ECOLOGICAL AND EVOLUTIONARY GENETICS OF *DROSOPHILA*

ECOLOGICAL AND EVOLUTIONARY GENETICS OF DROSOPHILA

Edited by

J. S. F. Barker

The University of New England
Armidale, Australia

William T. Starmer

Syracuse University
Syracuse, New York

and

Ross J. MacIntyre

Cornell University
Ithaca, New York

PLENUM PRESS • NEW YORK AND LONDON

Library of Congress Cataloging-in-Publication Data

Ecological and evolutionary genetics of Drosophila / edited by J.S.F.
  Barker, William T. Starmer, and Ross J. MacIntyre.
       p.   cm. -- (Monographs in evolutionary biology)
    "Proceedings of an international workshop on Ecological and
  Evolutionary Genetics of Drosophila, held January 5-10, 1989, in
  Armidale, Australia"--T.p. verso.
    Includes bibliographical references and index.
    ISBN 0-306-43671-X
    1. Drosophila--Genetics--Congresses. 2. Drosophila--Evolution-
  -Congresses. 3. Drosophila--Ecology--Congresses. 4. Ecological
  genetics--Congresses.   I. Barker, J. S. F. (James Stuart Flinton),
  1931-   . II. Starmer, W. T. (William T.) III. MacIntyre, Ross J.
  IV. Series.
  QL537.D76E24 1990
  595.77'4--dc20   90-45240
                                                              CIP

Proceedings of an international workshop on Ecological and Evolutionary
Genetics of *Drosophila*, held January 5–10, 1989, in Armidale, Australia

ISBN 0-306-43671-X

© 1990 Plenum Press, New York
A Division of Plenum Publishing Corporation
233 Spring Street, New York, N.Y. 10013

Printed in the United States of America

Preface

Ecological and evolutionary genetics span many disciplines and virtually all levels of biological investigation, from the genetic information itself to the principles governing the complex organization of living things. The ideas and information generated by ecological and evolutionary genetics provide the substance for strong inferences on the origins, changes and patterns of structural and functional organization in biological communities. It is the coordination of these ideas and thoughts that will provide the answers to many fundamental questions in biology.

There is no doubt that Drosophilids provide strong model systems amenable to experimental manipulation and useful for testing pertinent hypotheses in ecological and evolutionary genetics. The chapters in this volume represent efforts to use Drosophila species for such a purpose.

The volume consists of a dedication to William B. Heed, followed by four major sections: Ecological Genetics, Habitat Selection, Biochemical Genetics and Molecular Evolution. Each section is introduced by a short statement, and each chapter has an independent summary. The chapters contain the substance of talks given at a joint Australia-US workshop held January 5-10, 1989 at the University of New England, New South Wales, Australia.

We are indebted to the Division of International Programs of the National Science Foundation (USA) and to the Science and Technology Collaboration Section of the Department of Industry, Technology and Commerce (Australia) for the provision of financial support under the US/Australia Science and Technology Agreement.

Many people contributed to the preparation of this volume. We wish to thank Randi Starmer for the time consuming task of preparing the index. At Syracuse University, Bret Ingerman and Terry Jares were helpful in translating manuscripts from computer files prepared with a wide variety of software; as was Fay Hardingham at the University of New England. Alan Cowie provided technical assistance with international electronic mail transmission of manuscripts, and the entire volume was prepared to camera-ready copy by Mary Johnson and Ellen Klassen. We are greatly indebted to them all.

J. S. F. Barker
W. T. Starmer
R. J. MacIntyre

Contents

Chapter 3

**Fitness and Asymmetry Modification as an
Evolutionary Process.
A Study in the Australian Sheep Blowfly,
Lucilia cuprina and *Drosophila melanogaster***

John A. McKenzie, Philip Batterham, and Louise Baker

Chapter 4

**Extreme Environmental Stress: Asymmetry, Metabolic
Cost and Conservation**

P.A. Parsons

Chapter 5

The Effects of Natural and Artificial Selection on
Dysgenic Potential of a Wild Population of
Drosophila melanogaster

Michael Murphy, and John A. Sved

Chapter 6

P-Elements and Quantitative Variation in *Drosophila*

Chris Moran, and Adam Torkamanzehi

Chapter 7

Ecological and Evolutionary Importance of Host
Plant Chemistry

James C. Fogleman, and J. Ruben Abril

Chapter 8

The Nutritional Importance of Pure and Mixed Cultures
of Yeasts in the Development of *Drosophila mulleri*
Larvae in *Opuntia* Tissues and its Relationship to Host
Plant Shifts

William T. Starmer, and Virginia Aberdeen

Chapter 9

Experimental Analysis of Habitat Selection and
Maintenance of Genetic Variation

J.S.F. Barker

Chapter 10

Heritable Variation in Resource Use in *Drosophila*
in the Field

Ary A. Hoffmann, and Sharon O'Donnell

Chapter 11

Factors Maintaining Genetic Variation For Host Preference
in *Drosophila*

John Jaenike

Chapter 12

Theoretical Analysis of Habitat Selection and the
Maintenance of Genetic Variation

Philip W. Hedrick

Chapter 13

Alcohol Dehydrogenase and Alcohol Tolerance in
Drosophila melanogaster

Billy W. Geer, Pieter W.H. Heinstra, Ann M. Kapoun,
and Aleid van der Zel

Chapter 14

Environmental Modulation of Alpha-Glycerol-3-Phosphate
Oxidase (GPO) Activity in Larvae of *Drosophila
melanogaster*

Stephen W. McKechnie, Jennifer L. Ross, and
Kerrin L. Turney

Chapter 15

Physiology, Biochemistry and Molecular Biology of the Est-6 Locus in *Drosophila melanogaster*

Rollin C. Richmond, Karen M. Nielsen, James P. Brady, and Elizabeth M. Snella

Chapter 16

Insecticide Resistance as a Model System for Studying Molecular Evolution

Robyn J. Russell, Mira M. Dumancic, Geoffrey G. Foster, Gaye L. Weller, Marion J. Healy, and John G. Oakeshott

Chapter 17

Temperature Effects on Epicuticular Hydrocarbons and
Sexual Isolation in *Drosophila mojavensis*

Therese Ann Markow, and Eric C. Toolson

Chapter 18

Gene and Genome Structure in Diptera: Comparative
Molecular Analysis of an Eye Colour Gene in Three Species

Abigail Elizur, Ygal Haupt, Richard G. Tearle, and
Antony J. Howells

Chapter 19

Regulatory Evolution of ß-Carboxyl Esterases in
Drosophila

John G. Oakeshott, Marion J. Healy, and Anne Y. Game

Chapter 20

Molecular Isolation and Preliminary Characterisation
of a Duplicated Esterase Locus in *Drosophila buzzatii*

Peter East, Anne Graham, and Gillian Whitington

CHAPTER TWELVE

M ATTIWAZA, Tusratta's younger son, was not rewarded for having sought protection in Babylon. The Babylonians treated him as an enemy and robbed him of the riches which the old general, Aki-Tesup, had taken care to bring with them. As soon as this happened, Mattiwaza's supporters decided that all hope of winning back the Mitanni throne was lost and in future they could only expect gifts of sand and pebbles from their homeless sovereign. They dispersed accordingly, and Aki-Tesup, who at the age of seventy-eight still believed in the loyalty of vassals, died of a broken heart.

Mattiwaza, desperate, starving, torn by the thorns on paths so overgrown they weren't paths at all, and accompanied by a small group of servants and two Hurrite lords, finally reached the frontier of the Hittites, who had always been his dangerous hereditary enemies.

Then Suppiluliuma, the intelligent king of Khatti, knew that the moment to attack Egypt had come. Already, as much by force as by clever negotiation, he had succeeded in making inroads over the borders of the province of the land of Lower Kher*; he had also sounded out Abdashirti, the old Amorite prince, and Aziru his rebellious son, both worms in the splendid fruit of Egypt, and had made an alliance with them.

Now he gave his daughter in marriage to Mattiwaza and promised to put him back on his father's throne if he signed a treaty of vassaldom in which it was stipulated the Mittani would always support the Hittites against their enemies. Having done

* Land of Kher or of Rezen: Syrian province of Egypt which stretched from the fortress of Sile to the upper reaches of the Euphrates.

William B. Heed:
A Biography*

JAMES C. FOGLEMAN

"Anatomization, systematization, and classification are all
well enough, but these do not elicit conversation from the
specimen. The ... investigator must be ready to set his
method aside and sit down for an extended chat."

Heinrich Zimmer, *The King and the Corpse*, 1973,
Princeton University Press.

William B. Heed was born in 1926 in West Chester, Pennsylvania, the county seat of Chester County, 30 miles west of
Philadelphia. His father, John B. R. Heed, was a broker and a
respected member of the Council for the Borough of West
Chester. His mother, Margaret Lewis Heed, kept a garden that
was the envy of the neighborhood. She raised three children
and was an active member of the Methodist Church.

Heed's interest in science started very early with a pas-
sion for birds. At every opportunity in winter or summer, he
would gather his field glasses, trek out into the woods and
swamps below the local golf course, and remain for hours
searching for new and interesting species. He often waited
until dusk for a chance to hear a great horned owl. Owls of
any kind pique Heed's interest to this day. During high
school, these nature hikes and a concomitant interest in
scouting led to a friendship with Paul Martin, currently a
professor of geosciences at the University of Arizona.

Heed joined the U.S. Navy immediately after graduating from
high school in 1944 and was active for two years as a signal-
man on several tours of duty. This post permitted further
training in the ornithological world of birds of the open
ocean and far-off tropical isles. The G.I. Bill permitted
continued instruction for Heed, this time in the classrooms of
Ursinus College and subsequently Pennsylvania State University
where Heed graduated with a B.S. degree in 1950.

---

* This volume is dedicated to Professor Bill Heed for his many outstanding
contributions to the field of ecological genetics and evolution of
Drosophila. This biography originated with several taped interviews in
July, 1989, and was finalized by Heed at a later date.

Ecological and Evolutionary Genetics of Drosophila, Edited by
J.S.F. Barker *et al.*, Plenum Press, New York, 1990

When at Penn State, Heed again joined forces with Paul
Martin, then a student at Cornell, and Dick Robins, also a
Cornell undergraduate, on a four-month vertebrate collecting
trip to Tamaulipas, Mexico, for George Miksch Sutton and
others at the University of Michigan. Heed would report these
and other findings to the local biology club at Penn State
headed by Hubert Frings, a sensory physiologist of high cal-
iber and model instructor in the experimental method.

These were exciting times. Working with the bird collection at Michigan, spending a summer at the Michigan Biological
Station, writing papers, and interacting with many peers with
equal enthusiasm for ornithology and Mexico, were very positive experiences for Heed. Even so, things turned around
after a course in genetics from Jim Wright at Penn State and a
good look at Dobzhansky's book *Genetics and the Origin of
Species*. Suddenly chromosomes and genes became more fundamental than feathers and talons.

Finding the right graduate school became a primary concern.
The graduate catalogue of the University of Texas was most
reassuring for it mentioned something about evolutionary studies in the genus *Drosophila* and the availability of a truck
especially outfitted to collect species of *Drosophila* in the
U.S. and Mexico. Texas also offered Heed an assistantship.
It was not too many months after this offer that Charles
Walker, University of Michigan herpetologist, dropped Heed off
at Austin, Texas, on their way back from a successful collecting trip to "Frank's place" in an oak-sweet gum forest in the
Sierra Madre Oriental in Tamaulipas, Mexico. Heed was met in
the hallway of the old biology building by Wilson Stone who
ushered him into John Patterson's (Dr. Pat) office and a completely new adventure had begun. Dr. Pat pulled out "The
Drosophilidae of the Southwest" and leafed through the
delightfully painted portraits of a number of species, care-
fully explaining the history and characteristics of each one.

Although he became Patterson's last graduate student, Heed
was actually more influenced by Stone and Marshall Wheeler,
the systematist for the Texas group. Wheeler and Heed made
several long distance collecting trips to the western and
southeastern parts of the country during several summers.
According to Heed, it was Wheeler who showed him how to really
look at a fly and to appreciate what the different sets of
bristles mean in regard to the systematics of the group.

Heed recalls his experience at Texas as a particularly
special time. The camaraderie among the students on the fifth
floor of the new science building was part of it. Larry Mettler, Paul Moorhead, Marvin Wasserman, Herb Bruneau, Lynn
Throckmorton, Marvin Sieger, Tom Gregg and others arrived
within a few years of each other and formed a strong interac-
tive group which included several of Bob Wagner's students,
Jack Hubby and Jim Ragland. Many of these younger students
overlapped with the older entrenched veterans. T.C. Hsu,
Calvin Ward, Mary Alexander and Frances Clayton each had their
dissertations in various stages of completion.

Burke Judd arrived from Sturtevant's laboratory at Cal Tech as an assistant professor and Michael White was on board before he left for Australia. Th. Dobzhansky and Warren Spencer added greatly to the intellectual atmosphere as visiting professors. Visitors from South America included Crodowaldo Pavan and Danko Brncic. Shultz's beer garden became a focal point on Friday nights where heated discussions took place and bizarre revelations on matters scientific and political often came to light.

It was in the spring of 1952 that Heed was introduced to Sarah Roehr, an acquaintance of Marvin Sieger's. As a student, Sarah worked part time in the Speech and Hearing department on campus. She later became a graduate student and a practitioner in speech therapy. Sarah and Bill discovered they had a similar philosophy of life and that they were both on a quest for the truth. This is as true today as it was 37 years ago. The attraction was magnetic. They were married in Dallas in 1954 after a year's separation during which time Heed studied the Drosophilidae of El Salvador and other countries in Central America.

From the first collecting trip for *Drosophila*, Heed has had an abiding interest to discover the breeding sites of the various species since it was obvious to him that the genetics of natural populations would have little meaning until that information was known. These discoveries, according to Heed, would lay the foundation for an understanding of the actual structure of populations and permit experimentation under field conditions. His overriding interest in the ecology of *Drosophila* was met with a certain amount of skepticism among the leaders in ecological and population genetics, especially, they thought, if one planned to devote his entire career on this very elusive subject.

As things progressed at Texas, Adolph Meyer-Abich, a philosopher of biology from Germany, appeared on the scene through the suggestion of Th. Dobzhansky to Wilson Stone. Meyer-Abich had a deep interest in a research station entitled El Instituto Tropical de Investigaciones Cientificas which was located outside the city limits of San Salvador. Heed became a guest at that institute from October, 1953, to August, 1954, collecting flies from the tropical lowlands to the cloud forests at 7,000 feet. He also met the botanists, Paul Stanley and Carol Williams, at the experiment station at El Zamarano near Tegucigalpa, Honduras, and collected on the Caribbean coast at La Lima and Lancetilla. In Nicaragua, he made large collections in the sweet gum forests of the Matagalpa area and in the tropical lowland rain forests at the experiment station at El Recreo, eight hours upstream from Bluefields on the Rio Escondido. Many shipments of live material were sent to Austin and to Dobzhansky at Columbia during this time. Upon return from El Salvador, Heed worked with Wheeler's collection of pinned material as well as his own specimens and wrote up his notes in the form of a dissertation. The Ph.D. was awarded in 1955.

The very positive results from this effort in Central America convinced Stone that the whole circum-Caribbean area held
great promise for further evolutionary studies, and he lost
little time in organizing a larger force of interested investigators. Sarah and Bill were the first to circle the Caribbean for *Drosophila*, beginning a tradition of family collecting forays which continues to this day when his daughters
strive to get away from their busy schedules to join their
father in the deserts of the Southwest and Mexico.

The Caribbean project was a major activity at Texas from
1956 to the early 1960's. Heed collected with Hampton and
Meredith Carson and Marvin Wasserman on several extended expeditions in Central and South America and the West Indies.
These collections formed the basis for many research projects
for a number of investigators. The discovery of the parthenogenetic *D. mangabeirai* in El Salvador and later in the other
countries of Central America and in Trinidad, was one of the
highlights of the project. Several species groups were most
attractive to Heed because of their diversity and the relative
ease with which they could be bred in the laboratory.   The
systematics of the tripunctata group was pursued jointly with
Sarah Pipkin who was living in Panama at this time. Based on
their mutual interest, the two former students of Dr. Pat's
formed quite a team. It was the cardini species group, however, that became Heed's focus of attention for a number of
years. The major result of this work was the discovery of the
correlation between the accumulation and continued heterozygosity (or eventual fixation) of inversions in the polytene
chromosomes with their homologue of origin and with population
structure.   This work overlapped and took sustenance from
Wasserman's masterful chromosomal analysis of the repleta
species group.

In 1956, Heed joined the faculty at the University of Pennsylvania as an instructor when Charles Metz was in his final
year as head of the Department of Zoology.   There, he interacted with John Preer and Anna Racheal Whiting in teaching the
genetics laboratory. He also began his own course in organic
evolution which he still teaches today in, of course, a highly
modified form. Margaret Ellen was born in Dallas at the beginning of the second year at Penn.

In 1958, the Heed family left Pennsylvania and, after a
short stay in Austin, travelled to Tucson, where Heed took the
position of Assistant Professor at the University of Arizona.
He was soon funded by NSF to work in the Caribbean area and in
the Sonoran Desert on various aspects of *Drosophila* speciation
and ecology. Al Mead, Joe Marshall, Chuck Lowe and Paul
Martin were all interested in having Heed at Arizona because
he was a geneticist that was equally interested in ecology.

As things progressed in Tucson, Stone was making plans with
Elmo Hardy, the Dipteran taxonomist in Honolulu, and Herman
Spieth, a *Drosophila* behaviorist at Riverside, for an intensive study in the Hawaiian Islands. The Hawaiian *Drosophila*
where known to be high in species diversity, unusually large
in size, and exhibited very striking sexual dimorphisms.

The Hawaiian project commenced in 1963. Heed was invited
to join with eight other investigators as a member of the
original team. The first order of business was to collect and
culture as many species as possible for chromosomal analysis
and hybridization tests. Studies on the ecology, morphology
and behavior of the flies were to follow. Elmo Hardy was the
local coordinator of the expeditions to the outer islands of
Maui, Hawaii and Kauai. It soon became evident that these
Hawaiians were far different from those on the mainland, and a
major reorientation in thinking had to take place. It was not
too long, according to Heed, that everyone began to think like
an Hawaiian *Drosophila*, and the data started to accumulate.
Marshall Wheeler and Frances Clayton created a food medium
that was satisfactory for some of the species. Hampton Carson
and Harry Stalker discovered that the large picture-winged
species were the easiest to rear and have very photogenic
salivary chromosomes. Frances Clayton started a long career
of metaphase preparations of the Hawaiians. Herman Spieth commenced his many hours of observation on their mating behavior.
Lynn Throckmorton discovered that the internal anatomy of the
Hawaiians came in two major packages, the "drosophiloids" and
the "scaptomyzoids". Heed uncovered the breeding sites of a
host of species of Hawaiians including the rotting leaves of
many of the secondary trees in the rain forests. The leaves
of each species of tree had its own characteristic aroma so
that the flies could easily specialize on one or two kinds.
Heed eventually was able to categorize the proportion of
Hawaiian species that were host plant specialists (70%) compared to those that bred in a variety of substrates (30%).

Later, it became clear to all of us, according to Heed,
that the host plant specialization paralleled the systematic
groupings of the flies. It then became convenient to talk
about "bark breeders" or "fungus feeders" etc. as though each
group had its own common ancestor. This was true in the
majority of cases. Hawaii was somewhat of a microcosm showing
how the early evolution of the major groups of *Drosophila*
could have occurred.

Stone came to Honolulu to keep up on the proceedings, offer
suggestions, and clear the road of any obstacles for his
group. He was especially pleased to learn that the basic
chromosome number and configuration for the Hawaiians was the
same as those found in the mainland, and was surprised to
learn how invariant it really was. Heed spent a year's sabbatical leave and a number of summers and odd weeks in Hawaii.
By this time, Emily Robinson Heed and Meredith Anne Heed had
appeared on the scene, and Hawaii became part of their lives.

As the research team expanded, Heed interacted with Forbes
Robertson, Henry Kircher, Mike Kambysellis, Dick Richardson,
and Ken Kaneshiro on various projects designed to learn more
about the leaf breeders as well as the bark breeders and
flower breeders. The discovery of the correlation between the
nutritional capacity of the substrate and the number of eggs
that matured in the ovaries of the species that utilized them

was a major advance in understanding how more than 700 species
could coexist in this archipelago.

Heed held the position of Associate Entomologist in the
Department of Entomology, University of Hawaii, from 1963 to
1970.   The history of the Hawaiian *Drosophila* project was
reviewed by Herman Spieth in 1981 (Drosophila Information Service 56:6-14).

Heed's work in the Sonoran Desert spans three decades.
This subject has become known as the Cactus-Microorganism-
Drosophila Model System.   The early studies in the 1960's primarily involved the identification of the host plants and the
flies that bred in them.   This was also the beginning of the
studies on chemical ecology.   It was shown that *Drosophila*
pachea needed a piece of its host plant, senita cactus,
inserted into the medium in order to produce a second generation.   Early on, Heed was introduced to Henry Kircher, a natural products chemist, by Al Siegel, the molecular geneticist
in the Department of Agricultural Biochemistry.   Kircher was
in the same department.   The subsequent investigation on *D.*
pachea and senita cactus heightened the imagination of geneticists and entomologists alike and the desert work became
widely known.

The contact with Kircher blossomed into a 20-year collaboration which ended with Henry's untimely death in 1984.   This
collaboration extended to the Hawaiian project as well.
Through the work of Kircher and Heed, and (later) Fogleman, it
has become known that the chemistry of the various host plants
plays a primary role concerning the manner by which the cactus
niche, in its entirety, is subdivided in the desert.   Because
of these studies, Heed believes the list of differences
between saprophagous and phytophagous insects should be shortened in part because they obey the same "rules" in regard to
host plant selection and utilization.

Heed emphasized that an enormous amount of footwork and
scanning from the sky became characteristic of the desert
studies.   Discovering new and interesting places to collect
became an important activity.   Jean Russell, Heed's laboratory
assistant for many years, would travel with Ike, her bush
pilot husband, to many far away spots and return with bags
full of goodies, meaning odoriferous cactus rots neatly
wrapped in Mexican newspapers.   These trips became more extensive in later years when the Russell's flew Barney Ward and
Tom Starmer to Venezuela in search for *D. acanthoptera*.   They
did not find this species, but they did collect two new cactophilic species known today as *D. uniseta* and *D. starmeri*.
The Russells were key contributors to the productivity and
excitement of the Tucson laboratory.   Their friendship with
all of the Heeds over the years has been greatly valued.

During the early 70's, Heed became interested in the yeasts
which live in necrotic cactus tissue and serve as a source of
nutrition for adults and larvae of the cactophilic *Drosophila*.
He contacted Herman Phaff, a yeast expert at the University of
California (Davis), and ultimately spent a sabbatical year at

Davis (1972-73) learning how to work with yeasts and partici-
pating in the identification of over 300 yeast isolates from
cactus. Using both traditional taxonomic methods and determinations of GC content of representative strains, the yeast
flora of cactus rots eventually turned out to be a unique
group and about a dozen new species of yeasts have been
described. Although Heed initiated this study, Tom Starmer (a
former postdoctorate of Heed's) and Herman Phaff have maintained a collaborative research program on cactus yeasts to
the present day. In a recent autobiography (Annual Review of
Microbiology, Vol. 40:1-28, 1986), Phaff points out the utility of the research on cactus yeasts by stating: "The cactus-
Drosophila-yeast ecosystem has given us a wealth of information on the evolution and speciation of yeasts that no other
ecosystem has been able to supply."

The year in Davis was an especially busy time. Heed was a
guest in Dobzhansky's laboratory. Francisco Ayala's laboratory was next door. Herman Spieth's laboratory was in a separate building in the Zoology department while Herman Phaff and
Marty Miller were housed in the Food Sciences Department further down the mall. Heed spent research time in all four laboratories alternately throughout the year. He also took the
yeast course, with laboratory, offered by Phaff and Miller.
Field trips were taken to the Pacific northwest with Spieth
and to the big island of Hawaii with Ayala, Phaff and Miller.
It was at Davis that Heed became acquainted with other guests
and students of Ayala and Dobzhansky. Antonio Fontdevila from
Spain and Moritz Benado from Venezuela were two researchers
who later became involved in the cactophilic Drosophila work.
Marty Tracey and John McDonald helped Heed analyze the electrophoretic data on the nannoptera and repleta species groups
that were obtained in Ayala's laboratory. Bike riding through
the campus on the weekends with Sarah, Ellen, Emily and Anne
was always enthusiastically anticipated. It was a productive
year for everyone.

Over the years, Heed has made many contributions to the
scientific community in the form of service activities. He
has served as an Associate Editor for Evolution (1974), Vice
President for both the Society for the Study of Evolution
(1976) and the American Society of Naturalists (1978), Program
Director for Systematic and Evolutionary Biology at NSF
(1976), and Chairman of the National Policy Guidance Council
for the National Drosophila Species Resource Center (1975 to
the present).

According to Heed, the year in Washington was productive
but taxing. A departmental review was upcoming for the
National Science Board of NSF which took several months of
preparation. Heed reviewed the Hawaiian Drosophila program
for the board. Also, a new panel was inaugurated by Heed and
Fran James in order to accommodate the new thinking in population biology that was forthcoming. The systematics panel was
handling Drosophila proposals as best it could but relief was
vitally necessary. Heed worked with Jim Hickman in the systematics panel and with James in the population biology and
physiological ecology panel. Somehow, many grants were

funded, and the year came to a close thanks to the friendly and supportive help of John Brooks, Bill Sievers and Tom Callahan.

When asked to compare the Hawaiian project with the Sonoran Desert project, Heed quickly pointed out that the emphasis of the two programs has been significantly different. Cytological analysis of chromosomes, sexual behavior, and speciation eventually became the primary concerns of the Hawaiian program whereas the study of the desert system has emphasized the interactions between the flies and their environment (e.g., cactus chemistry and microorganisms) as they relate to insect-host plant relationships. Speciation of the cactophilic *Drosophila* is an extremely interesting topic, but speciation, as a process, has certainly occurred more often in Hawaii than in the desert. In fact, the basic mechanism of speciation in the two areas may be quite different. Heed's theory is that populations in the desert area had to speciate in order to survive increasingly stressful conditions, and this stands in stark contrast to the adaptive radiation of the Hawaiian flies which presumably occurred under relaxed selection. To distinguish these two modes of speciation, Heed coined the term "adaptive infiltration." Heed has been pursuing this concept for a number of years by studying the relatives of the four endemic desert species which are distributed outside or on the edge of the desert and the *Drosophila* of analogous environmental regions in the West Indies. Comparisons extending into the Tehuacan Desert of Mexico are projected to be the next step.

By 1982, work on the cactophilic *Drosophila* as a whole had reached an apogee, and a summit meeting was called by Stuart Barker and Tom Starmer. Investigators converged on the small town of Oracle, Arizona, from Australia, Spain and Brazil, as well as many parts of the United States. The results of the meeting were published as *Ecological Genetics and Evolution: The Cactus-Yeast-Drosophila Model System* by Academic Press in 1982. Barker and Starmer were the editors. According to Heed, it was during this meeting that plans were made for a collecting expedition to the West Indies on the ORV Cape Florida, an NSF research vessel stationed in Miami, Florida. The idea for the trip was communicated to Tom Starmer by Jack Fell, a yeast researcher. There was one trip in 1982 and a second one in 1983. Using the ORV Cape Florida as a floating laboratory and base from which to make collecting expeditions onto the islands was found to be highly successful. Several new and very significant species of yeasts and *Drosophila* were discovered.

The Tucson laboratory became heavy with cultures of undescribed species of *Drosophila* from Mexico and the West Indies. Heed spent many hours "training" the flies to laboratory food. Alfredo Ruiz arrived from the Autonomous University of Barcelona with his family and made significant contributions to the cytology and the measurements of fitness traits on the desert species. Bill Etges arrived from the University of Rochester and overlapped with Ruiz. Etges' expertise was life history evolution. He worked with *D. mojavensis* on this study which included elegant experiments to further document the tantalizing ability of the species, first discovered by Starmer, to store energy from the vapors of atmospheric alcohol. Etges, Ruiz and Heed made a successful trip to Sinaloa in 1985 in search for the rare cactus, *Stenocereus stanleyi*.

In order to take advantage of the growing number of cactophilic species in Tucson, Heed decided to join forces with Antonio Fontdevila's group who had been analyzing electrophoretically the cactophilics from South America and others sent over from the Tucson laboratory. Bill and Sarah spent the sabbatical year of 1986-87 in Spain with Antonio, Alfredo Ruiz, and others including a number of creative students. The results of that year's work are reported in this volume.

When Heed was questioned about the students he has supervised over the years he began listing them, their attributes and contributions beginning with the very first student. It was pretty obvious that Heed believed the students provided the "guts" for the intellectual life of the laboratory. Interestingly, not all the students worked on *Drosophila*. Jim Patton analyzed the chromosomes of several local rodents, George Conner worked on the genetics of *Mormoniella*, Penny Graf did the first bacterial analysis of cactus necroses, Don Vacek analyzed yeasts, and Dorothea Jurgenson collected saguaro seeds from Arizona and Mexico for an electrophoretic study. The list of students follows:

> Joy W. Cooper, M.S., 1964
> George W. Conner, Ph.D. 1965
> Penelope A. Graf, M.S., 1965
> James L. Patton, M.S., 1965, Ph.D., 1969
> John S. Grove, M.S., 1966
> Mary C. Truett, M.S., 1966
> Bernard L. Ward, M.S., 1969; Ph.D., 1975
> David P. Fellows, Ph.D., 1970
> J. Spencer Johnston, Ph.D., 1972
> William R. Johnson, M.S., 1973; Ph.D., 1980
> Susan Rockwood-Sluss, Ph.D., 1975
> Don C. Vacek, M.S., 1976; Ph.D., 1979
> Margaret C. Jefferson, Ph.D., 1977
> Robert L. Mangan, Ph.D., 1978
> Dave G. Baldwin, Ph.D., 1979
> Garry A. Duncan, Ph.D., 1979
> Dorothea E. Jurgenson, M.S., 1979
> Richard H. Thomas, Ph.D., 1985
> Pupulio S.N.A. Ssekimpi, Ph.D., 1986
> George L. Ball, M.A., 1987
> William T. Starmer, Postdoctoral Student, 1972-77
> James C. Foglemen, Postdoctoral Student, 1978-82
> William T. Etges, Postdoctoral Student, 1985-87

By way of a summary Heed attributes his continued interest in his kind of research as much to his family as to any other factor. According to Heed each individual member has been supportive in her own way. On the other hand, the family as a group has been extraordinarily close with the delightful wit and creative energy that has that strength-giving property not experienced in any other way. He is particularly indebted to Sarah for leading the way in her quest for the truth which lies outside the realm of science.

By way of a conclusion, what I had hoped to communicate is a view into the life of Bill Heed, a man who is more enthusiastic collecting flies out in the field than he is musing in his office. Bill has served as a mentor to many and has (perhaps unknowingly) dedicated his career to educating those who would believe that *Drosophila* only occur in half-pint milk bottles. He has certainly been the driving force for research in the desert *Drosophila* system which is considered by many, myself included, to be one of the most interesting and productive research programs in population biology. He has been my teacher, my colleague, and my friend.

William B. Heed: Publications

Robins, C. R., and Heed, W. B., 1951, Bird notes from la Joya de Salas, Tamaulipas, *Wilson Bull.* 63:263-270.

Martin, P. S., Robins, C. R., and Heed, W. B., 1954, Birds and biogeography of the Sierra de Tamaulipas, an isolated pine-oak habitat, *Wilson Bull.* 66:38-57.

Heed, W. B., 1956, Apuntes sobre la ecologia y la dispersion de los Drosophilidae (Diptera) de El Salvador, *Commun. Inst. Trop. Invest. Cient.* No. 2-3:59-74.

Carson, H. L., Wheeler, M. R., and Heed, W. B., 1957, A parthenogenetic strain of *Drosophila managabeirai* Malogolowkin, *Univ. Tex. Publs* 5721:115-122.

Heed, W. B., 1957a, Intraspecific relationships of *Drosophila crocina* Patterson and Mainland from three localities, *Univ. Tex. Publs* 5721:15-16.

Heed, W. B., 1957b, Ecological and distributional notes on the Drosophilidae (Diptera) of El Salvador, *Univ. Tex. Publs* 5721:62-78.

Heed, W. B., 1957c, A preliminary note on the *cardini* group of *Drosophila* in the Lesser Antilles, *Univ. Tex. Publs* 5721:123-124.

Heed, W. B., 1957d, An attempt to detect hybrid matings between *D. mulleri* and *D. aldrichi* under natural conditions, *Univ. Tex. Publs* 5721:182-185.

Heed, W. B., and Wheeler, M. R., 1957, Thirteen new species in the genus *Drosophila* from the Neotropical region, *Univ. Tex. Publs* 5721:17-38.

Heed, W. B., and Krishnamurthy, N. B., 1959, Genetic studies on the *cardini* group of *Drosophila* in the West Indies, *Univ. Tex. Publs* 5914:155-179.

Heed, W. B., 1960a, Ecology and abnormal karyotypes of *D. orbospiracula*, *Drosoph. Inf. Serv.* 34:84.

Heed, W. B., 1960b, Genetic, cytological and morphological clines in island populations of the *cardini* species group of *Drosophila* in the West Indies, *Records of Genetics Soc. of Amer.* No. 29:73 (Abstr.).

Heed, W. B., Carson, H. L., and Carson, M. S., 1960, A list of flowers utilized by drosophilids in the Bogota region of Columbia, *Drosoph. Inf. Serv.* 34:84-85.

Heed, W. B., 1962a, Genetic characteristics of island populations. *Univ. Tex. Publs* 6205:173-206.

Heed, W. B., 1962b, Review of B. Rensch's *Evolution Above the Species Level*, *Amer. J. phys. Anthrop.* 20:408-410.

Heed, W. B., Russell, J., and Harrington, D., 1962, Diversity and density of *Drosophila* in the immediate vicinity of Tucson with special reference to *D. pseudoobscura*, *Drosoph. Inf. Serv.* **36**:73-74.

Heed, W. B., 1963, Density and distribution of *Drosophila polymorpha* and its color alleles in South America, *Evolution* **17**:502-518.

Heed, W. B., and Blake, P., 1963, A new color allele at the e locus of *Drosophila polymorpha* from northern South America, *Genetics* **48**:217-234.

Heed, W. B., and Russell, J., 1963, Cytological studies in the *cardini* superspecies of *Drosophila*, *Proc. XI Int. Cong. Genet.* **1**:139 (Abstr.).

Carson, H. L., and Heed, W. B., 1964, Structural homozygosity in marginal populations of Nearctic and Neotropical species of *Drosophila* in Florida, *Proc. natn. Acad. Sci. USA* **52**:427-430.

Pipkin, S. B., and Heed, W. B., 1964, Nine new members of the *Drosophila tripunctata* species group (Diptera:Drosophilidae), *Pacif. Insects* **6**:256-273.

Heed, W. B., and Kircher, H. W., 1965, A unique sterol in the ecology and nutrition of *Drosophila pachea*, *Science* **149**:758-761.

Heed, W. B., and Russell, J. S., 1965, Inversion differences between island and continental species in the *cardini* species group of *Drosophila*, *Genetics* **52**:s447-s448 (Abstr.).

Patton, J. L., and Heed, W. B., 1965, Elevational differences in gene arrangements of *D. pseudoobscura* in the Santa Catalina Mountains, Tucson, *Drosoph. Inf. Serv.* **40**:69.

Heed, W. B., and Jensen, R. W., 1966, Drosophila ecology of the senita cactus, *Lophocereus schottii*, *Drosoph. Inf. Serv.* **41**:100.

Patton, J. L., Heed, W. B., and Lowe, C. H., 1966, Inversion frequency analysis of *Drosophila pseudoobscura* in the Santa Catalina Mountains, Arizona, *J. Ariz. Acad. Sci.* **4**:105-117.

Kircher, H. W., Heed, W. B., Russell, J. S., and Grove, J., 1967, Senita cactus alkaloids: their significance to Sonoran Desert *Drosophila* ecology, *J. Insect Physiol.* **13**:1869-1874.

Lowe, C. H., Heed, W. B., and Halpern, E. A., 1967, Supercooling of the saguaro species *Drosophila nigrospiracula* in the Sonoran Desert, *Ecology* **48**:984-985.

Heed, W. B., 1968, Ecology of the Hawaiian Drosophilidae. *Univ. Tex. Publs* **6618**:387-419.

Heed, W. B, and Russell, J. S. , 1968a, Inability of *D. pachea* to breed in cereus cacti other than senita, *Drosoph. Inf. Serv.* **43**:94-96.

Heed, W. B., and Russell, J. S., 1968b, Inversion replacement vs. fixation in the *cardini* species group of *Drosophila*, *Proc. XII Int. Cong. Genet.* **1**:321 (Abstr.).

Heed, W. B., Russell, J. S., and Ward, B. L., 1968, Host specificity of cactiphilic *Drosophila* in the Sonoran Desert, *Drosoph. Inf. Serv.* **43**:94-95.

Heed, W. B., Crumpacker, W., and Ehrman, L., 1969, *Drosophila lowei* a new american member of the *obscura* species group, *Ann. ent. Soc. Am.* **62**:388-393.

Kircher, H. W., and Heed, W. B., 1970, Phytochemistry and host-plant specificity in *Drosophila*, *Adv. Phytochem.* **3**:191-209.

Rockwood, E. S., Johnston, J. S., and Heed, W. B., 1970, Allozyme variation in natural populations of *Drosophila pachea*, *Genetics* **64**:s53 (Abstr.).

Ward, B. L., and Heed, W. B., 1970, Chromosome phylogeny of *Drosophila pachea* and related species, *J. Hered.* **61**:248-258.

Heed, W. B., 1971, Host plant specificity and speciation in Hawaiian *Drosophila*, *Taxon* **20**:115-121.

Heed, W. B., and Russell, J. S., 1971, Phylogeny and population structure in island and continental species of the *cardini* groups of *Drosophila* studied by inversion analysis, *Univ. Tex. Publs* **7103**:91-130.

Johnston, J. S., and Heed, W. B., 1971, Comparison of banana and rotted cactus as a bait for desert *Drosophila, Drosoph. Inf. Serv.* **46**:96.

Kambysellis, M. P., and Heed, W. B., 1971, Studies of oogenesis in natural populations of Drosophilidae I. Relation of ovarian development and ecological habits of the Hawaiian species, *Am. Nat.* **105**:31–49.

Fellows, D. P., and Heed, W. B., 1972, Factors affecting host plant selection in desert-adapted cactiphilic *Drosophila, Ecology* **53**:850–858.

Heed, W. B., 1972, Distribution extensions and gene arrangements for *D. parthenogenetica* and *D. montana, Drosoph. Inf. Serv.* **48**:100.

Heed, W. B., and Heed, S. R., 1972, Ecology, weather, and dispersal of *Drosophila* on an island mountain, *Drosoph. Inf. Serv.* **48**:100–101.

Spieth, H. T., and Heed, W. B., 1972, Ecological and experimental systematics in *Drosophila, Annu. Rev. Ecol. & Syst.* **3**:269–288.

Heed, W. B., 1973, Ecology and dispersal in Hawaiian *Drosophila, Proc. XIII. Int. Cong. Genet., Genetics* **74**:s113.

Kaneshiro, K. Y., Carson, H. L., Clayton, F. E., and Heed, W. B., 1973, Niche separation in a pair of homosequential *Drosophila* species from the island of Hawaii, *Am. Nat.* **107**:766–774.

Rockwood-Sluss, S., Johnston, J. S., and Heed, W. B., 1973, Allozyme genotype-environment relationships I. Variability in natural populations of *D. pachea, Genetics* **73**:135–146.

Baldwin, D. G., and Heed, W. B., 1974, A newly-discovered inversion in the second chromosome of *D. pseudoobscura. Drosoph. Inf. Serv.* **51**:128.

Jefferson, M. C., Johnson, W. R., Baldwin, D. G., and Heed, W. B., 1974, Ecology and comparative cytology of *Drosophila* on San Pedro Nolasco Island, *Drosoph. Inf. Serv.* **51**:65.

Kambysellis, M. P., and Heed, W. B., 1974, Juvenile hormone induces ovarian development in diapausing cave-dwelling *Drosophila* species, *J. Insect Physiol.* **20**:1779–1786.

McDonald, J. F., Heed, W. B., and Miranda, M., 1974, The larval nutrition of *Minettia flaveola* and *Phaonia parviceps* and its significance to the Hawaiian leaf-breeding *Drosophila, Pan-Pacif. Ent.* **50**:78–82.

Miller, M. W., Phaff, H. J., Heed, W. B., Starmer, W. T., and Miranda, M., 1974, Yeasts associated with *Drosophila* breeding sites in various species of cactus in desert regions of Arizona and northern Mexico, *Fourth International Symp. on Yeasts,* pp. 257–258 (Abstr.).

Phaff, H. J., Miller, M. W., Miranda, M., Heed, W. B., and Starmer, W. T., 1974, *Cryptococcus cereanus,* a new species of the genus *Cryptococcus, Int. J. Syst. Bact.* **24**:486–490.

Vacek, D. C., Ward, O. G., and Heed, W. B., 1974, Karyotype of *D. micromelanica* reared from Gambel Oak in Arizona, *Drosoph. Inf. Serv.* **51**:60.

Johnston, J. S., and Heed, W. B., 1975, Dispersal of *Drosophila:* the effect of baiting on the behavior and distribution of natural populations, *Am. Nat.* **109**:207–216.

Spieth, H. T., and Heed, W. B., 1975, The *Drosophila pinicola* species group, *Pan-Pacif. Ent.* **51**:287–295.

Starmer, W. T., Heed, W. B., Miller, M. W., Phaff, H. J., and Miranda, M., 1975, The diversity of yeasts associated with cactiphilic *Drosophila, J. Ariz.-Nev. Acad. Sci.* **10**:12–13.

Ward, B. L., Starmer, W. T., Russell, J. S., and Heed, W. B., 1975, The correlation of climate and host plant morphology with a geographic gradient of an inversion polymorphism in *Drosophila pachea, Am. Nat.* **109**:207–216.

Heed, W. B., Starmer, W. T., Miranda, M., Miller, M. W., and Phaff, H. J., 1976, An analysis of the yeast flora associated with cactiphilic *Drosophila* and their host plants in the Sonoran Desert and its relations to temperate and tropical associations, *Ecology* **57**:151–160.

Johnston, J. S., and Heed, W. B., 1976, Dispersal of desert-adapted

Drosophila: the saguaro-breeding *D. nigrospiracula*, *Am. Nat.* 110:629-651.

Miller, M. W., Phaff, H. J., Miranda, M., Heed, W. B., and Starmer, W. T., 1976, *Torulopsis sonorensis*, a new species of the genus *Torulopsis*, *Int. J. Syst. Bact.* 26:88-91.

Starmer, W. T., Heed, W. B., Miranda, M., Miller, M. W., and Phaff, H. J., 1976, The ecology of yeast flora associated with cactiphilic *Drosophila* and their host plants in the Sonoran Desert, *Microb. Ecol.* 3:11-30.

Fontdevila, A., Starmer, W. T., Heed, W. B., and Russell, J. S., 1977, Migrant selection in a natural population of *Drosophila*, *Experientia* 33:1447-1448.

Heed, W. B., 1977, A new cactus-feeding but soil-breeding species of *Drosophila* (Diptera: Drosophilidae), *Proc. ent. Soc. Wash.* 79:649-654.

Russell, J. S., Ward, B. L., and Heed, W. B., 1977a, Inversion polymorphism in *D. nannoptera*, *Drosoph. Inf. Serv.* 52:112.

Russell, J. S., Ward, B. L., and Heed, W. B., 1977b, Sperm storage and hybridization in *D. nannoptera* and related species, *Drosoph. Inf. Serv.* 52:70.

Starmer, W. T., and Heed, W. B., 1977, The infection of *Drosophila* cultures by species of the genus *Prototheca*, *Drosoph. Inf. Serv.* 52:12.

Starmer, W. T., Heed, W. B., and Rockwood-Sluss, E.S., 1977, Extension of longevity in *Drosophila mojavensis* by environmental ethanol: Differences between subraces, *Proc. natn. Acad. Sci. USA* 74:387-391.

Vacek, D. C., Starmer, W. T., and Heed, W. B., 1977, The Y-2 strain of yeast at the Texas stock center is *Candida krusei*, *Drosoph. Inf. Serv.* 52:80.

Heed, W. B., 1978, Ecology and Genetics of Sonoran Desert *Drosophila*, in: *Ecological Genetics: The Interface* (P. F. Brussard, ed.), Springer-Verlag, New York, pp.109-126.

Vacek, D. C., and Heed, W. B., 1979, The relevance of the ecology of citrus yeasts to the diet of *Drosophila*, *Microb. Ecol.* 5:43-49.

Fontdevila, A., Starmer, W. T., Heed, W. B., and Russell, J. S., 1980, Differential mating activity in two coexisting species of cactophilic *Drosophila*, *Drosoph. Inf. Serv.* 55:165-166.

Kambysellis, M. T., Starmer, W. T., Smathers, G., and Heed, W. B., 1980, Studies in oogenesis in natural populations of Drosophilidae. II. Significance of microclimatic changes on oogenesis of *D. mimica*, *Am. Nat.* 115:67-91.

Fogleman, J. C., and Heed, W. B., 1981, A comparison of the yeast flora in the larval substrates of *D. nigrospiracula* and *D. mettleri*, *Drosoph. Inf. Serv.* 56:38-39.

Fogleman, J. C., Hackbarth, K. R., and Heed, W. B., 1981, Behavioral differentiation between two species of cactophilic *Drosophila*. III. Oviposition site preference, *Am. Nat.* 118:541-548.

Fogleman, J. C., Starmer, W. T., and Heed, W. B., 1981a, Utilization of food resources by *Drosophila* larvae, *Genetics* 97:s36-37 (Abstr.).

Fogleman, J. C., Starmer, W. T., and Heed, W. B., 1981b, Larval selectivity for yeast species by *Drosophila mojavensis* in natural substrates, *Proc. natn. Acad. Sci. USA* 78:4435-4439.

Heed, W. B., 1981, Central and marginal populations revisited, *Drosoph. Inf. Serv.* 56:60-61.

Young, D. J., Vacek, D. C., and Heed, W. B., 1981, The facultative anaerobic bacteria as a source of alcohols in three breeding substrates of cactophilic *Drosophila*, *Drosoph. Inf. Serv.* 56:38-39.

Fogleman, J. C., Heed, W. B., and Kircher, H. W., 1982, *Drosophila mettleri* and senita alkaloids: fitness measurements and their ecological significance, *Comp. Biochem. Physiol.* 71A:413-417.

Fogleman, J. C., Starmer, W. T., and Heed W. B., 1982, Comparisons of yeast florae from natural substrates and larval guts of southwestern *Drosophila*, *Oecologia* **52**:187-191.

Heed, W. B., 1982, The origin of *Drosophila* in the Sonoran Desert, in: *Ecological Genetics and Evolution: The Cactus-Yeast-Drosophila Model System* (J. S. F. Barker, and W. T. Starmer, eds), Academic Press Australia, Sydney, pp. 65-80.

Starmer, W. T., Phaff, H. J., Heed, W. B., Miranda, M., and Miller, M. W., 1982, The yeast flora associated with the decaying stems of columnar cacti and *Drosophila* in North America, *Evol. Biol.* **14**:269-295.

Carson, H. L., and Heed, W. B., 1983, Methods of collecting *Drosophila*, in: *The Genetics and Biology of Drosophila*, Vol. 3d (M. Ashburner, H. L. Carson, and J. N. Thompson, eds), Academic Press, London, pp. 1-28.

Heed, W. B., 1983, Review of: Vol. 3b *The Genetics and Biology of Drosophila* (M. Ashburner, H. L. Carson, and J. N. Thompson, eds), *Amer. Sci.* **71**:315.

Markow, T. A., Fogleman, J. C., and Heed W. B., 1983, Reproductive isolation in Sonoran Desert *Drosophila*, *Evolution* **37**:649-652.

Heed, W. B., and Mangan, R. L. 1986, Community ecology of the Sonoran Desert *Drosophila*, in: *The Genetics and Biology of Drosophila*, Vol. 3e (M. Ashburner, H. L. Carson, and J. N. Thompson, eds), Academic Press, London, pp. 311-345.

Armengol, R., Sanchez, A., Ruiz, A., Heed, W. B., and Fontdevila, A., 1987, New data on the phylogenetic relationships in the *Drosophila mulleri* cluster, *Tenth European Drosophila Research Conference* p. 45 (Abstr.).

Etges, W. J., and Heed, W. B., 1987, Sensitivity to larval density in populations of *Drosophila mojavensis*: Influences of host plant variation on components of fitness, *Oecologia* **71**:375-381.

Sanchez, A., Armengol, R., Heed, W. B., Cerda, H., Benado, M., and Fontdevila, A., 1987, Allozyme relationships in the *mulleri* complex (Repleta Group) of *Drosophila*, *Tenth European Drosophila Research Conference* p. 221 (Abstr.)

Ruiz, A., and Heed, W. B., 1988, Host-plant specificity in the cactophilic *Drosophila mulleri* species complex, *J. Anim. Ecol.* **57**:237-249.

Heed, W. B., 1989, Origin of *Drosophila* of the Sonoran Desert II. In search for a founder event, in: *Genetics, Speciation, and the Founder Principle* (L. V. Giddings, K. Y. Kaneshiro, and W. W. Anderson, eds), Oxford Univ. Press, New York, pp. 253-278.

Fogleman, J. C., and Heed, W. B., 1989, Columnar cacti and desert *Drosophila*: the chemistry of host plant specificity, in: *Special Biotic Relationships in the Arid Southwest* (J. Schmidt, ed.), Univ. of New Mexico Press, Albuquerque, pp. 1-24.

Ruiz, A., Heed, W. B., and Wasserman, M., The evolution of the *mojavensis* cluster of cactophilic *Drosophila* with descriptions of two new species, *J. Hered.* (in press).

Starmer, W. T., Lachance, M., Phaff, H. J., and Heed, W. B., The biogeography of yeasts associated with decaying cactus in North America, the Caribbean, and northern Venezuela, *Evol. Biol.* (in press).

Heed, W. B., Sanchez, A., Armengol, R., and Fontdevila, A., Genetic differentiation among island populations and species of cactophilic *Drosophila* in the West Indies, this volume.

Ecological Genetics: Introduction

Ecological genetics concerns the adaptation of the individuals in natural populations to their habitats. Thus, in order to fully understand the adaptations, the phenotypes of individuals must be related to the ecological context of the population and to the underlying genetic bases of those phenotypes. The interplay between ecology, phenotypes and genotypes is complex. For example, the structure of a population (effective size, system of mating, subdivision and gene flow) can influence not only the speed of a genetic response to an environmental change, but can constrain the nature of that response as well. If the latter is true, the genetic basis of adaptations may not always represent "optimal" solutions. Understanding the genetic basis of traits associated with fitness differences is therefore a fascinating challenge for ecological geneticists. Unfortunately in normal environmental conditions, differences in fitness traits are often associated with low heritabilities. However, in situations of environmental stress, adaptations have been studied which have a more clearly detectable genetic basis.

There are two basic designs for analyzing the genetic basis of fitness differences. In the first or "quantitative trait loci mapping (QTL)", major or minor genes are mapped relative to defined chromosomal markers, such as allozyme loci or restriction fragment length polymorphisms (RFLPs). A clear disadvantage of this approach is that such maps of markers do not exist for most ecologically interesting organisms. Moreover, the construction of pedigrees to generate such maps inevitably involves breeding the organisms in the laboratory. Thus, in a new and surely very different environment, the phenotypic response and its inferred genetic basis may be altered.

The second basic design, the candidate locus approach, does not suffer from these disadvantages due to recent advances in molecular biology. Following this strategy, single genes with known metabolic functions are selected and studied for their effects on phenotypes which are likely to have ecological significance. With cloned probes and/or polymerase chain reaction (PCR) technology, allelic variation both at candidate loci and at interacting loci can be assessed without bringing the organisms into the laboratory for extended genetic analysis. The species which can be studied need not be restricted to those which can be bred in the laboratory.

Ecological and Evolutionary Genetics of Drosophila, Edited by
J.S.F. Barker *et al.*, Plenum Press, New York, 1990

The chapters in this section exemplify one or more of the above considerations. William Etges' analysis of genetic variation in *Drosophila mojavensis* populations from the Sonoran desert shows that coadapted gene complexes may not be involved in the adaptations of *D. mojavensis* to new host plants, despite numerous genetic differences in most life history traits associated with host plant shifts. Peter Parsons, in emphasizing the importance of environmental stress in uncovering adaptive genetic variation, has shown for dessication resistance in *D. melanogaster* that the selective response can be rapid and realized heritability can be high.

The nature of the genetic variants affecting fitness, particularly those variants acting in an additive fashion, are poorly understood. In recent years, the role of mobile genetic elements, or transposons, in generating these variants has become an important question. Michael Murphy and John Sved review the biology of the P factor in *Drosophila melanogaster* and describe how genetically marked elements can be used to potentially "tag" new fitness variants in population cage experiments. That the mobilization of the P factor can indeed generate additive genetic variance, is shown in the elegant experiments of Chris Moran and Adam Torkamanzehi. The advantage of this approach is that the loci contributing to the response to selection for bristle number could be marked with P factor insertions and ultimately be amenable to molecular cloning.

The candidate locus approach to identifying genes affecting fitness is exemplified by Alan Templeton and John McKenzie and their respective colleagues. In both cases, involving *Drosophila mercatorum* and *Lucilia cuprina*, single locus effects are shown most clearly in response to stress, as either a fluctuating environment (in Hawaii) or to the presence of an insecticide (in Australia). Both studies are also noteworthy because interacting, nonallelic loci have been identified in the phenotypic responses. In the analysis of abnormal abdomen in *D. mercatorum*, the use of cloned probes made direct field studies possible, demonstrating the power of the combination of the candidate locus approach and molecular biology.

It is hoped that these studies will serve as paradigms for ecological geneticists, both those working with *Drosophila* and with other organisms. As shown in these chapters, fitness differences, particularly those amplified in stressful environments, can be analyzed genetically and the molecular basis of the variation can now be understood. Future efforts should be directed towards candidate loci or major genes which can be or have been cloned, and for which interacting genetic loci have been identified (such as the heat shock genes). In this way, a fuller understanding of what Th. Dobzhansky referred to as "coadapted gene complexes" may eventually emerge.

The Ecological Genetics of Abnormal Abdomen in Drosophila mercatorum

ALAN R. TEMPLETON, HOPE HOLLOCHER, SUSAN LAWLER, AND
J. SPENCER JOHNSTON

1. INTRODUCTION

Natural selection and adaptation are central to theories of organismal evolution. In order to study natural selection in the field, four basic problem areas must be investigated as part of an ecological genetics research program. The first problem is to document the genetic basis of the phenotypic variation observed in nature. Without genetic variation underlying the phenotypic variation, it is impossible to have an evolutionary response to selective forces. The second problem is to determine the phenotypic expression of this genetic variation under field conditions. Phenotypes are the result of genotype-by-environment interactions, and there is no guarantee that phenotypic expression in the laboratory corresponds to that found in nature. The third problem is to relate the phenotypic variation to variation in life history. Natural selection requires fitness differences, and the potential for selection cannot be evaluated in a predictive fashion unless differences in life history attributes can be inferred. The fourth problem is to estimate the structure of the natural population. Evolutionary responses to natural selection represent an interaction between genetic architecture, fitness differences, and population structure (Templeton, 1982b). Hence, even if the genetic basis and fitness effects of a phenotypic syndrome were known, it is still not possible to make predictions about how allele or genotype frequencies will shift in nature in response to ecological conditions unless the relevant features of the population structure are also known.

Molecular genetic techniques are increasing our ability to perform such ecological genetic studies. Many ecological genetic investigations deal with continuous phenotypic variation

ALAN R. TEMPLETON, HOPE HOLLOCHER, AND SUSAN LAWLER * Department of Biology, Washington University, St. Louis, Missouri 63130, USA; J. SPENCER JOHNSTON * Department of Entomology, Texas A & M University, College Station, Texas 77843, USA.

Ecological and Evolutionary Genetics of Drosophila, Edited by
J.S.F. Barker *et al.*, Plenum Press, New York, 1990

that has traditionally been studied using quantitative genetics. However, an alternative is the candidate locus or measured genotype approach (Templeton and Johnston, 1988). With this approach, one begins with phenotypic variation. Picking a phenotypic syndrome that can be related to individual life history differences will maximize the chances for successfully studying the third problem noted above. One next uses accumulated knowledge from biochemistry, physiology, and/or development to identify loci whose known function can be directly related to the phenotypic variation of interest. These loci are called candidate loci. Molecular genetic techniques can often be used to see if genetic variation at the candidate loci can explain any of the observed phenotypic variation. If so, one has at least partially solved the first problem mentioned in the previous paragraph. However, unlike the usual quantitative genetic approach, one now has specific loci and variants, and these can often be measured independently of the phenotype in wild-caught individuals. This allows one to study genotype-by-environment interactions in nature in a more straightforward fashion, and hence increases the feasibility of successfully studying problem two. Finally, molecular genetic techniques can also be used in a powerful fashion to estimate many features of population structure. When all this information is available, it is possible to make specific predictions about adaptive response in nature. These can then be tested by making long-term field studies on the natural population.

The purpose of this paper is to describe our attempts to implement such a candidate locus approach to study natural selection in the cactophilic *Drosophila*, *D. mercatorum*.

2. THE *aa* SYNDROME IN *D. MERCATORUM*

In 1975, a polymorphic syndrome known as abnormal abdomen (*aa*) was discovered in natural populations of *Drosophila mercatorum* living near Kamuela, Hawaii. The syndrome is associated with a slow down in the egg-to-adult developmental time of females, the retention of juvenile abdominal cuticle in the adult, increased early fecundity in adult females, and decreased longevity of adult females (Templeton and Rankin, 1978; Templeton, 1982a, 1983). Hence, this syndrome directly affects life history and was an excellent phenotypic candidate for an ecological genetic program.

Many of the phenotypes described above are also associated with the bobbed syndrome found in *D. melanogaster*. This phenotypic syndrome is known to be caused by a deficiency of the 18S/28S ribosomal DNA (rDNA), which are found in tandem clusters on the X and Y chromosomes of most *Drosophila*. Hence, the rDNA was identified as a candidate locus for the *aa* syndrome. The rDNA was further implicated by a Mendelian genetic analysis that revealed that the *aa* syndrome depends upon two major elements that are about half a map unit apart and located near the heterochromatic end of the the X chromosome (Templeton *et al.*, 1985). In addition, a Y-linked element controls the morphological expression in males (Templeton *et*

al., 1985). Hence, the genetic locations for the major determinants of *aa* are the same as those for the 18S/28S rDNA.

Accordingly, molecular studies were undertaken on the
18S/28S rDNA of *D. mercatorum*. DeSalle *et al.* (1986) discovered that *aa* lines all had at least a third or more of their
28S genes bearing a 5kb insert. This insert is located in the
ß portion of the 28S genes and disrupts normal transcription
(DeSalle *et al.*, 1986). Hence, 18S/28S units bearing the insert are functionally inactivated, and X chromosomes bearing a
large proportion of inserted 28S genes have a deficiency of
functional rDNA genes even though the total number of rDNA
genes may be normal. However, we subsequently found that
strains having up to 90% of their 28S genes with the insert
will not express the syndrome unless a second criterion is
satisfied: there must be no preferential underreplication of
inserted repeats in polytene tissues such as the larval fat
body (DeSalle and Templeton, 1986). Euchromatic DNA is
greatly amplified in polytene tissues, but the rDNA tends to
be underreplicated relative to the euchromatin. In *aa* flies
this underreplication is uniform across the rDNA cluster, but
in non-*aa* flies with a large portion of their 28S genes bearing the insert, there is preferential underreplication of the
inserted 28S repeats (DeSalle and Templeton, 1986). Because
of this preferential underreplication, the noninserted functional 28S repeats are effectively overreplicated relative to
the nonfunctional inserted repeats. The presence or absence
of preferential underreplication is controlled by an X-linked
locus, with the allele coding for no preferential underreplication (hereafter referred to as *aa*) acting as a recessive for
the morphological effects of the syndrome. The dominant allele causes preferential underreplication for both cis and
trans rDNA repeats (DeSalle and Templeton, 1986).

In summary, the *aa* syndrome in *D. mercatorum* requires two
X-linked molecular events: 1) about a third or more of the
28S rDNA genes must bear a 5 kb insert, and 2) there must be
no preferential underreplication of inserted 28S genes in the
larval fat body.

2.1. Components of Genetic Variation in *aa* in Natural
Populations

To examine the contributions of the two X-linked components
to the genetic architecture of *aa* in natural populations, we
simultaneously characterized X chromosomes derived from nature
for their underreplication properties and for their proportion
of inserted 28S genes. It is impractical to do population
screening for the underreplication polymorphism using the procedures given in DeSalle and Templeton (1986), which involve
the isolation of DNA from the larval brains and fatbodies of
the same individuals. Accordingly, an alternative procedure
was developed. A standard *aa* stock had been bred which has
extremely high penetrance of juvenilized cuticle (Templeton *et
al.*, 1985). This *aa* stock has 80% of its 28S genes bearing
the insert and is homozygous for the recessive allele leading
to no preferential underreplication. When a wild caught male

(or son of a wild caught female) is crossed to an *aa* female, the resulting female offspring will have at least 40% of their 28S genes bearing the insert even if the X chromosome from the wild caught male bears no inserts at all. Hence, the resulting female progeny will always have sufficient inserted 28S genes to support the morphological expression of *aa*. This expression will only occur if these females are homozygous for the allele causing no preferential underreplication. Consequently, if the female offspring display juvenilized abdominal cuticle, the X chromosome extracted from the wild caught male must bear the *aa* allele for no preferential underreplication. Controlled experiments indicate that this morphological scoring procedure does indeed identify underreplication properties of X chromosomes (Templeton *et al.*, 1989). Finally, we score 50 female offspring to obtain quantitative estimates of penetrance (the percentage of flies with juvenilized cuticle) and expressivity (the average amount of abdominal cuticle that is juvenilized, given expression).

After the wild caught males have mated to *aa* females, we extract their DNA, restrict it with EcoRI, perform a Southern, and probe with a rDNA clone. The proportion of the 28S genes bearing the 5 kb insert can then be estimated by densitometry (Templeton *et al.*, 1989).

Figure 1 shows the results obtained by scanning 425 X chromosomes extracted from the natural population near Kamuela, Hawaii. Almost all X chromosomes have more than a third of their 28S genes bearing the insert. This means that insert proportion is not a limiting factor for *aa* expression in nature. In contrast, the test cross results reveal extensive polymorphism for the underreplication trait, with the fre-

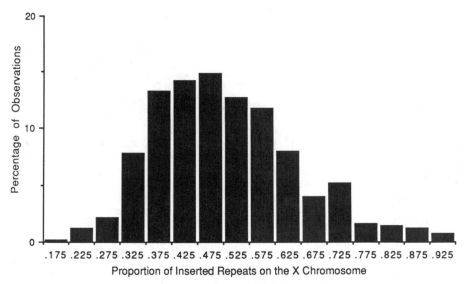

Figure 1. The proportion of 28S rDNA repeats that bear the *aa* insert in 425 X chromosomes extracted from a natural population of *D. mercatorum* living in the vicinity of Kamuela, Hawaii.

quency of the *aa* allele being 0.41 in this sample. These results imply that the natural population is indeed polymorphic for the presence or absence of *aa*, but that allelic variation at the underreplication locus is the primary source of variation in the presence or absence of the syndrome in *D. mercatorum*.

Combining the DNA and testcross data, we found that X chromosomes bearing the *aa* allele at the underreplication locus have a mean and standard deviation of insert proportion of 0.50 and 0.14 respectively, while + (non-*aa* at the underreplication locus) bearing X chromosomes have values of 0.51 and 0.14 respectively. There are no significant differences between these values. Using the total sample to define decatiles for insert proportions, there is also no significant difference between *aa* and non-*aa* X chromosomes in their distributions over these decatile intervals (chi-square = 5.54, 9 degrees of freedom). This lack of association does not imply that insert proportion plays no role in the expression of the *aa* syndrome. When we restrict our inference only to those X-chromosomes bearing the *aa* allele at the underreplication locus, there is no correlation between the insert proportion and *aa* penetrance in the testcross offspring, but there is a significant (5% level) correlation with expressivity of 0.16. This correlation is low, so quantitative variation in insert proportion is behaving only as a minor modifier of expression, given the presence of the *aa* allele at the underreplication locus.

2.2. Phenotypic Expression of *aa* in Nature

An understanding of the evolutionary significance of *aa* requires knowledge of the translation from genotype to phenotype. Our previous work involved inbred laboratory strains that display the morphological effects of the syndrome. However, morphologically abnormal flies are almost never observed in nature. There are probably at least two reasons for this. First, good morphological expression is obtained at 25°C, but not at lower temperatures. Our field studies indicate that the temperatures larvae experience would typically be lower than 25°C, so morphological expression should be reduced. Second, the natural population contains suppressors of the morphological expression. However, the significance of *aa* as a model system for natural selection does not lie in its morphological effects, but instead in its life history effects. Therefore, we wanted to know if *aa* expresses its life history effects in the naturally occurring genetic and environmental backgrounds that suppress the morphological effects.

To examine the effects of genetic background, we established two lines that differ at the *aa* locus from a December, 1986 collection as follows. By testcrossing several sons, several females caught in the same cactus patch were classified as being *aa/aa*, *aa/+*, and *+/+* at the underreplication locus. The male offspring of four *aa/+* were saved, and four females of each homozygous genotype were allowed to exhaust their sperm. Males bearing the *aa* allele were mass mated to the *aa/aa* females, and matched sib males bearing the + allele were mated to the +/+ females. In this way, we established two homozygous lines with no inbreeding and similar genetic backgrounds. The strains are designated K+X and K*aa*X. The life history studies were done shortly after the lines were established in order to minimize alterations in the genetic background associated with laboratory adaptation.

A detailed analysis of the experiments performed with these stocks will be presented elsewhere, so only a brief description of the results follows. There was no occurrence of juvenilized cuticle in the adults of the K+X strain among 100 females and 100 males, as expected. Of 100 males from the K*aa*X strain, none had juvenilized cuticle, and only 4 of 100 females showed juvenilized cuticle. Of these 4 females, 3 had only a small patch of juvenilized cuticle, and the fourth had a large patch that covered less than 50% of the abdominal cuticle. Hence, there is little morphological expression of *aa* in the K*aa*X stock. Of greater interest are life history phenotypes. Figure 2 presents the adult female survivorship results. The two stocks were significantly different, with the *aa* stock having a significantly lower adult survivorship (male survivorship was the same). The daily fecundity of these two strains was also measured for the first two weeks after eclosion. The K*aa*X strain had significantly higher fecundity for the first week after eclosion, but not the second week. Finally, the two strains showed a significant difference in eclosion time, with the K+X strain having a mean egg-to-adult time of 10.45 days, and K*aa*X 12.18 days. Hence, all the expected life history phenotypes are expressed on this genetic background that suppresses morphological expression.

Since phenotypes represent genotype-by-environment interactions, these laboratory studies have one major drawback - the ultimate significance of this syndrome is determined by its expression in the field environment, not the laboratory envi-

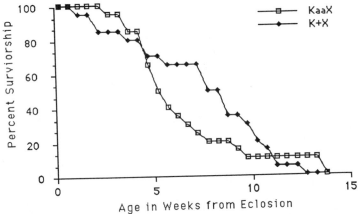

Figure 2. The survivorship schedules for the K*aa*X and K+x stocks. Starting with 20 females and 20 males, survivorship of males and mated females was monitored twice a week. This figure only shows the results for females.

Table I
Emergence patterns of *aa/-* females versus *+/+* females from natural rots[a]

Rot	D	N	n	Rank
B-2 '82	3.00	2	1	16.0
F-9 '82	1.75	14	2	18.5
C-4 '83	-1.50	5	2	-13.5
C-6 '83	0.67	9	3	6.0
F-3 '84	3.66	27	6	21.0
IV #4 '84	-1.12	14	3	-10.0
IV #7 '84	0.59	79	22	3.5
F #2 '85	-0.07	8	1	-1.0
IV #4 '85	1.37	47	20	12.0
B-1 #1 '87	0.73	41	18	8.0
B-1 #2 '87	2.63	65	30	17.0
B-1 #3 '87	-3.13	10	2	-20.0
B-1 #4 '87	0.59	96	42	3.5
F #2 '87	0.00	2	1	–
IV #1 '87	1.18	15	4	11.0
IV #2 '87	3.00	3	1	18.5
B-1 #1 '89	4.69	68	16	22.0
B-1 #2 '89	1.64	118	48	15.0
B-1 #3 '89	-0.35	23	5	-2.0
B-1 #4 '89	-0.63	5	1	-5.0
B-1 #5 '89	-1.50	2	1	-13.5
F #3 '89	0.70	12	2	7.0
F #5 '89	0.88	5	1	9.0

Average D: 1.44			Sum of - Ranks: -65.0	

[a]The first column gives the bagged rot designation, indicating both the cactus site and year of bagging. D is the observed difference in days to emergence after bagging between *aa/-* females and *+/+* females. N is the total number of females that emerged from the rot, and n is the minimum of the number of *aa/-* or *+/+* females. The average D over all rots is weighted by the n's. Finally, the last column gives the rank of the magnitudes of the D's, with the ranks then given the sign of the corresponding D.

ronment. To overcome this drawback, we needed to measure the life history phenotypes in the field. These field studies are most easily performed in the Hawaiian populations of *D. mercatorum*. Like most other repleta group flies, *D. mercatorum* is native to the American mainland, but populations have been introduced to Hawaii where their only larval food resource is rotting cladodes of *Opuntia megacantha* (Johnston and Templeton, 1982). We have placed mesh net bags over such rots and aspirated the emerging progeny every day. These progeny were then testcrossed to determine their genotype. Table I shows the average difference in days to emergence between females that were *aa/-* (where "-" can be either "+" or "*aa*" since the testcross results do not allow us to discriminate between females homozygous and heterozygous for *aa*) versus *+/+* for several bagged rots. Because the progeny emerging from any one rot may be the offspring of the same parents, the progeny cannot be regarded as independent observations. Accordingly, we will regard each cladode as the sampling unit

Table II

Emergence patterns of *aa* males versus + males from
natural rots[a]

Rot	D	N	n	Rank
F-9 '82	0.87	30	3	6.0
C-4 '83	4.00	3	1	17.0
C-6 '83	-0.22	13	3	-3.0
F-3 '84	4.64	33	10	19.0
IV #4 '84	2.44	13	6	12.0
IV #7 '84	-0.27	78	22	-4.0
B-1 #3 '85	3.17	4	1	13.0
F #2 '85	-1.38	9	1	-8.0
IV #4 '85	-3.26	36	7	-14.0
B-1 #1 '87	-0.05	35	14	-1.5
B-1 #2 '87	4.25	51	24	18.0
B-1 #3 '87	5.87	6	2	20.0
B-1 #4 '87	-1.13	97	48	-7.0
F #2 '87	0.00	3	1	–
IV #1 '87	0.00	9	3	–
IV #2 '87	1.50	3	1	10.0
B-1 #1 '89	-0.72	45	10	-5.0
B-1 #2 '89	0.05	140	55	1.5
B-1 #3 '89	1.41	24	8	9.0
B-1 #4 '89	-3.50	2	1	-15.0
B-1 #5 '89	2.00	2	1	11.0
F #3 '89	-1.20	6	1	-16.0
F #5 '89	0.00	3	1	–

Average D: 0.46   Sum of - Ranks: -73.5

[a]The first column gives the bagged rot designation, indicating both the cactus site and year of bagging. D is the observed difference in days to emergence after bagging between *aa* males and + males. N is the total number of males that emerged from the rot, and n is the minimum of the number of *aa* or + males. The average D over all rots is weighted by the n's. Finally, the last column gives the rank of the magnitudes of the D's, with the ranks then given the sign of the corresponding D.

and test for trends across cladodes. We test for such a trend using the Wilcoxon signed rank test to test the null hypothesis that *aa*/- and +/+ females have the same egg-to-adult developmental times under field conditions. The observed emergence differences are ranked irrespective of sign, and then the sums of the negative ranks are taken as a test of the null hypothesis. The sum of the negative ranks is only -65, which results in a rejection of the null hypothesis at the 5% level using a two-tailed test. Hence, the *aa*/- females are emerging significantly later than the +/+ females in natural rots. The average emergence time difference (weighted across rots by the minimum of the number of *aa*/- or +/+ females from a rot) is 1.44 days.

This emergence time difference could be due to egg-to-adult developmental time differences or due to *aa*/- females ovipositing later than +/+ females. Fortunately, there is a built in control to this experiment that discriminates between these two explanations; namely, the emerging males. Males only rarely express *aa* because most Y chromosomes suppress its expression (Templeton *et al.*, 1985). As shown in Table II, males bearing *aa* X chromosomes do not emerge from natural rots significantly later than males bearing + X chromosomes (the negative Wilcoxon rank sum is -73.5, with P > 0.2), and the average observed emergence time difference is only 0.46 days. This indicates that females bearing *aa* alleles tend to oviposit at the same time as +/+ females. Hence, females bearing *aa* alleles display an egg-to-adult developmental slow-down under natural environmental conditions.

The evolutionary significance of an allele is determined by what phenotypic effects can be transmitted through haploid gametes to future generations, and this is measured by the average excess (Templeton, 1982b). The average excess in this case is simply the average developmental time of females carrying the *aa* allele minus the mean developmental time for the entire population of females. Both of these quantities are easily calculated, so the average excess of egg-to-adult developmental time of the *aa* allele under natural conditions is 0.94 days; that is, on the average, females bearing an *aa* allele have an egg-to-adult developmental time that is about one day longer than the average developmental time for the entire female population.

Information about fecundity is also available from the bagged rot experiments. Over the years, we have been able to rather consistently pick rots such that most of the adults emerged from the bagged rots 1.5 to 3 weeks after bagging. This time period corresponds to the egg-to-adult developmental time for *D. mercatorum* in the laboratory and implies that we bagged the rots about the same time that females were actively ovipositing upon them. While bagging the rots, we often collected a sample of adult females from the same cactus patch. These females were brought into the laboratory and their male offspring were testcrossed to determine their frequency of *aa* X chromosomes. Similarly, the frequency of *aa* X chromosomes in the males emerging from bagged rots was determined by testcrossing. Finally, a second sample of flies from the same cactus was aged using the techniques described in Johnston and Ellison (1982) and/or the ecological conditions under which the collection was made were noted and related to our previous studies on the ecological determinants of age structure (Johnston and Templeton, 1982).

All of these data can be combined to estimate the age-specific fecundity effects of *aa* X chromosomes under natural conditions. Under neutrality, the frequency of *aa* X's in the sons of wild-caught females should be the same as that from the males emerging from bagged rots. The design insures that there is no opportunity for selection to alter the *aa* X frequencies observed in the sons from that found in their wild-caught female parents, so this frequency estimates the frequency of *aa* found in wild-caught females present in the cactus at the time of bagging. The frequency in bagged rot males is an estimate of the frequency of *aa* X chromosomes borne by

Table III

Estimates of the average excess of female fecundity caused by *aa* under
field conditions[a]

Cactus site	Year	Ecological conditions	Age structure	*aa* freq. in sons	*aa* freq. in rots	$\Delta p/p$
B	1982	Humid	>2 weeks	0.318	0.000	-0.318
IV	1982	Humid	>2 weeks	0.320	0.308	-0.038
C	1983	Dessicating	1 week	0.375	0.667	+0.779
F	1984	Dessicating	1 week	0.565	0.697	+0.233
IV	1984	Humid	>2 weeks	0.364	0.310	-0.146
B	1987	Kona Storm	1 week	0.311	0.526	+0.692[b]
	Pooled:		>2 weeks	0.329	0.304	-0.075
			1 week	0.376	0.559	+0.488[c]

[a]The cactus sites are indicated in Figure 3. The age structure was either
estimated directly or inferred from the ecological conditions. $\Delta p/p$ is
the frequency of *aa* in males emerging from bagged rots minus the frequency
in sons of wild-caught females, divided by the frequency in sons of wild-
caught females. This should be equal to the average excess in fecundity
association with the *aa* allele under the convention that average relative
fecundity is equal to 1.
[b]Significantly different frequencies in sons vs. males from rots at the 5%
level using the arcsin, square root test given in Templeton and Johnston
(1982).
[c]Significantly different frequencies in sons vs. males from rots at the 1%
level using the arcsin, square root test given in Templeton and Johnston
(1982).

ovipositing females weighted by their fecundities and then potentially modified by further selection on male larval/pupal
viability and developmental time differences. However, recall
that *aa* is only rarely expressed in males. Hence, by scoring
aa frequencies only through the males emerging from the bagged
rots, we should eliminate larval and pupal fitness differences. This leaves female fecundity as the principal source
of any observed frequency difference.

These frequency differences can be used to estimate the average excess of fecundity associated with *aa* under natural
conditions. From the equations in Templeton (1982b) and using
the convention that the average relative fecundity in the population is unity, the average excess in fecundity of *aa* is
estimated by $\Delta p/p$, where p is the frequency of *aa* in the sons
of wild-caught females, and Δp is the frequency in bagged rot
males minus the frequency in sons.

The data obtained from these types of experiments over several years are presented in Table III. Because of small sample sizes, only one Δp is significant at the 5% level, and it
represents an excess frequency in bagged rot males over sons
under ecological conditions associated with a young age
structure (Templeton *et al.*, 1989). However, the pattern for
all the other samples is remarkably consistent when classified
according to age pattern: all years associated with an old

age structure (most adult flies being greater than two weeks old as measured from eclosion) have negative Δp's; all years with young age structures have a positive Δp. Pooling over years with similar age structures, the estimated average excess for fecundity caused by aa chromosomes is +0.488 when the age structure is young (significant at the 1% level). When the age structure is old, the average excess is slightly negative (-0.075) but not significantly different from 0. These results are consistent with laboratory studies that indicate that aa increases first week female fecundity but has no significant effect on later fecundity.

2.3. The Adaptive Significance of aa

These studies show that the aa syndrome has the potential for a fitness trade-off between developmental time and early fecundity under natural conditions. Standard life history theory can be used to quantify the potential effects of these trade-offs on individual fitness (Templeton and Johnston, 1982). Let x be the age from hatching, e the age at eclosion, a=x-e the adult age from eclosion, n the maximal adult age, l_x the probability of survival to age x, $l_a = l_{a+e}/l_e$ the adult survivorship to age a, b_a the expected number of eggs laid by an adult female of age a, and r the Malthusian parameter, which is the overall fitness measure. Then, from the Euler-lotka equation,

$$l_e \sum_{a=1}^{n} l_a b_a e^{-r(a+e)} = 1 \qquad (1)$$

$$r \approx \frac{\sum_{a=1}^{n} l_a b_a - 1/l_e}{\sum_{a=1}^{n} l_a b_a (a+e)} \qquad (2)$$

with approximation (2) holding whenever r is close to 0. As can be seen from equation (2), the two phenotypic effects of aa measured in the field have opposite effects on fitness. Equation (2) is a decreasing function of e, so the delay in eclosion time observed in females bearing aa should decrease their fitness. In contrast, r is an increasing function of the b's, so the increased early fecundity of aa bearing females should increase their fitness. The net effect of these two antagonistic fitness components depends critically upon the adult age structure and other parameter values, as will now be illustrated.

We will use equation (2) to investigate the potential for aa to influence individual fitness using the phenotypic data gathered both in the laboratory and the field. We have only incomplete data on l_a from the field because the aging techniques are only reliable for about two weeks after eclosion. In the laboratory, adult flies can live several months, and in the field under humid conditions, almost all wild-caught flies are older than two weeks of age (Johnston and Templeton,

Table IV
The fecundity schedule used to evaluate equation (2) for *aa* and non-*aa*
flies as estimated from laboratory data[a]

Age in weeks since eclosion	Age specific fecundity (b_a)	
	aa	+
0.5	73	71
1.0	216	176
1.5	218	218
2.0	218	218
≥2.5	218	218

[a]The original data were obtained on a daily basis, but since longevity was
scored only on each third and seventh day of the week, the fecundities are
converted into half-week values. Further details are given in the text.

1982). Consequently, to evaluate the potential for fitness
differences under humid conditions, we will have to use laboratory determined l_a schedules which are compatible with the
field observations. It can be seen from Figure 2 that the
survivorship schedules in the first two weeks were high in
both the K+X and K*aa*X stocks, so these results are compatible
with the field observations obtained under humid conditions.
We will therefore use these survivorships in equation (2).

To evaluate equation (2), we also need age-specific fecundities for *aa* and + flies. The data in Table III indicate a
first week fecundity advantage but no significant difference
thereafter under field conditions. This is exactly the same
pattern as observed in the laboratory tests on K+X and K*aa*X.
We will therefore use the laboratory b_a schedules as shown in
Table IV. Since there were no significant differences after
the first week, this table gives the average over the two
lines for ages over two weeks. These experiments were not
continued after two weeks, so no information on late fecundity
is available. Hence, we simply continue the second week values
throughout the rest of the adult lifespan.

Equation (2) also depends on e, the egg-to-adult time. In
the laboratory studies, the difference between *aa* and + was
1.68 days, which is not significantly different from the 1.48
days estimated from the bagged rot experiments. Therefore, we
will use e values obtained from the laboratory strains of
e=1.74 weeks for *aa* and 1.50 weeks for +. Finally, to evaluate
equation (2), we also need information on larval viability
(l_e). No information exists on this in the field, and in the
absence of information we will assume that there are no differences in this trait. To insure that r is close to 0 for
both genotypes, we also assume that

$$1/l_e = [R_a(aa) + R_a(+)]/2 \qquad (3)$$

where

$$R_a(j) = \sum_{a=1}^{n} l_a(j)b_a(j) \qquad j = aa, + \quad (4)$$

Using these values, we obtain $r(aa) = -0.0122$ and $r(+) = 0.0100$. Hence, under adult survivorship conditions that are compatible with those observed in nature under humid conditions, aa is at a selective disadvantage.

The adult age structure changes drastically under desiccation (Johnston and Templeton, 1982). Under these environmental conditions, the adult mortality rate is 19% per day (Templeton and Johnston, 1982), and almost all the adult population dies within the first week after eclosion (Johnston and Templeton, 1982). To estimate the potential fitness effects of aa under desiccation, we calculated l_a assuming a 19% daily mortality rate, keeping all other parameter values as before. Under this age structure, $r(aa) = +0.0123$ and $r(+) = -0.0143$.

As these calculations show, there is not only a great potential for strong selection on aa, but the direction of selection is very sensitive to adult age structure. In particular, when the adult age structure is skewed towards the young ages, the early fecundity advantage of aa dominates and aa is selectively favored. When the adults are very long-lived, the delay in larval developmental time dominates and aa is at a selective disadvantage. These are the same fitness predictions made by Templeton and Johnston (1982) using a different modelling approach. In both cases, we conclude that aa is basically adaptive to the demographic environment of a young adult age structure. Note that the favorable selection for aa under desiccation is not due to desiccation resistance, but rather follows indirectly from the demographic implications of desiccation upon age structure.

3. POPULATION STRUCTURE

The natural population in Hawaii is found on patches of *Opuntia megacantha* that grow on the slopes of the Kohala Mountains and in the saddle between Kohala and Mauna Kea. Under normal weather conditions, there is a very steep humidity and rainfall gradient in this region, with the upper elevations of the cacti distribution being very wet and humid and the lower elevations very dry. The saddle region can be very wet or dry depending upon the winds that blow through the saddle. A transect on the mountainside near Puu Kawaiwai was set up because it covers an extreme range of humidity conditions under normal weather conditions. An additional collecting site was established in the saddle near this transect, as shown in Figure 3. These collecting sites are within 3 km of one another, so before any predictions can be made concerning adaptive response to this spatial heterogeneity in the environment, it is critical to estimate the amount of gene flow between the collecting sites.

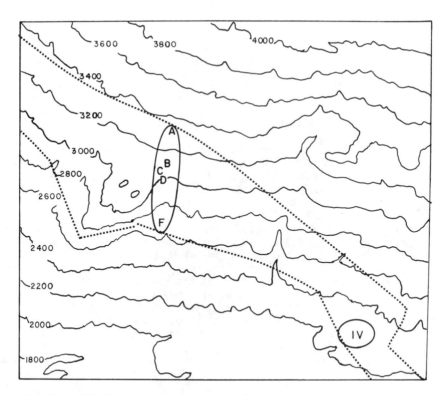

Figure 3. The collecting sites near Kamuela, Hawaii. The contour lines show elevation above sea level in feet in 200-ft increments. The dotted line shows the approximate distribution of the cactus *Opuntia megacantha*. Sites A through F represent a transect on the hillside. Site IV lies in the saddle between Kohala and Mauna Kea.

Johnston and Templeton (1982) studied dispersal using marked capture/recapture techniques and concluded that dispersal is very limited, primarily because the strong trade winds in this area prevent movement between cactus patches most days of the year. Even when dispersal does occur, it is primarily between adjacent patches, leading to the potential for much isolation by distance. They further showed that these dispersal rates were sufficiently low that local adaptation on the transect should be possible. DeSalle *et al.* (1987) examined population structure by utilizing genetic surveys of nuclear and mitochondrial DNA variability. By combining these two types of DNA studies, they showed that the product of the inbreeding effective size with the gene flow rate, Nm, was bounded between 4 and 8. This value is sufficiently large to insure virtual panmixia for neutral nuclear markers, but it is sufficiently small to allow local differentiation for selected markers. Consequently, if the environmental heterogeneity discussed above induces strong changes in the fitnesses associated with *aa*, the natural population has the potential for showing local adaptation despite the small geographical distances involved.

4. TESTING THE ADAPTIVE RESPONSE PREDICTIONS

From the results of the previous sections, we can make the prediction that *aa* is adaptive under the ecological conditions that favor a young adult age structure, and that patterns of local adaptation to heterogeneity in these conditions should be observable in our study area. To test these predictions (which were first made in Templeton and Johnston, 1982), we have monitored the frequency of the *aa* allele at the under-replication locus by testcrossing wild-caught males and sons of wild-caught, inseminated females, as described earlier. These frequency data are then combined with data on spatial and temporal variation in the relevant ecological parameters in order to test our predictions about adaptive response.

As shown in Templeton *et al.* (1987) and Templeton and Johnston (1988), there was considerable spatial and temporal variation in the frequency of *aa* from 1980 through 1985. Throughout this period, *aa* tended to be in its highest frequencies under dry conditions, which are associated with young adult age structures (Johnston and Templeton, 1982). Because humidity has shown extreme spatial and temporal fluctuations, the tracking of the *aa* frequency to these fluctuations is strong support for the idea that the fitness trade-offs outlined earlier are real and that natural selection is operating upon this syndrome.

These results are consistent with either the hypothesis that *aa* is an adaptation to desiccation or that it is an adaptation to the young adult age structure associated with desiccation. The collections made in December of 1986 and 1987 allowed us to discriminate between these possibilities. During the winter, "Kona" storms are more common. These storms produce widespread and prolonged rainfall. In both of these December collections, Kona storms preceded our collecting period, producing a brief period of extremely wet conditions. At the top of our transect, the cacti grow in large clumps that accumulate many fallen pads. The wet conditions caused these accumulated pads to rot, creating a temporary superabundance of larval food resource. This in turn led to a population explosion and a young age structure, even though the conditions were very humid (Templeton *et al.*, 1989). In contrast, the increase in larval food resource at the bottom of the hill (where the cacti are small and do not accumulate many fallen pads) was very modest and the age structure is older (Templeton *et al.*, 1989). Table V presents the *aa* frequency data from 1986 and 1987, and as can be seen the frequencies at the top of the hill are significantly higher than those at the bottom. This is exactly the opposite of the normal pattern (see Templeton *et al.*, 1987; Templeton and Johnston, 1982, 1988). This reversal of the cline is strong evidence that *aa* is truly a demographic adaptation and not a desiccation adaptation.

Starting in 1986, we also began monitoring the other genetic component of the *aa* syndrome, the proportion of inserted 28S rRNA genes. As noted earlier, insert proportion only af-

Table V
Frequency of the *aa* allele in *D. mercatorum* as a function of collecting
site in 1986 and 1987[a]

	Site	Altitude	'86 freq. of *aa*	Sample size	'87 freq. of *aa*	Sample size
Large	A	1036m	-	-	0.33	6
cacti	B	950m	0.43	80	0.47	266
	C	930m	0.43	46	0.47	47
	D	880m	0.43	40	0.33	18
Totals for top sites:			0.43	166	0.46	337
Small	F	795m	0.38	26	0.26	31
cacti	IV	670m	0.32	60	0.34	58
Totals for bottom sites:			0.34	86	0.31	89

[a]In both years, there is no significant heterogeneity among sites within
the top and bottom areas, but the frequency for the top sites is significantly higher (at the 5% level) than the frequency at the bottom sites.

fects the degree of expression given the presence of *aa* at the
underreplication locus in the laboratory testcrosses. We have
finished our survey of the insert proportions for the 1986
collection, and the results indicate that the contribution of
insert proportion to *aa* expression in nature is also
conditional upon the underreplication locus. An ANOVA of all
X chromosome insert proportions revealed a significant effect
of site of capture (F=3.62 with 5 and 232 degrees of freedom,
P < 0.01). However, when the analysis was redone by using
seperate ANOVA's on the X chromosomes characterized as *aa* vs.
non-*aa* based upon the testcross results, a dramatic contrast
was uncovered. For the non-*aa* X chromosomes, the resulting F
statistic was 0.88 with 4 and 129 degrees of freedom, which is
not significant. However, for the *aa* X chromosomes, F=3.94
with 4 and 85 degrees of freedom, which is significant at the
1% level. Both Tukey and Bonferonni multiple comparison tests
revealed significant differences between insert proportions in
aa chromosomes, with the insert proportion decreasing with decreasing altitude, as shown in Table VI. This cline parallels
the frequencies of the *aa* allele at the underreplication locus
during 1986, which were also decreasing with decreasing altitude (Table V). Recall that there is no overall association
between insert proportion and *aa* alleles, so this result is
not explicable as a hitchhiking effect. Rather, it suggests
that natural selection is indeed operating upon the insert
proportions, but only on those X chromosomes that also bear
the *aa* allele at the underreplication locus. Hence, both
molecular components of the *aa* syndrome respond to natural selection, but the response of insert proportions is conditional
upon the genetic state at the underreplication locus. Hence,
this is an example of coadaptation at the molecular level.

Table VI

The insert proportions in *aa* X chromosomes from wild caught *D. mercatorum*
as a function of collecting site[a]

Site	Altitude	Proportion	Sample size
B	950m	0.59	28
C	930m	0.50	18
D	880m	0.59	16
Totals for top sites:		0.56	62
F	795m	0.46	20
IV	670m	0.49	8
Totals for bottom sites:		0.47	28

[a]Sites that share a vertical line in a column were not significantly different by either the Tukey or Bonferoni multiple comparison procedure. The
average for the top sites is also signficantly different from the average
for the bottom sites.

5. CONCLUSION

The results summarized above strongly indicate that natural
selection is operating on both of the major genetic components
underlying the abnormal abdomen phenotypic syndrome. Moreover, this selection is mediated through the interaction of
the life history effects of *aa* with the ecological factors determining adult age structure. These strong conclusions are
possible because our studies allowed us to make *a priori* predictions about the expected adaptive responses to both spatial
and temporal environmental heterogeneity. These predictions
could then be tested directly in the field through long-term
genetic and ecological monitoring of the natural population.
This is perhaps the greatest strength of the candidate locus
approach. The naturally occurring variations in environmental
parameters create the analogue of test treatments in empirical
studies, but to utilize these potential treatment effects in
an analytical design, we must have an easily measured response
variable. In the case of adaptive responses caused by natural
selection, this response must be a genetically based one that
is manifest at the local population level. The candidate locus approach greatly simplifies the long-term genetic monitoring of natural populations that is necessary for making ecological genetics into an empirical, rather than just a
descriptive, science.

6. SUMMARY

Natural populations of *Drosophila mercatorum* are variable
for the number of X-linked 28S ribosomal genes bearing a 5 kb
insert. A separate polymorphic X-linked gene controls whether
or not 28S repeats bearing the insert are preferentially underreplicated during the formation of polytene tissue. Female
flies having at least a third of their 28S genes bearing the
insert and lacking the ability to preferentially underrepli-
cate inserted repeats display the abnormal abdomen (*aa*) syn-

drome. The syndrome is characterized by retention of juvenile abdominal cuticle in the adult, a slowdown in larval developmental time, and an increase in early female fecundity. The life history traits are expressed in nature, and provide a basis for strong natural selection. The *aa* syndrome should be favored whenever the adult age structure is skewed towards young individuals, and field studies confirm this prediction.

ACKNOWLEDGEMENTS. We wish to thank Rose Leightner for her excellent technical help that was so essential to this work and Ary Hoffmann for his suggestions for improving an earlier draft of this manuscript. This work was supported in part by NIH grant R01 AG02246 and in part by a National Science Foundation Graduate Fellowship awarded to Susan Lawler and a Genetics Training Grant (GM08036) awarded to Hope Hollocher.

References

DeSalle, R., Slightom, J. and Zimmer, E., 1986, The molecular through ecological genetics of abnormal abdomen in *Drosophila mercatorum*. II. Ribosomal DNA polymorphism is associated with the abnormal abdomen syndrome in *Drosophila mercatorum*, *Genetics* 112:861-875.

DeSalle, R., and Templeton, A. R., 1986, The molecular through ecological genetics of abnormal abdomen in *Drosophila mercatorum*. III. Tissue-specific differential replication of ribosomal genes modulates the abnormal abdomen phenotype in *Drosophila mercatorum*, *Genetics* 112:877-886.

DeSalle, R., Templeton, A., Mori, I., Pletscher, S., and Johnston, J. S., 1987, Temporal and spatial heterogeneity of mtDNA polymorphisms in natural populations of *Drosophila mercatorum*, *Genetics* 116:215-223.

Johnston, J. S. and Ellison, J. R., 1982, Exact age determination in laboratory and field-caught *Drosophila*, *J. Insect Physiol.* 28:773-780.

Johnston, J. S. and Templeton, A. R., 1982, Dispersal and clines in *Opuntia* breeding *Drosophila mercatorum* and *D. hydei* at Kamuela, Hawaii, in: *Ecological Genetics and Evolution: The Cactus-Yeast-Drosophila Model System* (J. S. F. Barker, and W. T. Starmer, eds), Academic Press Australia, Sydney, pp. 241-256.

Templeton, A. R., 1982a, The prophecies of parthenogenesis, in: *Evolution and Genetics of Life Histories* (H. Dingle, and J. P. Hegmann, eds), Springer-Verlag, New York, pp. 75-101.

Templeton, A. R, 1982b, Adaptation and the integration of evolutionary forces, in: *Perspectives on Evolution* (R. Milkman, ed.), Sinauer Associates, Sunderland, Massachusetts, pp. 15-31.

Templeton, A. R., 1983, Natural and experimental parthenogenesis, in: *The Genetics and Biology of Drosophila*, Vol. 3C (M. Ashburner, H. L. Carson, and J. N. Thompson, eds), Academic Press, London, pp. 343-398.

Templeton, A. R., Crease, T. J., and Shah, F., 1985, The molecular through ecological *genetics* of abnormal abdomen in *Drosophila mercatorum*. I. Basic genetics, *Genetics* 111:805-818.

Templeton, A. R., Hollocher, H., Lawler, S., and Johnston, J. S., 1989, Natural selection and ribosomal DNA in *Drosophila*, *Genome* 31:296-303.

Templeton, A. R. and Johnston, J. S., 1982, Life history evolution under pleiotropy and K-selection in a natural population of *Drosophila mercatorum*, in: *Ecological Genetics and Evolution: The Cactus-Yeast-Drosophila Model System* (J. S. F. Barker, and W. T. Starmer, eds), Academic Press Australia, Sydney, pp. 241-256.

Templeton, A. R., Johnston, J. S., and Sing, C. F., 1987, The proximate and ultimate control of aging in *Drosophila* and humans, in: *Evolution of Longevity in Animals* (A. D. Woodhead, and K. H. Thompson, eds), Plenum Press, New York, pp. 123-133.

Templeton, A. R. and Johnston, J. S., 1988, The measured genotype approach to ecological genetics, in: *Population Genetics and Evolution* (G. de Jong, ed.), Springer-Verlag, Berlin, pp. 138-146.

Templeton, A. R. and Rankin, M. A., 1978, Genetic revolutions and control of insect populations, in: *The Screwworm Problem* (R. H. Richardson, ed.), University of Texas Press, Austin, pp. 83-112.

CHAPTER 2

Direction of Life History Evolution in Drosophila mojavensis

WILLIAM J. ETGES

1. INTRODUCTION

Studies of the genetic basis for life history evolution involving analysis of demographic change in response to patterns of environmental variability have yet to provide a general explanation for the diversity of life histories often observed among species. Part of this problem is due to the lack of information about the forces actually responsible for causing the genetic variation observed in natural populations, without which we cannot evaluate the significance of the variation measured or the precise outcome in long-term life history evolution. Adaptation to environmental variability can lead to different equilibrium life histories all with equivalent fitnesses (Schaffer and Rosenzweig, 1977). The form of life history expected will depend on the pattern of environmental variation and degree of correlation among life history traits (Tuljapurkar, 1988; Orzack and Tuljapurkar, 1989). Only when observed genetic variation and covariation in components of fitness can be associated with the causes in nature responsible for their maintenance will understanding of the microevolutionary processes directing life history evolution be possible (Istock *et al.*, 1976; Reznick and Endler, 1982; Etges, 1989a).

The process of adaptation to different environments leading to life history evolution can also cause reproductive isolation in allopatry arising from pleiotropy or linkage. Genetic differentiation caused by adaptation to new environments may secondarily cause sexual isolation because those genes involved in adaptation may influence sexual isolation (Muller 1939, 1942; Rice, 1987). In concert with genetic drift, adaptive divergence in allopatry may be a potent factor in speciation due to disruptive selection causing fixation of incompatibility genes producing sexual and/or postmating isolation upon secondary contact (Dobzhansky, 1940; Mayr, 1963; Carson,

WILLIAM J. ETGES * Department of Ecology and Evolutionary Biology, University of Arizona, Tucson, Arizona 85721, USA. Present address: Department of Zoology, University of Arkansas, Fayetteville, Arkansas 72701, USA.

Ecological and Evolutionary Genetics of Drosophila, Edited by
J.S.F. Barker *et al.*, Plenum Press, New York, 1990

1971, 1975, 1982; Nei, 1976; Nei *et al.*, 1983; Kaneshiro, 1980).

The microevolutionary connections between changes in life history and reproductive isolation are central to arguments about species formation. In its most simplistic form, evolution may be viewed as a long process of allelic substitutions at many loci of small effect or frequency changes of alleles at loci controlling regulatory functions or "major genes". At some crucial point, genetic change becomes irreversible (Dobzhansky, 1970) whether reinforced by speciation or not. Such irreversible evolutionary change may be achieved in any number of ways including classical divergence in allopatry (Mayr, 1963; Futuyma and Mayer, 1980) and genetic drift in structured populations (Wright, 1932, 1982; Templeton, 1980; Carson and Templeton, 1984). Since the form of genetic change must at some point be expressed through variation in fitness, genetic analysis of variation in integrated developmental and physiological pathways, i.e., life history traits, will reveal something about the nature of shifts involved in large-scale genetic reorganizations during evolutionary transitions. When functionally related sets of life history traits are reshaped together, the process probably involves many genes and irreversibility may occur quickly (Istock, 1982).

Natural populations of *Drosophila mojavensis* are ideal for the study of causal relationships between adaptive divergence in life histories and incipient speciation. Not only does premating isolation exist among certain geographically isolated populations (Wasserman and Koepfer, 1977; Zouros and D'Entremont, 1980; Markow *et al.*, 1983), but also the origin of this species and history of range expansion, mediated by shifts to alternate host cacti, have been described in detail (Johnson, 1980; Heed, 1982; Heed and Mangan, 1986; Ehrman and Wasserman, 1987; Etges and Heed, 1987; Etges, 1989b). Populations of *D. mojavensis* from Baja California are considered ancestral and invaded mainland Mexico by switching host plants. The Gulf of California now forms a major geographical barrier to gene flow as evidenced by both inversion and allozymic differentiation among mainland and Baja populations (Zouros, 1973; Johnson, 1980). A central-marginal pattern of inversion polymorphism closely follows the distribution of host plants (Johnson, 1980). Polymorphism is mainly restricted to agria cactus, *Stenocereus gummosus*, in Baja California and one small patch in coastal Sonora (Fig. 1). A rare ancestral chromosome is found only in central Baja, and agria is the preferred host plant even when other cacti are present (Heed, 1982). Mainland populations are typically chromosomally monomorphic throughout Sonora, northern Sinaloa, and Southern Arizona and use organ pipe cactus, *S. thurberi*, with occasional use of cina, *S. alamosensis*, a major host plant of *D. arizonensis* in Sonora and Sinaloa (Heed, 1982). In southern California, all populations of *D. mojavensis* are fixed for an alternate gene arrangement where barrel cactus, *Ferocactus acanthodes*, is the sole host plant.

Incipient speciation among Baja and mainland Sonora populations has been suggested in studies of premating isolation

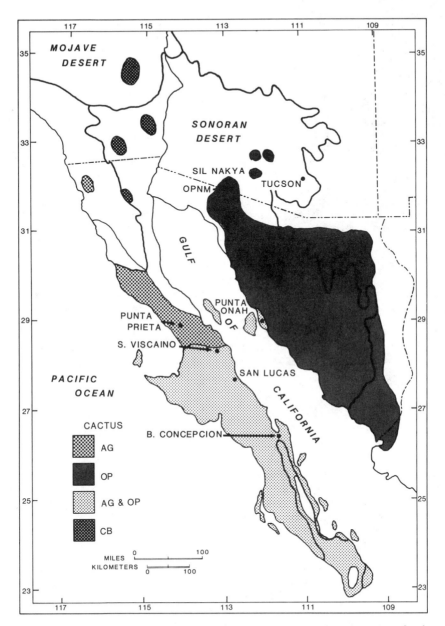

Figure 1. Map of the Sonoran Desert showing the ranges of the major host cacti of *D. mojavensis* and the collecting sites described in the text. Cactus types are: AG, agria; OP, organ pipe; CB, California barrel cactus.

between Sonora females and Baja males (Wasserman and Koepfer, 1977; Zouros and D'Entremont, 1980; Zouros, 1981; Koepfer, 1987a, 1987b). Sonoran females discriminate against Baja males with apparently little postmating isolation (Zouros and D'Entremont, 1980). Male mating behavior is influenced by genes on the X and Y chromosomes and interactions with other chromosomes while female behavior is influenced mainly by genes on the second and fifth chromosomes (Zouros, 1981). Because this genetic basis of sexual isolation does not correlate well with the pattern of inversion polymorphism on the second and third chromosomes, Zouros (1981) concluded that incipient speciation in *D. mojavensis* may not be related to "the process of adaptation to local environments" (p. 716) adding to the view that inversions *per se* have little to do with speciation events (Carson, 1975, 1978, 1982; Craddock, 1974; Paterson, 1981; Walsh, 1983; but see White, 1978).

However, considerable life history variation exists among those populations which show behavioral isolation. Flies from Baja express shorter egg to adult development times, higher viabilities, smaller thorax sizes, lower lifetime fecundities, and slower rates of sexual maturation than mainland flies (Etges and Heed, 1987; Etges and Klassen, 1989). Detailed analysis of the magnitude and kind of genetic transitions that have produced life history differences will be necessary if connections to sexual isolation are to made, as will within-population studies of expression of genetic variances and covariances of life history traits and sexual isolation on fermenting cactus substrates like those used in nature. There is no evidence for hybrid sterility between Baja and mainland Sonora populations (Zouros, 1973), yet some variation in male genital morphology has been noted (Heed, unpublished data). The present study was undertaken to further clarify the magnitude of genetic differentiation in life history now present among ancestral Baja populations and derived mainland populations by means of population crosses cultured on both agria and organ pipe tissues.

Changes in the means and variances among parental, F_1, and F_2 generations in the expression of these fitness components were of particular interest in examining the extent of genetic coadaptation within populations (Vetukhiv 1953, 1954; McFarquhar and Robertson, 1963; Anderson, 1968). Locally adapted populations may exhibit genetic integration or epistasis among groups of genes that have been shaped by natural selection. Crosses among such populations may show heterosis in the F_1s or F_2 breakdown, or loss of fitness in the F_2 generations caused by the breaking up of coadapted gene complexes by recombination. As McFarquhar and Robertson (1963) noted, "differences in co-adaptation should be assessed in terms of fitness with respect to the normal environment". In attempting to measure variation in components of fitness under nearly natural conditions, I assayed the extent of coadaptation in Baja and mainland populations by comparing generation means and variances for the fitness components studied on fermenting cactus substrates.

Table I
Collection records of *D. mojavensis* used for this study. Cactus refers to the host plant from which imagoes were reared, and record ID refers to the collection numbers of William B. Heed at the University of Arizona

Location[a]	Record ID	Cactus	# Rots	# Emerged
Baja California				
Punta Prieta (P)	A916	Agria	11	3918[b]
South Viscaino (V)	A917	Agria	4	1699[c]
		Organ Pipe	3	14
South of Bahia Concepcion (B)	A918	Agria	3	107
San Lucas (L)	A920	Agria	2	217
Mainland Sonora and Arizona				
Punta Onah (O)	FOG2	Agria	2	200[d]
Organ Pipe National Monument, AZ (M)	OPNM	Organ Pipe	1	1074
Santa Rosa Mountains, AZ(S)	A911	Organ Pipe	2	155

[a]Abbreviated population designation in parentheses.
[b]Includes 233 adults aspirated from rots in the field.
[c]Includes 38 adults aspirated from rots in the field.
[d]Estimated number.

2. MATERIALS AND METHODS

2.1. Origin of Stocks

Seven populations of *D. mojavensis* were sampled from nature 3-5 generations prior to the start of the first set of crosses (Table I). In addition to adults aspirated from rots in the field, all adults eclosing from rots returned to the lab were used to start laboratory populations. All flies were reared in mass culture in large numbers on banana-yeast-malt-Karo-agar food in shell vials until crosses were made.

2.2. Cactus Rearing Conditions

Artificial rots were made with 25 g pieces of fresh cactus placed on 75 g of sand in half-pint milk bottles sealed with cotton. After autoclaving, each was inoculated with a pecti-nolytic bacterium and seven species of yeast common to natural rots of both cactus species (Starmer 1982a; Fogleman and Starmer, 1985). Yeasts used were: *Pichia cactophila, P. mexi-*

cana, *P. amethionina* var. *amethionina*, *Cryptococcus cereanus*, *Candida valida*, *C. ingens* and *C. sonorensis*.

2.3. Population Crosses - I

For each cross, about 200 virgin adults of each sex were placed in polyethylene freezer boxes with a hole on one side large enough for a removable petri dish containing an oviposition medium (1% agar-agria rot juice). First, four crosses were made: both parental populations and both reciprocal F_1 crosses. Four second generation crosses included one parental population, F_1 backcrosses to each parental population, and an F_2 population made by intercrossing half of the adults from each reciprocal. Combinations of populations were chosen for crossing to represent a variety of genetic differences within and among geographical regions in the species range of *D. mojavensis*.

2.4. Population Crosses - II

In order to assess the degree to which experimental error due to rearing generations at different times may have influenced the expression of life history variation, a second set of crosses using populations A924, a more recently collected Santa Rosa Mountains stock, and A916 were performed in the same fashion as above, but all replicated generations were cultured side-by-side (Etges, 1989c). Replicates of 50 or 400 eggs per 35 g cactus tissue were started for these crosses to determine how larval competition might influence gene expression of life history traits.

For both sets of population crosses, eggs were collected from small petri dishes on each population cage over 6 hr intervals, surface sterilized by washing in sterile water, 70 % ethanol and again in sterile water. Groups of 100, or 50 and 400 eggs for the second experiment, were counted out onto sterile filter paper strips and placed on freshly inoculated agria or organ pipe cactus. For each generation, eggs from each cross were counted onto 3 (2 replicates for the second experiment) replicates of each cactus species each day. All cultures were grown at 25°C in a 14:10 LD photoperiod.

Egg to adult viability and development time were measured by collecting eclosing adults every day. The number of unhatched eggs was recorded after removing the filter paper strips from the culture bottles and subtracted from the total number of eggs. Adult thorax size was measured with an ocular micrometer on aged adults. Three adults per replicate per day of development time per sex were scored for thorax size.

2.5. Statistical Analysis

Development time data were rescaled by log transformation, viability data were arc sin transformed, and thorax length

Table II

Nested ANOVA results for all parental, control populations cultured in this study on both host cacti after reclassifying each population as either Baja or mainland depending upon geographical origin. Means for each variable (1 SE) showing significant Cactus x Region interaction terms are shown

Source of variation	df	TYPE IV SS	F	P
Egg to Adult Development Time				
Region	1	0.0265	164.83	0.0001
Population(Region)[a]	32	0.0272	5.29	0.0001
Cactus	1	0.0080	49.95	0.0001
Cactus x Region	1	0.0027	16.97	0.0001
Cactus x Pop(Region)[b]	32	0.0173	3.36	0.0001
Error	136	0.0219		

	Baja	n[c]	Mainland	n
Organ Pipe	12.29[d] (0.077)	54	12.74 (0.101)	48
Agria	11.78 (0.039)	54	12.59 (0.061)	48

Source of variation	df	TYPE IV SS	F	P
Egg to Adult Viability				
Region	1	0.5595	19.80	0.0001
Population(Region)	32	5.4715	6.05	0.0001
Cactus	1	0.3676	13.01	0.0004
Cactus x Region	1	0.0000	0.00	0.9769
Cactus x Pop(Region)	32	1.9415	2.15	0.0013
Error	136	3.8419		
Female Thorax Size				
Region	1	1.3306	124.19	0.0001
Population(Region)	32	0.7505	2.19	0.0010
Cactus	1	0.0000	0.00	0.9622
Cactus x Region	1	0.0234	2.19	0.1413
Cactus x Pop(Region)	32	0.2858	0.83	0.7201
Error	139	1.4893		
Male Thorax Size				
Region	1	0.6249	138.50	0.0001
Population(Region)	32	0.6530	4.52	0.0001
Cactus	1	0.0004	0.09	0.7621
Cactus x Region	1	0.0479	10.61	0.0014
Cactus x Pop(Region)	32	0.2192	1.52	0.0524
Error	139	0.6272		

	Baja	n	Mainland	n
Organ Pipe				
(females)	4.21[e] (0.013)	57	4.38 (0.013)	48
(males)	3.87 (0.010)	57	3.99 (0.010)	48
Agria				
(females)	4.18 (0.003)	58	4.38 (0.017)	47
(males)	3.86 (0.014)	57	4.02 (0.011)	48

[a]Population(Region) = populations nested within geographical region.
[b]Cactus x Pop(Region) = interaction of cactus with populations nested within geographical region.
[c]n refers to the total number of cultures per cactus for each region. The number of cultures for each parental population were: S- 9, M- 21, O- 18, P- 15, V- 18, L- 15, and B- 6. Population abbreviations are given in Table I.
[d]Development time in days.
[e]Thorax size in micrometer units X 10.

data were rescaled to ocular micrometer units x 10 (4 units =1 mm). Maternal effects were assessed by t-tests of the means of the reciprocal crosses. Differences between pooled Baja and mainland populations for each trait were assayed with nested ANOVA. Differences among mid-parent and F_1 (MP-F_1) means, mid-parent and F_2 (MP-F_2) means, and F_1 and F_2 (F_1-F_2) means were subjected to t-tests.

3. RESULTS

3.1. Life History Differences between Baja and Mainland Populations

With a few exceptions, development times were shorter, viabilities were higher, and thorax sizes were smaller for the Baja populations than for the mainland populations (Fig. 2). Development times of Baja populations were shorter on agria than on organ pipe cactus accounting for a significant Cactus x Population interaction (Table II). Thus, Baja *D. mojavensis* developed faster on their resident host cactus signifying regional adaptation to agria in Baja California. Egg to adult viability differed among mainland and Baja populations, but no host plant adaptation was evident because there was no Cactus x Region interaction; however, different populations within regions expressed differences in viability due to cactus substrates suggesting geographical variation in egg to adult viability (Table II). Mainland adults were larger than Baja adults on both cacti, but Baja males were smaller on agria generating another Region x Cactus interaction (Table II).

Discriminant function analyses were performed with the replicated parental populations using the four fitness components as variables. On agria and organ pipe, these life history differences significantly discriminated among Baja and mainland populations (agria - Wilk's Lambda = 0.318, χ^2 = 115.58, P > 0.0001, df =4; organ pipe - Wilk's Lambda = 0.620, χ^2 = 48.26, P > 0.0001, df = 4). Thus, significant differences exist between Baja and mainland populations in these life history traits.

Latitudinal clines in development time and thorax size, but not viability, were evident among populations. The total number of replicates cultured for each parental population was averaged and regressed on latitude, so only seven data points were analyzed for each character. Development time on agria ($r = 0.726$, $P = 0.065$), female thorax size on organ pipe ($r = 0.725$, $P = 0.065$), female thorax size on agria ($r = 0.754$, $P = 0.05$), male thorax size on organ pipe ($r = 0.770$, $P = 0.043$), and male thorax size on agria ($r = 0.812$, $P = 0.026$) were all positively associated with increasing latitude. Such clines suggest that climatic variation in addition to host plant variation may have influenced development time and thorax size; however, life history differences characterizing Baja and mainland populations were particularly evident in the Punta Prieta (Baja) and Punta Onah (mainland) populations that are located at similar latitudes (Figs 1, 2). Further analy-

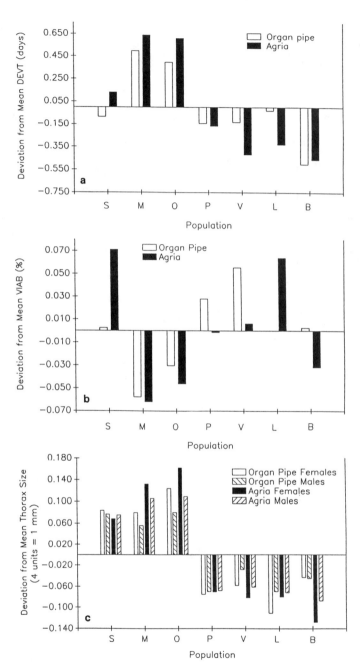

Figure 2. Comparisons of (a) egg to adult development time, (b) viability, and (c) adult thorax size among the 7 populations used in this study. Differences are expressed as deviations from the overall mean of all populations for each trait. Populations are arrayed from north to south (see Fig. 1). Population labels are defined in Table I.

Table III

Differences in $\log_{10}$ (egg to adult development time) among populations for Baja California and mainland populations and their crosses[a] of *D. mojavensis* cultured on organ pipe (op) and agria (ag) cacti

Cross		Contrast		
		$MP-F_1$	$MP-F_2$	F_1-F_2
Across the Gulf				
PS and SP	op	-0.007	0.006	0.014
	ag	-0.002	-0.002	0.000
PM and MP	op	-0.001	0.001	0.002
	ag	0.002	0.018*	0.020**
PO	op	0.013	-0.013	-0.026*
OP	op	-0.010	–	-0.003
PO	ag	0.014	-0.012	-0.025+
OP	ag	-0.031**	–	0.019
VO and OV	op	-0.029	0.004	0.033*
VO	ag	-0.014	0.007	0.021*
OV	ag	0.008	–	-0.001
LM and ML	op	-0.019+	-0.037**	-0.018
	ag	-0.033**	-0.012	0.022*
LO	op	-0.003	-0.005	-0.002
OL	op	0.006	–	-0.011*
LO and OL	ag	-0.004	0.014	0.017*
Within Baja				
VP	op	-0.050**	-0.009	0.041+
PV	op	0.013	–	-0.022
VP	ag	-0.018*	0.000	0.018*
PV	ag	0.004	–	-0.004
LV and VL	op	0.008	0.004	-0.005
	ag	-0.002	0.013	0.015*
BL and LB	op	0.022*	0.015	-0.007
BL	ag	-0.033*	0.017*	0.050*
LB	ag	-0.002	–	0.018*
Within the mainland				
OM and MO	op	-0.010	-0.004	0.021
	ag	0.022**	-0.003	-0.025*
SM and MS	op	0.005	-0.009+	-0.014
SM	ag	-0.045**	-0.010	0.035*
MS	ag	0.008	–	-0.018

+ $0.1 < P < 0.05$, * $P < 0.05$, ** $P < 0.01$, *** $P < 0.001$.
[a]Maternal population label precedes paternal label.

sis of more Baja and mainland populations from similar latitudes is warranted.

3.2. Population Crosses

Eleven crosses were performed among populations: six involved populations on either side of the Gulf of California, with three within-Baja crosses and two within-mainland

crosses.   Means   of   the F_1 and F_2 generation   crosses   were
adjusted for the differences between the parental populations
cultured with them to remove the average effects of variations
among cactus tissue quality that can influence larval growth
and development (Etges, unpublished data).   When differences
among reciprocal crosses were significant, replicates were not
pooled.   Contrasts among generation means were constructed
including mid-parent-F_1 differences as indicators of overdominance or underdominance and mid-parent-F_2 comparisons as indicators of F_2 breakdown.

Considering all crosses, development time of the F_1s tended
to be greater than that of the mid-parents with little indication of a cactus effect; maternal influences were significant
in some cases (Table III).   A majority of these significant
differences involved within-region crosses indicating as much,
or more, genetic differentiation within Baja and the mainland
as between regions.   Few differences were significant among
mid-parent-F_2 comparisons, but development time was significantly greater in a majority of the F_1s than in the F_2s.
Recombination of genotypic arrays influencing development time
in these populations thus caused decreases in development
time.

Egg to adult viability increased in those F_1s showing significant differences, suggesting some heterosis (Table IV).
More evidence of maternal effects were found than with development time.   However, a majority of the F_2 comparisons were
negative, clearly showing increases in F_2 viability or lack of
coadaptation.

Maternal effects were not apparent for female or male thorax size nor was there much evidence for heterosis in the F_1s
(Table V).   Of the 17 significant differences between mid-parent means and F_2s, seven were positive, and six of these seven
were from within region crosses. Most F_2s from across the gulf
crosses were larger than the mid-parents, but there was no
trend in comparisons of F_1s and F_2s.   Little evidence for coadaptation was thus apparent for thorax size, except for the
results from the within-region crosses.

Virtually no evidence for either decreased variance among
the F_1s or increased variance among the F_2s was found in comparisons of pooled within-bottle variances among generations
(Table VI).   Some increased variance within the parental mainland populations was suspected, which could have obscured
comparisons among generations, but this was significant in
only two of eight cases (Table VI).   Similar results were
obtained from a separate set of independent crosses using high
and low larval densities on both host cacti (Etges, 1989c;
Table VII).   Variability in development time was greater in
the Arizona population at high densities and the F_2s were more
variable than the F_1s, but only on organ pipe. Increased variability among F_2s was consistent with coadaptation within populations, but expression of such genetic variation depended
upon breeding substrates. Variation in thorax size did not
provide strong evidence for coadaptation of genes influencing

Table IV

Differences in arcsin (egg to adult viability) among populations for Baja
California and mainland populations and their crosses[a] of *D. mojavensis*
cultured on organ pipe (op) and agria (ag) cacti

Cross			Contrast		
			$MP-F_1$	$MP-F_2$	F_1-F_2
Across the Gulf					
	PS	op	0.123	-0.087+	-0.230*
	SP	op	-0.106	-	0.019
	PS	ag	-0.081	-0.141	-0.059
	SP	ag	0.020	-	-0.160
	PM	op	-0.045	0.123	0.167
	MP	op	-0.050	-	0.073
PM and	MP	ag	-0.049	0.110	0.061
	PO	op	-0.393+	-0.106	0.286*
	OP	op	0.102	-	-0.208
PO and	OP	ag	-0.250+	-0.203	0.047
VO and	OV	op	0.171	0.067	-0.103
		ag	0.042	-0.117+	-0.158
	LM	op	-0.209*	-0.159	-0.368+
	ML	op	0.393*	-	-0.552+
LM and	ML	ag	0.301*	-0.159	-0.460*
LO and	OL	op	-0.140+	-0.438**	-0.298*
	LO	ag	0.007	-0.102	-0.108
	OL	ag	-0.390*	-0.102	0.090
Within Baja					
	VP	op	0.113	-0.157*	-0.270**
	PV	op	-0.235+	-	0.078
	VP	ag	0.180+	-0.323*	-0.511*
	PV	ag	-0.080	-	-0.243+
LV and	VL	op	-0.110	0.207+	0.317*
		ag	0.011	-0.116	-0.127
BL and	LB	op	-0.477**	-0.087	0.390*
		ag	-0.044	0.134	0.179
Within the mainland					
OM and	MO	op	0.006	0.047	0.005
	OM	ag	-0.107	-0.019	0.087
	MO	ag	0.134	-	-0.153*
SM and	MS	op	-0.191*	-0.049	0.142
	SM	ag	-0.096	-0.270**	-0.173
	MS	ag	-0.181	-	-0.089

+ 0.1 < P < 0.05, * P < 0.05, ** P < 0.01, *** P < 0.001.
[a]See footnote - Table III.

Table V
Differences in thorax size (1 unit = 0.25 mm) among populations for Baja California and mainland populations and their crosses (see footnote - Table III) of *D. mojavensis* cultured on organ pipe (op) and agria (ag) cacti

		Contrast					
		MP-F$_1$		MP-F$_2$		F$_1$-F$_2$	
Cross		Females	Males	Females	Males	Females	Males
Across the Gulf							
PS and SP	op	0.090+	0.060	0.105	-0.235***	0.015	-0.195**
	ag	0.022	-0.006	-0.092	-0.142**	-0.114	-0.136*
PM and MP	op	0.055	0.065	0.063	0.007	0.118*	0.072
	ag	0.065	0.064	-0.022	-0.047	0.043	0.017
PO and OP	op	-0.022	-0.019	0.047	-0.012	0.069	0.007
	ag	-0.090	-0.044	0.129	-0.056	0.220**	-0.013
VO and OV	op	0.088	0.020	-0.054	0.008	-0.142	-0.012
	ag	-0.006	-0.036	-0.221**	-0.159*	-0.216**	-0.123*
LM and ML	op	-0.087	-0.041	-0.216**	0.027	-0.128+	0.069
	ag	0.006	-0.022	-0.157+	-0.120+	-0.163	-0.098
LO and OL	op	-0.051	-0.073+	0.081	0.161**	0.132*	0.234***
	ag	0.000	-0.013	0.036	-0.049	0.036	-0.036
Within Baja							
VP and PV	op	-0.070	0.002	-0.184*	-0.215*	-0.115	-0.217*
	ag	-0.100	-0.084	-0.458**	0.233**	-0.358**	0.317**
LV and VL	op	-0.005	0.078*	-0.028	-0.009	-0.024	-0.087*
	ag	0.172*	0.025	0.067+	0.198**	-0.105	0.172*
BL and LB	op	0.065	0.025	-0.080	0.000	-0.145	-0.026
	ag	-0.001	-0.080+	0.214+	0.046	0.215+	0.126*
Within the mainland							
OM and MO	op	-0.015	0.005	0.105*	0.026	0.121*	-0.156***
	ag	-0.020	0.018	0.090	0.025	0.110**	0.007
SM and MS	op	-0.033	-0.036	-0.048	-0.037	-0.014	0.000
	ag	0.102	0.067	0.121+	0.015	0.019	-0.052

+ 0.1 < P < 0.05, * P < 0.05, ** P < 0.01, *** P < 0.001.

body size (Table V). Overall, recombination did not generally increase variability among the F$_2$s from interpopulation crosses of *D. mojavensis*.

4. DISCUSSION

The expansion of the geographical range of *D. mojavensis* has involved shifts to alternate host cacti and genetic changes in life history. An historical explanation for this pattern involves geographic isolation in Baja California and subsequent divergence from a mainland form, now the sibling species, *D. arizonensis* (Ehrman and Wasserman, 1987). Gastil *et al.* (1975) have postulated that tectonic drift transported the Baja peninsula from its connection to mainland Mexico to its present location, effectively isolating *D. mojavensis*. Following migration across the islands of the Gulf of California, *D. mojavensis* invaded the mainland by switching to organ pipe in Sonora and Arizona, and cina in southern Sonora and Sinaloa. Origins of the southern California barrel cactus-breeding and Santa Catalina Island *Opuntia*-breeding populations are discussed elsewhere (Johnson, 1980; Heed, 1982).

Table VI

Ratios of pooled within-group variances for parental, F_1, and F_2 generations from all crosses in this study cultured on both organ pipe, op, and agria, ag, cacti for egg to adult development time, DEVT, female thorax size, THFM, and male thorax size, THML. F ratios for egg to adult viability were formed using replicate averages

Generations Compared	Cactus	df	MS Ratio	F
1. P_2/P_1				
DEVT	op	2052	0.4264	0.784
		2917	0.5440	
	ag	2232	0.6232	1.872 ***
		3092	0.3329	
VIAB	op	97	0.0516	0.671
		64	0.0769	
	ag	97	0.0505	1.208
		65	0.0418	
THFM	op	215	0.0273	1.270 +
		282	0.0215	
	ag	215	0.0203	0.832
		253	0.0244	
THML	op	216	0.0195	1.224
		293	0.0159	
	ag	243	0.0152	1.076
		266	0.0141	
2. Mid-Parents/F_1				
DEVT	op	4969	0.5697	0.908
		4303	0.6271	
	ag	5324	0.4546	0.857
		4869	0.5305	
VIAB	op	97	0.0516	0.671
		64	0.0769	
	ag	97	0.0505	0.208
		65	0.0418	
THFM	op	497	0.0240	1.087
		427	0.0221	
	ag	468	0.0225	0.842
		413	0.0267	
THML	op	509	0.0175	0.951
		437	0.0184	
	ag	509	0.0146	1.139
		444	0.0128	
3. F_2/F_1				
DEVT	op	2297	0.4058	0.647
		4303	0.6271	
	ag	2392	0.4064	0.766
		4869	0.5305	
VIAB	op	33	0.0637	0.828
		66	0.0769	
	ag	33	0.0581	1.390
		68	0.0418	
THFM	op	76	0.0172	0.779
		427	0.0221	
	ag	206	0.0172	0.644
		413	0.0267	
THML	op	180	0.0172	0.935
		437	0.0184	
	ag	203	0.0142	1.108
		444	0.0128	
4. F_2/Mid-Parents				
DEVT	op	2468	0.4252	1.270 *
		2151	0.3349	
	ag	2588	0.4234	0.689
		2118	0.6146	
VIAB	op	33	0.0637	1.234
		102	0.0516	
	ag	33	0.0581	1.151
		101	0.0505	
THFM	op	196	0.0166	0.926
		177	0.0179	
	ag	226	0.0173	1.033
		182	0.0168	
THML	op	204	0.0164	1.089
		169	0.0151	
	ag	225	0.0137	0.950
		215	0.0144	

+ $0.1 < P < 0.05$, * $P < 0.05$, *** $P < 0.001$.

Table VII
Across generation F ratios of pooled within-bottle variances of egg
to adult development time and thorax size in *D. mojavensis*

a. Development time

		Organ Pipe		Agria	
Variance Ratio[a]	Density[b]	F	df	F	df
1. P_2/P_1	L	0.630	618/631	0.854	614/640
	H	2.314***	4915/4833	1.285*	4979/4942
2. Mid-parents/F_1	L	1.173	1249/1165	1.154	1254/1202
	H	1.133	9748/9842	0.943	9921/10339
3. F_2/Mid-parents	L	1.171	1181/1249	0.903	1367/1254
	H	1.035	9784/9748	1.145	10203/9921
4. F_2/F_1	L	1.374**	1181/1165	1.041	1367/1202
	H	3.122***	9784/9842	1.080	10203/10339

b. Thorax Size

Variance Ratio	Sex	Density	F	df	F	df
1. P_2/P_1	F	L	1.312	78/75	0.593	71/78
	F	H	1.127	319/243	0.996	475/423
	M	L	0.834	70/73	0.950	67/72
	M	H	1.205	325/228	0.956	489/422
2. Mid-parents/F_1	F	L	1.128	153/138	0.957	149/148
	F	H	0.948	562/576	1.103	898/933
	M	L	1.914***	143/144	1.238	139/144
	M	H	1.134	553/585	0.874	911/942
3. F_2/Mid-parents	F	L	0.839	159/153	1.138	161/149
	F	H	1.191	606/562	1.073	915/898
	M	L	0.725	164/143	1.425**	164/139
	M	H	0.972	599/553	1.049	941/911
4. F_2/F_1	F	L	0.947	159/138	1.090	161/148
	F	H	1.129	606/576	1.184	915/933
	M	L	1.387*	164/144	1.764***	164/144
	M	H	0.943	599/585	0.960	941/942

* $P < 0.05$, ** $P < 0.01$, *** $P < 0.001$.
[a]Generation designations are defined in the text.
[b]L = low density, H = high density.

Life history evolution can thus be attributed, at least partly, to the shift to different host cacti along with the morphological, cytological, behavioral, and genic differentiation among populations that has prompted description of geographical subspecies, races, and subraces of *D. mojavensis* (Mettler, 1963; Zouros, 1973; Zouros and D'Entremont, 1980).

Regional differences in development time and viability can be considered adaptations to the host cacti used. Baja populations express shorter development times and smaller thorax sizes on their resident host plant, agria, than do mainland populations on their host plant, organ pipe cactus. Unraveling the causes for the smaller adult size of Baja flies and the larger adult sizes of the mainland flies will require field studies of dispersal and estimation of genetic correlations with other traits, particularly development time. If smaller thorax size is genetically correlated with shorter development time, small thorax size may not be adaptive.

The characteristics of agria and organ pipe rots that have been implicated with these life history shifts concern rot duration, abundance, and dispersion. Agria cactus is considered a more predictable resource for adult feeding and oviposition, but unpredictable for larval survivorship because stem diameters are smaller and tissues decompose faster leading to faster rot desiccation than organ pipe (Johnson, 1980; Heed, 1981; Etges and Heed, 1987; Etges, 1989c). Thus, agria is a more ephemeral breeding substrate but more abundant as a feeding site than organ pipe, and those populations that use agria are more viable, develop faster, and are smaller as adults.

Invasion of mainland Sonora and Arizona was possible because of the presence of organ pipe, and to the south, cina and organ pipe. Little is known of the ecology or genetics of the cina-breeding populations, except that they sometimes coexist with *D. arizonensis* (Ruiz and Heed, 1988). In Sonora and Arizona, organ pipe-breeding *D. mojavensis* populations are faced with greater trophic unpredictability for adults because organ pipe rots are less frequent (rots per plant per hectare) and further apart (Johnson, 1980; Mangan, 1982; Heed and Mangan, 1986; Ruiz and Heed, 1988). However, organ pipe rots are usually much larger than agria rots because stem diameters are larger providing temporally more stable breeding sites. Longer lasting rots could have allowed increases in larval development time and adult body size in organ pipe-breeding populations of *D. mojavensis* with a corresponding increase in ovariole numbers (Heed, unpublished data) and lifetime fecundity (Etges and Klassen, 1989). Dispersal ability must be greater in organ pipe-breeding populations because of the greater distance between rots than in agria-breeding populations (Mangan, 1982), which may also explain the body size increases. Thorax size is known to be correlated with dispersal ability (Roff, 1981) and rot-to-rot distances among all Sonoran Desert Drosophila (Johnston and Heed, 1976; Heed and Mangan, 1986).

Despite adaptation to such very different breeding and feeding environments, little reorganization of the gene pool influencing life history differences among populations has occurred since invasion of the mainland. Most of the genetic differences among populations are additive or nearly so (Etges, 1989c). The lack of coadaptation among population crosses suggests that new epistatic gene complexes did not evolve in the face of considerable gene frequency change. Wright (1977) predicted that microevolutionary divergence would be most rapid when small populations or demes were separated in space or time, gene flow was reduced or absent, and demes were exposed to differing environments. It is unlikely that the lack of genetic restructuring or incompatibility is due to gene flow with the ancestral Baja populations given the degree of allozymic differentiation and the abrupt loss of inversion polymorphism outside of the small agria patch in coastal Sonora (Heed, 1978; Johnson, 1980). The mainland populations have clearly begun a new evolutionary trajectory, yet no postmating reproductive barriers have yet evolved.

Thus, the mainland populations have not undergone irreversible genetic changes in life history traits and only the one-way sexual isolation between mainland females and Baja males suggests any evolution of reproductive isolation. Further study of the quantitative genetic basis of within-population variation and covariation of life history traits and premating isolation may provide insight into the genetic processes of species formation before isolation is complete.

5. SUMMARY

Baja California and mainland populations of *D. mojavensis* exhibit geographic differences in components of fitness, egg to adult development time, viability, and adult thorax size that are genetically based. These regional differences are consistent, and are greater between regions than within regions. The expression of life history variation was influenced by the type cactus used for larval growth and development. Baja populations of *D. mojavensis* exhibited shorter development times and smaller thorax sizes on their resident cactus, agria, than the mainland populations did on their host plant, organ pipe cactus, suggesting host plant adaptation. Smaller thorax sizes of Baja adults may not be adaptive if thorax size is genetically correlated with development time. Results of population crosses within and between Baja and mainland populations for these life history characters showed few instances of overdominance in the F_1s or F_2 breakdown, and many cases of maternal effects for development time and viability. The lack of coadaptation suggests that selection has not produced epistatic gene complexes in spite of considerable gene frequency change since *D. mojavensis* shifted to new host cacti. Since Baja and mainland populations are fully interfertile and exhibit only one-way premating isolation, further genetic analysis of the population differences in life history should yield insight into the process of incipient speciation.

ACKNOWLEDGEMENTS. Tom Starmer provided the yeast species used and Stan Alcorn lent strains of a pectolytic bacterium. C. S.

Klassen provided expert technical assistance. Felix Breden made valuable statistical suggestions. I thank the Tohono O'odham Nation, formerly the Papago Tribe of Arizona, for permission to collect on their reservation and Bill Heed for allowing me to use his unpublished data. Funding was supplied by NSF grant BSR-8503472, NIH grant BRSG 2 SO7 RR07101-09, and the Department of Zoology at the University of Arkansas.

References

Anderson, W. W., 1968, Further evidence for coadaptation in crosses between geographic populations of *Drosophila pseudoobscura*, *Genet. Res.* **12**:317-330.

Carson, H. L., 1971, Speciation and the founder principle, *Stadler Genet. Symp.* **3**:51-70.

Carson, H. L., 1975, The genetics of speciation at the diploid level, *Am. Nat.* **109**:83-92.

Carson, H. L., 1978, Speciation and sexual selection in Hawaiian *Drosophila*, in: *Ecological Genetics: The Interface* (P. F. Brussard, ed.), Springer-Verlag, New York, pp. 93-107.

Carson, H. L., 1982, Speciation as a major reorganization of polygenic balances, in: *Mechanisms of Speciation* (C. Barigozzi, ed.), A. R. Liss, New York, pp. 411-433.

Carson, H. L., and Templeton A. R., 1984, Genetic revolutions in relation to speciation phenomena: The founding of new populations, *Annu. Rev. Ecol. & Syst.* **15**:97-131.

Craddock, E. M., 1974, Reproductive relationships between homosequential species of Hawaiian *Drosophila*, *Evolution* **28**:593-606.

Dobzhansky, Th., 1940, Speciation as a stage in evolutionary divergence, *Am. Nat.* **74**:312-321.

Dobzhansky, Th., 1970, *Genetics of the Evolutionary Process*, Columbia Univ. Press, New York.

Ehrman, L., and Wasserman, M., 1987, The significance of asymmetrical sexual isolation, *Evol. Biol.* **21**:1-20.

Etges, W. J., 1989a, Chromosomal influences on life-history variation along an altitudinal transect in *Drosophila robusta*, *Am. Nat.* **133**:83-110.

Etges, W. J., 1989b, Divergence in cactophilic *Drosophila*: The evolutionary significance of adult ethanol metabolism, *Evolution* **43**:1316-1319.

Etges, W. J., 1989c, Evolution of developmental homeostasis in *Drosophila mojavensis*, *Evol. Ecol.* **3**:189-201.

Etges, W. J., and Heed. W. B., 1987, Sensitivity to larval density in populations of *Drosophila mojavensis*: Influences of host plant variation on components of fitness, *Oecologia* **71**:375-381.

Etges, W. J., and Klassen C. S., 1989, Influences of atmospheric ethanol on adult *Drosophila mojavensis*: Altered metabolic rates and increases in fitness among populations, *Physiol. Zool.* **62**:170-193.

Fogleman, J. C., and Starmer W. T., 1985, Analysis of community structure of yeasts associated with decaying stems of cactus. III. *Stenocereus thurberi*, *Microb. Ecol.* **11**:165-173.

Futuyma, D. J. and Mayer G. C., 1980, Non-allopatric speciation in animals, *Syst. Zool.* **29**:254-271.

Gastil, R. G., Phillips, R. P., and Allison, E. C., 1975, Reconnaissance geology of the state of Baja California, in: The Geological Society of America, Inc., Memoir 140, Boulder, Colorado, pp. 139-143.

Heed, W. B., 1978, Ecology and genetics of Sonoran Desert *Drosophila*, in: *Ecological Genetics: The Interface* (P. F. Brussard, ed.), Springer-Verlag, New York, pp. 109-126.

Heed, W. B., 1981, Central and marginal populations revisited, *Drosoph. Inf. Serv.* **56**:60-61.

Heed, W. B., 1982, The origin of *Drosophila* in the Sonoran desert, in: *Ecological Genetics and Evolution: The Cactus-Yeast-Drosophila Model System* (J.S.F. Barker, and W. T. Starmer, eds), Academic Press Australia, Sydney, pp. 65-80.

Heed, W. B., and Mangan R. L., 1986, Community ecology of the Sonoran Desert *Drosophila*, in: *The Genetics and Biology of Drosophila*, Vol. 3E (M. Ashburner, H. L. Carson, and J.N. Thompson Jr., eds), Academic Press, New York, pp. 311-345.

Istock, C. A., 1982, Some theoretical considerations concerning life history evolution, in: *Evolution and Genetics of Life Histories* (H. Dingle and J. P. Hegmann, eds), Springer-Verlag, New York, pp. 7-20.

Istock, C. A., Zisfein, J., and Vavra, K., 1976, Ecology and evolution of the pitcher-plant mosquito. 2. The substructure of fitness, *Evolution* **30**:535-547.

Johnson, W. R., 1980, Chromosomal polymorphism in natural populations of the desert adapted species, *Drosophila mojavensis*, PhD Dissertation, University of Arizona.

Johnston, J. S., and Heed. W. B., 1976, Dispersal of desert-adapted *Drosophila*: the Saguaro-breeding *Drosophila nigrospiracula*, *Am. Nat.* **110**:629-651.

Kaneshiro, K. Y., 1980, Sexual isolation, speciation and the direction of evolution, *Evolution* **34**:437-444.

Koepfer, H. R., 1987a, Selection for sexual isolation between geographic forms of *Drosophila mojavensis*. I. Interactions between the selected forms, *Evolution* **41**:37-48.

Koepfer, H. R., 1987b, Selection for sexual isolation between geographic forms of *Drosophila mojavensis*. II. Effects of selection on mating preference and propensity, *Evolution* **41**:1409-1412.

McFarquhar, A. M., and Robertson, F. W., 1963, The lack of evidence for co-adaptation in crosses between geographical races of *Drosophila subobscura*, *Coll., Genet. Res.* **4**:104-131.

Mangan, R. L., 1982, Adaptations to competition in cactus breeding *Drosophila*, in: *Ecological Genetics and Evolution: The Cactus-Yeast-Drosophila Model System* (J. S. F. Barker, and W. T. Starmer, eds), Academic Press Australia, Sydney, pp. 257-272.

Markow, T. A., Fogleman, J. C., and Heed, W. B., 1983, Reproductive isolation in Sonoran Desert *Drosophila*, *Evolution* **37**:649-652.

Mayr, E., 1963, *Animal Species and Evolution*, Harvard Univ. Press, Cambridge, Massachusetts.

Mettler, L. E., 1963, *D. mojavensis* baja, a new form in the *mulleri* complex, *Drosoph. Inf. Serv.* **28**:57-58.

Muller, H. J., 1939, Reversibility in evolution considered from the standpoint of genetics, *Biol. Rev.* **14**:261-280.

Muller, H. J., 1942, Isolating mechanisms, evolution, and temperature, *Biol. Symp.* **6**:71-125.

Nei, M., 1976, Mathematical models of speciation and genetic distance, in: *Population Genetics and Ecology* (S. Karlin and E. Nevo, eds), Academic Press, New York, pp. 723-765.

Nei, M., Maruyama, T., and Wu, C.-I., 1983, Models of evolution of reproductive isolation, *Genetics* **103**:557-579.

Orzack, S. H., and Tuljapurkar, S. D., 1989, Population dynamics in variable environments VII. The demography and evolution of iteroparity, *Am. Nat.* **133**:901-923.

Paterson, H. E. H., 1981, The continuing search for the unknown and unknowable: A critique of contemporary ideas on speciation, *S. Afr. J. Sci.* **77**:113-119.

Rice, W. R., 1987, Speciation via habitat specialization: the evolution of reproductive isolation as a correlated character, *Evol. Ecol.* **1**:301-314.

Reznick, D,. and Endler, J. A., 1982, The impact of predation on life history evolution in Trinidadian guppies, *Evolution* **36**:160-177.

Roff, D. A., 1981, On being the right size, *Am. Nat.* **118**:405-422.

Ruiz, A., and Heed, W. B., 1988, Host plant specificity in the cactophilic *Drosophila mulleri* species complex, *J. Anim. Ecol.* **57**:237-249.

Schaffer, W. M., and Rosenzweig, M. L., 1977, Selection for optimal life histories. II. Multiple equilibria and the evolution of alternative reproductive strategies, *Ecology* **58**:60-72.

Starmer, W. T., 1982, Analysis of community structure of yeasts associated with decaying stems of cactus. I. *Stenocereus gummosus*, *Microb. Ecol.* **8**:71-81.

Templeton, A. R., 1980, The theory of speciation via the founder principle, *Genetics* **94**:1011-1038.

Tuljapurkar, S. D., 1988, An uncertain life: Demography in random environments, Working Paper Series #10, The Stanford Institute for Population and Resource Studies, 97 pp.

Vetukhiv, M., 1953, Viability of hybrids between local populations of *Drosophila pseudoobscura*, *Proc. natn. Acad. Sci. USA* **39**:30-34.

Vetukhiv, M., 1954, Integration of the genotype in local populations of three species of *Drosophila*, *Evolution* **8**:241-251.

Walsh, J. B., 1983, Rate of accumulation of reproductive isolation in chromosome arrangements, *Am. Nat.* **120**:510-532.

Wasserman, M., and Koepfer, R. H., 1977, Character displacement for sexual isolation between *Drosophila mojavensis* and *Drosophila arizonensis*, *Evolution* **31**:812-823.

White, M. J. D., 1978, *Modes of Speciation,* W. H. Freeman. USA.

Wright, S., 1932, The roles of mutation, inbreeding, crossbreeding and selection in evolution, Proc. 6th Internatl. Cong. Genetics **1**:356-366.

Wright, S., 1977, *Evolution and the Genetics of Populations*, Vol. 3. Univ. Chicago Press, Chicago.

Wright, S., 1982, Character change, speciation, and the higher taxa, *Evolution* **36**:427-443.

Zouros, E., 1973, Genic differentiation associated with the early stages of speciation in the *mulleri* subgroup of *Drosophila*, *Evolution* **27**:601-621.

Zouros, E., 1981, The chromosomal basis of sexual isolation in two sibling species of *Drosophila*: *D. arizonensis* and *D. mojavensis*, *Genetics* **97**:703-718.

Zouros, E., and D'Entremont, C. J., 1980, Sexual isolation among populations of *Drosophila mojavensis*: response to pressure from a related species, *Evolution* **34**:421-430.

CHAPTER 3

Fitness and Asymmetry Modification as an Evolutionary Process
A Study in the Australian Sheep Blowfly, Lucilia cuprina and Drosophila melanogaster

JOHN A. MCKENZIE, PHILIP BATTERHAM, AND LOUISE BAKER

1. INTRODUCTION

Biologists have usually considered resistance to pesticides as an applied problem, albeit one of considerable importance. An ever increasing range of organisms have become resistant to chemical agents used in their control, adding to the costs of a number of industries (Georghiou, 1986). The future availability of effective new control agents is also a cause of concern (Metcalf, 1980; Hotson, 1985).

Each of these factors has been important in emphasizing the need for the implementation of strategies to manage resistance (Georghiou, 1983; Tabashnik and Croft, 1982) or, ideally, susceptibility (Daly and McKenzie, 1986). Such strategies will be most successful if based on data generated within a fundamental evolutionary framework (Roush and McKenzie, 1987). The benefits of this approach are relevant not only in an applied context. Clarke (1975) has defined an approach to demonstrate the action of selection in the evolutionary process and has emphasized that a limiting constraint is often the unambiguous identification of the selective agent.

Studies of the evolution of pesticide resistance are not hindered by this; the pesticide is the selective agent in the transition of a population from susceptibility to resistance. Furthermore, while in many laboratory studies and in some natural systems (e.g., anthelmintic resistance in nematodes (Martin *et al.*, 1988)) resistance is polygenically determined (Crow, 1957; Brown and Pal, 1971 and Roush and McKenzie, 1987 for reviews), in the field, resistance of economic importance is generally determined by allelic substitution at one or two loci (Roush and McKenzie, 1987). In each case the genetic mechanism is in accord with that expected on the basis of how the phenotypic variation is selectively channelled (Whitten and McKenzie, 1982; McKenzie, 1985). However, in the present

JOHN A. MCKENZIE, PHILIP BATTERHAM, AND LOUISE BAKER * Department of Genetics, University of Melbourne, Parkville 3052, Australia.

Ecological and Evolutionary Genetics of Drosophila, Edited by
J.S.F. Barker *et al.*, Plenum Press, New York, 1990

context, it is most important to note that resistance to insecticides in the field has a simple genetic basis. When taken
in conjunction with the identification of the selective agent,
it is suprising that the number of studies that have estimated
the relative fitness of resistance phenotypes in the presence
of the insecticide has been restricted (Muggleton, 1986; Roush
and McKenzie, 1987 for a review). Models of the evolution of
resistance have thus been based on generalized relationships
in which susceptible individuals (genotypically SS) are at a
selective advantage in the absence of the pesticide and
selected against when it is present.

It must be emphasized that the dominance, or recessiveness,
of resistance in these models is defined with respect to fitness. Inheritance of resistance in insects is generally partially or co-dominant (Whitten and McKenzie, 1982). The relative advantage of susceptibles in the absence of the insecticide is a consequence of the R allele disrupting biochemical
and physiological developmental processes (Brown and Pal,
1971) resulting in that allele being extremely rare in populations before the insecticide is introduced. The selective disadvantage of individuals carrying the R allele may, however,
be relatively small in some cases (Roush and McKenzie, 1987)
and may be further ameliorated if subsequent selection, in the
presence of the insecticide, integrates the R allele into an
appropriate genetic background (Abedi and Brown, 1960;
Georghiou, 1972).

The investigation of the evolution of insecticide resistance therefore affords the opportunity of studying the interface between relative fitness, co-adaptation and development
which is of key importance in defining constraints placed upon
more general aspects of evolutionary change (Maynard Smith *et
al.*, 1985). This paper considers this interface by reference
to the evolution of resistance to the insecticides dieldrin
and diazinon by the Australian sheep blowfly, *Lucilia cuprina*.

2. RESISTANCE TO INSECTICIDES BY THE AUSTRALIAN SHEEP BLOWFLY

An historical overview of the evolution of resistance to
dieldrin and diazinon is given in Table I. Resistance, with
respect to fitness, is partially dominant for dieldrin and
recessive for diazinon in the presence of the respective chem-

Table I
Dieldrin and diazinon resistance in the Australian sheep blowfly

Chemical	Group	Period of major use	Resistance detected	Genetics
Dieldrin	Cyclodiene	1955-1958	1957	Single gene (chromosome V)
Diazinon	Organo-phosphate	1958-1980 (still used)	1967	Basically single gene (chromosome IV)

icals (Whitten *et al.*, 1980; McKenzie and Whitten, 1982, 1984), results in accord with the expectation of the general model. These expectations also held for genotypes of the dieldrin resistance locus in trials on untreated sheep and on artificial medium lacking the insecticide. Susceptible individuals were selectively favored; the number of F_2 adults from an initial cross between a resistant and a susceptible strain was SS 524, RS 847 and RR 402 (x^2_2 = 20.31; P < 0.001). However, when diazinon resistance genotypes were compared, in similar trials, differences were insignificant (SS 379, RS 813 and RR 373; x^2_2 = 2.42; 0.30 > P > 0.20).

The genotypes used in these trials were derived from strains collected in the field in 1979. When diazinon resistance genotypes were compared by Arnold and Whitten (unpublished), using strains derived soon after resistance first evolved, susceptible individuals were at a selective advantage in the absence of diazinon (McKenzie, 1987). These results suggested that the fitness of genotypes resistant to diazinon had been modified during the period the chemical was used after resistance had developed, resulting in an increase in fitness relative to that of susceptible individuals. Modification of this kind had not occurred for genotypes resistant to dieldrin.

2.1. Selection of Fitness Modifiers

The modification of the initially deleterious effects of the R allele through its integration into the genome in a coadaptive process is often implied to be commonplace (Abedi and

Table II
The importance of genetic background and co-adaptation to the relative fitness of resistant and susceptible phenotypes

Species	Genetic background	Co-adaptation of R allele	Reference
Tetranychus urticae	Yes	No	Helle, 1965
Culex quinquefasciatus	Yes	No	Amin and White, 1984
Musca domestica	No	No	Whitehead *et al.*, 1985
Tribolium castaneum	No	No	Beeman and Nanis, 1986
Lucilia cuprina Dieldrin resistance	No	No	McKenzie and Gatford (unpubl.)
Diazinon resistance	Yes	Yes	McKenzie *et al.*, 1982

Brown, 1960; Georghiou, 1972; McEnroe and Naegele, 1968). This may not be the case (Roush and McKenzie, 1987). An effective way of demonstrating the effect is to disrupt the genetic background of the resistant strain by repeated backcrossing and selection, and to test the relative fitness of segregant genotypes at regular stages of the backcross program (Roush and McKenzie, 1987). The six studies using this approach have yielded species and/or insecticide specific results (Table II).

Genetic background was important to the fitness estimates in three cases but there was evidence for co-adaptation in only one, the diazinon resistance system of *Lucilia cuprina*. In this system, diazinon has been used for a significant period subsequent to the evolution of resistance (Table I) and after the R allele was close to fixation in many populations (McKenzie *et al.*, 1980). This evolutionary situation provides ideal circumstances for the selection of modifiers (Fisher, 1958; Charlesworth, 1979; Lenski, 1988), an opportunity that has not existed in the other systems.

2.2. Mapping of a Fitness Modifier in *Lucilia cuprina*

McKenzie *et al.* (1982) demonstrated fitness modification of diazinon resistance genotypes by comparing changes in the proportions of susceptibles, over generations, in discrete generation population cages started with RS flies. These flies resulted from an initial cross between a field-derived RR strain and a laboratory SS strain and after 3, 6 and 9 generations of backcrossing of the field strain to the SS strain. The proportion of SS flies in the population cages remained approximately constant, over generations, in the 0 and 3 generations of backcrossing comparisons, suggesting a fitness relationship of SS=RS=RR. In the 6 and 9 generations of backcrossing cages the proportion of SS increased, indicating a fitness set of SS>RS>RR. After 9 generations of backcrossing this fitness relationship was similar to that when resistance first evolved (McKenzie, 1987).

Subsequent genetical analysis has mapped the gene (gene complex) responsible for the modification to the *w* region of chromosome III (McKenzie and Purvis, 1984; McKenzie and Game, 1987). That is, the modifier is unlinked to the diazinon resistance locus on chromosome IV.

Comparisons of percentage egg hatch, the percentage of first instar larvae reaching adulthood and egg to adult time of development for combinations of modifier and resistance genotypes show the modifier affects only the latter (McKenzie and Game, 1987). Developmental time is decreased for RS and RR genotypes. The effect is dominant in these cases but the developmental time of SS genotypes is unaffected by modifier genotype (Fig. 1).

3. DEPARTURES FROM SYMMETRY: THE INFLUENCE OF A RESISTANT ALLELE

The introduction of a resistant allele into the genome may result in a developmental challenge to the organism (Brown and Pal, 1971) leading to an initially lower fitness of RS (and RR) individuals relative to SS. It may be predicted that the newly evolved phenotypes will be more variable than long-established ones if the developmental constraints stabilizing phenotypic expression are disrupted (Waddington, 1957; Maynard Smith et al., 1985). Departures from developmental homeostasis and the impact of environmental, or genetical, stress on the developmental process have been assessed by deviation from symmetry of a normally bilaterally symmetrical organism (Van Valen, 1962; Soulé, 1967; Maynard Smith et al., 1985; Palmer and Strobeck, 1986 for a review). Therefore, the working hypothesis is that when resistance first evolves resistant phenotypes will exhibit greater asymmetry than susceptibles.

This hypothesis was tested in *L. cuprina*. Asymmetry was estimated from the absolute difference ($|L-R|$) in bristle count between the left and right sides of the frontal head stripe, the outer wing margin and the R_{4+5} wing vein (Clarke and McKenzie, 1987). The mean asymmetry of the dieldrin resistant strain was significantly greater than that of susceptible strains, the asymmetry of which was similar to that of the diazinon resistant strain. When the field background of the resistant strains was disrupted by repeated backcrossing to a susceptible strain, the mean asymmetry of dieldrin resistant flies did not vary, but, the mean asymmetry of diazinon resistant segregants increased with each generation of backcrossing until a plateau was reached after seven generations. The influence of the resistant allele on asymmetry was shown to be dominant in both resistant systems (Clarke and McKenzie, 1987).

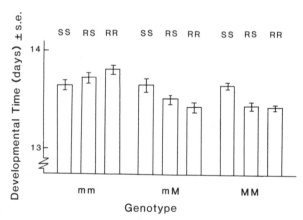

Figure 1. Developmental times (egg to adult) of individuals of specified modifier (M, m) and diazinon resistance (R, S) genotype (based on 30 trials, after McKenzie and Game, 1987).

These results are in accord with the hypothesis that the presence of a resistant allele in the genome initially increases mean asymmetry. Subsequent selection for a modified genetic background has occurred for the diazinon resistance system of *L. cuprina* resulting in a return to base line levels of asymmetry (Clarke and McKenzie, 1987), presumably as a result of selection for a return to normal developmental home-ostasis (Waddington, 1957; Maynard Smith *et al.*, 1985).

The gene(s) responsible for this modification of asymmetry mapped to chromosome III (Clarke and McKenzie, 1987), the same chromosome to which the fitness modifier was mapped (McKenzie and Game, 1987), suggesting the same gene (gene complex) may be responsible in each case. This presented a novel experimental opportunity to investigate the associations between changes in asymmetry and fitness, a topic of much debate and of data open to different interpretations (Palmer and Strobeck, 1985; Jones, 1987).

3.1. Mapping of a Fitness/Asymmetry Modifier

The genetic basis of asymmetry modification was assessed using the resistant strain of McKenzie and Game (1987), and a dieldrin resistant strain was analysed for comparison. Resistant flies were crossed to a susceptible strain in which chromosomes II-VI carried a recessive marker (*bp; ru; gl; mv; y*, respectively). F_1 males were testcrossed and asymmetry scored for each class (McKenzie and Clarke, 1988). The results are unambiguous (Fig. 2).

The data of both dieldrin and diazinon resistant strains fell into 2 classes, one of base level asymmetry ($x \approx 2$, a similar asymmetry value to that of susceptibles), the other with elevated asymmetry values ($x \geq 3$). In the cross with the dieldrin resistant strain whenever the R allele was present (Parents (RR), F_1 and +*mv* progeny) asymmetry levels were in the latter class. When progeny were genotypically *mv mv*, SS at the dieldrin resistance locus, base levels of asymmetry were observed. The results were consistent with the dieldrin resistant allele having a disruptive effect. The effect was dominant (see also Clarke and McKenzie, 1987).

Parents, F_1, *gl gl* and +*gl*; +*ru* genotypes of the diazinon cross gave phenotypes with base level asymmetries. Flies of genotype +*gl*; *ru ru*, RS without chromosome III material from the resistant strain, have elevated asymmetries (Fig. 2). Therefore, the effect of the diazinon resistant allele was disruptive, the modifier's influence in ameliorating this was dominant and the modifier mapped to chromosome III (McKenzie and Clarke, 1988). This result was consistent with that of the initial analysis using a different diazinon resistant strain (Clarke and McKenzie, 1987). The asymmetry modifier maps to the *w* region of chromosome III and available data are consistent with the fitness/asymmetry modifier being the same gene (gene complex) (McKenzie and Clarke, 1988).

3.2. The Fitness/Asymmetry Association

The discussion thus far has not attempted to partition departures from symmetry into the three recognised asymmetry classes (fluctuating asymmetry, directional asymmetry, and antisymmetry, see Palmer and Strobeck, 1986 for definitions). Fluctuating asymmetry is due to environmental influences on development (developmental "noise", Waddington, 1957), whereas directional asymmetry and antisymmetry reflect genetical influences. The presentation of data in many studies makes it difficult to interpret the robustness of allocation to a specific class of asymmetry and the field has a tendency to

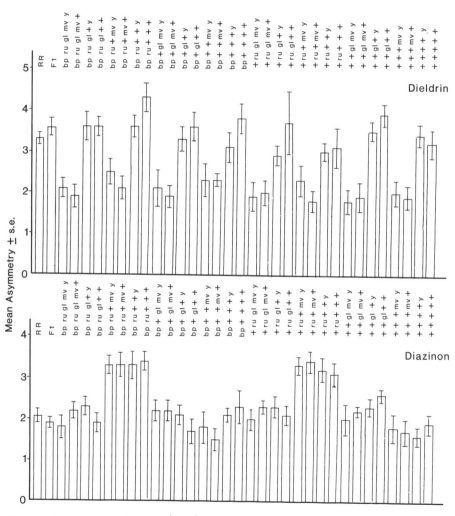

Figure 2. Mean asymmetry ($|L-R|$ ± s.e.) of resistant (RR, sample size, 100), F_1 (30) and the 32 phenotypic classes of testcross progeny (10 per class) of M_15 x RR for dieldrin and diazinon resistant strains (diazinon data after McKenzie and Clarke, 1988).

become a minefield of semantical and statistical debate, particularly in the area of population phenetics (Palmer and
Strobeck, 1986; Wayne *et al.*, 1986; Willig and Owen, 1987;
Modi *et al.*, 1987; Swain, 1987). Some of these difficulties
would be overcome if the distributions of signed (R-L)
differences were described (Palmer and Strobeck, 1986). Data
of the diazinon resistance system emphasize this point.

Fluctuating asymmetry was observed for susceptible flies
and also for resistant flies carrying the modifier. Antisymmetry was observed for resistant phenotypes without the modifier (McKenzie and Clarke, 1988). The evolution of resistance
and subsequent selection of the modifier therefore represent a
transition from fluctuating asymmetry through antisymmetry to
fluctuating asymmetry; a perturbation and re-establishment of
developmental stability that parallels the observed transition
in the relative fitness of resistant individuals. It is this
association that is of particular importance as, while often
assumed, convincing examples in the literature are rare (Leary
et al., 1984; Palmer and Strobeck, 1986; Jones, 1987). That
the clarity of the association occurs in an instance where the
genotypes involved in the evolutionary change are defined and
presumptive developmental changes occur is unlikely to be
serendipitous. Developmental changes are likely to influence
fitness (and changes in asymmetry); the converse does not necessarily follow.

The influence of a single gene (gene complex) on developmental stability is also of special interest. There is some
evidence that co-adapted gene complexes may have significant
influences (Leary *et al.*, 1984), however, most models are
based on whole genome heterotic interactions (Lerner, 1954;
Maynard Smith *et al.*, 1985; Palmer and Strobeck, 1986).

4. THE MECHANISM OF MODIFIER ACTION: *SCL* AND *NOTCH* HOMOLOGY?

Diazinon resistance in *L. cuprina* can be explained by the
more efficient hydrolysis of the insecticide by the altered
gene product of an esterase locus (Hughes and Devonshire,
1982; Hughes and Raftos, 1985). The increased levels of
asymmetry and lowered fitness of resistant phenotypes, when
resistance first evolved, suggest that the altered gene product disrupts normal developmental patterns. The modifier ame-
liorates these effects.

Organophosphorus insecticides influence the nervous system,
therefore it is possible to suggest a model involving interaction between the gene products of a neurogenic locus and the
resistance locus. Under this model the modifier would be an
allele of the neurogenic locus. It is therefore of interest to
note that Scalloped wings, *Scl*, of *L. cuprina* which, on the
basis of phenotype, recessive lethality and the conservation
of the linkage maps of *L. cuprina* and *Drosophila melanogaster*
(Foster *et al.*, 1981), is considered homologous to *Notch* of *D.
melanogaster*, is closely linked to the *w* locus (Maddern *et
al.*, 1986), the region to which the fitness/asymmetry modi-

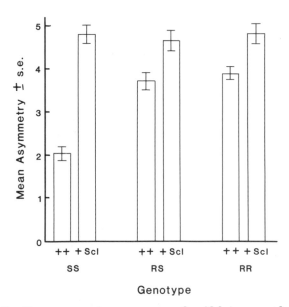

Figure 3. Mean asymmetry ± s.e. of wild-type and Scalloped wings flies of the three diazinon resistance classes. (Methods follow Clarke and McKenzie, 1987; sample size, 50).

fier maps (McKenzie and Clarke, 1988). *Notch* is known to play a key role in neurogenesis and development (Yedvobnick *et al.*, 1985; Artavanis-Tsakonas, 1988) and interacts with the *E*(*spl*) gene which may function in signal transduction (Hartley *et al.*, 1988).

The model remains speculative, but two predictions flow from it. Firstly, mutations at the *Scl* and *Notch* loci should generate phenotypes of increased asymmetry, relative to wild-type, and, secondly, the modifier should lower the asymmetry and increase the fitness of Scalloped wings phenotypes. Each of these expectations is met.

4.1. The Asymmetry of *Scl* and *Notch* Phenotypes

The asymmetry of wild-type and Scalloped wings segregants within each diazinon resistance class is shown in Figure 3. Within each resistance class it is clear that higher asymmetries are observed for the Scl/+ genotypes. The increased asymmetry level of phenotypes of resistant genotypes RS and RR, relative to SS, in the wild-type class demonstrates the dominant disruptive effect of the R allele and that the *Scl* strains do not carry the modifier. Scalloped wings phenotypes are antisymmetric for each of the characters scored by Clarke and McKenzie (1987).

Strains of *D. melanogaster* can be divided into two classes of asymmetry values, although it should be noted that there is

Table III
Mean asymmetry (x ± s.e.) of wild-type and *Notch*
mutation strains pooled over the orbital (1), scutellar
(2), thoracic (3), sternopleural (4), abdominal (5),
palp (6) and vibrissal (7) bristle characters
(sample size, 120)

Strain	x ± s.e.
Slankumen	5.08 ± 0.15
Canton S	5.09 ± 0.17
facet^{swbBG}	5.46 ± 0.18
Umen 79	5.67 ± 0.14
facet	7.21 ± 0.21
Abruptex	7.26 ± 0.21
Notch-8	7.75 ± 0.23
split	8.85 ± 0.23

considerable variation within each class (Table III). The
Notch mutants *fa*, *Ax*, N^8 and *spl* have significantly higher
asymmetry values than the three wild-type stocks and the *fa*^{swbBG}
mutant. Outcrossing a *Notch* strain to the Canton-S wild-type
stock for five generations produced no difference in mean
asymmetry. Therefore the asymmetry phenotype segregated with
the *Notch* allele, a conclusion supported by the observation
that other mutations in the region have wild-type asymmetry
levels (w^a 5.20 ± 0.20; *dnc* 5.20 ± 0.16).

The wild-type asymmetry phenotype exhibited by *fa*^{swbBG} is
noteworthy given the molecular basis of this mutation. The *fa*
and *fa*^{swbBG} mutations result from the insertion of different
retrotransposons in precisely the same region of the second
intron of the *Notch* gene; *opus* in *fa* and *flea* in *fa*^{swbBG}. In
each case the retrotransposon is inserted in reverse orientation to the *Notch* gene suggesting that the mutant phenotype
may be due to transcriptional interference resulting in per-
turbed levels of *Notch* product (Kidd and Young, 1986). Given
that the retrotransposons *opus* and *flea* show distinct patterns
of transcription during *Drosophila* development, it would seem
likely that a difference in levels of *Notch* product at a specific stage in development may be responsible for the differing asymmetry phenotypes of *fa* and *fa*^{swbBG}. Results for the
remaining mutants confirm the importance of the wild-type
Notch gene product in the maintenance of symmetry.

The four strains of the "wild-type class" display fluctuating asymmetry for each of the characters scored. The strains
with increased asymmetry display character specific responses
relative to a wild-type comparison. Fluctuating asymmetry
increases for characters 1-4 when signed (L-R) distributions
are compared, suggesting that the mutation rendered the developmental processes associated with those characters more susceptible to environmental influences. The (L-R) distributions
of other characters (5-7) are bimodal, about mean zero, indicating antisymmetry where departures from symmetry are genetically based (Palmer and Strobeck, 1986).

The *Notch* gene has recently been described as "toporhythmic" given its role in apportioning cells to different fates during differentiation and thus playing an essential role in the fine tuning of pattern formation (Artavanis-Tsakonas, 1988). The bristle asymmetry phenotypes support this designation and are consistent with the suggested fine tuning role of *Notch* in development.

Artavanis-Tsakonas (1988) proposed that the *Notch* gene functions during development for the differentiation of individual cells. One cell within a partially committed cell cluster responding to a cue acquires a specific fate, resulting in the recruitment of surrounding cells through cell interactions. The cueing and/or recruitment processes would be sensitive to both genetic and environmental factors, thus leading to the antisymmetry and fluctuating asymmetry observed for bristle characters in *Notch* mutants (Table III).

If *Notch* and *Scl* are indeed homologous, and the modifier is allelic to *Scl*, the opportunity exists to consider cellular interactions in a system where normal, disrupted and modified developmental variants are available and in which the evolutionary significance of the variation is known. Some support for the modifier being an allele of the *Scl* locus comes from its influence on the asymmetry and fitness of the Scalloped wings phenotype, although it should be noted that other interpretations are possible. The notation used below assumes allelism and follows that of McKenzie and Clarke (1988).

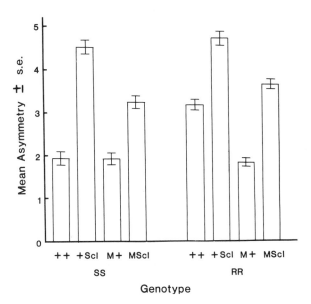

Figure 4. Mean asymmetry ± s.e. of flies of the genotypes specified (sample size, 50).

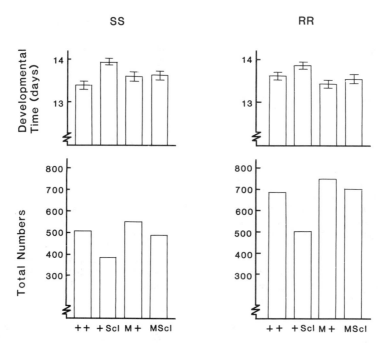

Figure 5. Developmental times ± s.e. and the total number of flies emerging for each of the specified genotypes (based on 30 trials).

4.2. The Influence of the Modifier on *Scl* Phenotypes

Pure-breeding modifier and non-modifier susceptible strains were crossed to *Scl* +; SS flies and developmental time estimated as in McKenzie and Game (1987). The number of wild-type and Scalloped winged flies emerging in each cross was recorded and flies of each class scored for asymmetry following Clarke and McKenzie (1987). The procedure was repeated using resistant strains.

In each case the modifier reduces the asymmetry of Scalloped wings phenotypes in a partially dominant manner, relative to its dominant influence on the asymmetry of resistant flies with normal wings (Fig. 4). *Scl* and the modifier also have a significant impact on the number of flies emerging and their developmental time (Fig. 5).

Comparison of the number of Scalloped wings and wild-type phenotypes of each cross, against the 1:1 ratio expected with equal viability of each phenotype, shows an excess of wild-type flies in SS (x^2_1 = 16.98, P < 0.001) and RR (x^2_1 = 27.74, P < 0.001) backgrounds in the absence of the modifier. The 1:1 ratio is observed in the RR flies (x^2_1 = 1.52, 0.30 > P > 0.20), and approximated in SS flies (x^2_1 = 3.93, 0.04 < P < 0.05), in the presence of the modifier (Fig. 5). The expression of the Scalloped wings phenotype is similar in each cross, suggesting that it is the viability of the Scalloped wings phenotype that is influenced by the modifier. Developmental time comparisons give similar results (Fig. 5).

In the absence of the modifier, the developmental time of Scalloped wings flies is significantly greater than wild-type for both SS (t_{58} = 2.47, P < 0.02) and RR (t_{58} = 2.06, P < 0.05) genotypes. However, for the same comparisons, the difference between the developmental times is insignificant (SS, t_{58} = 0.28, 0.8 > P > 0.7; RR, t_{58} = 0.68, 0.6 > P > 0.5) in the presence of the modifier. It is of interest to note that while the developmental times of ++; RR and M+; RR genotypes are not significantly different (t_{58} = 1.60; 0.2 > P > 0.1) in these trials, the result is in the same direction as that observed by McKenzie and Game (1987); the trend is towards the modifier enhancing the developmental rate of RR flies.

5. CONCLUSIONS AND FUTURE DIRECTIONS

The studies presented in this paper have shown that the evolution of resistance has associated fitness and developmental "costs". In the absence of conditions that specifically favor resistant individuals, the presence of the insecticide, they are selected against although the selection of modifiers may alter this if the insecticide is used for a sufficient period after the R allele is common in the population (Roush and McKenzie, 1987). Thus, selection for modifiers occurs in the presence of the insecticide and it is therefore important to note that, in the *L. cuprina* system, all measurements of the modification of fitness of resistant individuals have been against a susceptible standard in the absence of the insecticide (McKenzie *et al.*, 1982; McKenzie and Purvis, 1984; McKenzie and Game, 1987). The results from these studies suggest that MM; RR flies will have greater relative fitness than ++; RR flies in the presence of diazinon. While there have been some difficulties in defining the appropriate concentrations of the insecticide to make the comparison, the hypothesis is currently being tested.

More generally, it will be of interest to assess the influence of resistant alleles on relative fitness and development across a range of resistance systems and organisms with measurements of asymmetry being a useful tool. There is no shortage of candidates for such studies (Georghiou, 1986)! Ideally, the comparisons between susceptible and resistant phenotypes should be made with individuals of a common genetic background and, while the evolutionary opportunity for modification of fitness and development is not likely to be common, in appropriate systems, changes in fitness with continued pesticide use should be investigated within a similar framework (Roush and McKenzie, 1987).

The study of resistance systems may be seen as an extension of the field in which the influence of environmentally imposed stress on departures from symmetry and, by implication, on development has been considered. The results have been somewhat ambiguous (Palmer and Strobeck, 1986) perhaps because subtle developmental changes may depend on whether the stress is "general" or "specific" (Parsons, 1961). Extreme caution is therefore necessary in the interpretation of population phenetic data in the absence of detailed ecological and genetical information. Where such data are available, measures of developmental symmetry may be particularly powerful tools in tracking evolutionary transitions and defining developmental constraints. For example, asymmetry estimates from museum specimens of *Biston betularia*, over the period of changes of morph frequency, would allow any disruptive effect of the *carbonaria* gene to be assessed and any subsequent selection of modifiers to be considered.

The distributions of departures from symmetry in the neurogenic genes *Notch* and, assuming homology, *Scl* are of interest. To test whether more general parallels may be drawn between toporhythmic genes and asymmetry phenotypes it would be necessary to first examine the symmetry of mutants for the other neurogenic genes which display similar phenotypes to those observed in Notch mutants (i.e., *Delta*, *Enhancer of split*, *mastermind*, *big brain* and *neuralized*). Should the comparison hold, the relationship could be further tested in other species and may prove to be an initial screen of presumptive developmental variants.

Each of the above directions offers the chance of extending the data base of tantalising projections, the generality of which requires confirmation. Indeed, there is an equally speculative framework in the reductionalist approach to an understanding of the mechanistic basis of fitness/asymmetry modification of diazinon resistant phenotypes of *L. cuprina*. However, preliminary Southern analysis suggests this species does have a gene that is homologous to *Notch*. If the working hypotheses that *Scl* is homologous to *Notch* and that the modifier is allelic to *Scl* are confirmed, this system may have much to offer at the interface of molecular, developmental and evolutionary genetics. In such circumstances speculation, and experimentation, seems warranted.

6. SUMMARY·

The evolution of alleles that confer resistance to the insecticides dieldrin and diazinon had a deleterious effect on development of the Australian sheep blowfly, *L. cuprina*. The effect was demonstrated by lowered fitness and increased bilateral asymmetry of resistant, relative to susceptible, phenotypes. Use of diazinon, to control the blowfly, for a significant period after resistance had evolved selected a dominant modifier to ameliorate these effects in diazinon-resistant phenotypes providing unusually convincing evidence of an association between changes in fitness and asymmetry, an area of considerable debate. Data are consistent with the modifier being an allele of the *Scl* locus. *Scl* is considered homologous to *Notch* of *D. melanogaster*, a locus known to play a key role in development and neurogenesis.

ACKNOWLEDGEMENTS. The work on which this paper is based was made possible by the support of the Australian Research Council and the Research Trust Fund of the Australian Wool Corporation.

References

Abedi, Z. H., and Brown, A. W. A., 1960, Development and reversion of DDT-resistance in *Aedes aegypti*, *Can. J. Genet. Cytol.* **2**:252-261.

Amin, A. M., and White, G. B., 1984, Relative fitness of organophosphate-resistant and susceptible strains of *Culex quinquefasciatus* Say (Diptera: Culicidae), *Bull. ent. Res.* **74**:591-598.

Artavanis-Tsakonas, S., 1988, The molecular biology of the *Notch* locus and the fine tuning of differentiation in *Drosophila*, *Trends Genet.* **4**:95-100.

Beeman, R. W., and Nanis, S. M., 1986, Malathionin resistance alleles and their fitness in the red flour beetle (Coleoptera: Tenebrionidae), *J. econ. Ent.* **79**:580-587.

Brown, A. W. A., and Pal, R., 1971, *Insecticide Resistance in Arthropods*, W.H.O., Geneva.

Charlesworth, B., 1979, Evidence against Fisher's theory of dominance, *Nature, Lond.* **278**:848-849.

Clarke, B., 1975, The contribution of ecological genetics to evolutionary theory: detecting the direct effects of natural selection on particular polymorphic loci, *Genetics* **79**:101-113.

Clarke, G. M., and McKenzie, J. A., 1987, Developmental stability of insecticide resistant phenotypes in blowfly; a result of canalizing natural selection, *Nature, Lond.* **325**:345-346.

Crow, J. F., 1957, Genetics of insect resistance to chemicals, *A. Rev. Ent.* **2**:227-246.

Daly, J., and McKenzie, J. A., 1986, Resistance management strategies in Australia: The *Heliothis* and "Wormkill" programmes, in: *Pests and Diseases*, BCPC, Brighton (U.K.), pp. 951-959.

Fisher, R. A., 1958, *The Genetical Theory of Natural Selection*, 2nd ed., Dover, New York.

Foster, G. G., Whitten, M. J., Konovalov, C., Arnold, J. T. A., and Maffi, G., 1981, Autosomal genetic maps of the Australian sheep blowfly, *Lucilia cuprina dorsalis* R-D (Diptera: Calliphoridae) and possible correlations with the linkage maps of *Musca domestica* L. and *Drosophila melanogaster* (Mg.), *Genet. Res.* **37**:55-69.

Georghiou, G. P., 1972, The evolution of resistance to pesticides, *Annu. Rev. Ecol. & Syst.* **3**:133-168.

Georghiou, G. P., 1983, Management of resistance in arthropods, *in: Pest Resistance to Pesticides* (G. P. Georghiou, and T. Saito, eds), Plenum, New York, pp. 769-792.

Georghiou, G. P., 1986, The magnitude of the resistance problem, in: *Pesticide Resistance: Strategies and Tactics for Management*, National Academy Press, Washington, D.C, pp. 14-43.

Hartley, D. A., Preiss, A., and Artavanis-Tsakonas, S., 1988, A deduced gene product from the *Drosophila* neurogenic locus, *Enhancer of split*, shows homology to mammalian G-protein B subunit, *Cell* **55**:785-795.

Helle, W., 1965, Resistance in the acarina: mites, *Adv. Acarol.* **2**:71-93.

Hotson, I. K., 1985, New developments in nematode control: the role of the animal health products industry, in: *Resistance in Nematodes to Anthelmintic Drugs* (N. Anderson, and P. J. Walker, eds), CSIRO, Aust. Wool Corp., Sydney, pp. 117-125.

Hughes, P. B., and Devonshire, A. L., 1982, The biochemical basis of resistance to organophosphorus insecticides in the sheep blowfly, *Lucilia cuprina*, *Pestic. Biochem. & Physiol.* **18**:289-297.

Hughes, P. B., and Raftos, D. A., 1985, Genetics of an esterase associated with resistance to organophosphorus insecticides in the sheep blowfly, *Lucilia cuprina* (Wiedemann)(Diptera: Calliphoridae), *Bull. ent. Res.* **75**:535-544.

Jones, J. S., 1987, An asymmetrical view of fitness, *Nature, Lond.* 325:298-299.

Kidd, S., and Young, M. W., 1986, Transposon-dependent mutant phenotypes at the *Notch* locus of *Drosophila, Nature, Lond.* 323:89-91.

Leary, R. F., Allendorf, F. W., and Knudsen, K. L., 1984, Superior developmental stability of heterozygotes at enzyme loci in salmonid fishes, *Am. Nat.* 124:540-551.

Lenski, R. E., 1988, Experimental studies of pleiotropy and epistasis in *Escherichia coli.* II. Compensation for maladaptive effects associated with resistance to virus T4, *Evolution* 42:433-440.

Lerner, I. M., 1954, *Genetic Homeostasis*, Wiley, New York.

Maddern, R. J., Foster, G. G., Whitten, M. J., Clarke, G. M., Konovalov, C. A., Arnold, J. T. A., and Maffi, G., 1986, The genetic mutations of *Lucilia cuprina* R.-D. (Diptera: Calliphoridae), CSIRO, Div. Entomol. Report No. 37, Canberra.

Martin, P. J., McKenzie, J. A., and Stone, R. A., 1988, The inheritance of thiabendazole resistance in *Trichostrongylus colubriformis, Int. J. Parasitol.* 18:703-709.

Maynard Smith, J., Burian, R., Kauffman, S., Alberch, P., Campbell, J., Goodwin, B., Lande, R., Raup, D., and Wolpert, L., 1985, Developmental constraints and evolution, *Q. Rev. Biol.* 60:266-287.

McEnroe, W. D., and Naegele, J. A., 1968, The coadaptive process in an organophosphorus-resistant strain of the two-spotted spider mite, *Tetranychus urticae, Ann. ent. Soc. Am.* 61:1055-1059.

McKenzie, J. A., 1985, Genetics of resistance to chemotherapeutic agents, in: *Resistance in Nematodes to Anthelmintic Drugs* (N. Anderson, and P. J. Waller, eds), CSIRO, Aust. Wool Corp., Sydney, pp. 89-95.

McKenzie, J. A., 1987, Insecticide resistance in the Australian sheep blowfly - messages for pesticide usage, *Chem. Ind.* 8:266-269.

McKenzie, J. A., and Clarke, G. M., 1988, Diazinon resistance, fluctuating asymmetry and fitness in the Australian sheep blowfly, *Lucilia cuprina, Genetics* 120:213-220.

McKenzie, J. A., and Game, A. Y., 1987, Diazinon resistance in *Lucilia cuprina*: mapping of a fitness modifier, *Heredity* 59:381-391.

McKenzie, J. A., and Purvis, A., 1984, Chromosomal localisation of fitness modifiers of diazinon resistance genotypes of *Lucilia cuprina, Heredity* 53:625-634.

McKenzie, J. A., and Whitten, M. J., 1982, Selection for insecticide resistance in the Australian sheep blowfly, *Lucilia cuprina, Experientia* 38:84-85.

McKenzie, J. A., and Whitten, M. J., 1984, Estimation of relative viabilities of insecticide resistance genotypes of the Australian sheep blowfly, *Lucilia cuprina, Aust. J. biol. Sci.* 37:45-52.

McKenzie, J. A., Dearn, J. M., and Whitten, M. J., 1980, Genetic basis of resistance to diazinon in Victorian populations of the Australian sheep blowfly, *Lucilia cuprina, Aust. J. biol. Sci.* 33:85-95.

McKenzie, J. A., Whitten, M. J., and Adena, M. A., 1982, The effect of genetic background on the fitness of diazinon resistance genotypes of the Australian sheep blowfly, *Lucilia cuprina, Heredity* 49:1-9.

Metcalf, R. L., 1980, Changing role of insecticides in crop protection, *A. Rev. Ent.* 25:219-256.

Modi, W. S., Wayne, R. K., and O'Brien, S. J., 1987, Analysis of fluctuating asymmetry in cheetahs, *Evolution* 41:227-228.

Muggleton, J., 1986, Selection for malathion resistance in *Oryzaephilus surinamensis* (L.) (Coleoptera: Silvanidae): fitness values of resistant and susceptible phenotypes and their inclusion in a general model describing the spread of resistance, *Bull. ent. Res.* 76:469-480.

Palmer, A. R., and Strobeck, C., 1986, Fluctuating asymmetry: measurement, analysis, patterns, *Annu. Rev. Ecol. & Syst.* **17**:391-421.

Parsons, P. A., 1961, Fly size, emergence time and sternopleural chaeta number in *Drosophila*, *Heredity* **16**:455-473.

Roush, R. T., and McKenzie, J. A., 1987, Ecological genetics of insecticide and acaricide resistance, *A. Rev. Ent.* **32**:361-380.

Soulé, M., 1967, Phenetics of natural populations. II. Asymmetry and evolution in a lizard, *Am. Nat.* **101**:141-160.

Swain, D. P., 1987, A problem with the use of meristic characters to estimate developmental stability, *Am. Nat.* **129**:761-768.

Tabashnik, B. E., and Croft, B. A., 1982, Managing pesticide resistance crop-arthropod complexes: interactions between biological and operational factors, *Environ. Entomol.* **11**:1137-1144.

Van Valen, L., 1962, A study of fluctuating asymmetry, *Evolution* **16**:125-142.

Waddington, C. H., 1957, *The Strategy of the Genes*, Allen and Unwin, London.

Wayne, R. K., Modi, W. S., and O'Brien, S. J., 1986, Morphological variability and asymmetry in the cheetah (*Acinonyx jubatus*), a genetically uniform species, *Evolution* **40**:78-85.

Whitehead, J. R., Roush, R. T., and Norment, B. R., 1985, Resistance stability and coadaptation in diazinon-resistant house flies (Diptera: Muscidae), *J. econ. Ent.* **78**:25-29.

Willig, M. R., and Owen, R. D., 1987, Fluctuating asymmetry in the cheetah: methodological and interpretive concerns, *Evolution* **41**:225-227.

Whitten, M. J., and McKenzie, J. A., 1982, The genetic basis for pesticide resistance, Proc. 3rd Aust. Conf. Grassland Invertebrate Ecol., South Aust. Gov. Printer, Adelaide, pp. 1-16.

Whitten, M. J., Dearn, J. M., and McKenzie, J. A., 1980, Field studies on insecticide resistance in the Australian sheep blowfly, *Lucilia cuprina*, *Aust. J. biol. Sci.* **33**:725-735.

Yedvobnick, B., Muskavitch, M. A. T., Wharton, K. A., Halpern, P. E., Grimwade, B. G., and Artavanis-Tsakonas, S., 1985, Molecular genetics of *Drosophila* neurogenesis, *Cold Spring Harb. Symp. Quant. Biol.* **50**:841-854.

CHAPTER 4

Extreme Environmental Stress: Asymmetry, Metabolic Cost and Conservation

P.A. PARSONS

1. INTRODUCTION

Fisher (1930) in *The Genetical Theory of Natural Selection* wrote "If therefore an organism be really in any high degree adapted to the place it fills in the environment, this adaptation will be constantly menaced by any undirected agencies liable to cause changes to either party in the adaptation". In using this quote in an article on *Genetics of resistance to environmental stresses in Drosophila populations*, (Parsons, 1973), I then added that "Environmental stresses of a man-made type are going to assume progressively more prominence with time, especially as they can be regarded as rather more 'directed' than Fisher's 'undirected' agencies." At that time, the main 'directed' agency consisted of various chemicals such as insecticides. Now we have a far more insidious scene comprising increased concentration of CO_2 and less-abundant atmospheric gases including chloro-fluorocarbons (CFCs), all implying substantial exchange of materials between terrestrial systems and the atmosphere (Mooney *et al.*, 1987). The impact of CFCs is a likely depletion of stratospheric O_3 so diminishing its role as a protective absorber of short-wave radiation (Cicerone, 1987). It is now well-known that collectively, these changes could increase world temperature by up to 5°C within the next 100 years. This is <u>exceedingly</u> rapid as compared with prehistoric changes of a similar magnitude, so that conditions of severe environmental stress on the world's biota are likely.

The literature on variation under stress was reviewed extensively by Parsons (1987) in a wide range of organisms, with the conclusion that both phenotypic and genotypic variability increase under conditions of severe stress imposed by the physical and biological environments. This applies directly to quantitative traits of importance in determining survival, and more indirectly at the level of genes control-

---

P.A. PARSONS * Division of Science and Technology, Griffith University, Brisbane. Present address: Department of Zoology, University of Adelaide, Adelaide, S.A. 5000, Australia.

Ecological and Evolutionary Genetics of Drosophila, Edited by
J.S.F. Barker *et al.*, Plenum Press, New York, 1990

ling protein variation. This conclusion assumes that populations have not been previously directionally selected for phenotypic extremes for the trait in question, which could ultimately produce lowered variability (Parsons, 1980).

Concentrating upon *Drosophila* evidence, this article has three purposes:

<u>Firstly</u>, adaptation to stress at the developmental level will be considered, and it will be shown that stress increases variability. Following Waddington (1953, 1956), severe environmental perturbations applied at certain critical times in development may greatly increase developmental variability of morphological structures. In relation to natural habitats, however, these experiments can be regarded as rather contrived since a trait that can be studied on all individuals in a wide range of taxa is needed. Fluctuating asymmetry (FA) of bilateral structures (Palmer and Strobeck, 1986) is an appropriate trait, and it will be shown that environmental (and genetic) changes can increase FA, representing a deterioration in developmental homeostasis expressed by adult morphology.

<u>Secondly</u>, the developmental cost of stress implied above is generalized to show that stress has a metabolic cost. This is demonstrated using desiccation resistance which can be regarded as an environmental probe for magnifying genetic variability. Lines selected for increased desiccation resistance have decreased behavioral activity and metabolic rate, and are resistant to an array of generalized stresses. This is leading to an integrated approach to an understanding of the basis of environmental stress resistance (Hoffmann and Parsons, 1989a).

<u>Thirdly</u>, the background of available experimental data on stress resistance in *Drosophila* permits questions concerning the possibility of adaptation to a rapid temperature increase as predicted by some for the greenhouse scenario. It is important, furthermore, to consider other stresses simultaneously such as pollutants and drought, since stresses do not normally occur in isolation. Metabolic cost arguments can, in principle, provide predictions, which could be investigated experimentally in a species such as *D. melanogaster*.

2. DEVELOPMENTAL STRESS - VARIABILITY AND ASYMMETRY

There is a substantial literature on life span in *Drosophila* indicating little response to directional selection when larvae are at a low density (Lints *et al.*, 1979) but significant responses to selection when larval density is not controlled (Rose, 1984). Luckinbill and Clare (1985) and Clare and Luckinbill (1985) selected for adult life span and found that populations with intense competition responded strongly to selection for later reproduction, but when populations of developing larvae were at a low density, there was little overall response. They concluded that the genes controlling longevity are conventional in every sense, showing near additivity in the uncontrolled-density treatment. That is, it is the expression of genes that differs between density

treatments, such that some stress is needed for the maintenance of reasonable levels of genetic variability. It is not
therefore surprising that the assessment of correlations among
life-history traits has proved on many occasions to be such a
difficult problem, leading to so much discussion in the literature.

Turning to larval resource utilization under stress,
Robertson (1960) studied the size of *D. melanogaster* on different diets. For within culture variance of body size, representing the combined effects of genetic segregation and
unidentified environmental variation, the lowest variance was
found on live yeast medium, while on diets deficient in fructose, RNA, and protein, the variance was greater, representing
an increase in the phenotypic expression of gene differences.
It is of course well known that as stresses become increasingly severe, especially if applied early in ontogeny, phenotypic variation in the form of phenocopies is a distinct possibility (Waddington, 1953). However, such extreme circumstances may be of minor significance in the present evolutionary context, because the stresses would not normally be
applied for sufficient generations to be assimilated genetically.

Figure 1 gives results of an experiment on an Oregon-R
stock of *D. melanogaster* grown at 25°C and 30°C, indicating

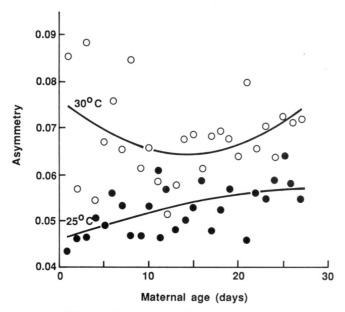

Figure 1. Fluctuating asymmetry, FA, of sternopleural
chaeta number plotted against maternal age for the
Oregon-R stock grown at 25°C(·) and 30°C(o), from Parsons
(1962). FA was computed as the sum of the absolute
differences between the left and right hand sides of
flies, A, divided by the total number of chaetae, T, on
60 flies scored daily at each temperature.

far more fluctuating bilateral asymmetry (FA) of steropleural chaeta number at 30°C over a range of maternal ages especially in flies derived from young females, when developmental stability appears to be particularly high (Parsons, 1962). At 30°C, there is a great deal of variability from day to day which tends to obscure trends but a fitted parabola borders on significance for maternal age (P <0.10). The extreme variability at 30°C might be expected since this temperature is an extreme stress when slight changes in micro-environment may have major effects. At 25°C, FA increases with maternal age (P <0.01), a result consistent with many observations on increased developmental instability with maternal age, especially in mice and our own species.

In another series of experiments, Parsons (1961) found that sternopleural chaeta number and FA were higher at 30°C than at 25°C. The conclusion that morphological homeostasis was poorer at 30°C was confirmed by the common occurrence of flies with other abnormalities including crumpled and curled wings. Fly weight was only reduced slightly, indicating the possibility of death due to general poor morphological homeostasis of the fly, presumably resulting from an upset in underlying metabolic processes by the extreme temperature.

A chemical stress, phenyl-thio-urea (PTU), was also used in these experiments. In contrast with 30°C, PTU reduced fly weight substantially and increased its variability. Emergence time was delayed by PTU and its variability increased. PTU, however, had little effect upon chaeta number variability and FA. It is therefore a stress with little effect on the developmental stability of flies, being a stress affecting fly weight but not the variability of the morphological structure of the fly. Presumably, the contrast resides in the point that temperature is a generalized stress, and at certain limits, the basic physiology of the organism will become restrictive with multiple consequences, whereas PTU is a specific chemical inhibitor involving far fewer metabolic pathways than temperature.

An organism can, however, adapt morphologically to extreme temperatures at least to some extent. This was demonstrated in a study of sternopleural chaeta number and FA over 30 generations at 25°C, the normal culture temperature, and three 'novel' environments, 20°C, 30°C and one fluctuating between 20°C and 30°C in a smooth diurnal cycle (Thoday, 1958). Adaptation to the new environments was indicated by a chaeta number change, and regressions of chaeta number over generations were greater in the new environments than in the old, indicating that chaeta number measures something of adaptive significance. In particular, chaeta number increased over generations at 20°C and decreased at 30°C, which is consistent with the well-known observation that flies tend to have fewer chaetae when cultured at high as compared with low temperatures (Parsons, 1961).

The results also provided direct evidence that sternopleural FA is a trait of adaptive significance, since there is no indication of regression on generations in the 25°C population, but each of the other populations (in new environments) showed a negative regression of FA on generations, which is significant in the 30°C and the 20/30°C population, and close to significance in the 20°C population. This is reasonable, since 20°C is certainly not a major environmental perturbation by comparison with 30°C which is at the limits of survival of *D. melanogaster.*

Comparing inbred strains and hybrids, FA is normally reduced significantly in the F1 and F2 generations (Mather, 1953; Reeve, 1960) which is consistent with FA level as a fitness measure, assuming the inbreds to have inferior fitness to the hybrid generation. Artificial selection often reaches limits, not because variability is exhausted, but because the fitness of the organisms so produced is lower than in unselected organisms. Thoday (1958) analyzed FA in lines selected for high and low chaeta number, and found that FA increased equivalently in both lines. Hence directional selection can cause deterioration of the systems responsible for the developmental homeostasis of organisms. Finally, Reeve (1960) found increased FA in major mutants of the chaeta system, implying that in the mutant genetic background, developmental homeostasis is reduced, a conclusion consistent with data in the sheep blowfly, *Lucilia cuprina* (McKenzie and Clarke, 1988). Additional suggestive evidence for FA as a useful monitor of fitness changes in a wide array of organisms is presented by Palmer and Strobeck (1986).

Therefore, for traits not showing strong canalization, FA assessments have a role as an epigenetic measure of stress. The expectation is for increasing FA away from an optimum, especially at stress levels approaching lethality where major abnormalities may also occur. At these levels, normal development is upset at the molecular, chromosomal and epigenetic levels (Parsons, 1987, 1988). Even so, in a new environment, adaptation can occur as shown by a progressive fall in FA over a number of generations in *Drosophila* exposed to temperature extremes.

3. AN INTEGRATED APPROACH VIA METABOLIC COST

A general feature of stressed systems is an increase in energy expenditure, since following a severe perturbation, energy is diverted from maintenance and production to repair and recovery (Odum *et al.*, 1979). In other words, there is a metabolic cost in response to stress. One measure of the energy available to an organism at a given time is the adenylate energy charge (AEC) derived from relative concentrations of ATP, ADP and AMP, and which measures the availability of metabolic energy (Atkinson, 1977; Hochachka and Somero, 1984). A wide range of environmental perturbations affect AEC's (Ivanovici and Weibe, 1981); these include nutritional stress, oxygen depletion, desiccation and heat and chemical stresses. This implies that AEC has a role as a metabolic indicator of the relative severity of an ecological stress (Ivanovici and Weibe, 1981), although some consider that more work is needed to establish the generality of this conclusion, especially in marine organisms (Schafer and Hackney, 1987).

Desiccation is a stress for which substantial variation occurs in natural populations of various insect and plant species, and this is readily relatable to habitat (Parsons, 1973, 1987). This suggests that when a population is exposed to stressful desiccation levels in nature, there should be rapid phenotypic changes. Based upon this premise, selection for increased resistance to desiccation was carried out at an 85% mortality level in *D. melanogaster* (Hoffmann and Parsons, 1989a) i.e., each successive generation was based upon the 15% of flies that survived this stress. Responses to selection were exceedingly rapid. The mean of three selected lines increased from 18 hr to 28 hr after nine generations of selection (based upon females only). This increase exceeds the resistance of the control lines by more than 50%. The realized heritabilities were exceedingly high, averaging 0.65, and are consistent with the large responses to selection. The heritabilities are substantially higher than estimates for most morphological traits and especially for conventional life-history traits. Populations therefore have the potential to undergo extremely rapid genetic changes when exposed to dry habitats, and this is consistent with field observations.

However, there was a cost. One way that an organism may increase its resistance to a range of environmental stresses is by reducing its metabolic energy. In accord with this prediction, the desiccation-resistant strains had a lower metabolic rate than unselected sensitive strains (Table I). Predictably, these resistant strains showed less behavioral activity, i.e., they moved less, which is a reasonable association with metabolic rate, and in addition they had a lower fecundity measured as egg deposition over three days. Increased resistance to many environmental stresses should be genetically correlated because of a reduction in metabolic energy expenditure (Hoffmann and Parsons, 1989a,b). Table I shows the validity of this prediction for starvation, toxic levels of ethanol and acetic acid, irradiation with exceedingly high doses of Co^{60}-γ rays, and high temperatures.

An integrated approach to environmental stress is therefore emerging by considering metabolic rate as the key trait, and using desiccation resistance as the environmental probe for magnifying genetic variability. Resistances to different stresses should often be correlated due to a common genetic mechanism relatable to metabolic rate. Genotypes with low rates of metabolism may then be favored under a range of stressful conditions, since the cost of stress is reduced in them. Examples in several other species, in particular mammals such as beef cattle (Frisch, 1981), will be presented in Hoffmann and Parsons (1990). Exceptions should include biochemical changes that are largely specific for a particular type of stress. For example, increased resistance to insecticides may involve mechanisms that are specific to a particular class of insecticides (Oppenoorth, 1985). As already noted in FA studies, PTU appears to be a stress of a specific kind; indeed major genes sensitive to PTU have been found in *D. melanogaster* (Parsons, 1973).

Table I

A comparison of lines of *D. melanogaster* selected for desiccation tolerance with control lines

	Metabolic rate[a] ($O_2mg^{-1}hr^{-1}$)	Activity[b]		Fecundity[c]	Starvation[d]	Ethanol[e]	Acetic Acid[e]	Heat[f]	Radiation[g]
		Low humidity	High humidity						
Selected	2.1	10.4	6.0	131	98	63	23	0.30	0.36
Control	2.6	12.6	9.0	162	79	28	12	0.55	0.75
Significance	<0.01	<0.05	<0.05	<0.01	<0.001	<0.01	<0.001	<0.01	<0.01

[a] $O_2mg^{-1}hr^{-1}$ based upon 20 flies.
[b] Number of flies (out of 15) that moved in a 30 second interval.
[c] Total eggs laid in 3 successive 24 hour periods at 3-5 days of age.
[d] LT_{50} in hours for 20 flies in a humid atmosphere.
[e] LT_{50} in hours for 20 flies
[f] Proportion of 20 flies knocked down in a vial after 1.5 hours at 37°C.
[g] Proportion of 20 flies dead 22 hours after exposure to $Co^{60}\gamma$ irradiation of dose 1.2 kGy.
Adapted from Hoffmann and Parsons (1989a,b).

4. IS ADAPTATION TO A RAPID TEMPERATURE INCREASE POSSIBLE?

Simplistically, this is a problem in quantitative genetics in understanding genetic responses to stress. Under laboratory conditions, it appears that the potential for genetic responses for major increases in temperature is not great (Parsons, 1989), as compared with more specific stresses such as desiccation. Changing the adaptive norm of a species (Schmalhausen, 1949) is not easy, and such reactions appear strongly canalized to the ranges of existing populations for temperature. All really new reactions of an organism are never adaptive, and as the environment moves an organism away from the norm, fitness tends to fall. In geographically widespread species, different populations may evolve different resistances initially by acclimation and then by assimilated genetic changes, but the limit is the set of conditions within which metabolism can occur. Since a major consequence of adaptation to stress is a fall in metabolic rate, the scope for a species to widen these limits is restricted because of the build up of metabolic costs.

Evidence is accumulating that organisms are subjected to periods of intense climatic stress in nature. For example, plant and animal distributions may be largely determined by climatic extremes. Much of this work derives from the successful application of calculated bioclimatic envelopes based upon climatic similarities with actual species distributions on the Australian continent (Busby, 1986). Since stress periods are a feature of natural habitats, it follows that if the environment were to become suddenly more extreme, mass extinctions would be likely, as observed periodically in the fossil record (Raup and Boyajian, 1988). Major range expansions involving shifts in resistances to stress should then be unlikely under the rapid climatic change envisaged under some greenhouse predictions.

Experimentally, Forman (1964) took populations of the moss, Tetraphis pellucida, of the moist coniferous zones of the temperate zone, and grew them in a microphytotron under various combinations of controlled temperature, relative humidity, pH and light intensity, and found that parameters relating to the mean monthly maximum and mean monthly minimum temperatures were the most useful in explaining distributions. Collectively, both the experimental approach and that based upon interpreting species distributions, emphasize the importance of the physical features of the environment incorporating extremes.

The experimental approach has been applied to those Drosophila species that can be cultured in the laboratory. Ten species from non-tropical, mainly temperate zone, habitats, showed much greater resistance to extreme temperatures and desiccation in the laboratory than nine tropical species (Parsons, 1981). The range of resistance of the tropical species was quite narrow in accord with the narrow range of tropical temperatures, especially those from within rainforests. Within the much wider range of the non-tropical species, broad associations with habitat were discernible, for example, *D. funebris*, a species found in Arctic Russia was extraordinarily cold resistant. At the intraspecific level, *D. melanogaster* populations from temperate regions were more tolerant of climatic extremes than those from more benign tropical regions (Stanley and Parsons, 1981). General discussions of plant distributions also emphasize stress physiology, in particular the two components of drought stress and low temperature stress (Osmond *et al.*, 1987) indicating striking parallels with the two major determinants of *Drosophila* distributions. Osmond *et al.* (1987) emphasize that stress affects most plants during part of their life cycle and Boyer (1982) estimates that environmental stress effects limit U.S. agricultural produc-tivity to 25% of its potential. Extrapolating to *Drosophila*, a genus intimately dependent upon plants for resources, is difficult, although there is evidence that flies are commonly starved under field conditions (Bouletreau, 1978).

Undoubtedly, the greenhouse changes will lead to range expansions of certain stress resistant species into habitats such as rainforests. The process of climatic matching (i.e., a comparison of the climatic characteristics of the regions of origin with those of the regions to be utilized - a process most successfully achieved using climatic characteristics emphasizing extremes) as advocated for economic accessions (Burt *et al.*, 1975) should be predictive in this regard. Equally, the process of climatic matching following greenhouse changes implies the elimination of rare stress-sensitive species since the climate will change beyond their resistance levels. In the humid tropics of Australia, this means that the common *Drosophila* species of heat-stressed lowland regions would tend to move into the floristically richest rainforests at higher altitudes, eliminating species therein in the process. Furthermore, the intimate relationship between *Drosophila* and plants means that parallel plant changes are likely since the species distributions of *Drosophila* correlate closely with floral types in Australia rainforests (Parsons, 1981). Indeed, *Drosophila* can be regarded as an indicator organism of habitat change (Parsons, 1985) especially as there are typical faunas within rainforests largely differing from faunas of more cosmopolitan species outside rainforests. The role of *Drosophila* distributions is emphasized by observations suggesting that a temperature increase as small as 2°C will affect desiccation- and temperature-sensitive species of the rainforests of the humid tropics, with the possibility of species extinctions (Parsons, 1989).

The metabolic cost argument means that generalized stresses should be at least partially cumulative, for example, global warming may be associated with simultaneous desiccation and pollution stress in some regions, and the potential and observed effects of multiple pollutants is becoming a matter of concern (Hinrichsen, 1986). Even partial genetic associations of pollution stress with metabolic rate would exacerbate the effects of temperature change. There is a clear need for the experimental study of various stresses singly, and in combination, in a species such as *D. melanogaster*, to provide a model of the potential effects of environmental stresses generally. In this context, parallel FA studies should provide a monitor of fitness changes as assessed morphologically.

5. SUMMARY

The norm of reaction in the sense of Schmalhausen (1949) can be upset by environmental (and genetic) perturbations. Here, environmental perturbations are emphasized which have costs at the developmental and metabolic levels.

This is illustrated morphologically in *D. melanogaster* using fluctuating asymmetry, FA, a fitness trait for which variability is increased substantially at 30°C, a high temperature close to lethality. Metabolically, an increase in desiccation resistance in *D. melanogaster* by directional selection had the cost of a reduction in the metabolic rate of resistant lines. Resistance to combinations of stresses should be correlated, and this was demonstrated for a range of generalized stresses including starvation, toxic levels of ethanol and acetic acid, irradiation, and high temperature.
Experimental work within and among species, and analyses of species distributions, indicate the importance of climatic stress in determining the range of species, and the difficulty of extending these ranges. The postulated greenhouse temperature change is therefore likely to cause species extinctions, especially in species-rich rainforest habitats, and based upon metabolic cost arguments, this may be exacerbated by interactions with other generalized stresses. FA levels may prove useful in assessing the intensity of environmental stresses.

References

Atkinson, D. E., 1977, *Cellular Energy Metabolism and its Regulation,* Academic Press, New York.
Bouletreau, J., 1978, Ovarian activity and reproductive potential in a natural population of *Drosophila melanogaster, Oecologia* 33:319-342.
Boyer, J. S., 1982, Plant productivity and environment, *Science* 218:443-448.
Burt, R. L., Reid, R., and Williams, W. T., 1975, Exploration for, and utilization of, collections of tropical pasture legumes, *Agro-Ecosystems* 2:293-307.
Busby, J. R., 1986, Bioclimate prediction system. Users Manual, Bureau of Flora and Fauna, Canberra.
Cicerone, R. J., 1987, Changes in stratospheric ozone, *Science* 237:35-42.
Clare, M. J., and Luckinbill, L. S., 1985, The effects of gene-environment interaction on the expression of longevity, *Heredity* 55:19-29.
Fisher, R. A., 1930, *The Genetical Theory of Natural Selection,* Clarendon Press, Oxford.
Forman, R. T. T., 1964, Growth under controlled conditions to explain the hierarchical distributions of a moss, *Tetraphis pellucida, Ecol. Monogr.* 34:1-25.
Frisch, J. E., 1981, Changes occurring in cattle as a consequence of selection for growth rate in a stressed environment, *J. agric. Sci., Camb.* 96:23-38.
Hinrichsen, D., 1986, Multiple pollutants and forest decline, *Ambio* 15:258-265.

Hochachka, P. W., and Somero, G. N., 1984, *Biochemical Adaptation*, Princeton University Press, Princeton.

Hoffmann, A. A., and Parsons, P. A., 1989a, An integrated approach to environmental stress tolerance and life-history variation: Desiccation tolerance in *Drosophila*, *Biol. J. Linn. Soc.* **37**:117-136.

Hoffmann, A. A., and Parsons, P. A., 1989b, Selection for increased desiccation resistance in *Drosophila melanogaster*: Additive genetic control and correlated responses to other stresses, *Genetics* (in press).

Hoffmann, A. A., and Parsons, P. A., 1990, *Evolutionary Genetics and Environmental Stress*, (in preparation).

Ivanovici, A. M., and Wiebe, W. J., 1981, Towards a working 'definition' of stress: a review and critique, in: *Stress Effects on Natural Ecosystems* (G. W. Barrett, and R. Rosenberg, eds), John Wiley, New York, pp. 13-27.

Lints, F. A., Stoll, J., Gruwez, G., and Lints, C. V., 1979, An attempt to select for increased longevity in *Drosophila melanogaster, Gerontology* **25**:192-204.

Luckinbill, L. S., and Clare, M. J., 1985, Selection for life span in *Drosophila melanogaster, Heredity* **55**:9-18.

Mather, K., 1953, Genetical control of stability in development, *Heredity* **7**:297-336.

McKenzie, J. A., and Clarke, G. M., 1988, Diazinon resistance, fluctuating asymmetry and fitness in the Australian sheep blowfly, *Lucilia cuprina, Genetics* **120**:213-220.

Mooney, H. A., Vitousek, P. M., and Matson, P. A., 1987, Exchange of materials between terrestrial ecosystems and atmosphere, *Science* **238**:926-932.

Odum, E. P., Finn, J. T., and Franz, E., 1979, Perturbation theory and the subsidy-stress gradient, *Bioscience* **29**:349-352.

Oppenoorth, F. J., 1985, Biochemistry and genetics of insecticide resistance, in: *Comprehensive Insect Physiology, Biochemistry and Pharmacology* (G. A. Kerkut, and I. L. Gilbert, eds), Pergamon Press, Oxford, pp. 731-773.

Osmond, C. B., Austin, M. P., Berry, J. A., Billings, W. D., Boyer, J. S., Dacey, J. W. H., Nobel, P. S., Smith, S. D., and Winner, W. E., 1987, Stress physiology and the distribution of plants, *Bioscience* **37**:38-48.

Palmer, A. R., and Strobeck, C., 1986, Fluctuating symmetry: measurement, analysis, patterns, *Annu. Rev. Ecol. & Syst.* **17**:391-421.

Parsons, P. A., 1961, Fly size, emergence time and sternopleural chaeta number in *Drosophila, Heredity* **16**:455-473.

Parsons, P. A., 1962, Maternal age and developmental variability, *J. exp. Biol.* **39**:251-260.

Parsons, P. A., 1973, Genetics of resistance to environmental stresses in *Drosophila* populations, *A. Rev. Genet.* **7**:239-265.

Parsons, P. A., 1980, Isofemale strains and evolutionary strategies in natural populations, *Evol. Biol.* **13**:175-217.

Parsons, P. A., 1981, The evolutionary ecology of Australian *Drosophila*: a species analysis, *Evol. Biol.* **14**:297-350.

Parsons, P. A., 1985, Tropical *Drosophila*: resistance to environmental stresses and species diversity, *Proc. Ecol. Soc. Aust.* **13**:43-49.

Parsons, P. A., 1987, Evolutionary rates under environmental stress, *Evol. Biol.* **21**:311-347.

Parsons, P. A., 1988, Evolutionary rates: effects of stress upon recombination, *Biol. J. Linn. Soc.* **35**:49-68.

Parsons, P. A., 1989, Environmental stresses and conservation of natural populations, *Annu. Rev. Ecol. & Syst.* (In press).

Raup, D. M., and Boyajian, G. E., 1988, Patterns of generic extinction in the fossil record, *Paleobiology* **14**:109-125.

Reeve, E. C. R., 1960, Some genetic tests on asymmetry of sternopleural chaeta number in *Drosophila*, *Genet. Res.* **1**:151-172.

Robertson, F. W., 1960, The ecological genetics of growth in *Drosophila* I. Body size and development time on different diets, *Genet. Res.* **1**:288-304.

Rose, M. R., 1984, Laboratory evolution of postponed senescence in *Drosophila melanogaster, Evolution* **38**:1004-1009.

Schafer, T. H., and Hackney, C. T., 1987, Variation in adenylate energy charge and phosphoadenylate pool size in estuarine organisms after an oil spill, *Bull. Environ. Contam. & Toxicol.* **38**:753-761.

Schmalhausen, I. I., 1949, *Factors of Evolution,* Blakiston Company, Philadelphia.

Stanley, S. M., and Parsons, P. A., 1981, The response of the cosmopolitan species, *Drosophila melanogaster* to ecological gradients, *Proc. Ecol. Soc. Aust.* **11**:121-130.

Thoday, J. M., 1958, Homeostasis in a selection experiment, *Heredity* **12**:401-415.

Waddington, C. H., 1953, Genetic assimilation of an acquired character, *Evolution* **7**:118-126.

Waddington, C. H., 1956, *Principles of Embryology*, George Allen and Unwin, London.

CHAPTER 5

The Effects of Natural and Artificial Selection on Dysgenic Potential of a Wild Population of Drosophila melanogaster

MICHAEL MURPHY, AND JOHN A. SVED

1. INTRODUCTION

It is now known that transposons occur ubiquitously in both eukaryotes and prokaryotes. A greater variety has been characterised in *Drosophila melanogaster* than in any other eukaryote, and more is known about selective effects associated with particular transposons. In particular, the dysgenic effects associated with two transposons, the "P factor" and "I factor" have been well documented (Engels, 1989; Finnegan, 1989). In each case, the effects are manifested when strains containing at least one copy of the transposon are crossed to strains which lack any copies.

The conditions under which transposons are maintained in natural populations have been studied by several authors (see Charlesworth, 1988). The possibility of replicative transposition to new sites in the genome means that the normal balance between mutation and selection expected for Mendelian genes is not expected to apply for transposons. The number of transposons in a population may rise in spite of detrimental effects. The factors responsible for maintaining copy number of different transposons are not presently clear, especially given the complication that many transposons, including the P and I factors, exist in complete and truncated forms.

The aim of our study is to determine whether seasonal factors in natural populations, or conditions of raising in laboratory populations, affect the distribution of transposons. Conceptually, this question can be answered either by studying the number of transposons, or by studying their overall effect. In this study, we have examined only the latter. A complete study of the number of transposons would require estimates of not only the number of transposons but also a description of which are complete and which are incomplete. Such a description has been achieved for P factors for one inbred strain (Simmons *et al.*, 1987). Much work would be

_____

MICHAEL MURPHY, AND JOHN A. SVED * School of Biological Sciences, University of Sydney, N.S.W. 2006, Australia.

involved in carrying out such a study on an outbred strain, in which variability between individuals is expected. A prohibitive amount of work would be necessary to characterise as many populations as we have tested in this study.

As discussed below, we used male recombination levels as the most convenient measure of dysgenic potential. It should be emphasised that we are using dysgenic effects in hybrid progeny to characterise the natural and artificial populations. These hybrid progeny are obtained from crosses between males from the strains in question and females from a standard laboratory strain. However, we are really interested in studying the distribution of transposons within the wild type strains themselves, and there is evidence (Engels and Preston, 1980) that these hybrid effects normally do not occur, or at least occur at a very low level, in such populations. The expected selective effects are those generated either by the disruption to genes caused by the insertion of transposons, by the extra metabolic processes associated with the replication of transposons and with the expression of their genes, or perhaps by indirect deleterious effects associated with the transposon gene products.

2. METHODS AND MATERIALS

The overall design of sampling and treatments in the experiment is shown in Figure 1. More complete details are available in Murphy (1986). Four wineries were sampled at each of three sampling times indicated in the figure. These samples were taken respectively in November 1981, March 1982 and November 1982. The November samples were taken at a time when

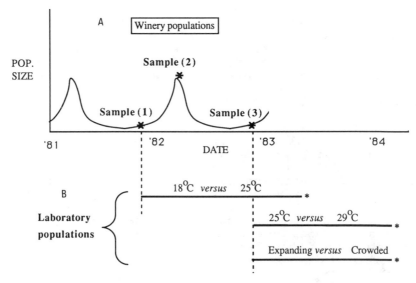

Figure 1. A. Sampling times in the experiment, in relation to the expected size of the winery populations. B. Times of initiation and final sampling of the laboratory populations.

the population was in the first phases of recovery from the winter trough, while the March sample was taken just after vintage, when populations could be expected to be at a maximum.

Figure 1 also shows the course of three laboratory population comparisons, made starting with flies taken from the two November samples. Two of these comparisons were made to test the effects of rearing at low (18°C) and high (29°C) temperatures respectively. The third comparison was made to simulate the effect of expanding versus stationary populations. The tipover periods in this case were 11 and 19 days respectively. The 11-day regime places a premium on early egg-lay and fast development, while the 19-day regime presumably selects for maximum survival and fertility under conditions of crowding. It should be noted that the two types of populations were held for the same amount of time, so that the 11-day populations in fact underwent nearly twice as many generations as the 19-day populations. Similar differences also existed between the populations maintained at 18°C, which had a 21-day cycle, 25°C which had a 14-day cycle, and 29°C which had a 10-day cycle.

2.1. The Measurement of Dysgenesis

Figure 2 shows our standard procedure for testing for dysgenesis. The same tester strain, *al cn bw, e* was used throughout the experiments. The markers *al* (0.0), *cn* (57.5) and *bw* (104.5) span most of chromosome 2, while the marker *e* is on chromosome 3. As detailed below, the tester strain *al cn bw, e* was classified as M in the P-M system and R in the I-R system (Bregliano and Kidwell, 1983).

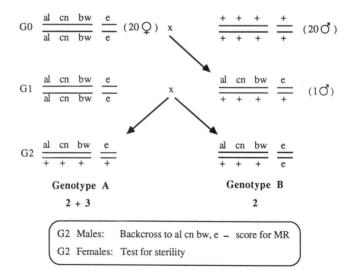

Figure 2. Procedure for producing G2 males and females used to measure dysgenesis.

It should be noted that the procedure for testing for dysgenesis differs from the usual tests, which involve hybrid progeny from a single backcross (see e.g., Bregliano and Kidwell, 1983). Use of a second backcross allowed the production of two genotypes, which were (A) heterozygous for chromosomes 2 and 3, and (B) heterozygous for chromosome 2. Thus separate estimates of recombination on chromosome 2 were obtained in the presence and absence of a wild-type (P) third chromosome. As related later, some individuals homozygous for just chromosome 3 were also tested for sterility. We assume throughout the discussion that chromosomal genotypes are involved, although this is only true if male recombination is absent. In an estimated 1% or less of cases, the third chromosome genotypes are expected to contain recombinant regions. The design allowed also for the estimation of segregation ratio distortion. Independent estimates were made for chromosomes 2 and 3 in genotype A, and for chromosome 2 in genotype B.

For each winery population, 10 G1 males from each of the A and B genotypes were individually backcrossed to *al cn bw, e*. Three males from each genotype A progeny group and five males from each genotype B progeny group were backcrossed individually to *al cn bw, e*. females. All progeny in the tests were raised at 25°C.

The analysis of male recombination data is complicated by the occurrence of clusters of recombinant products. Occasional large progeny groups are found which distort the analysis of recombination frequencies. For this reason we have used the "minimum recombination" statistic (Kidwell, Kidwell and Sved, 1977) in the analysis below, in which the observation of any number of recombinants found in a given region for a particular male is scored as a single event. When recombination frequencies are high, information is lost through the use of such a statistic, but low frequencies were found throughout the present experiment.

In testing for female sterility, we scored for both the presence or absence of eggs, and the presence or absence of larvae. The absence of eggs shows gonadal (GD) sterility (Kidwell and Novy, 1979). Although no attempt was made to count the number of eggs, we also scored the number of cases where less than 10 eggs were laid. As seen below, we found very little GD sterility in our experiments by either measure.

In cases where eggs were laid, we scored for absence of larvae, and also for cases where very few (less than 10) larvae hatched. High levels of sterility were detected in this way. The sterility appears to correspond to SF sterility (Picard, 1976), which is attributable to the I-R dysgenesis system. When females from the tester strain *al cn bw, e* were crossed to the laboratory strain Canton-S, an apparently identical kind of sterility was found, suggesting that the *al cn bw, e* strain is an R strain, and that the P-M system is not primarily responsible for the sterility. Furthermore the fertility of aged females was found to revert to wild type levels, a characteristic of SF sterility (Picard *et al.*, 1977).

3. RESULTS

3.1. Wild Populations

3.1.1. Male Recombination

Figure 3 shows the distribution of mean recombination percentages for the three series of winery collections, and Table I shows the analysis of variance of the results. The left-hand portion of Figure 3 shows the distribution of male recombination values in the four winery populations. The male recombination percentages have been averaged from the A and B genotypes (Fig. 1) to derive these values. Consideration of the totals shows that there is no systematic difference between the collection made at a high population density (March '82) and those made at the low population densities, although the difference between collections is marginally significant (Table I, line 1).

The two right-most distributions of Figure 3 compare the male recombination percentages in the presence and absence of wild type chromosome 3, i.e., genotype A versus B of Figure 1. These values have been averaged over all four wineries. The presence of a wild type chromosome 3 clearly results in a large increase in the incidence of crossing-over on chromosome 2. The analysis of variance between chromosome classes (Table I, line 4) shows that this difference is highly significant.

Table I also shows that the difference between wineries is highly significant. We also carried out a chi-square analysis of the differences between chromosomes within wineries (results not shown), and many of the series show significant heterogeneity, as expected if there is a heterogeneous distribution of P factors over the different chromosomes

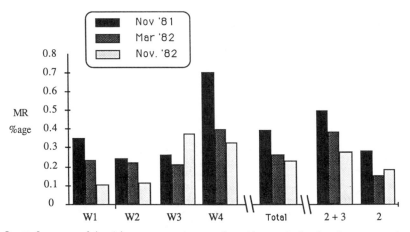

Figure 3. Male recombination percentages for the original winery series.

(e.g., Kidwell and Kidwell, 1976; Sinclair and Grigliatti, 1985). The significant differences between wineries shows that there is an added tendency for high recombination chromosomes to be concentrated in some wineries more than others. The non-significance of the series x winery interaction suggests that the differences between wineries persist over collections.

3.1.2. Segregation Ratio Distortion

In the progeny of a hybrid in the P-M dysgenesis system, the frequency of the P chromosome is often reduced compared to expectation (Kidwell, Kidwell and Sved, 1977). The measurement of segregation ratio distortion is frequently complicated by a counter-acting inviability of individuals carrying the M chromosome. Thus in the present series of crosses, progeny carrying the P chromosome are phenotypically wild type, while progeny carrying the M chromosome are homozygous for three markers, al, cn and bw, each of which may expected to affect viability to some extent. Note that it is an assumption that it is thdistribution of P factors over the different chromosomese P factors, rather than some other transposon or chromosomal property, which are responsible for the segregation ratio distortion.

Averaging over all measurements made in the experiments of this paper, the frequency of the P-bearing offspring comes to 0.551 for chromosome 2 and 0.521 for chromosome 3, indicating that any segregation ratio distortion is counterbalanced by viability effects. However, we have shown that there is a negative correlation between the segregation ratio and the amount of male recombination. Each series of measurements is based on a set of 10 replicate chromosomes. We have calculated the correlation coefficient in each such case and find that out of a total of 36 correlation coefficients, 32 are negative. A similar negative correlation was found by Hiraizumi (1971).

Table I
Analysis of Variance for the Original Winery Collections[a]

Source	df	Sum of Squares	Mean Square	F-test Square	P value
Series	2	1.101	0.551	3.484	.0325
Wineries	3	2.874	0.958	6.061	.0006
S x W	6	1.515	0.253	1.598	.1490
Chromosomes	1	1.963	1.963	12.419	.0005
S x C	2	0.232	0.116	0.733	.4816
W x C	3	0.249	0.083	0.534	.6659
S x W x C	6	0.880	0.147	0.928	.4755
Error	211	33.347	0.158		

[a]The recombination frequencies have been transformed using an arcsin transformation. The number of degrees of freedom for the Error term, expected to be 24 x 9 = 216, is reduced because of five missing values.

The significant correlation shows that segregation ratio distortion and male recombination are related measures of dysgenic potential. We attempted to use the segregation ratio distortion to test the differences between populations, as variability in individual measurements is such that this statistic is not as useful a measure in discriminating between populations.

3.1.3. Female Sterility

The total amount of gonadal sterility, as measured by the complete absence of eggs, reached 10% in one series, but was othersise extremely low in all experiments. All progeny were raised at 25°C, whereas GD sterility is maximised at 29°C (Kidwell, Kidwell and Sved, 1977), which may be a contributing factor to the lack of GD sterility observed.

In contrast to gonadal dysgenesis, high levels of SF sterility were observed. Figure 4 summarises the sterility data in an identical format to that used for male recombination in Figure 3. Once again there is little evidence for a difference between collection series, as can be seen from the totals. Considerable heterogeneity was observed between wineries in the third collection, although little in the first two.

Perhaps the most striking feature of the results from Figure 4 is the consistently lower sterility level found when two chromosomes were heterozygous as compared with a single chromosome heterozygous. This pattern is the opposite to the cumulative inter-chromosomal effect found in the P-M dysgenesis system (e.g., Voelker, 1974; see also Figure 3), and has not previously been emphasised for the I-R system (e.g., Bucheton and Picard, 1978). It is not clear from the results presented in Figure 4 whether the effect is chromosome specific. However, we tested sterility levels for individuals

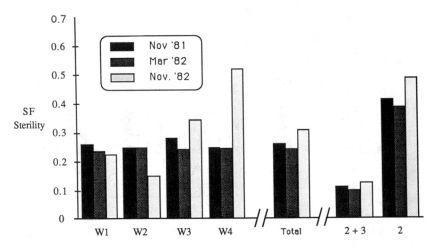

Figure 4. SF sterility observed in winery collections.

heterozygous for just chromosome 3 in some of the laboratory-held series, and found sterility levels comparable to those for chromosome 2.

The mechanism responsible for this reduction in sterility when more than one wild type chromosome is involved is not clear. One hypothesis is that there is an optimum level of sterility determined by an intermediate number of factors, presumably I factors. A second hypothesis is that the sterility is affected by the number of "reactive" chromosome homozygotes (i.e., chromosome pairs not containing the I factor) present in the hybrid. Thus genotype A females (Fig. 1) have only the X chromosome homozygous, while genotype B have the X chromosome and chromosome 3 homozygous, and therefore have a higher level of sterility. It should be noted that we also tested the sterility of females with X, 2 and 3 heterozygous, and found sterility levels slightly lower than but comparable to those with chromosomes 2 and 3 heterozygous, indicating that the relationship between number of reactive chromosome pairs and sterility could not be a simple one.

Under either of these hypotheses, the overall level of sterility is likely to be a poor predictor of the number of factors responsible for creating the sterility. It means that measurements of sterility obtained from hybrids are unlikely to be influenced in a simple way by conditions which affect the wild type population. Fortunately, measurements of male recombination do not seem to suffer from this problem.

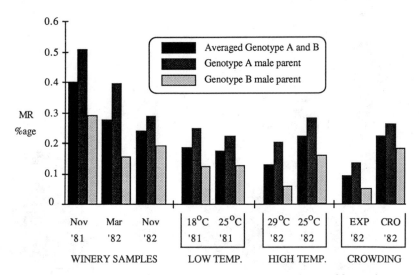

Figure 5. Male recombination percentages from all series.

3.2. Laboratory Populations

3.2.1. Male Recombination

Mean recombination frequencies for all laboratory samples are shown in Figure 5. The first three samples repeat, for comparison, results from the three collection series. Matched pairs of populations are shown grouped.

The most striking result from Figure 5 is that male recombination has fallen in all populations kept in the laboratory as compared with the original samples. This result refers to the total recombination (black bars) averaged over both A and B genotypes. There are altogether 12 original winery samples, and 24 laboratory population samples, each with ten counts. The means of the two sets are 0.30% and 0.17% respectively, and this difference is significant at the 1% level based on a t-test comparing the two sets (t_{34} = 3.51).

Significant differences between series must be interpreted with some caution. One possible cause of differences between the laboratory and natural population samples is the tester stock used. Although this stock is, and has been, kept under constant conditions for a number of years, subtle changes in the stock over the course of the experiment could conceivably contribute to changes in dysgenesis.

The complication of possible changes in the tester stock, and also of differences in the starting populations, does not affect the interpretation of results from the matched pairs of laboratory populations. Of the three pairs, the most striking difference is that between the "expanding" and "crowded" populations. Table II gives the analysis of this pair of populations, and shows that the difference between the two series is significant at the 1% level. In interpreting this result, it must be borne in mind that the "expanding" populations have been through 34 generations, as opposed to the 17 generations for the "crowded" populations.

Table II
Analysis of variance for comparison of male recombination between "expanding" and "crowded" populations

Source	df	Sum of squares	Mean square	F-test	P value
Series	1	0.650	0.650	10.85	.001
Wineries	3	0.443	0.148	2.46	.065
S x W	3	0.668	0.223	3.71	.013
Chromosomes	1	0.264	0.264	4.40	.038
S x C	1	0.0004	0.0004	0.007	.932
W x C	3	0.179	0.060	1.00	.396
S x W x C	3	0.074	0.025	0.41	.746
Error	142	8.510	0.060		

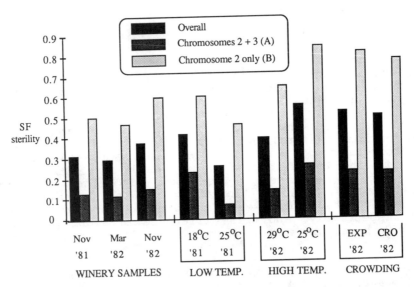

Figure 6. Sterility values observed in all samples.

The difference between the two populations kept at 29°C and 25°C is also significant, at the 5% level. The differences between the 18°C and 25°C populations are not significant.

3.2.2. Sterility

The results from all populations are summarised in Figure 6. Significant differences were found in some comparisons, but there was no overall trend comparable to that found for male recombination. In view of the comparisons between the single- and double-heterozygous genotypes, the lack of any simple trends is not surprising.

4. DISCUSSION

Two potential forces need to be distinguished in any discussion of the distribution of transposons in natural and selected populations. The first is conventional change in frequency, independent of any transposition effects. The second is the change in numbers due to replicative transposition and to excision, processes associated with dysgenesis. We would expect that conventional changes in frequency may be important in our experiments, since they are carried out in pure populations in which dysgenic effects seem to be minimised (Engels and Preston, 1980). However, the possibility of a low background of dysgenic effects is hard to rule out.

The most striking effect observed in our experiments is the reduction in dysgenic potential in laboratory populations, as measured by male recombination, particularly in the simulation of an "expanding" population. Although we have no proof that the male recombination is due to P factors, this seems the most likely explanation.   The crosses involved also I-R dysgenesis, but male recombination has not been reported for this system (Bregliano and Kidwell, 1983).   The fall in male recombination levels parallels the results found by Engels and Preston (1980) for GD sterility, although in the study reported by these authors there was also a minority of strains which showed a rise in dysgenic potential.

Why should dysgenic potential fall in laboratory populations? This question could of course only be answered if we had some clear idea of what factors are responsible for maintaining dysgenic potentials in natural populations at their present level. For example, the wild type distribution could be due to a balance between deleterious effects caused by the transposons and a tendency to increase by replicative transposition.   The reduction in dysgenic potential under laboratory conditions, particularly those populations maintained under the fast passage regime, could be due to enhancement of the deleterious effects under these conditions. An alternative explanation can be put forward in terms of some favourable effects of the transposons in natural populations, and a reduction in the laboratory populations.

Both of the above are selective explanations, which seem to us the simplest way of accounting for the observed effects. However, the possibility also exists that transposition rates are modified in laboratory populations as compared with wild populations.   If this is the case, the results could be explained without postulating any effects on the fitness of the flies.

5. SUMMARY

Flies were sampled from four different winery populations at times of low and high population density.   There were significant differences in dysgenic potential between populations, as measured by the level of male recombination induced in hybrids between marker stock females and males from the populations.   No differences were discernable, however, between low and high density populations.   Levels of sterility were also measured. There was very little of the gonadal sterility that would be expected from strong reactions in the P-M system of hybrid dysgenesis. Substantial levels of egg hatch sterility were found, as expected from the I-R system.   However, we found that the levels of sterility induced by a single chromosome were much higher than the levels induced by two chromosomes, an observation which suggests that the level of sterility may be a poor predictor of the number of factors on wild chromosomes which contribute to the sterility.

Laboratory populations initiated from the wild-caught flies were maintained under various conditions of temperature and crowding for 20-30 generations. Dysgenic potential in these populations, as measured by male recombination, fell in all cases. It fell most strongly in the populations which were held under a regime of early passage. These observations suggest that there is selection against certain transposons, even in pure populations in which dysgenic effects are not expected to be high, and that selection for fast development maximises this selective pressure.

ACKNOWLEDGEMENTS. The work in this study was supported by a grant from the Australian Research Grants Scheme.

References

Bregliano, J. C., and Kidwell, M. G., 1983, Hybrid dysgenesis determinants, in: *Mobile Genetic Elements* (J.A. Shapiro, ed.), Academic Press, New York, pp. 363-410.

Bucheton, A., and Picard, G., 1978, Non-Mendelian female sterility in *Drosophila melanogaster*: hereditary transmission of reactivity levels, *Heredity* **40**:207-223.

Charlesworth, B., 1988, The maintenance of transposable elements in natural populations, in: *Plant Transposable Elements* (Oliver Nelson, ed.), Plenum Publishing Co., New York, pp. 189-212.

Engels, W. R., 1989, P elements in Drosophila., in: *Mobile DNA* (D. E. Berg and M. M. Howe, eds), American Soc. for Microbiology, pp. 437-484.

Engels, W. R., and Preston, C. R., 1980, Components of hybrid dysgenesis in a wild population of *Drosophila melanogaster*, *Genetics* **95**:111-128.

Finnegan, D. F., 1989, The I factor and I-R hybrid dysgenesis in *Drosophila melanogaster*, in: *Mobile DNA* (D. E. Berg, and M. M. Howe, eds), American Soc. for Microbiology, pp. 503-518.

Hiraizumi, Y., 1971, Spontaneous recombination in *Drosophila melanogaster* males, *Proc. natn. Acad. Sci. USA* **68**:268-270.

Kidwell, M. G., and Kidwell, J. F., 1976, Selection for male recombination in *Drosophila melanogaster*, *Genetics* **84**:333-351.

Kidwell, M. G., Kidwell, J. F., and Sved, J. A., 1977, Hybrid dysgenesis in *Drosophila melanogaster*: A syndrome of aberrant traits including mutation, sterility and male recombination, *Genetics* **86**:813-833.

Kidwell, M. G., and Novy, J. B., 1979, Hybrid dysgenesis in *Drosophila melanogaster*: sterility resulting from gonadal dysgenesis in the P-M system, *Genetics* **92**:1127-1140.

Murphy, M., 1986, The effects of environmental conditions on dysgenic potential of *Drosophila melanogaster*, MSc Thesis, University of Sydney.

Picard, G., 1976, Non-Mendelian female sterility in *Drosophila melanogaster*: hereditary transmission of I factor, *Genetics* **83**:107-123.

Picard, G., Lavige, J. M., Bucheton, A., and Bregliano, J. C., 1977, Non-Mendelian female sterility in *Drosophila melanogaster*: physiological pattern of embryo lethality, *Biol. Cell.* **29**:89-98.

Simmons, M. J., Raymond, J. D., Boedigheimer, M. J., and Zunt, J. R., 1987, The influence of nonautonomous P elements on hybrid dysgenesis in *Drosophila melanogaster*, *Genetics* **117**:671-685.

Sinclair, D. A. R., and Grigliatti, T. A., 1985, Investigation of the nature of P-induced male recombination in *Drosophila melanogaster*, *Genetics* **110**:257-279.

Voelker, R. A., 1974, The genetics and cytology of a mutator factor in *Drosophila melanogaster*, *Mutat. Res.* **22**:265-276.

CHAPTER 6

P-Elements and Quantitative Variation in Drosophila

CHRIS MORAN, AND ADAM TORKAMANZEHI

1. INTRODUCTION

1.1. Mutation and Evolution

Mutation is ultimately the source of all evolutionary change. For this reason, much effort has been expended in understanding the nature of mutational changes and in attempting to measure and predict the consequences of variation in rates of mutation on levels of genetic variation and consequent evolutionary rates. The early studies of Clayton and Robertson (1955) suggested that mutation has little effect on response to selection, as did other attempts at manipulating mutation rates using radiation as the mutagenic agent (Clayton and Robertson, 1964; Hollingdale and Barker, 1971; Kitagawa, 1967) in order to experimentally modify evolutionary rates. In general, such studies were relatively disappointing in that the levels of enhancement of genetic variation and rates of evolution were quite small, presumably because of the deleterious effects associated with the radiation-induced mutations. As a practical means of enhancing rates of evolution and genetic improvement in domestic plants and animals, "mutation breeding" using radiation as the mutagenic source has fallen into disfavour.

Frankham (1980) has shown that unequal crossing over is a substantial source of genetic variation for characters affected by changes in copy number of the tandemly repeated genes coding for ribosomal RNA. Newly arising copy-number mutations are an important source of genetic variation within lines selected for bristle number. However, because of the unique relationship between the tandemly repeated rRNA genes and variation in this character, the general significance of these results is not clear.

This paper will be concerned with another potentially more effective and more general source of mutation than radiation

CHRIS MORAN, and ADAM TORKAMANZEHI * Department of Animal Husbandry, University of Sydney, NSW 2006, Australia.

Ecological and Evolutionary Genetics of Drosophila, Edited by
J.S.F. Barker *et al.*, Plenum Press, New York, 1990

or even unequal crossing over. The very existence of this category of mutagen, recognition of its potential as an evolutionary force and to some extent even the analysis of its
effect have depended on advances in molecular biology which
have revealed the structure and to a lesser extent the function of transposable elements.

1.2. Transposable Elements

The original discoverer of transposable elements, Barbara
McClintock, was well aware of their evolutionary potential.
However, her views on the existence of feedback mechanisms
which couple rates of transposition and hence mutation to
levels of so called "genomic stress", which may be internal or
external in origin (McClintock, 1978), are still rather con-
troversial and as yet unsupported by convincing evidence.
Junakovic et al. (1986) have shown that rates of transposition
of copia-like elements can be manipulated by heat shock and
there is evidence of a temperature effect on transposition
rate for P elements in dysgenic crosses (Engels, 1983; Torkamanzehi et al., 1988). In these cases, temperature shock or
stress may be considered as a form of external "genomic
stress". In any case, McClintock must be credited not only
with the discovery of transposable elements but with some
interesting and provocative suggestions concerning their evolutionary significance.

It was the "rediscovery" of the existence of transposable
elements, the discovery of their molecular nature and the
realisation of their ubiquity among plant and animal species
that has helped to generate the latest round of enthusiasm for
studies of mutation as an evolutionarily significant force in
the medium to short term. (There has of course never been any
doubt as to its long term significance.) This realisation of
the existence and commonness of endogenous mutagens, with the
capacity to lie dormant for long periods and then suddenly
create bursts of mutation, has led to a rethinking of previous
conclusions concerning the influence of mutation on rates of
evolution. Indeed it has led to a reinterpretation of experimental results from long term selection experiments, where the
absence of or escape from a selection plateau or the obvious
occurence of mutations within inbred selection lines has frequently been noted.

The discovery of the molecular nature and definition of the
role of the P element as an endogenous mutagen was very recent
(Bingham et al., 1982; Rubin et al., 1982; Spradling and
Rubin, 1982; O'Hare and Rubin, 1983). Genetically defined P
factors were known and the properties of the P-M system of
hybrid dysgenesis (Kidwell et al., 1977; Bregliano and Kidwell, 1983; Engels, 1979) were quite well described prior to
this, but the molecular breakthroughs mentioned previously
have been essential for the development of our understanding
of the evolutionary consequences of these elements.

1.3. Quantifying the Influence of New Mutations

Encouraged by the molecular and genetic analyses of transposable element (TE) occurrence and behaviour, there has been
a blossoming of theoretical and experimental studies aimed at
examining the influence of concurrent mutation on evolutionary
responses. At the theoretical level, many studies (Lande,
1975; Hill, 1982a, 1982b; Keightley and Hill, 1983, 1987,
1988; Lynch, 1988) have examined the influence of mutation
with various models of selection and gene action on medium
term or equilibrium levels of genetic variation and hence of
response to selection. In general, such studies suggest that
mutations occurring more or less contemporaneously with the
selection applied can contribute significantly to selection
response in the long term and with appropriate selection
regimes may permit an indefinitely continuous response to
selection. The parameter which is usually calculated to indicate the relative importance of the contribution of mutation
is V_m/V_E (Lande, 1975), where V_m is the rate of input of new
genetic variance due to mutation and V_E is the environmental
variance. A generally quoted figure for the value of V_m/V_F is
10^{-3} (Lande, 1975) for spontaneous mutations not related to
viability, although it is very variable over the range 10^{-4} to
5×10^{-2} (Lynch, 1988). Mackay (1985, 1987a, 1987b) has performed a number of experiments to assess the effect of transposition on genetic variance for quantitative characters. Her
data produce estimates of V_m/V_F of 0.1 to 0.2 during periods of
hybrid dysgenesis (Lynch, 1988), implying a one hundred to one
thousand-fold increase in the input of mutational variance due
to P element transposition. The enormity of this increase
clearly requires verification, given the implications for evolutionary rates in *Drosophila*, and will be discussed in detail
later in this paper.

1.4. Transposition and Evolution

The intrinsic revertibility, propensity to create chromosomal rearrangements, ability to introduce new controlling
regions which influence gene expression, tendency for nonrandom occurrence in time and chromosomal space are all features
of TE-mutation which have suggested evolutionary possibilities
beyond those arising from simple models of nucleotide sequence
change. Indeed Kimura (1985), whose neutral theory presupposes
that most nucleotide substitutions are selectively neutral,
has suggested that visible and lethal mutations are primarily
caused by transposons, thus implying that TE-mediated mutation
provides the principal source material of adaptive evolution.
On the other hand, Engels (1986) has argued that "mutations
caused by transposable element insertions are not being fixed
in natural populations, and are thus not important for evolution." Engels believes that most evidence favours the interpretation that TEs are parasitic or selfish DNA sequences
which are more likely to harm rather than help the prospects
of the species in which they occur.

1.5. Transposition and Quantitative Variation

Several experimental studies have directly examined the role of transposable-element-mediated mutation in governing selection response for quantitative characters. These studies have used the P-element system in *Drosophila melanogaster*, which has been chosen because it is straightforward to manipulate and control transposition rates in the progeny of certain categories of crosses in this system. In particular, high levels of transposition will be expected to occur in the germline of F1 and some subsequent generations of dysgenic crosses, that is in lines initiated by crossing M stock females with P stock males. It is also known that no or infrequent transposition will occur in the germ-line of F1 progeny from crosses of P stock females with M stock males, that is in lines begun with a nondysgenic cross. In the absence of information to the contrary Mackay (1984, 1985, 1987a, 1987b), and Torkamanzehi *et al.* (1988 and unpublished) have made the fundamental assumption that no transposition will occur in subsequent generations of nondysgenic lines as well. Thus they have attempted to examine the influence of P-element transposition by comparing selection responses between transposition positive dysgenic lines and and putatively transposition negative nondysgenic lines. The results of these selection experiments are rather equivocal in that Mackay, particularly in her earlier experiments, consistently found enormously enhanced responses in her dysgenic lines, which were attributed to transposition. However Torkamanzehi *et al.* (*ibid*) have found no consistent evidence favouring the influence of transposition using the same experimental design. In fact, Torkamanzehi *et al.* (unpublished) and Moran *et al.* (unpublished) have found clear evidence for the derepression of transposition from analyses of cytotype as well as direct evidence for transposition in the nondysgenic control lines. Clearly the experimental design employed by Mackay in her studies and also used by Torkamanzehi *et al.* is faulty. Indeed Mackay (1986) herself has produced evidence of the inadequacy of the reciprocal crossing protocol used to establish the putatively transposition-positive and negative lines, since she found similar frequencies of recessive lethals and distrubutions of fitness in lines derived from dysgenic and nondysgenic crosses.

1.6. Transposition and Fitness

Using a variety of experimental procedures, numerous authors (Mukai and Yukuhiro, 1983; Yukuhiro *et al.*, 1985; Simmons, Raymond, Culbert, and Laverty, 1984; Simmons, Raymond, Johnson, and Fahey, 1984; Fitzpatrick and Sved, 1986; Mackay, 1986; Eanes *et al.*, 1988) have examined the influence of P-element transposition on either viability or total fitness. All studies have found a very substantial impact of transposition on fitness. By introducing P-elements into X chromosomes which previously lacked them, Eanes *et al.* (1988) were able to show that most of the impact on fitness observed in these studies could be attributed to mobilisation of elements already in place. Imprecise excisions are known to occur at a frequency of about 35% under dysgenic conditions (Voelker *et al.*, 1984). Gross chromosomal rearrangements also appear to be much more common (Engels and Preston, 1984) than may have been suggested by the original molecular studies of O'Hare and Rubin (1983) which stressed the precision of the molecular events involved in insertion and excision. This imprecision almost certainly accounts for the greater deleterious effects of mobilisation of elements already in place as compared with the introduction of new elements. Novel insertions undoubtedly have an effect on viability and fitness, estimated to be an average decrease of 1% per insertion (Eanes *et al.*, 1988), but at least in terms of overall impact, excisions appear to be more important. Unfortunately it is not possible to gauge the average effect of an excision, as none of the studies previously mentioned monitored excision events. The apparently more adverse effect of excision as compared with insertion might suggest that insertions have the more important role in adaptive evolution, but a satisfactory assessment of the relative importance of the two will have to await further studies.

1.7. Regulation of P Element Transposition

Our level of understanding of regulation of P element transposition is still inadequate. For example, Engels' (1979) theory of cytotype is adequate for predicting accurately the results in the F1 of crosses between P and M stocks of *Drosophila*, but is clearly inadequate for making accurate predictions for later generations. The main reason for this inadequacy is that the theory is based on empirical observations of particular crosses and is essentially descriptive. Without a firm molecular basis for the concept of cytotype, it is not possible to extrapolate beyond the restricted set of circumstances under which the concept was derived and in particular, it is impossible to make predictions about the evolution of regulation of transposition in the medium to long term.

However, there has been a number of recent advances in understanding of P element regulation, which suggest that a full understanding of the process(es) involved is not far away. First it has been shown that the restriction of P element transposition to the germ line alone is due to differential splicing of the transposase transcript (Laski *et al.*, 1986), with an 87kD active protein being made from the germline mRNA from which all three introns have been removed. Somatic mRNA on the other hand retains the third intron, which has within it a stop codon and as a result a 66kD inactive protein is produced. It has been speculated (Rio, 1988) that this inactive protein may be responsible for negatively regulating transposition, either by forming inactive heteropolymers or by competitive binding to sites within DNA. Robertson (reported in Snyder and Doolittle, 1988) has produced direct evidence for a regulatory role for this truncated transposase, based on comparisons of the properties of mutations caused by P element insertions at the *singed* locus in either P cytotype or M cytotype backgrounds. Nitasaka *et al.* (1987) also have produced evidence for a regulatory role for truncated transposase, based on analysis of a single internally deleted P element in a Q strain of *Drosophila*. This element confers P cytotype and would, if transcribed and translated, produce a protein very similar to the somatic truncated transposase. Black *et al*. (1987) have also produced evidence favouring a regulatory role for truncated transposase, based on studies of truncated elements called "KP" elements. A much smaller truncated transposase protein would be produced in this case and it has been speculated that it negatively regulates transposition by forming "poisoned" enzyme heteropolymers with the full length transposase polypeptide. Black *et al*. (1987) have suggested that natural selection has operated to cause the spread of KP elements throughout the world because of the moderating effect they have on P transposition. However, KP regulation is clearly not the same as P cytotype regulation as it is established rapidly and does not show the reciprocal difference between crosses. Clearly the molecular studies have established that truncated transposase is a mediator of repression, although its mode of action still remains to be fully elucidated.

2. EXPERIMENTAL STUDIES OF CREATION OF QUANTITATIVE VARIATION BY P TRANSPOSITION

2.1. Crossing Protocols

2.1.1. Response to selection for abdominal bristle number

Both Mackay (*ibid*) and Torkamanzehi *et al*. (*ibid*) have used reciprocal crossing procedures to establish selection lines in which transposition was either putatively repressed (nondysgenic lines) or derepressed (dysgenic lines). Figures 1 and 2 summarise the results produced by Mackay (1984, 1985) and Torkamanzehi *et al*. (1988 and unpublished) respectively for selection on abdominal bristle number. The selection regime was identical in each case and the experimental conditions were identical, except for the first experiment of Torkamanzehi *et al*. (1988), where the flies were maintained at 23-24°C rather than 20°C as in all other cases. The striking difference between these results is that Mackay has found substantially enhanced response in 8/8 comparisons between dysgenic and nondysgenic lines in two independent experiments. The phenotypic variance is also consistently enhanced in her dysgenic lines, even after selection plateaux have been reached.

By contrast, Torkamanzehi *et al*. have found no consistent evidence for enhanced response in dysgenic lines in three independent experiments which used the same crossing protocol. In their first experiment (Fig 2a), one of the nondysgenic lines displayed enhanced response with a transient increase in phenotypic variance as the response occurred. In their second experiment (Fig 2b), which used a different P stock (Torkamanzehi *et al*., unpublished), one dysgenic line showed a very slight enhancement of response and an increase in phenotypic variance which was both smaller and later than that observed by Mackay. In any case, one of the nondysgenic controls had a similarly elevated phenotypic variance, but without any apparent effect on response. In a final experiment

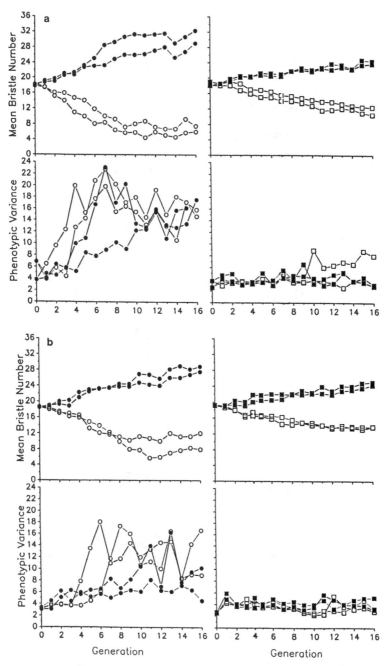

Figure 1. Selection responses and phenotypic variances of abdominal bristle number for Mackay's (1985, 1986, 1987) experimental tests of the effect of P element transposition. Results for dysgenic lines are shown on the left with circular symbols and for nondysgenic lines on the right with square symbols. Lines were initiated using either (a) non-inbred or (b) inbred Canton S (M) and Harwich (P) stocks.

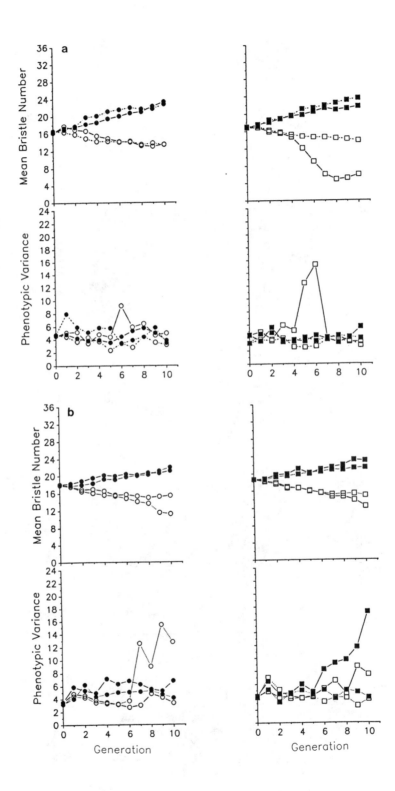

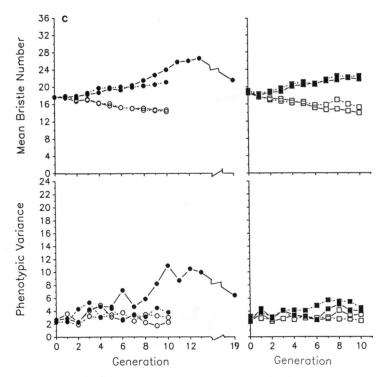

Figure 2. Selection responses and phenotypic variances of abdominal bristle number for the experiments of Torkamanzehi, Moran and Nicholas (1988 and unpublished). Results are presented in an identical format and scale to those of Mackay in Figure 1. Lines were initiated by crosses of (a) Canton S (M) and Harwich (P) stocks, provided by Dr John Sved; (b) the same Canton S stock and Parra Wirra, a strong P stock from Australia, also provided by Dr John Sved, and (c) Canton S and Harwich stocks provided by Dr Trudy Mackay.

(Fig 2c), in which both the P and M stock were identical to those used by Mackay and were in fact provided by her, there was some evidence of enhanced response in one of the four dysgenic lines. This line had slightly elevated response relative to the corresponding replicate dysgenic line and the nondysgenic control lines. Analysis of the total response in this line using compound and balancer chromosomes revealed that of the total upward responses of 9.5 and 7.2 bristles for the females and males respectively, approximately 20% resided on the X chromosome with 35% on chromosome 2 and 45% on chromosome 3 (Morton, 1988), with no evidence of mutations of large effect. Phenotypic variance was also inflated for this line, but not to the extent observed in Mackay's experiments. There were adverse fitness effects associated with the selection response, since with relaxation of selection for six generations the line mean and phenotypic variance both decreased to values near those observed prior to the burst of response (Morton, 1988).

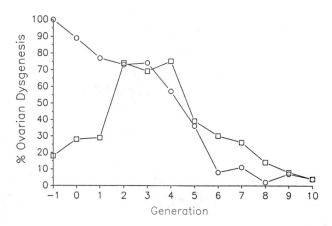

Figure 3. Ovarian dysgenesis tests of cytotype during selection in dysgenic and nondysgenic lines. Results are averaged
over the four dysgenic (circular symbols) and four nondysgenic (square symbols) replicates. Random females were taken
from the selection lines each generation, mated with Para
Wirra P stock males and F_1 progeny raised at 28.5°C . For each
replicate, 96 female progeny were tested for egg-laying ability on grape juice medium in Tissue Culture Microtest Plates.
The percentage of females laying no eggs is presented.

2.1.2. Cytotype change in dysgenic and nondysgenic lines

As mentioned previously, there is substantial evidence for
problems with the experimental design used in the previously
described selection experiments. In particular, it is now
clear that transposition is not repressed in lines derived
from nondysgenic crosses. This would clearly account for the
inability of Torkamanzehi et al. (ibid) to obtain consistently
enhanced response to selection in dysgenic lines, as well as
explaining the occurrence of occasional elevated response and
phenotypic variance in the nondysgenic controls. However, it
makes Mackay's results doubly difficult to explain, first
since they are not repeatable and secondly since the evidence
on lack of repression of transposition in the controls suggests that they should not have been obtained in the first
place.

By monitoring cytotype each generation during the second
and third of the experiments described previously, Torkamanzehi et al. (unpublished) were able to obtain independent
evidence for the derepression of transposition in the nondysgenic "control" lines. Cytotype was assayed using the ovarian
dysgenesis test which involved mating females from the selection lines with males of a P tester stock. Fertility of the F_1
daughters raised at 28.5°C was measured, with 100% fertility
indicating P cytotype in the parental females and zero fertility indicating M cytotype and no repression of transposition.
Results are shown averaged across replicates for the dysgenic
and nondysgenic lines in the second experiment in Figure 3. It
can be seen that nondysgenic lines initially have strong P

cytotype, which they transiently lose before regaining high
levels of repression of transposition. The dysgenic lines
behave as expected with a gradual increase in the ability to
repress transposition. The third experiment of Torkamanzehi *et
al*. (unpublished) produced qualitatively similar results to
these (not shown) for change in cytotype in both dysgenic and
nondysgenic lines.

2.1.3. Other evidence for derepression of transposition in nondysgenic lines

In Figure 2a, it can be seen that one of the nondysgenic
lines provided greatly enhanced response to selection relative
to the other nondysgenic lines and also relative to the dysgenic lines. It was soon apparent that the response was
attributable to a mutation of large effect located on the X
chromosome. Detailed analysis of this mutation (Harcourt,
1987; Moran *et al.*, unpublished) revealed that it is located
on the tip of the X, 0.4 cM proximal to the *yellow* locus. The
only known candidate locus for the mutation in this region is
scute, but it complements alleles and a deletion of *scute*.
Both in terms of phenotype and map position, the mutation does
not appear to have been described previously, at least in
Lindsley and Grell (1968), so provisionally it has been named
tonock. [*tonock* is a Farsi word meaning sparse and patchy].
There is strong genetic evidence that the *tonock* mutation is
due to a P element insertion, since it reverts to wild-type at
a frequency of 6.8% in dysgenic crosses, but is otherwise very
stable. Furthermore, *in situ* mapping using a P element probe
has revealed a new site of insertion at position 2A-C (Fig.
4a,b), which is consistent with the genetic mapping data. The
X chromosome from the *tonock* stock has at least six other
sites of P element hybridisation that were not present on the
parental P stock X chromosome. These results confirm the derepression of transposition that occurs in nondysgenic lines, as
there is no other way in which the new sites of hybridisation
could have been produced.

Ironically, the very results which confirm the unsuit-abil-
ity of the crossing protocol for assessing the contribution of
transposition-mediated mutation to quantitative genetic variation actually provide strong evidence that P transposition can
create genetic variation which has a very substantial effect
on response to selection and evolutionary rates. Clearly, however, it is not possible to quantify the overall contribution
of transposition-induced mutation with the crossing protocol
because of the lack of an appropriate control, which precludes
the possiblity of recognising the influence of mutations of
small effect. An alternative experimental design is essential.

Although it is not strictly relevant to the theme of this
paper, it is worth mentioning here that the *tonock* insertion
mutation has a disruptive effect on the developmental stability of abdominal bristle production. By comparing adjusted
coefficients of variation of left-right difference in abdominal bristle number, which had been calculated using the mean
of total bristle number rather than the mean of the left-right

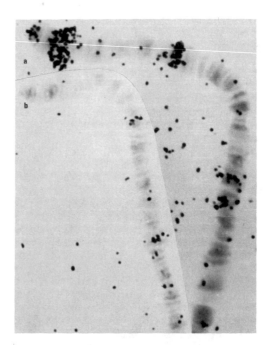

Figure 4. *In situ* hybridisation evidence of transposition in
a nondysgenic line. Hybridisation of the pπ25.1P element
probe to the tip of the X chromosome of (a) the *tonock*
nondysgenic derivative and (b) the Harwich parental stock.
Note the site at 2A-C, responsible for the *tonock* mutation
and the enhanced downwards response shown in Figure 2a.

difference, it was found that the *tonock* stock flies experienced very significantly higher levels of fluctuating
asymmetry {see Palmer and Strobeck (1986) for a recent review
of fluctuating asymmetry} than either of the parental stocks
from which it was derived (Harcourt, 1987; Moran *et al.*,
unpublished). This is not unexpected for a newly arisen mutation, although it is perhaps a novel observation for an insertion mutation.

2.2. Use of Transformed Stocks Produced by Microinjection of P Elements

2.2.1. The experimental design of selection experiments

Clearly the inconsistency between laboratories and the fundamental unsuitability of the crossing protocol for assessing
and quantifying the influence of transposition demand an
alternative experimental design. Such a design must take
account of the inadequacies of our understanding of cytotype
and other aspects of regulation of transposition, yet provide
unequivocal answers. Since it is possible to create P stocks
by microinjection of cloned P factors into M stocks (Spradling
and Rubin, 1982), it is clear that such stocks might provide

the necessary resource for these selection experiments, particularly as the parental M stock would provide an isogenic
control relative to the transformed P stock derivative. If
repression of transposition has already been established in
the P derivative, then a new round of transposition can be
induced by crossing the isogenic M and P stocks, prior to
selection. As there is no requirement for crossing of unrelated stocks, inflation of the genetic variance due to
hybridisation is avoided. Indeed if a suitably inbred M stock
is transformed, then the sensitivity of detection of transposition effects should be greatly enhanced since they do not
have to be disentangled from effects due to background genetic
variation.

2.2.2. Results using experimentally transformed stocks

Using a transformed stock, such an experiment is now under-
way in our laboratory (Torkamanzehi et al., unpublished). The
stock in use is the 88-4-1 transformed stock provided by Professor Margaret Kidwell (Daniels et al., 1987). It has approximately 15 intact P elements and approximately 40 degenerate
elements per haploid genome, as compared with 25 and 40
respectively for a standard, naturally-occurring strong P
stock, Harwich-77. Daniels et al. (1987) have also shown that
88-4-1 has both strong P factor activity in dysgenic crosses
and strong P cytotype. This has been confirmed in our laboratory. Although the selection experiment using 88-4-1 and its
ry^{506} M stock parent is still incomplete, the results (Fig. 5
and Table I) are quite obvious even at this stage. Lines initiated by dysgenically crossing the transformed P stock with
its M stock parent provide considerably enhanced response relative to the parental M stock control in three of the four
comparisons, and estimates of realised h^2 also are significantly different from zero. These estimates of h^2 can be
applied to the appropriate estimates of phenotypic variance to
obtain estimates of additive genetic variance, V_A. The replicate 2 up line (UR2) is problematical in that there was some
evidence of enhanced upwards response which disappeared as the
experiment progressed. The population mean for this line also
seems to fluctuate more than in the control lines. Nevertheless the phenotypic variance is only slightly if at all
inflated relative to the control variances.

The extent of induction of genetic variance can be estimated using the following reasoning. The ry^{506} M stock control
has shown no response to either upwards or downwards selection, with none of the estimates of realised heritability
being significantly different from zero. Therefore, all of the
additive genetic variance in the transformed lines must have
been induced by transposition. Further, since the observed
phenotypic variance in the control lines is entirely environmental in origin, it can be used as an estimate of the environmental variance, V_E of the transformed lines. As shown in
Table I, this permits partitioning of V_P in the transformed
lines into V_A, V_E and a non-additive genetic component, V_{NA},
which may however be overestimated due to increased

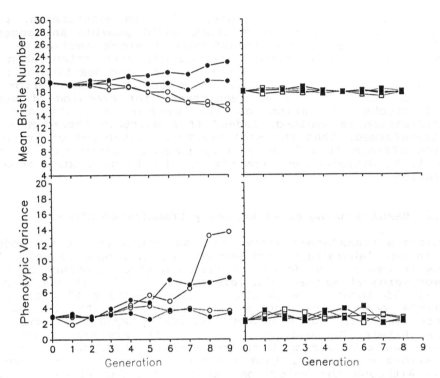

Figure 5. Unambiguous evidence for the influence of P element transposition on response to selection and genetic variance using an inbred M stock and a transformed P stock derivative. Selection response and phenotypic variance of abdominal bristle number for a transposition positive line initiated by a dysgenic cross between ry[506] and the 88-4-1 P stock derivative (circular symbols on left) and for the ry[506] M stock control (square symbols on right). Note that these stocks are co-isogenic apart from the P element insertions which have been introduced into 88-4-1 by microinjection.

environmental sensitivity. The enhancement of variance in the transposition positive transformed lines must be attributable to the original insertions when the stock was produced plus any other insertions and excisions that have resulted during the most recent round of transposition. The estimates of V_A in these lines can be treated as estimates of V_m, in which case V_m/V_E for the transformed stocks ranges from 0 to 0.45, averaging 0.32 for the three stocks where there was clear evidence of induction of variation. Compared with the generally quoted value of 10^{-3}, it can be seen that transposition has created approximately a three hundred fold increase in genetic variance above the standardly accepted spontaneous rate of introduction of mutational variance.

Whether one judges the influence of transposition from the response graphs or from V_m/V_E statistics, it is unambiguously clear that transposition has had a very substantial effect on

Table I
Preliminary analyses of selection responses attributable to P elements
introduced by microinjection[a]

	UR1	DR1	UR2	DR2
"Transformed" Lines				
h^2	0.15	0.20	-0.01	0.17
±SE	0.02	0.01	0.03	0.02
	***	***	n.s.	***
V_P	5.22	6.61	3.11	3.47
V_A	0.78	1.33	0.0	0.60
V_{NA}	1.64	2.48	0.31	0.07
Control Lines				
h^2	-0.04	-0.01	-0.01	0.02
±SE	0.02	0.01	0.02	0.08
	n.s.	n.s.	n.s.	n.s.
V_P	3.73	2.90	2.87	2.69
V_A	0.0	0.0	0.0	0.0

[a]Realised h^2 has been estimated by regression of cumulative response on cumulative selection differential, except for control DR2, where it is based on individual generation response and selection differential. Standard errors are appropriate for the regression coefficients, but underestimate the true standard error of realised heritability. Phenotypic variances are simple arithmetic means across generations and ignore obvious heterogeneity of variances particularly in the transformed lines, where there is an obvious trend for increasing variances in most lines. Other parameters are estimated as explained in the text. No attempt has been made to analyse changes in genetic and phenotypic parameters which may be occurring during the course of the experiment in this preliminary analysis.

response to selection even in the short term. Admittedly some of this response can be attributable to genetic variance that accumulated as a result of transposition in the transformed line between 1982 when it was produced and 1988 when it was used in this experiment. However, even this period of time is relatively short in evolutionary terms.

3. CONCLUSIONS

Experimental protocols based on comparisons of selection response between lines initiated with dysgenic and nondysgenic crosses have been discredited as a means of evaluating the evolutionary influence of transposition, due to the inadequacy of the negative controls employed. Nevertheless it is undoubtedly true that P element transposition can contribute significantly to selection response in the short to medium term. One category of evidence comprises selection responses that can be directly attributed to mutations of large effect, caused by insertion of transposable elements during the course of the experiment. Better evidence, of greater generality since it includes the influence of mutations of small effect, has been obtained from experiments employing stocks transformed by microinjection. The advantages of appropriate negative controls and the greater sensitivity of detection of mutational effects have been combined to demonstrate a very significant influence of transposition on response to selection. This can be generalised to encompass the influence of transposition on the evolutionary potential of all species bearing transposable elements. One area requiring further study concerns the relative importance of insertion versus excision events. It will be very interesting to discover whether the adaptive evolution simulated in these experiments can be attributed primarily to insertions which have a relatively low cost in terms of viability and fitness effects or whether imprecise excisions and chromosomal rearrangements also have an important role.

4. SUMMARY

The influence in the short term of P element transposition on quantitative genetic variation is examined. It is shown that experimental protocols which employ lines derived from dysgenic crosses (transposition positive) and nondysgenic crosses (putatively transposition negative controls) are unsuitable for analysing this problem. Cytotype assays demonstrate a transient loss of repression of transposition in the "negative" controls. Furthermore, *in situ* and genetic analysis of a mutation of large effect directly demonstrate the occurrence of transposition in the "negative" controls. An alternative experimental approach using M stock controls and co-isogenic artificially transformed P stock derivatives has unequivocally demonstrated the effect of P element transposition on quantitative variation, with an enormous increase in genetic variance in most but not all cases.

ACKNOWLEDGEMENTS. We wish to thank Associate Professor Frank Nicholas, Department of Animal Husbandry, University of Sydney for perceptive and helpful comments on the manuscript. Dr John Sved, School of Biological Sciences, University of Sydney provided fly stocks and helpful advice on the experiments and preparation of the manuscript. Dr Allan Lohe, Department of Genetics, School of Medicine, Case Western Reserve University, provided guidance and assistance with the *in situ* hybridisations. We wish to particularly thank Professor Margaret Kidwell, Department of Ecology and Evolutionary Biology, University of Arizona, who kindly provided the transformed stocks of flies. We are grateful to Merrilee Baglin and Stephen Brown for technical assistance and statistical advice respectively. A.T. acknowledges support of a scholarship provided by the Ministry of Education and Higher Degrees, Islamic Republic of Iran. Our reseach has been supported by maintenance funds provided by the Department of Animal Husbandry, University of Sydney.

References

Black, D.M., Jackson, M.S., Kidwell, M.G., and Dover, G.A., 1987, KP element accumulation represses hybrid dysgenesis in *D. melanogaster*, *EMBO J.* **6**:4125-4135.

Bingham, P.M., Kidwell, M.G., and Rubin, G.M., 1982, The molecular basis of P-M hybrid dysgenesis: the role of the P element, a P-strain-specific transposon family, *Cell* **29**:995-1004.

Bregliano, J.C., and Kidwell, M.G., 1983, Hybrid dysgenesis determinants, in: *Mobile Genetic Elements* (J.A. Shapiro, ed.), Academic Press, London and New York, pp. 363-410.

Clayton, G.A., and Robertson, A., 1955, Mutation and quantitative variation, *Am. Nat.* **89**:151-158.

Clayton, G.A., and Robertson, A., 1964, The effect of X-rays on quantitative characters, *Genet. Res.* **5**:410-422.

Daniels, S.B., Clark, S.H., Kidwell, M.G., and Chovnick, A., 1987, Genetic transformation of *Drosophila melanogaster* with an autonomous P element: phenotypic and molecular analyses of long established transformed lines, *Genetics* **115**:711-723.

Eanes, W.F., Wesley, C., Hey, J., Houle, D., and Ajioka, J.W., 1988, The fitness consequences of P element insertion in *Drosophila melanogaster*, *Genet. Res.* **52**:17-26.

Engels, W.R., 1979, Hybrid dysgenesis in *Drosophila melanogaster*: rules of inheritance of female sterility, *Genet. Res.* **33**:219-236.

Engels, W.R., 1983, The P family of transposable elements in *Drosophila*, *A. Rev. Genet.* **17**:315-344.

Engels, W.R., 1986, On the evolution and population genetics of hybrid-dysgenesis-causing transposable elements in *Drosophila*, *Phil. Trans. R. Soc.* B **312**:205-215.

Engels, W.R., and Preston, C.R., 1984, Formation of chromosomal rearrangements by P factors in *Drosophila melanogaster*, *Genetics* **107**:657-678.

Fitzpatrick, G.J., and Sved, J.A., 1986, High levels of fitness modifiers induced by hybrid dysgenesis in *Drosophila melanogaster*, *Genet. Res.* **48**:89-94.

Frankham, R., 1980, Origin of genetic variation in selection lines, in: *Selection Experiments in Laboratory and Domestic Animals* (A. Robertson, ed.), Commonwealth Agricultural Bureaux, Slough, pp. 56-68.

Harcourt, R., 1987, Molecular and genetic characterisation of a P element insertion in *Drosophila melanogaster*, Unpublished B.Sc.Agr. thesis, University of Sydney.

Hill, W.G., 1982a, Rate of change in quantitative traits from fixation of new mutations, *Proc. natn. Acad. Sci. USA* **79**:142-145.

Hill, W.G., 1982b, Prediction of response to artificial selection from new mutations, *Genet. Res.* **40**:225-278.

Hollingdale, B., and Barker, J.S.F., 1971, Selection for increased abdominal bristle number in *Drosophila melanogaster* with concurrent irradiation. I. Populations derived from an inbred line, *Theor. & Appl. Genet.* **41**:208-215.

Junakovic, N., Di Franco, C., Barsanti, P., and Palumbo, G., 1986, Transposition of Copia-like nomadic elements can be induced by heat shock, *J. Mol. Evol.* **24**:89-93.

Keightley, P.D., and Hill W.G., 1983, Effects of linkage on response to directional selection from new mutations, *Genet. Res.* **42**:193-206.

Keightley, P.D., and Hill W.G., 1987, Directional selection and variation in finite populations, *Genetics* **117**:573-582.

Keightley, P.D., and Hill W.G., 1988, Quantitative genetic variability maintained by mutation-stabilising selection balance in finite populations, *Genet. Res.* 52:33-43.

Kidwell, M.G., Kidwell, J.F., and Sved, J.A., 1977, Hybrid dysgenesis in *Drosophila melanogaster*: a syndrome of aberrant traits including mutation, sterility and male recombination, *Genetics* 86:813-833.

Kimura, M., 1985, The neutral theory of molecular evolution, *New Scientist* 107:43-46.

Kitagawa, O., 1967, The effects of X-ray irradiation on selection response in *Drosophila melanogaster*, *Jap. J. Genet.* 42:121-137.

Lande, R., 1975, The maintenance of genetic variability by mutation in a polygenic character with linked loci, *Genet. Res.* 26:221-235.

Laski, F.A., Rio, D.C., and Rubin, G.M., 1986, Tissue specificity of *Drosophila* P element transposition is regulated at the level of mRNA splicing, *Cell* 44:7-19.

Lindsley, D.L., and Grell, E.H., 1968, *Genetic Variations of* Drosophila melanogaster, Carnegie Institute of Washington, Publication no. 627.

Lynch, M., 1988, The rate of polygenic mutation, *Genet. Res.* 51:137-148.

Mackay, T.F.C., 1984, Jumping genes meet abdominal bristles: hybrid dysgenesis-induced quantitative variation in *Drosophila melanogaster*, *Genet. Res.* 44:231-237.

Mackay, T.F.C., 1985, Transposable element-induced response to artificial selection in *Drosophila melanogaster*, *Genetics* 111:351-374.

Mackay, T.F.C., 1986, Transposable element-induced fitness mutations in *Drosophila melanogaster*, *Genet. Res.* 48:77-87.

Mackay, T.F.C., 1987a, Transposable element-induced polygenic mutations in *Drosophila melanogaster*, *Genet. Res.* 49:225-233

Mackay, T.F.C., 1987b, Transposable element-induced quantitative genetic variation in *Drosophila*, in: *Proceedings of the Second International Conference on Quantitative Genetics* (B.S. Weir, E.J. Eisen, M.M. Goodman, and G. Namkoong, eds), Sinauer, Sunderland, Mass., pp. 219-235.

McClintock, B., 1978, Mechanisms that rapidly reorganise the genome, in: *Stadler Symposium, Vol. 10*, (G.P. Reder, ed.), Columbia, Mo., pp. 25-48.

Morton, R.L., 1988, Genetic characterisation of a high bristle line of *Drosophila melanogaster* produced under selection in a dysgenic cross, Unpublished B.Sc.Agr. thesis, University of Sydney.

Mukai, T., and Yukuhiro, K., 1983, An extremely high rate of deleterious viability mutations in *Drosophila* possibly caused by transposons in non-coding regions, *Jap. J. Genet.* 59:316-319.

Nitasaka, E., Mukai, T., and Yamazaki, T., 1987, Repressor of P elements in *Drosophila melanogaster*: Cytotype determination by a defective P element carrying only open reading frames 0 through 2, *Proc. natn. Acad. Sci. USA* 84:7605-7608.

O'Hare, K., and Rubin, G.M., 1983, Structures of P transposable elements and their sites of insertion and excision in the *Drosophila melanogaster* genome, *Cell* 34:25-35.

Palmer, A.R., and Strobeck, C., 1986, Fluctuating asymmetry: measurement, analysis, patterns, *Annu. Rev. Ecol. & Syst.* 17:391-421.

Rio, D.C., 1988, P elements in *Drosophila melanogaster*: their molecular biology and use as vector systems, in: *Transposition* (A. J. Kingsman, K. F. Chater, and S. M. Kingsman, eds), Cambridge University Press, Cambridge, pp. 287-300.

Rubin, G.M., Kidwell, M.G., and Bingham, P.M., 1982, The molecular basis of P-M hybrid dysgenesis: the nature of induced mutations, *Cell* 29:987-994.

Simmons, M.J., Raymond, J.D., Culbert, P., and Laverty, T.R., 1984, Analysis of dysgenesis induced lethal mutations on the X chromosome of a Q strain of *Drosophila melanogaster*, *Genetics* 107:49-63.

Simmons, M.J., Raymond, J.D., Johnson, N.A., and Fahey, T.M., 1984, A comparison of mutation rates for specific loci and chromosome regions in dysgenic hybrid males of *Drosophila melanogaster*, *Genetics* **106**:85-94.

Snyder, M., and Doolittle, W.F., 1988, P elements in *Drosophila*: selection at many levels, *Trends Genet.* **4**:147-149.

Spradling, A.C., and Rubin, G.M., 1982, Transposition of cloned P elements into *Drosophila* germ line chromosomes, *Science* **218**:341-347.

Torkamanzehi, A., Moran, C., and Nicholas, F.W., 1988, P element-induced mutation and quantitative variation in *Drosophila melanogaster*: lack of enhanced response to selection in lines derived from dysgenic crosses, *Genet. Res.*, **51**:231-238.

Voelker, R.A., Greenleaf, A., Gyurkovics, H., Wisely, G., Huang, S., and Searles, L., 1984, Frequent imprecise excision among reversions of P element-caused lethal mutations in *Drosophila*, *Genetics* **107**:279-294.

Yukuhiro, K., Harada, K., and Mukai, T., 1985, Viability mutations induced by P elements in *Drosophila melanogaster*, *Jap. J. Genet.* **60**:531-537.

Habitat Selection: Introduction

Habitat selection is a key component in the adaptive response of animals to their environments, and it can profoundly influence the selective regime to which the pre-adult and, in some cases, the adult stages of the organisms are exposed.

In insects such as *Drosophila*, habitat selection is directly determined by ovipositing females whose behavior is affected by visual, chemical and physical properties of the host. The genetic basis of oviposition site preference (OSP) is likely to be complex, yet it represents an obvious target for strong selection pressures which can lead to fundamental changes in the genetic architecture of a species. Laboratory studies have revealed some of the variation associated with OSP, but the extent to which this variation is manifested by wild flies in the field with naturally occurring resources remains to be determined. Most previous experiments have controlled the physiological condition of the flies, the distribution of resources they encounter, and the physical environment. It is not known whether variation in these factors in the field will swamp any underlying genetic differences among flies in their use of different breeding sites.

Very little work has been done on possible genetic correlations between adult host response and offspring performance in different breeding sites. If positive genetic correlations between response and offspring performance are found, it will be important to determine whether they are due to linkage disequilibrium or pleiotropy. In theory, genetic variation in habitat selection can greatly increase the likelihood of genetic polymorphism for fitness variants. Such polymorphisms should, particularly if OSP is associated with positive assortative mating, lead rapidly to reproductive isolation.

OSP behavior and larval performance will be limited by the properties of the host environments. These properties, even in a "simplified" ecological context such as the cactus-yeast-*Drosophila* system, will be complex, involving the interactions between host plant chemistry, microbial populations and their successions, the presence of parasites as well as resource abundance. The definition of these components, however, is an essential first step in understanding the relationship between OSP and larval performance.

Several of the chapters in this section examine the relationships between OSP and host components. James Fogleman and

Ecological and Evolutionary Genetics of Drosophila, Edited by
J.S.F. Barker *et al.*, Plenum Press, New York, 1990

Ruben Abril profile the volatile chemicals of cactus hosts and their role in attracting females to oviposit. They also describe the bacterial flora of the host and a variety of secondary plant compounds which directly affect larval survival. The role of yeast communities in determining the suitability of cacti as host plants for desert *Drosophila* is addressed by Tom Starmer and Virginia Aberdeen. They discuss the importance of the flies in vectoring yeast species to wounded cacti, thus accelerating the rot process and developing the site for larval survival. Yeasts provide concentrated sources of nutrients for the immature stages and appear to be involved in habitat detoxification, larval foraging behavior, as well as adult feeding, mating and oviposition behavior. Thus yeasts are likely to be important components in differential habitat use.

The remaining chapters concern, at least in part, the kind of genetic variation affecting habitat selection, and the factors maintaining the variation. Philip Hedrick develops the theoretical basis for the evolution and maintenance of genetic polymorphisms associated with differential OSP's and subsequent larval survival. Genetic variation in OSP in three different *Drosophila* species groups, is empirically analyzed by Stuart Barker with *D. buzzatii*, using cactophilic yeast species from their natural breeding site in *Opuntia* cacti as the oviposition site resources, by John Jaenike with *D. tripunctata*, using mushrooms and tomatoes, and by Ary Hoffmann and Sharon O'Donnell with *D. melanogaster* and *D. simulans*, using different types of fruit. In each case, polygenic variation was detected, which is associated with heterogeneity in the particular habitat. Importantly, these chapters also address questions of experimental strategy and design, and of mechanisms for the maintenance of this variation.

These experiments with different *Drosophila* species clearly show that the genes involved in habitat selection can be identified, and that the contributions of the various components of these habitats to selectively important genotype-environment interactions can be measured. Future studies on the relatively straightforward *Drosophila* host-plant or host-fungus interactions should provide a wealth of information on critical issues in habitat selection. Sophisticated genetic and molecular techniques (such as the QTL method discussed earlier) for mapping and analyzing the genes which play a key role in these interactions are now available. Indeed, these studies should eventually interface with the active research concerned with the genetic and molecular architecture of the *Drosophila* nervous system.

CHAPTER 7

Ecological and Evolutionary Importance of Host Plant Chemistry

JAMES C. FOGLEMAN, AND J. RUBEN ABRIL

1. INTRODUCTION

In the last two decades, there has been an increasing interest in the inter-disciplinary subject of the chemical ecology of insects. A large part of this subject concerns the chemistry of the interactions between insects and their host plants. The cactus-microorganism-*Drosophila* model system of the Sonoran Desert provides an excellent opportunity to pursue the subject of chemical ecology in a system which is also amenable to the study of evolutionary and ecological genetics. By examining the chemical interactions between the desert *Drosophila* and their cactus host plants, insights into aspects of the habitat that impact on the fitness of the flies can be gained. The cactophilic *Drosophila* in this model system feed and breed in necrotic stems of columnar cacti. In order to do this, the flies must be able to locate suitable rot pockets, assimilate required nutrients, and be able to tolerate whatever toxic compounds might be present in the cactus tissue. Microorganisms which grow in the developing rot serve as a food source for the *Drosophila* as well as modify the cactus tissue both physically and chemically. Therefore, the chemical interactions between the cacti, the microorganisms, and the flies are of major importance in determining the *Drosophila*-host plant relationships which exist in the Sonoran Desert.

2. HOST PLANT RELATIONSHIPS

The cactus-microorganism-*Drosophila* model system involves four species of *Drosophila* and five species of columnar cacti. The four *Drosophila* species are: *D. nigrospiracula*, *D. mettleri*, *D. mojavensis*, and *D. pachea*. The first three species are members of the *repleta* species group while the last belongs to the *nannoptera* group. Several recent reviews have discussed the genetics, ecology, and phylogenetic relationships of these four species (Heed, 1978 and 1982; Heed and

JAMES C. FOGLEMAN, AND J. RUBEN ABRIL * Department of Biological Sciences, University of Denver, Denver, CO 80208, USA.

Ecological and Evolutionary Genetics of Drosophila, Edited by
J.S.F. Barker *et al.*, Plenum Press, New York, 1990

Mangan, 1986). The summary of their evolutionary histories is twofold: first, each species appears to have independently evolved its own set of adaptations to the desert. This idea is based on the fact that they are not closely related phylogenetically and because they inhabit different cacti. Second, *D. pachea* and *D. mojavensis* show evidence of evolutionary origin derived from their respective ancestors south of the desert along the west coast of Mexico. In contrast, *D. nigrospiracula* and *D. mettleri* are not such clearly derived forms and are not as well known with respect to their place of origin.

The columnar cacti of Mexico are highly specialized plants, and the five species that are involved in the model system have been phylogenetically placed in two subtribes within the tribe Pachycereeae (Gibson and Horak, 1978; Gibson and Nobel, 1986). Saguaro (*Carnegiea gigantea*), cardón (*Pachycereus pringlei*), and senita (*Lophocereus schottii*) are included in the the subtribe Pachycereinae, while pitaya agria (*Stenocereus gummosus*) and organ pipe (*S. thurberi*) are included in the subtribe Stenocereinae. All five of the cacti are chemically differentiated, and some of these chemical differences are considered to be subtribe characteristics.

There is essentially a one-to-one relationship between the fly species and the cactus species such that, in any particular locality, each *Drosophila* species utilizes necrotic tissue from only one cactus species. The fly-cactus associations are presented in Table I. Due to differences in the geographic distributions of the cacti, and one case of behavioral preference, three of the four *Drosophila* species shift host plants or substrates between the Baja Penisula and the mainland part of the Sonoran Desert. For example, *D. nigrospiracula* uses necrotic saguaro on the mainland and cardón, on the Baja Peninsula. Saguaro does not occur on the peninsula and cardón

Table I
Host plant relationships and distribution of host plants in the Sonoran Desert

Resident species	Major host plant on peninsula	Major host plant on mainland	Host plant distribution	
D. nigrospiracula	Cardón	Saguaro	Saguaro Cardón	M[a] MP[b]
D. mettleri	Cardón soaked soil	Saguaro soaked soil	Saguaro Cardón	M MP
D. mojavensis	Agria	Organ pipe	Agria Organ pipe	MP M & P[c]
D. pachea	Senita	Senita	Senita	M & P

[a]Mainland, [b]Mostly peninsula, [c]Mainland and peninsula.

is relatively infrequent on the mainland. *Drosophila mettleri* breeds primarily in soils which have been soaked with rot exudate from saguaro or cardón, so it exhibits the same host shift as *D. nigrospiracula*. Although these two species share the same host plants, they are completely separate with respect to their breeding niche. The host plants for *D. mojavensis*, agria and organ pipe, are sympatric on the Baja Peninsula, but the preference of *D. mojavensis* for agria leaves organ pipe virtually unused in that region. Organ pipe is the major host on the mainland where agria is all but absent. Senita cactus is present in both areas of the Sonoran Desert and is used monophagously by *D. pachea*. Extensive rearing records, together with records of the adults collected directly from natural substrates, indicate that the specificity of the relationships in Table I is very high, though not absolute. In general, 95% or more of the flies caught on or reared out of a rot on a particular species of cactus are the resident *Drosophila* for that cactus. The one exception to this trend is *D. mojavensis*. Recent rearing records from naturally occurring agria and organ pipe rots show that *D. arizonensis*, a sibling species of *D. mojavensis*, can represent a significant proportion (up to 50%) of the flies which eclose from certain individual rots (Ruiz and Heed, 1988).

The chemistry involved in the fly-cactus associations readily falls into two categories: the chemistry of host plant utilization and the chemistry of host selection (Fogleman and Heed, 1989). Utilization, in the model system, involves the physiological ability of a *Drosophila* species to use the necrotic tissue of a cactus species as a substrate. Selection is defined as the behavioral preference of a fly species for a particular cactus species.

3. HOST PLANT UTILIZATION

Chemical aspects of host utilization by *Drosophila* include tolerance to the toxic phytochemicals which may be present in necrotic cactus tissue and strict requirements for nutritional factors which may not occur in all potential cactus species. Cactus compounds which have been identified as toxic to the cactophilic *Drosophila* are listed in Table II. This information has been gathered from a large number of separate experiments and is recently reviewed by Fogleman and Heed (1989). Typically, the toxic effects of the compounds in Table II have been characterized by extracting the compounds of interest from fresh tissue, adding the compounds to an otherwise innocuous medium, and measuring its effect on fitness components of flies. The standard fitness components are viability (either egg-to-adult or larva-to-adult), larval development time, adult longevity, and/or thorax size. The last parameter has been shown to be positively correlated with female fecundity in cactophilic *Drosophila* (Heed and Mangan, 1986) and flying ability in dipterans (Roff, 1977). The putative mode of action of the compounds is speculative since no one has actually investigated their biochemical effects in the cactophilic flies. However, experiments with other insects have

Table II
Toxic compounds in cactus tissue

Compounds	Present in	Toxic to	Putative mode of action
Complex alkaloids	Senita	*D. nigrospiracula* *D. mojavensis*	Blocks steroid metabolism or phytosterol assimilation
Simple alkaloids	Saguaro Cardón	Toxic in elevated concentrations	Blocks steroid metabolism or phytosterol assimilation
Sterol diols	Agria Organ pipe	*D. nigrospiracula*	Unknown toxic effect Insufficient as dietary sterol
Medium chain fatty acids	Agria Organ pipe	*D. mojavensis* *D. nigrospiracula*	Inhibition of oxidative phosphorylation
Triterpene glycosides	Agria Organ pipe	*D. mojavensis* *D. nigrospiracula* microbes	Feeding deterrent Unknown toxic effect
All natural products	Soaked soils (Elevated conc.)	*D. nigrospiracula* *D. mojavensis* *D. pachea*	See above

either shown or suggested that similar compounds manifest toxicity in this manner.

Several investigators have suggested that alkaloids, like the isoquinoline alkaloids in saguaro and senita, affect insects by blocking steroid metabolism or phytosterol assimilation (Schreiber, 1958; Harley and Thorsteinson, 1967). Unlike senita, the normal concentration of alkaloids in saguaro is apparently not sufficient to cause a fitness reduction in any of the cactophilic species. The actual physiological mechanism of the toxic effect of the sterol diols found in agria and organ pipe cacti on *D. nigrospiracula* is unknown. Neither *D. nigrospiracula* nor *D. mojavensis* can survive if sterol diols are the sole source of dietary sterols (Fogleman et al., 1986), but this dietary insufficiency does not appear to be related to the toxic effect. The insecticidal properties of medium chain fatty acids have long been appreciated and are effective against a large variety of insects and other arthropods (Shepard, 1951; Binder et al., 1979). Triterpene glycosides have been reported as having an incredible variety of biological activities. It has been demonstrated that the triterpene glycosides found in cacti reduce the fitness of both *D. nigrospiracula* and *D. mojavensis* as well as inhibit the the growth of some yeasts and bacteria (Kircher, 1977;

Starmer *et al.*, 1980; Phaff *et al.*, 1985; Fogleman and Armstrong, 1989).

Drosophila pachea has not been included in most of these studies because of the strict nutritional dependency of this species for the unusual sterols that are only found in senita (Fogleman *et al.*, 1986). Consequently, *D. pachea* could not utilize any other cacti, even if it could tolerate their toxic compounds.

Soil, which has been soaked by rot juices, are rather unusual oviposition substrates for *Drosophila*. Soaked soils represent a situation where all of the cactus solutes become more concentrated due to the evaporation of water. Although *D. nigrospiracula*, *D. mojavensis*, and *D. pachea* can tolerate the toxic compounds in the tissues of their respective host plants, they cannot tolerate the elevated concentrations present in soaked soil (Fogleman, 1984). For example, the alkaloids in saguaro comprise about 1% of the dry weight of the plant and are not toxic at this concentration. However, the concentration of alkaloids in soil which has been soaked by saguaro juice can be up to 25 times that in the tissue and is toxic to *D. nigrospiracula* at this level (Meyer and Fogleman, 1987).

Given the five cactus species plus soaked soils and the four species of *Drosophila*, there are 24 possible substrate-fly combinations. Six of these combinations represent the normal *Drosophila*-host plant relationships leaving 18 combinations which do not typically occur in nature. Eleven of these 18 exclusions or approximately 61% are primarily due to chemistry. The biological and chemical factors known to be involved in host utilization in the model system are presented in Table III along with the appropriate references.

4. ANALYSIS OF FRESH CACTUS TISSUE: CARBOHYDRATES

Research on the natural products of cacti, such as those listed in Table II, and their effect on fitness components of the desert *Drosophila* has progressed to the point that these interactions are reasonably well understood. However, the cacti have never been specifically analyzed for carbohydrates. Ecologically, carbohydrates are very important. In this system, sugars serve as the primary carbon source or energy source for microbial growth. They have also been implicated as contributing to the fitness of the flies (Brazner *et al.*, 1984).

In analyzing cactus tissue for carbohydrates, a slightly different extraction procedure was employed compared to that used by Kircher (1982) in his analysis of the chemical composition of columnar cacti. Briefly, the protocol starts with lyophilization (which yields the percent dry weight of the tissue) followed by sequential extraction with ether, methanol, and water. The ether soluble or lipid fraction would contain sterols, fatty acids, neutral triterpenes (if present), and alkaloids (if present). The methanol soluble

Table III

Chemical and biological factors involved in host plant utilization

			Substrates			
Drosophila	Agria	Organ pipe	Saguaro	Cardón	Senita	Soils
D. pachea	◄-------	nutritionally deficient[a] (lack required sterols)	-------►		normal host	physical[b] and/or chemical
D. mojavensis	normal host	normal host	◄--interspecific[c]--► competition		intolerance[d] to alkaloids	physical and/or chemical
D. nigrospiracula	◄---intolerance[e]---► to fatty acids and sterol diols			normal host	intolerance[d] to alkaloids	physical[b] and/or chemical
D. mettleri	◄--------------		behavior[f]		--------------►	normal host

[a]Heed and Kircher, 1965; Goodnight and Kircher, 1971; Campbell and Kircher, 1980; and Fogleman et al., 1986.
[b]Fogleman et al., 1981; Fogleman and Heed, 1981; Fogleman, 1984; Meyer and Fogleman, 1987; and Fogleman and Newman, unpublished.
[c]Fellows and Heed, 1972; Mangan, 1982; Brazner et al., 1984; and Heed and Mangan, 1986.
[d]Kircher et al., 1967; and Fogleman et al., 1982.
[e]Fellows and Heed, 1972; Fogleman et al., 1986; and Fogleman and Kircher, 1986.
[f]Fogleman et al., 1981; Fogleman et al., 1982; Mangan, 1982; Fogleman, 1984; and Heed and Mangan, 1986.

fraction contains triterpene glycosides (if present), poly- or oligosaccharides, and some simple sugars. The water soluble fraction would contain simple sugars, proteins, and salts. Most of the toxic compounds are lipophilic and would be found in the ether extract. What is not soluble in any of these solvents, is considered insoluble residue and is composed mainly of cell wall material. The methanol and water soluble fractions along with the insoluble residue were then acid hydrolyzed, sugars were reduced and acetylated into alditol acetates, and a capillary gas chromatograph was used to quantify the sugars that were present.

The results on the composition of cacti with respect to the fractions mentioned above are shown in Table IV. Several interesting *Opuntia* species (prickly pear cactus) and two types of *Opuntia* fruits have been included along with four of the five columnar cacti of the Sonoran Desert. Each value given in this table is the average of two separate determinations of tissue from a single stem. Variance between replicates was generally less than 1% dry weight which indicates that our technique was very good, but, unfortunately, no indication of the variance between plants within a species can be derived from these data. It should also be noted that the *O. phaeacantha* fruit, although red in color, was hard and probably not ripe.

The data in the first two columns of Table IV are not appreciably different from the values previously published by Kircher (1982). However, the same cannot be said for the remaining fractions. The insoluble residue fraction in Table IV is, on the average, approximately half of values previously reported, and the methanol and water soluble fractions are correspondingly higher. The major change in the extraction procedure has been the addition of a final water extraction. It is obvious that there is a substantial fraction of water soluble material in cacti.

Table IV
Chemical composition of fresh cactus tissue

| Cactus | Percent water | | Percent of dry weight | | |
		Lipids	Methanol soluble	Water soluble	Insoluble residue
Saguaro	87.0	1.8	16.7	27.4	54.1
Senita	88.9	4.6	32.8	38.6	24.0
Agria	88.1	7.2	36.4	32.1	24.3
Organ pipe	72.2	11.1	20.7	21.4	46.8
Opuntia stricta	93.3	1.1	19.2	31.4	48.3
Opuntia phaeacantha	82.5	1.8	13.4	19.4	65.4
Opuntia ficus-indica	92.0	1.9	13.1	33.7	51.3
O. phaeacantha fruit	71.3	2.4	63.5	9.3	24.8
O. ficus-indica fruit	90.7	0.9	76.9	9.3	12.9

Free sugars and some low molecular weight oligosaccharides are expected to be mainly in the water soluble fraction while complex carbohydrates, such as certain polysaccharides, are not soluble in either methanol or water and would occur in the insoluble residue. The methanol soluble fraction should also contain some oligosaccharides as well as glycosides. The sugars that were detected in the methanol fraction are presented in Table V. Quantities of sugars are expressed in percent of the dry weight of the plants. From the data in this table, it can be seen that saguaro, senita, and the *Opuntia* species do not have much sugar in this fraction, and what they do have is mainly glucose. Agria and organ pipe have comparably more sugar, presumably because of the presence of triterpene glycosides which are methanol soluble. The sugar moieties of these triterpene glycosides are composed of glucose and rhamnose, and the relatively high concentrations of rhamnose in these cacti can be seen in the data. Not surprisingly, the fruits have the highest total concentration of sugars. About 95% of the sugar in *Opuntia* fruits was detected in the form of glucose. It should be noted that the presence of mannose is an artifact of the process of making alditol acetates. Fructose, either by itself or as a component of sucrose, is converted by this process into glucose and mannose. The detection of mannose in *Opuntia* fruits, therefore, indicates that at least some fructose was present.

Table VI presents the sugars that were detected in the water soluble fraction of the cacti. This fraction contains the relatively low concentrations of sugars, such as arabinose, xylose, and galactose, that are typically found in plants. It appears that most of the fruit sugars have already been extracted. Within cactus tissue, agria and organ pipe contain the most water soluble sugars, followed by the *Opuntia* species, and saguaro and senita have the least.

Hydrolysis of the insoluble residue and subsequent analysis by gas chromatography did not detect any of the sugars listed in Tables V and VI. A colorimetric test for uronic acids (Ahmed and Labavitch, 1977), however, yielded an extremely strong positive result. It is hypothesized, therefore, that the insoluble residue is composed primarily of uronic acid, probably in the form of poly-galacturonic acid (pectin) which is known to be a component of plant cell wall material.

Adding the data of Tables V and VI together yields data on the total sugar content of these substrates that is potentially available for microorganism growth and *Drosophila* nutrition. These data are presented in Table VII. The highest sugar content is found in cactus fruits, but, compared to commercial fruits, even these values are relatively low. For example, the average total sugar content of apples and bananas is approximately 77.1% and 72% dry weight respectively (Whiting, 1970). Due to their high triterpene glycoside content, agria and organ pipe cacti are next highest in sugar content. Evidence that the sugars of these glycosides are, in fact, consumed during the rotting process has been presented by Kircher (1982). The *Opuntia* species, as a group, are lower

Table V

Quantitative analysis of methanol soluble carbohydrates

Cactus	Sugars in the methanol soluble fraction (% dry weight)								
	Rham	Rib	Arab	Xyl	Mann	Gal	Glu	Unk	Total
Saguaro	0.09	0.02	0.00	0.01	0.00	0.03	0.65	0.01	0.81
Senita	0.17	0.10	0.00	0.04	0.00	0.10	1.54	0.00	1.95
Agria	1.24	0.04	0.00	0.09	0.06	0.05	5.37	0.14	6.99
Organ pipe	1.61	0.08	0.00	0.08	0.00	0.05	2.75	0.08	4.65
Opuntia stricta	0.01	0.10	0.02	0.01	0.01	0.06	0.99	0.00	1.20
Opuntia phaeacantha	0.06	0.03	0.03	0.02	0.03	0.10	1.77	0.05	2.09
Opuntia ficus-indica	0.03	0.04	0.01	0.02	0.03	0.07	1.80	0.03	2.03
O. phaeacantha fruit	0.04	0.11	0.05	0.00	0.18	0.20	14.89	0.14	15.61
O. ficus-indica fruit	0.05	0.40	0.06	0.06	0.41	0.34	23.73	0.23	25.28

Sugars: Rham = Rhamnose; Rib = Ribose; Arab = Arabinose; Xyl = Xylose; Mann = Mannose;
Gal = Galactose; Glu = Glucose; Unk = sum of all unknown peaks.

Table VI

Quantitative analysis of water soluble carbohydrates

Cactus	Rham	Rib	Arab	Xyl	Fuc	Mann	Gal	Glu	Unk	Total
						Sugars in the water soluble fraction (% dry weight)				
Saguaro	0.73	0.12	0.25	0.25	0.12	0.00	1.26	0.59	0.00	3.32
Senita	0.07	0.47	0.11	0.28	0.00	0.00	0.29	0.36	0.28	1.86
Agria	0.80	0.64	0.51	0.54	0.00	0.00	1.10	1.61	0.00	5.20
Organ pipe	0.61	0.18	2.68	0.17	0.00	0.03	2.68	0.61	0.00	6.96
Opuntia stricta	0.16	0.22	1.68	0.74	0.00	0.00	1.17	0.21	0.00	4.18
Opuntia phaeacantha	0.16	0.45	0.26	0.06	0.00	0.00	0.58	0.69	0.00	2.20
Opuntia ficus-indica	0.26	2.38	0.42	0.20	0.00	0.00	0.95	1.62	0.00	5.83
O. phaeacantha fruit	0.28	0.06	0.06	0.02	0.00	0.03	0.21	0.12	0.00	0.78
O. ficus-indica fruit	0.24	0.14	0.29	0.06	0.00	0.05	0.54	0.13	0.00	1.45

Sugars: Rham = Rhamnose;   Rib = Ribose;   Arab = Arabinose;   Xyl = Xylose;   Fuc = Fucose;
          Mann = Mannose;   Gal = Galactose;   Glu = Glucose;   Unk = sum of all unknown peaks.

Table VII

Total sugar content of fresh cactus tissue

Cactus	Total sugar content (% dry weight)									
	Rham	Rib	Arab	Xyl	Fuc	Mann	Gal	Glu	Unk	Total
Saguaro	0.82	0.14	0.25	0.26	0.12	0.00	1.29	1.24	0.01	4.13
Senita	0.24	0.57	0.11	0.32	0.00	0.00	0.39	1.90	0.28	3.81
Agria	2.04	0.68	0.51	0.63	0.00	0.06	1.15	6.98	0.14	12.19
Organ pipe	2.22	0.26	2.68	0.25	0.00	0.03	2.73	3.36	0.08	11.61
Opuntia stricta	0.17	0.32	1.70	0.75	0.00	0.01	1.23	1.20	0.00	5.38
Opuntia phaeacantha	0.22	0.48	0.29	0.08	0.00	0.03	0.68	2.46	0.05	4.29
Opuntia ficus-indica	0.29	2.42	0.43	0.22	0.00	0.03	1.02	3.42	0.03	7.86
O. phaeacantha fruit	0.32	0.17	0.11	0.02	0.00	0.21	0.41	15.01	0.14	16.39
O. ficus-indica fruit	0.29	0.54	0.35	0.12	0.00	0.46	0.88	23.86	0.23	26.73

Sugars: Rham = Rhamnose; Rib = Ribose; Arab = Arabinose; Xyl = Xylose; Fuc = Fucose;
Mann = Mannose; Gal = Galactose; Glu = Glucose; Unk = sum of all unknown peaks.

in sugar content than agria and organ pipe, and saguaro and senita have the lowest total sugar content.

The above analysis of carbohydrates does not address the question of the availability of sugars in these substrates. Most of the sugars in commercial fruits (e.g., apples and bananas) exist as mono- and disaccharides (Whiting, 1970) while the sugars present in stems of the columnar cacti are, in general, tied up in more complex molecules such as polysaccharides and glycosides (see Starmer and Aberdeen, Chapter 8, this volume). The apparent evolutionary progression that both *Drosophila* and yeasts have made in invading the desert niche, e.g., from general fruits to *Opuntia* fruits to *Opuntia* cladodes to columnar stems, has been accompanied by a decrease in the availability of free sugars and an increase in the concentration and complexity of potentially toxic secondary plant compounds.

5. ANALYSIS OF NECROTIC CACTUS TISSUE: VOLATILE COMPOUNDS

Not only are sugars the primary energy source for microbial growth, but the result of microbial action on sugars is the production of volatile fermentation products which are putatively involved in host plant selection in the desert system. As previously mentioned, all four of the cactophilic *Drosophila* species show high host plant specificity. It is also well established that *Drosophila*, in general, are attracted to the products of microbial fermentation (Fogleman, 1982). Fermentation products include low molecular weight alcohols, esters, and volatile fatty acids. The question that will be addressed in this section is: Do the differences in sugars which are present in cactus tissue translate into cactus species specific volatile patterns?

Some data have already been published on the volatiles detected in naturally occurring agria, organ pipe, senita, saguaro, saguaro soaked soil, and *Opuntia stricta* substrates (Starmer *et al.*, 1986; Fogleman and Heed, 1989). The methods used in these gas chromatographic (GC) analyses are given in Starmer *et al.* (1986). The data set on the volatiles present in necrotic tissue of the columnar cacti has been recently updated and is provided in Table VIII.

Before discussing the differences between cactus species with respect to volatile patterns, several caveats should be mentioned. First, within any particular rot, volatile patterns change over time. The stages of rotting which are most attractive to the cactophilic *Drosophila* were not necessarily the only stages that were sampled to obtain these data. A wide variety of rot stages were sampled, yet several lines of evidence suggest that *Drosophila* are attracted to the initial or early stages of the rotting process (Starmer, 1982; Fogleman and Foster, 1989). As a result, the differences in volatiles that are important to the flies may not be reflected in these data. Second, experiments using *Drosophila melanogaster* have demonstrated that individual volatiles are

Table VIII
Concentrations of volatiles in naturally occurring *Drosophila* substrates (in mM)

Volatile	Agria (N=24)			Organ pipe (N=43)			Senita (N=29)			Saguaro (N=58)		
	Avg ± SD		Max	Avg ± SD		Max	Avg ± SD		Max	Avg ± SD		Max
Methanol	8.3 ±	6.7	33.1	2.7 ±	2.7	14.1	5.6 ±	4.7	22.9	6.7 ±	5.0	20.9
Acetone	3.8 ±	6.9	24.8	0.1 ±	0.3	1.3	0.1 ±	0.1	0.3	0.5 ±	1.4	7.3
Ethanol	4.6 ±	8.3	25.1	2.2 ±	5.6	28.3	1.4 ±	3.7	20.3	0.9 ±	1.1	4.9
2-Propanol	11.0 ±	20.9	75.4	0.6 ±	2.8	18.4	0.5 ±	0.9	3.5	0.9 ±	2.0	9.4
1-Propanol	2.8 ±	3.9	11.9	0.9 ±	2.1	12.2	1.1 ±	1.5	5.5	0.3 ±	0.8	4.1
2-Propyl Acetate	0.9 ±	1.6	5.6	0.1 ±	0.5	2.9	0.0 ±	0.1	0.4	0.3 ±	0.9	5.6
1-Propyl Acetate	0.0 ±	0.1	0.5	0.0 ±	0.1	0.5	0.0 ±	0.0	0.1	0.0 ±	0.1	0.5
Acetoin	0.7 ±	2.2	10.2	0.9 ±	1.7	7.6	0.1 ±	0.1	0.6	0.5 ±	1.2	7.4
Acetic Acid	28.4 ±	22.9	82.1	33.2 ±	36.8	175.8	34.0 ±	38.8	111.9	39.1 ±	30.2	115.4
Propionic Acid	11.8 ±	9.6	39.6	8.1 ±	8.6	44.7	5.0 ±	4.3	14.6	5.3 ±	6.9	44.5
2,3-Butanediol	4.0 ±	5.4	19.9	3.2 ±	5.8	25.7	2.2 ±	5.2	20.7	1.8 ±	3.7	23.9
iso-Butyric Acid	0.1 ±	0.2	0.6	0.0 ±	0.0	0.2	0.1 ±	0.3	1.6	0.1 ±	0.5	2.9
n-Butyric Acid	2.5 ±	2.3	8.8	0.7 ±	1.1	3.8	3.9 ±	5.3	20.6	3.0 ±	3.8	15.4

Volatile	Saguaro soil (N=21)			Cardón (N=5)			Cardón soil (N=3)		
	Avg ± SD		Max	Avg ± SD		Max	Avg ± SD		Max
Methanol	5.0 ±	5.6	24.9	0.9 ±	0.4	1.6	0.7 ±	1.2	2.1
Acetone	0.1 ±	0.2	0.7	0.2 ±	0.3	0.7	0.0 ±	0.1	0.1
Ethanol	0.5 ±	1.1	4.9	1.2 ±	0.6	1.9	0.3 ±	0.5	0.9
2-Propanol	0.2 ±	0.6	2.3	0.0 ±	0.0	0.0	0.0 ±	0.0	0.0
1-Propanol	0.1 ±	0.4	1.3	0.0 ±	0.0	0.0	0.0 ±	0.0	0.0
2-Propyl Acetate	0.1 ±	0.3	1.3	0.0 ±	0.0	0.0	0.0 ±	0.0	0.0
1-Propyl Acetate	0.0 ±	0.0	0.0	0.0 ±	0.0	0.0	0.0 ±	0.0	0.0
Acetoin	1.0 ±	1.4	5.1	0.2 ±	0.1	0.3	0.0 ±	0.0	0.0
Acetic Acid	24.7 ±	25.9	85.4	6.0 ±	3.5	15.8	5.0 ±	1.1	5.7
Propionic Acid	4.4 ±	5.8	20.3	0.8 ±	0.8	1.8	1.3 ±	1.2	2.3
2,3-Butanediol	2.6 ±	3.7	13.0	0.3 ±	0.4	0.7	0.0 ±	0.0	0.0
iso-Butyric Acid	0.1 ±	0.3	1.3	0.0 ±	0.0	0.0	0.0 ±	0.0	0.0
n-Butyric Acid	2.0 ±	3.8	16.4	0.0 ±	0.0	0.0	0.0 ±	0.0	0.0

Table IX
Results of analysis of variance tests (F_s) for within rot and geographical variation in volatile concentrations

| Volatile | Within rot variation | | Geographical variation | |
| | 2-way ANOVA for organ pipe rots | | 1-way ANOVA for saguaro rots | 1-way ANOVA for organ pipe rots |
	Between	Within	2 localities	3 localities
Methanol	1.111	0.431	4.007	1.768
Acetone	---	---	1.774	0.614
Ethanol	0.985	0.695	1.374	1.898
2-Propanol	0.660	2.472	0.391	0.258
1-Propanol	0.905	0.411	2.045	0.353
2-Propyl Acetate	---	---	0.607	0.395
Acetoin	0.941	0.823	0.075	1.271
Acetic Acid	4.633**	0.498	12.975**	2.881
Propionic Acid	5.775**	0.172	0.183	3.253*
2,3-Butanediol	3.138	1.328	0.839	2.951
iso-Butyric Acid	---	---	---	---
n-Butyric Acid	1.704	0.773	1.063	0.169

*P < 0.05, **P < 0.01.

not as important with respect to attraction as the overall pattern or combination of volatiles (Hutner *et al.*, 1937). This probably applies to the cactophilic species as well.

In addition to the above cautions, other sources of variation may affect volatile patterns. Within rot variation was examined by taking multiple samples from the same organ pipe rot for several plants within a locality. The results of two-way analysis of variance (ANOVA) tests are given in Table IX. Although the ranges were quite high, no significant variation was attributable to differences within individual necroses. Also shown in Table IX are two, one-way ANOVA's for differences between localities as compared to within localities for volatiles in saguaro and organ pipe. It does not appear that geographic locality is a significant source of variation in volatile pattern either. Some of the samples of agria and saguaro used to generate Table VIII were collected during different seasons. When winter (October to March) samples were separated from summer (April to September) samples from the same locality, several interesting differences in volatile patterns were noticed. These data are shown in Table X and the results of the statistical analyses (one-way ANOVA's) of the data are presented in Table XI. For agria, the winter samples contain significantly higher average concentrations of acetone, ethanol, 2-propanol, and 2-propyl acetate and significantly lower concentrations of propionic acid and iso-butyric acid. From the trends in agria, one might think that the higher summer temperatures are depleting summer rots of volatiles with low boiling points. However, the trends in saguaro are reversed, in that it is the summer samples of saguaro that contain higher concentration of acetone, ethanol,

Table X
Seasonal differences in volatile concentrations in two *Drosophila* substrates (in mM)

	Agria							Saguaro						
	Winter (N=14)			Summer (N=10)				Winter (N=18)			Summer (N=25)			
Volatile	Avg ±	SD	Max	Avg ±	SD	Max		Avg ±	SD	Max	Avg ±	SD	Max	
Methanol	7.5 ±	4.0	14.5	9.4 ±	9.5	33.1		6.7 ±	5.7	20.9	8.5 ±	4.9	20.7	
Acetone	6.1 ±	8.4	24.8	0.5 ±	1.0	3.2		0.0 ±	0.0	0.1	1.2 ±	1.9	7.3	
Ethanol	7.7 ±	9.8	25.1	0.3 ±	0.4	1.2		0.6 ±	1.2	4.9	1.2 ±	1.1	4.1	
2-Propanol	18.0 ±	25.4	75.4	1.1 ±	0.9	2.5		0.2 ±	0.5	1.7	1.8 ±	2.8	9.4	
1-Propanol	4.0 ±	4.8	11.9	1.1 ±	1.0	3.1		0.3 ±	0.6	2.3	0.7 ±	1.1	4.1	
2-Propyl Acetate	1.4 ±	1.9	5.6	0.1 ±	0.3	0.9		0.1 ±	0.2	0.6	0.6 ±	1.2	5.6	
1-Propyl Acetate	0.1 ±	0.1	0.5	0.0 ±	0.0	0.0		0.0 ±	0.0	0.0	0.0 ±	0.1	0.5	
Acetoin	1.0 ±	2.8	10.2	0.3 ±	0.7	2.1		0.5 ±	1.0	4.2	0.5 ±	1.5	7.4	
Acetic Acid	23.2 ±	20.1	78.2	35.6 ±	25.6	82.1		19.6 ±	28.0	115.4	44.3 ±	27.6	96.9	
Propionic Acid	7.9 ±	5.6	18.9	17.2 ±	11.4	39.6		3.8 ±	10.4	44.5	6.5 ±	5.2	20.5	
2,3-Butanediol	3.5 ±	5.3	19.9	4.7 ±	5.7	18.3		3.2 ±	6.2	23.9	0.9 ±	1.5	5.6	
iso-Butyric Acid	0.0 ±	0.1	0.3	0.2 ±	0.2	0.6		0.1 ±	0.2	0.7	0.3 ±	0.7	2.9	
n-Butyric Acid	2.1 ±	2.3	6.9	3.0 ±	2.3	8.8		1.4 ±	3.7	15.4	4.4 ±	4.2	14.1	

both propanols, and 2-propyl acetate (although only acetone and 2-propanol were statistically significant). Obviously, other parameters besides temperature must be affecting the volatile patterns. One possibility is that the microorganism community composition may vary both seasonally and between cactus species.

Because of the seasonal differences in volatile patterns, the statistical analysis of the differences between cactus species (Table XI) was based on winter samples only. The one-way ANOVA's on each volatile compound also excluded those cactus species which did not contain the volatile being tested. Statistically significant differences were found between cactus species for methanol, acetone, ethanol, 2-propanol, 1-propanol, 2-propyl acetate, and n-butyric acid. The conclusion is that the volatile patterns present in the different substrates are demonstrably different. It appears from examining the data in Tables VIII and X that necrotic agria tissue is the most distinctive with respect to volatiles. This substrate typically contains much higher concentrations of acetone, ethanol, both propanols, and 2-propyl acetate than the other substrate types. The few cardón samples that have been analyzed did not show high concentrations of any volatile compound, were not associated with large fly populations, and may not be typical for substrates of this cactus species.

The data in Tables VII, VIII, X, and Table 2 in Starmer *et al.* (1986) provide the opportunity to examine correlations between the total sugar content in fresh tissue of the cacti and the volatile concentrations in necrotic tissue. This

Table XI
Results of analysis of variance tests (F_s) for seasonal variation and differences between all cactus species with respect to volatile concentrations

	Seasonal variation		
	1-way ANOVA for saguaro rots	1-way ANOVA for agria rots	Between all cactus species
Volatile	Winter - Summer	Winter - Summer	species
Methanol	1.232	0.454	4.522***
Acetone	7.118*	4.345*	12.767***
Ethanol	2.886	5.622*	3.929**
2-Propanol	5.709*	4.367*	13.164***
1-Propanol	1.953	3.498	7.969***
2-Propyl Acetate	3.044	4.543*	10.345***
Acetoin	0.000	0.591	0.444
Acetic Acid	8.281**	1.770	1.282
Propionic Acid	1.257	7.037*	1.945
2,3-Butanediol	3.208	0.281	0.424
iso-Butyric Acid	1.380	10.476**	0.000
n-Butyric Acid	5.886	0.893	3.849**

*$P < 0.05$, **$P < 0.01$, ***$P < 0.001$.

Table XII

Sugar – volatile correlation matrix

Volatile	Rhamnose	Ribose	Arabinose	Xylose	Galactose	Glucose	Total sugar
			Correlation coefficients[a]				
Methanol	0.196	0.489	-0.809	-0.051	-0.485	0.540	0.086
Acetone	0.487	0.656	-0.295	0.510	-0.158	0.888*	0.594
Ethanol	0.470	0.635	0.148	0.746	0.056	0.821	0.718
2-Propanol	0.509	0.692	-0.260	0.511	-0.142	0.912*	0.629
1-Propanol	0.601	0.807	-0.301	0.202	-0.137	0.955*	0.662
2-Propyl Acetate	0.593*	0.495	-0.336	0.272	-0.055	0.866	0.595
Acetoin	0.958*	-0.129	0.381	-0.349	0.760	0.597	0.808
Acetic Acid	0.284	-0.163	-0.362	-0.901*	0.069	0.021	-0.048
Propionic Acid	0.873*	0.512	-0.038	-0.061	0.258	0.944*	0.838*
2,3-Butanediol	0.883*	0.568	0.097*	-0.028	0.312	0.947*	0.894*
n-Butyric Acid	-0.670	0.363	-0.931*	-0.008	-0.957	-0.223	-0.701

[a] $r_{c(df=3; \ alpha=0.05)} = 0.878$

analysis yields information on which sugars are primarily responsible for the production of specific volatiles. Using data from agria, organ pipe, saguaro, senita, and *Opuntia stricta*, a correlation matrix was generated and is presented in Table XII. The concentrations of both acetoin (3-hydroxy-2-butanone) and 2,3-butanediol were significantly correlated with rhamnose. Since these two volatiles are chemically related (acetoin is an oxidized form of the diol), this result was not unexpected. Propionic acid was marginally correlated with rhamnose (P = 0.052). Ribose was not significantly correlated with any of the volatiles. The group, arabinose, xylose, and galactose, does not appear to have much to do with volatile production since the majority of the correlation coefficients are negative. Glucose, on the other hand, does appear to be heavily involved in volatile production since significant correlations were found between the amount of glucose present in the tissue and the concentration of acetone, 2-propanol, 1-propanol, propionic acid, and 2,3-butanediol. Several additional volatiles show marginally significant coefficients (ethanol, P = 0.087 and 2-propyl acetate, P = 0.056). In general, higher concentrations of glucose produce greater concentrations of most volatiles, and glucose and rhamnose are the most important sugars with respect to overall volatile production.

6. HOST PLANT SELECTION

The remaining question is: Do the flies respond to the volatile patterns of necrotic cactus tissue in a manner which explains their host specificity? This question has only been addressed in experiments using *D. mojavensis*. The field experiments described by Fellows and Heed (1972) certainly suggest that this species exhibits a preference hierarchy for necrotic tissue of the columnar cacti. The observed preference hierarchy was: agria > organ pipe > saguaro > senita. Extensive field observations indicate that *D. mojavensis* has a strong preference for agria rots even when suitable organ pipe rots are available. In a series of substrate choice experiments performed in the laboratory, Downing (1985) confirmed the preference hierarchy of *D. mojavensis* (see Fogleman and Heed, 1989 for methods). Control experiments demonstrated that *D. mojavensis* does not distinguish between agria and organ pipe when fresh rather than rotted tissue is used.

Table XIII shows some of the results of Downing's experiments. If given a choice between necrotic tissue of agria and organ pipe, *D. mojavensis* strongly responds to agria. The average response to agria over four replicates was 78% and is statistically significant (compared to a null hypothesis of equal response to both substrates). Downing then characterized the volatile patterns for each of his substrates by GC analysis and synthesized two volatile solutions. One solution had the volatile pattern which mimicked the agria rot and the pattern of the other mimicked organ pipe. Table XIII shows the results of behavioral response tests using the volatile solutions instead of the actual substrates. *Drosophila*

Table XIII
Behavioral response of *D. mojavensis* to cactus substrates and volatile
solutions

Replicate	N	Number trapped by necrotic tissue		N	Number trapped by volatile solution	
		Agria No. (%)	Organ pipe No. (%)		Agria No. (%)	Organ pipe No. (%)
1	420	291 (79)	79 (21)	403	290 (90)	34 (10)
2	402	275 (76)	87 (24)	419	229 (94)	15 (6)
3	370	227 (81)	55 (19)	416	295 (78)	85 (22)
4	406	292 (76)	94 (24)	417	340 (88)	46 (12)
Average		271 (78)	79 (22)		289 (87)	45 (13)

mojavensis exhibited the same response when presented with the
volatile solutions as when given a choice between the rots
themselves. The average response to the volatile pattern of
agria was 87% and is also statistically significant. These
results provide strong support for the contention that
volatiles are the chemical basis of host plant selection for
D. mojavensis. Obviously, similar experiments with the other
three cactophilic species must be performed before a general
statement can be made regarding host selection in desert
flies.

Several additional experiments have been performed in order
to elucidate the contribution of single volatiles to the
behavioral response of *D. mojavensis* (Fogleman and Davis, unpublished). These experiments were designed to test for
attraction response to an increased concentration of a single
volatile compound when all of the volatiles which might be
important to attraction were present. Control traps contained
12 volatiles mixed into sterilized (autoclaved) saguaro
homogenate at the concentrations given in Table XIV. Experimental traps contained the same 12 volatiles at control concentrations with the exception that one volatile was present
at five times the control level. Saguaro homogenate was used
as an innocuous matrix because it contains none of the
volatiles in question and is more realistic than water due to
its viscosity. The results of two replicate tests for each
volatile are presented in Table XIV. Ethanol, 2-propyl
acetate, propionic acid, and 2,3-butanediol were shown to be
significantly more attractive at the higher concentration,
while no significant differences were detected between control
and experimental traps for methanol, acetone, and acetic acid.
Significantly more flies responded to the control traps when
experimental traps contain elevated concentrations of 2-
propanol, 1-propanol, 1-propyl acetate, acetoin, and n-butyric
acid. It is unknown, however, whether the last five compounds
are just less attractive at elevated concentrations or actually repulsive. The fact that *Drosophila* adults respond to a
range of volatile concentrations above which the volatile is
repellent has been reported in the literature (Reed, 1938).

Table XIV

Behavioral response of *D. mojavensis* to single volatiles. Experimental traps contained the control level of volatiles plus 5x the control concentration of the single volatile being tested

Volatiles	Vol. conc. in control traps (mM)	Total flies caught	Average percent		χ^2	p	Response to elevated concentration
			Control traps	Exp. traps			
Methanol	5.0	411	49	51	0.29	n.s.	No change
Acetone	3.0	1062	48	52	1.09	n.s.	No change
Ethanol	5.0	299	28	72	57.40	<0.001	More attractive
2-Propanol	3.0	513	60	40	19.89	<0.001	Less attractive
1-Propanol	3.0	529	63	37	36.52	<0.001	Less attractive
2-Propyl Acetate	1.0	538	37	63	38.54	<0.001	More attractive
1-Propyl Acetate	1.0	530	61	39	27.17	<0.001	Less attractive
Acetoin	0.5	497	69	31	70.17	<0.001	Less attractive
Acetic Acid	10.0	535	51	49	0.23	n.s.	No change
Propionic Acid	5.0	553	32	68	71.61	<0.001	More attractive
2,3-Butanediol	2.0	382	42	58	10.06	<0.001	More attractive
n-Butyric Acid	2.0	1150	53	47	5.57	<0.05	Less attractive

n.s. = not significant.

The general conclusion which can be derived from the results in Table XIV is that not only is the pattern or combination of volatiles important in attraction, but the concentration of each volatile is important as well.

7. SUMMARY

The experiments reported herein summarize the chemical interactions that are important in determining the relationships between desert-adapted *Drosophila* and columnar cacti. The majority of the cacti used by flies in the Sonoran Desert contain toxic compounds that effectively exclude all but the resident species. Since, in most cases, a substrate mistake is fatal, it is not surprising that the flies exhibit high host plant specificity. The behavioral discrimination of the *Drosophila* for particular species of cacti is thought to be due to another set of chemical interactions between microorganisms and plant chemistry. Microbial fermentation of carbohydrates produces a pattern of volatile compounds to which the *Drosophila* respond in a species-specific manner. Host plant chemistry, therefore, is intrinsically involved in both host utilization and host selection.

In invading the desert, the evolutionary pattern of host usage suggests that both the *Drosophila* and the cactophilic yeasts have had to contend with an apparent decrease in the concentration of free sugars and an increase in the concentration of toxic plant compounds. Thus, host plant chemistry is also important from an evolutionary perspective. Knowledge of the chemical interactions between cacti, microorganisms, and *Drosophila* has significantly contributed to the understanding of this interesting and useful model system.

ACKNOWLEDGEMENTS. The authors would like to thank Robert Downing and Daniel Davis for the use of their unpublished data and Joan Foster for critically reviewing the manuscript. This work was supported by NIH grant GM34820 to JCF. Much of the research described in this chapter is based on the pioneering work of the late Prof. Henry Kircher. The many contributions that he made in the field of chemical ecology of the cactus-microorganism-*Drosophila* model system and the training that he gave to both authors are gratefully acknowledged.

References

Ahmed, el-.R. A., and Labavitch, J. M., 1977, A simplified method for accurate determination of cell wall uronide content, *J. Food Biochem.* 1:361-365.

Binder, R. G., Chan, B. G., and Elliger, C. A., 1979, Antibiotic effects of C10-C12 fatty acid esters on pink bollworm, bollworm, and tobacco budworm, *Agric. Biol. Chem.* 43:2467-2471.

Brazner, J., Aberdeen, V., and Starmer, W. T., 1984, Host-plant shifts and adult survival in the cactus breeding *Drosophila mojavensis*, *Ecol. Entomol.* 9:375-381.

Campbell, C. E., and Kircher, H. W., 1980, Senita cactus: a plant with interrupted sterol biosynthetic pathways, *Phytochemistry* **19**:2777-2779.

Downing, R. J., 1985, The chemical basis for host plant selection in *Drosophila mojavensis*, MS Thesis, Univ. of Denver, Denver, CO.

Fellows, D. P., and Heed, W. B., 1972, Factors affecting host plant selection in desert-adapted *Drosophila*, *Ecology* **53**:850-858.

Fogleman, J. C., 1982, The role of volatiles in the ecology of cactophilic Drosophila, in: *Ecological Genetics and Evolution: The Cactus-Yeast-Drosophila Model System* (J. S. F. Barker, and W. T. Starmer, eds), Academic Press Australia, Sydney, pp. 191-206.

Fogleman, J. C., 1984, The ability of cactophilic *Drosophila* to utilize soaked soil as larval substrates, *Drosoph. Inf. Serv.* **60**:105-107.

Fogleman, J. C., and Armstrong, L., 1989, Ecological aspects of cactus triterpene glycosides. I. Their effect on fitness components of *Drosophila mojavensis*, *J. Chem. Ecol.* **15**:663-676.

Fogleman, J. C., and Foster, J. L. M., 1989, Microbial colonization of injured cactus tissue and its relationship to the ecology of desert-adapted *Drosophila*, *Appl. & Environ. Microbiol.* **55**:100-105.

Fogleman, J. C., and Heed, W. B., 1981, A comparison of the yeast flora in the larval substrates of *D. nigrospiracula* and *D. mettleri*, *Drosoph. Inf. Serv.* **56**:38-39.

Fogleman, J. C., and Heed, W. B., 1989, Columnar cacti and desert *Drosophila*: the chemistry of host plant specificity, in: *Special Biotic Relationships in the Arid Southwest* (J. Schmidt, ed.), Univ. of New Mexico Press, Albuquerque, pp. 1-24.

Fogleman, J. C., and Kircher, H. W., 1986, Differential effects of fatty acid chain length on the viability of two species of cactophilic *Drosophila*, *Comp. Biochem. Physiol.* **83A**:761-764.

Fogleman, J. C., Duperret, S. M., and Kircher, H. W., 1986, The role of phytosterols in host plant utilization by cactophilic *Drosophila*, *Lipids* **21**:92-96.

Fogleman, J. C., Hackbarth, K. R., and Heed, W. B., 1981, Behavioral differentiation between two species of cactophilic *Drosophila*. III. Oviposition site preference, *Am. Nat.* **118**:541-548.

Fogleman, J. C., Heed, W. B., and Kircher, H. W., 1982, *Drosophila mettleri* and senita cactus alkaloids: fitness measurements and their ecological significance, *Comp. Biochem. Physiol.* **71A**:413-417.

Gibson, A. C., and Horak, K. E., 1978, Systematic anatomy and phylogeny of mexican columnar cacti, *Ann. Missouri Bot. Gard.* **65**:999-1057.

Gibson, A. C., and Nobel, P. S., 1986, *The Cactus Primer*, Harvard Univ. Press, Cambridge.

Goodnight, K. C., and Kircher, H. W., 1971, Metabolism of lathosterol by *Drosophila pachea*, *Lipids* **6**:166-169.

Harley, K. L. S., and Thorsteinson, A. J., 1967, The influence of plant chemicals on the feeding behavior, development and survival of the two-striped grasshopper, *Melanoplus bivattatus* (Say), Acrididae: Orthoptera, *Can. J. Zool.* **45**:305.

Heed, W. B., 1978, Ecology and genetics of Sonoran Desert *Drosophila*, in: *Ecological Genetics: The Interface* (P. F. Brussard, ed.), Springer-Verlag, New York, pp.109-126.

Heed, W. B., 1982, The origin of *Drosophila* in the Sonoran Desert, in: *Ecological Genetics and Evolution: The Cactus-Yeast-Drosophila Model System* (J. S. F. Barker, and W. T. Starmer, eds), Academic Press Australia, Sydney, pp. 65-80.

Heed, W. B., and Kircher, H. W., 1965, Unique sterol in the ecology and nutrition of *Drosophila pachea*, *Science* **149**:758-761.

Heed, W. B., and Mangan, R. L., 1986, Community ecology of the Sonoran Desert *Drosophila*, in: *The Genetics and Biology of Drosophila*, Vol. 3e (M. Ashburner, H. L. Carson, and J. N. Thompson Jr., eds), Academic

Press, London, pp. 311-345.

Hutner, S. H., Kaplan, H. M., and Enzmann, E. V., 1937, Chemicals attracting *Drosophila*, *Am. Nat.* **71**:575-581.

Kircher, H. W., 1977, Triterpene glycosides and queretaroic acid in organ pipe cactus, *Phytochemistry*, **16**:1078-1080.

Kircher, H. W., 1982, Chemical composition of cacti and its relationship to Sonoran Desert *Drosophila*, in: *Ecological Genetics and Evolution: The Cactus-Yeast-Drosophila Model System* (J. S. F. Barker, and W. T. Starmer, eds), Academic Press Australia, Sydney, pp. 143-158.

Kircher, H. W., Heed, W. B., Russell, J. S., and Grove, J., 1967, Senita cactus alkaloids: their significance to Sonoran Desert ecology, *J. Insect Physiol.* **13**:1869-1874.

Mangan, R. L., 1982, Adaptations to competition in cactus breeding *Drosophila*, in: *Ecological Genetics and Evolution: The Cactus-Yeast-Drosophila Model System* (J. S. F. Barker, and W. T. Starmer, eds), Academic Press Australia, Sydney, pp. 257-272.

Meyer, J. M., and Fogleman, J. C., 1987, Significance of saguaro cactus alkaloids in the ecology of *Drosophila mettleri*, a soil-breeding, cactophilic drosophilid, *J. Chem. Ecol.* **13**:2069-2081.

Phaff, H. J., Starmer, W. T., Tredick, J., and Miranda, M., 1985, *Pichia deserticola* and *Candida deserticola*, two new species of yeasts associated with necrotic stems of cacti, *Int. J. Syst. Bacteriol.* **35**:211-216.

Reed, M. R., 1938, The olfactory reactions of *Drosophila melanogaster* Meigen to the products of fermenting bananas, *Physiol Zool.* **11**:317-325.

Roff, D. C., 1977, Dispersal in dipterans: Its costs and consequences, *J. Anim. Ecol.* **46**:443-456.

Ruiz, A., and Heed, W. B., 1988, Host-plant specificity in the cactophilic *Drosophila mulleri* species complex, *J. Anim. Ecol.* **57**:237-249.

Schreiber, K., 1958, Über einige Inhaltsstoffe der Solanaceen und ihre Bedeutung fur die Kartoffelkaferresistenz, *Entomol. exp. appl.* **1**:28-37.

Shepard, H. H., 1951, *The Chemistry and Action of Insecticides*, McGraw-Hill, New York.

Starmer, W. T., 1982, Analysis of the community structure of yeasts associated with the decaying stems of cactus. I. *Stenocereus gummosus*, *Microb. Ecol.* **8**:71-81.

Starmer, W. T., Barker, J. S. F., Phaff, H. J., and Fogleman, J. C., 1986, Adaptations of *Drosophila* and yeasts:their interactions with the volatile 2-propanol in the cactus-microorganism-*Drosophila* model system, *Aust. J. biol. Sci.* **39**:69-77.

Starmer, W. T., Kircher, H. W., and Phaff, H. J., 1980, Evolution and speciation of host plant specific yeasts, *Evolution* **34**:137-146.

Whiting, G. C., 1970, Sugars, in: *The Biochemistry of Fruits and their Products*, Vol 1 (A. C. Hulme, ed.), Academic Press, New York, pp. 1-31.

CHAPTER 8

The Nutritional Importance of Pure and Mixed Cultures of Yeasts in the Development of Drosophila mulleri Larvae in Opuntia Tissues and its Relationship to Host Plant Shifts

WILLIAM T. STARMER, AND VIRGINIA ABERDEEN

1. INTRODUCTION

The prevalence of yeasts in the diets of Drosophila species has stimulated studies on the role of yeasts in almost all aspects of the biology of Drosophila. The early work of Delcourt and Guyenot (1910) and Baumberger (1919) on the nutritional importance of yeasts for Drosophila was followed in the middle of the 20th century by two major research efforts. One was primarily concerned with the nutritional value of yeasts to Drosophila in the laboratory (reviewed by Sang, 1978) while the other was directed toward understanding the evolutionary biology of natural populations of Drosophila species by studying their natural food, yeasts (reviewed by Begon, 1982). Both lines of investigation suggested that the two organisms depend on one another for survival. The dependencies are basically mutualistic. The yeasts rely on insects such as Drosophila species for dispersal to new habitats while the Drosophila rely on the yeasts for nutrition during larval development and egg maturation.

Investigations into the yeasts associated with desert adapted drosophilids was started in the early 1970s by W. B. Heed and H. J. Phaff. Most of the early work was done in the Sonoran Desert where Drosophila species breed and feed in the decaying tissue of cactus species (Heed 1977, 1982; Heed and Mangan, 1986; Fogleman and Heed, 1989). Numerous investigations on the taxonomy (reviewed by Lachance et al., 1988), ecology (reviewed in Starmer and Fogleman, 1986) and biogeography (Starmer et al., 1989) of yeasts that live in the decaying tissues of cactus have revealed an interesting constellation of interactions between cactus yeasts and cactus inhabiting drosophilids. Other than direct nutritional benefits of yeasts to Drosophila these studies have shown that:

WILLIAM T. STARMER, and VIRGINIA ABERDEEN * Biology Department, Syracuse University, Syracuse, New York 13210, USA.

Ecological and Evolutionary Genetics of Drosophila, Edited by
J.S.F. Barker et al., Plenum Press, New York, 1990

1. *Drosophila* adults serve as an important vector of yeasts between host plants (Ganter, 1988; Starmer *et al.*, 1988a). Although other flying insects such as syrphids, nereids, muscoids, dolichopodids and wasps also feed and some breed on the decaying cactus tissue, along with the drosophilids, they do not carry yeasts as often nor do they carry yeasts at the same level of abundance and diversity as do the drosophilids. This is apparently true for *Drosophila* species that live in temperate woodlands (Gilbert, 1980).

2. Some cactus living yeasts are capable of detoxifying the larval environment for the benefit of the fly. In one case the host tissue (*Stenocereus thurberi*, organ pipe cactus) contains potentially toxic medium chain length fatty acids bound as esters. When the ester bonds are broken, the free fatty acids can be lethal to the larvae of *Drosophila* species (Fogleman *et al.*, 1986) and can slow the larval development of *Drosophila mojavensis*, which lives in decaying organ pipe stem tissue. A common yeast in this system, *Candida ingens*, is capable of metabolizing the fatty acids and thus alleviates the detrimental effects of the toxic compounds (Starmer, 1982). Another case of environmental modification by yeasts that results in benefits to the fly involves the degradation of toxic alcohols. Decaying stems of *Stenocereus* species often have 2-propanol and acetone as major compounds of their volatile profiles (Starmer *et al.*, 1986). These alcohols at moderate to high concentrations are harmful to both larval and adult stages of *Drosophila*, however, several yeasts, including the common cactus yeasts *Candida sonorensis* and *Cryptococcus cereanus* can utilize 2-propanol and acetone as a source of carbon. In so doing, the potential toxic effect of the volatiles is at least partly removed (Starmer *et al.*, 1986).

3. Larvae of cactus breeding *Drosophila* selectively feed on the yeasts found in their natural habitat (Fogleman *et al.*, 1981, 1982). Vacek (1982) found that larvae of *D. buzzatii* breeding in *Opuntia stricta* tissue in eastern Australia tend to prefer four to five common yeasts equally well, while the work of Fogleman and colleagues indicates that Sonoran Desert *Drosophila* species have a more distinct preference that may be dependent on the host plant tissue. In general it appears that larval preferences are mediated by the volatiles produced by the yeasts and that common cactus specific yeasts such as *Pichia cactophila* are capable of producing relatively high amounts of attractive volatiles (Fogleman, 1982).

4. Adult cactophilic drosophilids use yeasts and their products as cues to the environment in several ways. There is ample evidence that feeding and oviposition site preferences are influenced by yeast species. *Pichia cactophila*, one of the most common cactus associated yeasts, is often the yeast preferred in oviposition choice tests (Vacek, 1982, Barker *et al.*, 1986)

Adult females of *D. buzzatii* also appear to make oviposition choices based on their genotypes (Barker *et al.*, 1986). In addition, studies with *D. buzzatii* have shown that yeasts are transferred between the sexes during courtship and mating (Starmer *et al.*, 1988b). This phenomenon, coupled with the observation that previous yeast diet can influence the choices of mates, implicates yeasts as potentially important components of behavioral modification in *Drosophila*.

In summary, it appears that yeasts are involved in the direct nutrition of larvae and adults, the modification of the environment (especially toxins) in which the larvae feed, the chemical cues used by larval and adult stages for finding food, the chemical cues used in finding suitable sites for depositing eggs and influencing the behavior of adults during courtship and mating. Yeasts are thus significant microorganisms in almost all aspects of the the biology of cactophilic drosophilids.

Studies on the nutritional sufficiency of cactus inhabiting microorganisms for cactophilic *Drosophila* have shown that yeasts are in general sufficient for larval growth and that some bacteria are also sufficient (Wagner, 1944, 1949; Vacek, 1982; Starmer, 1982). Additional studies with mixed cultures of yeasts have revealed that even though monocultures of yeasts on cactus tissues are nutritionally sufficient, bicultures of different yeast species can provide additional benefits to the larvae in terms of faster development, larger adult size or increased viability (Starmer and Fogleman, 1986; Starmer and Barker, 1986). The biculture effect was detected during studies of *D. mojavensis* and its columnar cactus host *Stenocereus gummosus* (agria) in conjunction with agria yeasts and studies using *D. buzzatii* and its prickly pear cactus host *Opuntia stricta* in conjunction with *O. stricta* yeasts. In the former case, the larvae of *D. mojavensis* experience increased viability when given bicultures of yeasts, while in the latter case *D. buzzatii* grew faster and attained larger sizes when fed bicultures of yeasts.

In this paper we will present evidence from an extended study of the biculture effect. Many of the questions that are addressed in this study are based on the notion that the most likely evolutionary sequence of host use for cactophilic *Drosophila* is a progression from using fruit to using cladodes to using columnar stems. Five questions that are pertinent to understanding the role of mixed cultures of yeasts in the nutrition of cactus breeding drosophilids are:

1. The biculture effect appears to occur in the stems of columnar cacti (agria) and the cladodes of *Opuntia,* but does it occur in the decaying tissue of cactus fruit?

2. Do yeasts that produce the biculture effect in one tissue also show the effect when trans- or cross-inoculated onto another tissue (e.g., cactus cladode yeasts inoculated onto fruit tissue)?

3. Is the biculture effect dependent on the strains or species of yeasts?

4. What is the most probable proximal mechanism responsible for the effect when it does occur? and

5. What does the pattern of effects within the different systems (cactus fruit, *Opuntia* cladode and columnar stem) reveal about the evolution of the *Drosophila*-yeast interactions?

2. EXPERIMENTAL ANALYSIS

In order to answer these questions we chose *Drosophila mulleri* as an appropriate species because this species uses both fruit and cladode tissue of *Opuntia* species throughout its distribution. This species thus represents an intermediate in the transition between sole use of decaying fruit and the restricted use of necrotic cladodes in nature. A strain of *D. mulleri* (ORV-25) collected in Jamaica was used in all of the experiments. Experiments were conducted with axenic larvae obtained by dechorionation of eggs with Chlorox (Starmer and Gilbert, 1982). The cladode tissue used was from *Opuntia stricta* collected from Caribbean islands, while the fruit tissue used was from *Opuntia ficus-indica*. In all experiments the experimental unit was a 30 ml shell vial with 10 g of sand and 5 g of tissue (cladode or fruit) seeded with a given number of larvae (20 or 25) and inoculated with a particular yeast or nutrient treatment. Vials were incubated in an environmentally controlled room at 26°C. The vials with sand and tissue were plugged with cotton, covered with a plastic cap, then sterilized by autoclaving (15 min, 15 psi) before being subjected to treatments. The number of adults and

Table I

Yeast strains used in the experimental test along with their sources and localities of isolation

Yeast species	Strain number	Source	Locality
Pichia cactophila	88-303.1	*Cephalocereus*	Big Pine Key, Florida
P. cactophila	86-2.1	*Opuntia*	Kamuela, Hawaii
P. cactophila	87-474.1	*Nopalea*	Zanatepec, Oaxaca
Cr. cereanus	86-304.1	*Cephalocereus*	Big Pine Key, Florida
Cr. cereanus	86-8.4	*Opuntia*	Kamuela, Hawaii
Cr. cereanus	87-475.1	*Nopalea*	Zanatepec, Oaxaca
Pichia barkeri	86-929.2	*Opuntia* fruit	Canaveral National Seashore, Florida
Kloeckera apiculata	86-926.1	*Opuntia* fruit	Canaveral National Seashore, Florida
Issatchenkia terricola	86-928.1	*Opuntia* fruit	Canaveral National Seashore, Florida

Table II

The mono- versus biculture effect of different strains of *P. cactophila* and *Cr. cereanus* on the development and viability of *D. mulleri* larvae growing in *O. stricta* cladode tissue

Source	DF	Development (Days)		Viability	
		SS	F	SS	F
Corrected Total	29	14.028		0.7368	
Mono vs Biculture	1	3.326	8.7**	0.0211	0.83
Error	28	10.702		0.7156	

** P<0.01.

their day of emergence were recorded from which mean days until emergence was calculated on a per vial basis. These means were used along with the vial viability (number surviving/number seeded) in analyses of variance for each experimental group.

2.1. Strain Effects Experiment

Earlier experiments that detected the biculture effect were done with single strains as representative of different yeast species. In order to evaluate the potential variability of strains from within a species, an experiment was conducted with three different strains of *Pichia cactophila* and *Cryptococcus cereanus*. These strains were chosen from widely separated geographic regions and from different host plants (Table I, *Opuntia* in Hawaii, *Cepaholcereus* in Florida, and *Nopalea* in southern Mexico). They were inoculated onto *O. stricta* tissue as six monocultures (three of each species) and all nine possible between species bicultures. These 15 treatment combinations were replicated twice. Each of the 30 vials were seeded with 20 axenic larvae of *D. mulleri*.

Table II contains the analysis of variance for both development time and viability. This analysis indicates that there was a difference in development time that is due to the type of culture that was inoculated (mono or bi). Figure 1 shows the means for the development on each monoculture versus development on the corresponding biculture. It is clear from this figure that larvae developed faster in bicultures, regardless of the the source of the strains for each species (i.e., 17 of the 18 possible comparisons develop faster on bicultures ; they are above the line of equivalent development time for mono and bicultures).

2.2. Cross-Inoculation Experiments

Four related experiments were conducted using axenic first instar larvae of *D. mulleri*. Each experiment was designed to evaluate the effects of yeasts on the development of larvae.

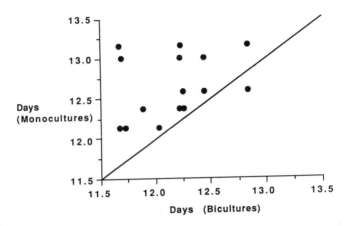

Figure 1. Development time of *D. mulleri* larvae on *O. stricta* cladode tissue that was inoculated with bicultures and monocultures of cactus-specific yeasts. Treatment means (average of vial means) of each biculture treatment are plotted against the two corresponding monoculture treatment means. Three strains of *P. cactophila* and three strains of *Cr. cereanus* were used and only between species biculture combinations were used in the experiment. The diagonal line represents the line of equivalent development time.

Each experiment consisted of a host tissue type (*O. stricta* cladode or fruit) and a set of three host specific yeast species (*P. cactophila*, *C. sonorensis* and *Cr. cereanus* were used as typical representatives of *Opuntia* cladode yeasts; while *P. barkeri*, *K. apiculata* and *I. terricola* were used as typical *Opuntia* fruit yeasts). The four combinations of host tissue type and host yeast type (fruit tissue with fruit yeasts, fruit tissue with cladode yeasts, cladode tissue with fruit yeasts and cladode tissue with cladode yeasts) formed the four experimental groups. The treatment combinations for each experiment consisted of inoculations of mono and all possible bicultures of the three yeasts (i.e., six yeast treatments). Each treatment was seeded with 20 larvae and replicated twice except for the fruit tissue with fruit yeast, for which each treatment was replicated four times.

An analysis of variance for each host x yeast combination is given for development time and viability in Table III. Yeast effects on days of development were significant in the fruit yeast with fruit tissue combination and in the cladode yeast with cladode tissue combination. The means of development time for each monoculture-biculture combination of each host x yeast experimental group are plotted in Figures 2 and 3. These figures shows that the biculture effect (i.e., larvae develop faster on bicultures) was only present in the combination of cladode host with cladode yeasts. The cross-inoculations did not produce an effect, while bicultures of fruit yeasts on fruit tissues appeared to retard development of the larvae but increased the viability of the larvae (data not shown).

Table III

Analysis of variance for days until emergence and viability of *D. mulleri* larvae growing on pieces of *Opuntia* fruit or cladode that were inoculated with mono- or bicultures of fruit and cladode yeasts. One degree of freedom is partitioned out of the Yeast effects to evaluate the effect of mono- versus bicultures of yeast on the development of the larvae

Host Type Yeast Type		Fruit Fruit			Fruit Cladode		Cladode Fruit		Cladode Cladode	
Source	DF	SS	F	DF	SS	F	SS	F	SS	F
Days										
Corrected Total	23	3.63		11	1.85		72.2		25.1	
Yeast	5	2.99	16.7***	5	1.23	2.36	28.5	0.78	24.8	117.1***
Mono-Biculture	1	0.43	3.0	1	0.01	0.06	10.7	1.73	10.8	7.5*
Error	18	0.64		6	0.62		43.7		0.25	
Viability										
Corrected Total	23	0.27		11	0.27		0.32		0.17	
Yeast	5	0.15	4.69**	5	0.14	1.35	0.18	1.56	0.05	0.52
Mono-Biculture	1	0.05	4.83**	1	0.02	0.92	0.01	0.27	0.02	1.03
Error	18	0.12		6	0.13		0.14		0.12	

* $P < 0.05$, ** $P < 0.01$, *** $P < 0.001$.

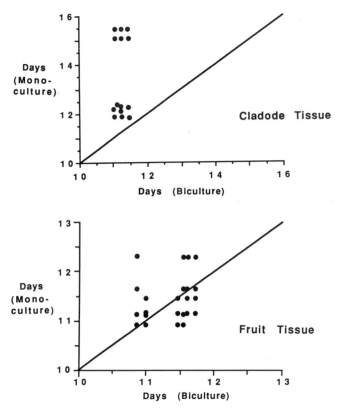

Figure 2. Development time of *D. mulleri* larvae on cladode-specific yeasts inoculated onto either *Opuntia* cladode or fruit tissue. Vial means for each biculture treatment are plotted against the vial means of each of the corresponding monoculture treatments. The diagonal line represents the line of equivalent development time.

2.3. Nutritional Supplement Experiment

In order to evaluate the phenomenon of faster larval development of *Drosophila mulleri* when given bicultures of yeasts growing on *Opuntia stricta* tissue, a two way factorial experiment was conducted. One factor was yeast treatment consisting of an inoculation of *P. cactophila* and *Cr. cereanus* as either monocultures or together as a biculture. The second factor was nutrient treatments of: no additional nutrients, a supplement of yeast extract (0.5%), or the addition of glucose (5.0%) to 5 g of homogenized *Opuntia stricta* cladode tissue. Twenty axenic first instar larvae of *D. mulleri* were seeded onto the appropriate treatment combination (3 yeast treatments by 3 nutrient treatments). Four replicates of each treatment combination were prepared.

Table IV presents a two-way analysis of variance for this experiment and shows that there was a yeast-by-nutrient interaction with regard to the developmental time of the larvae.

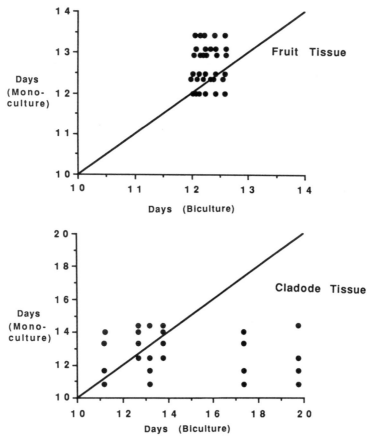

Figure 3. Development time of *D. mulleri* larvae on fruit-specific yeasts inoculated onto either *Opuntia* cladode or fruit tissue. Vial means for each biculture treatment are plotted against the vial means of each of the corresponding monoculture treatments. The diagonal line represents the line of equivalent development time.

Mean development time for each yeast-by-nutrient combination is also listed in Table IV. These means show that:

1. the biculture effect was not present in this experiment (compare both yeasts to each yeast when no nutrients were added),

2. the addition of yeast extract appeared to provide the fastest larval development time overall, and

3. *P. cactophila* reared larvae responded best to the addition of nutrients.

Table IV

Analysis of the effects and interaction between yeasts (*P. cactophila* and *Cr. cereanus*) and nutrients (none added, yeast extract added or glucose added) on the length of development (days) and viability (number surviving/20) of *D. mulleri* growing in 5 g of homogenized *Opuntia stricta* cladode tissue

ANOVA

Source	DF	Days SS	Days F	Viability SS	Viability F
Corrected Total	35	17.40		0.32	
Yeast (Y)	2	3.05	4.74*	0.05	2.70
Nutrients (N)	2	1.41	2.20	0.01	0.63
Y x N	4	4.24	3.29*	0.007	0.19
Error	27	8.70		0.254	

* $P < 0.05$

Interaction (means in days)

		Yeast *P. cactophila*	Yeast *Cr. cereanus*	Yeast Both
	None	12.39	10.54	11.22
Nutrient:	Yeast Extract	10.95	10.92	10.92
	Glucose	11.35	11.09	11.48

3. DISCUSSION

Many fungal-insect interactions that are mutualistic are ultimately based on relatively nutrient-poor substrates that often have low levels of nitrogen and vitamins. This appears to be the case for wood-inhabiting fungus-beetle associations, interactions between scale insects and their mutualistic fungi, associations of wood wasps and their fungal symbionts, as well as the interactions between ants and the saprotrophic fungi that they culture on cut leaf fragments (Cooke, 1977). The endosymbiotic yeasts or yeast-like fungi that form in the lymph between fat body cells or occupy specialized host cells (mycetomes) in insects of the orders Homoptera and Coleoptera also relate to the habit of feeding on nutrient-poor substrates, low in nitrogen and deficient in vitamins (Buchner, 1965).

3.1. Nitrogen Limits

Larval grazing on yeasts growing in decaying cactus fruits, *Opuntia* cladodes or columnar cactus stems is an effective means of concentrating nitrogen in tissues where nitrogen is diffuse and at low levels. Nitrogen content of columnar cactus tissue is approximately 0.3-0.5% (dry weight, Fogleman, personal communication) while yeasts are about 8%. Thus lar-

ically increase their nitrogen intake by as much as 16-26 times. Nitrogen in the waste products can also be a limiting factor. Urea and uric acid are both known to delay development, decrease survival and cause larval stop in crowded cultures of *Drosophila melanogaster* (Botella *et al.*, 1985). It has been suggested that some fungi utilize their host's waste products for the synthesis of protein, thus alleviating toxic metabolites by recycling them (Cooke, 1977). This may also be pertinent to the mutualistic interactions between yeasts and drosophilids since most yeasts are able to use uric acid as a source of nitrogen (LaRue and Spencer, 1968). Nitrogen sources and amounts may be critical to the development and maturation of drosophilids but this is most likely a universal problem that has been solved by yeasts in general, and like other insect-fungus mutualistic interactions mentioned above, could form the basis for the non-specific mutualism between drosophilids and yeasts.

3.2. Sugars and Carbohydrates

In the evolution of host use for cactophilic *Drosophila* species there has been a discernable shift in the availability of sugars and in the complexity of carbohydrates available in the tissues. In a broad perspective, fruits of all types have relatively high levels of free sugars. The average total sugar content of many fruits range from 5-10% (fresh weight) and oligosaccharides other than the disaccharide sucrose as well as polysaccharides are relatively rare (Whiting, 1970). *Opuntia* cladode tissue also has free sugars present but at a lower concentration (0.5-2% fresh weight in mature cladodes of *O. ficus-indica,* Kircher, personal communication; 0.8% in three different *Opuntia* species, Rodriques-Felix and Cantwell, 1988) than most fruits. However, cladode tissue contains far more polysaccharides (0.25-1.75% fresh weight is water soluble polysaccharides in *O. ficus-indica*, Kircher personal communication; 4.6% fresh weight is complex polysaccharides in several species, Rodriques-Felix and Cantwell, 1988) as compared to fruit. Columnar cactus stems have little free sugars available because most sugars are either bound as glycosides or are components of oligo- and polysaccharides. It thus appears that the shift in host plant use has been accompanied by a shift in the availability of free sugars. This shift has also changed the types of microorganism present as well as their physiological abilities in the decaying tissue. The yeasts normally recovered from decaying cactus tissue grow readily on sterile fruit and cladode tissue but poorly on sterile columnar stem tissue. If columnar stem tissue is inoculated with a bacterium (e.g., soft-rot bacteria of the genus *Erwinia*) and yeasts, then the yeasts grow in abundance.

3.3. Host Shifts

The host shift from cactus fruit - to *Opuntia* cladode - to columnar stem tissue is also accompanied by a change in the complexity and abundance of inhibitory compounds. Cactus species in the columnar subtribe Stenocereinae contain copious

quantities of triterpene glycosides, which are toxic to some
yeasts (Starmer *et al.*, 1980). In addition, unusual lipids
and their components found in the stem tissues of some *Stenocereus* species can be detrimental to the development of resident as well as non-resident *Drosophila* species. Some cactus
species in the columnar subtribe Pachycereinae contain alkaloids that are inhibitory to some *Drosophila* species but
apparently do not affect yeast growth (Fogleman and Kircher,
1983).

The overall picture for chemical changes experienced during
the evolution from fruit to cladode to columnar stem use is a
decrease in the availability of free sugars, an increase in
the amount of sugars that are tied up in polysaccharides and
glycosides and an increase in the presence of toxic compounds.
The suitability and requirements for growth appear more severe
and demanding as the chemical shifts of the habitat progress.
Brazner *et al.* (1984) and Etges and Klassen (1989) have suggested that a successful host shift to the columnar cacti for
D. mojavensis has required a greater reliance on the alcohols
produced during the decay process as sources of energy. This
also may be true for the yeast communities since their dominant physiological abilities appear to be in the realm of
alcohol metabolism (Heed *et al.*, 1976; Lachance and Starmer,
1986).

3.4. Biculture Effect

Experiments reported here for the effect of yeast bicultures on the development of larvae also seem to follow a pattern that fits the shift in host use. Three points are worth
noting:

1. The strength of the biculture effect appears to increase
   as the shift from fruit to cladode to stem is made.
   Evidence on this point comes from the earlier work on *D.
   mojavensis* in agria stems (Starmer and Fogleman, 1986)
   and *D. buzzatii* in *Opuntia* cladodes (Starmer and Barker,
   1986), and from the evidence given in Table III and Figures 2 and 3. The difference in magnitude of the biculture effect for fruit and cladode tissue is not clear
   since both showed effects but of different types
   (viability increases in fruit tissue and decreases in
   development time in cladode tissue). In addition the
   benefits of bicultures to the viability of the larvae in
   fruit is reduced by the increased development time experienced by larvae under these culture conditions. The
   progression from fly-yeast mutualism to a fly-yeast community mutualism could be driven by the increasing
   severity of the host environment.

2. The biculture effect may be specialized in the sense
   that it appears to be host dependent but diffuse in that
   several different combinations of species within a particular host allow larvae to develop faster or experience greater survival. However, there does not seem to
   be much within-species variation for the presence or

absence of the biculture effect between species. Cultures from widely different geographic areas and host plants produce the same results (Table II, Fig. 1). Until more experiments are conducted using different *Drosophila* species under several possibly important conditions (e.g., conditions that test for larval density dependence, abiotic dependence such as temperature, consistency and source of the cactus, or associated soft-rot bacteria) this point is open to doubt. We have conducted other experiments in which the biculture effect is not present (e.g., Table IV) even though conditions were similar to the other experiments reported here. In these cases the apparent contradiction is probably due to our lack of understanding of the mechanism and lack of control of the conditions under which the effect is most apparent.

3. The mechanism for the biculture effect is unknown but several possible mechanisms that involve yeast-yeast cross-feeding are possible. It is known that the initial growth rates for several pairs of cactus yeasts (i.e., yeasts of agria communities) are mutually enhanced when grown together and that early in the growth of larvae more yeast cells are available due to this mutualistic growth of the yeasts (Starmer and Fogleman, 1986). Preliminary work by Ganter (unpublished) indicates that mutual enhancement of growth rates for pairs of yeasts may be host tissue dependent. Cross-feeding between bacterial species in the hindgut of termites appears to facilitate the overall fermentation process important to termite survival (Schultz and Breznak, 1979). In the case of the cactus yeasts in the larval environment we do not know if cross-feeding does take place but the difference in the physiological profiles and vitamin requirements of these yeasts provides suggestive evidence that it could occur.

4. SUMMARY

The mutualistic relationship between *Drosophila* and yeasts involve a number of interesting interactions. The primary benefit received by the yeast is transportation to new habitats by the adult drosophilids and transport within the habitat by foraging larvae. The yeasts in turn provide nutritional benefits to the larvae and adult drosophilids by providing essential nutrients. In some cases the yeast can detoxify an otherwise toxic environment for the larval or adult stages. Yeasts also appear to be involved in differential habitat use (i.e., larval foraging behavior, adult feeding and oviposition behavior) and mating behavior of adults.

Results presented in this paper show that mixed cultures of yeasts in the larval environment provide greater benefits in terms of developmental speed or survival as compared to single cultures of yeasts. This biculture effect was shown to be dependent on the type of host plant tissue and the yeast community specific for that tissue. Observations on the strength of the biculture effect suggest that tissues with greater chemical complexity and deficiencies result in stronger biculture effects.

References

Barker, J. S. F., Vacek, D. C., East P. D., and Starmer, W. T., 1986, Allozyme genotypes of *Drosophila buzzatii:* Feeding and oviposition preferences for microbial species, and habitat selection, *Aust. J. biol. Sci.* **39**:47-58.

Baumberger, J. D., 1919, A nutritional study of insects, with special reference to microorganisms and their substrata, *J. exp. Zool.* **28**:1-81.

Begon, M., 1982, Yeasts and *Drosophila*, in: *The Genetics and Biology of Drosophila*, Vol 3B (M. Ashburner, H. L. Carson, and J. N.Thompson Jr., eds), Academic Press, New York, pp. 345-384.

Botella, L. M., Moya, A., Ganzalez M. C., and Mensua, J. L., 1985, Larval stop, delayed development and survival in overcrowded cultures of *Drosophila melanogaster:* effect of urea and uric acid, *J. Insect Physiol.* **31**:179-185.

Brazner, J., Aberdeen, V., and Starmer, W. T., 1984, Host-plant shifts and adult survival in the cactus breeding *Drosophila mojavensis*, *Ecol. Entomol.* **9**:375-381.

Buchner, P., 1965, *Endosymbiosis of Animals with Plant Microorganisms*, New York Intersciences Publishers, John Wiley and Sons, New York.

Cooke, R., 1977, *The Biology of Symbiotic Fungi*, John Wiley and Sons, New York.

Delcourt, A., and Guyenot E., 1910, De la possibilite d'etudier certain Dipteres en milien definite, *C. r. hebd. Séance. Acad. Sci., Paris* **151**:255-257.

Etges, W. J., and Klassen C. S., 1989, Influences of atmospheric ethanol on adult *Drosophila mojavensis*: Altered metabolic rates and increases in fitness among populations, *Physiol. Zool.* **62**:170-193.

Fogleman, J. C., 1982, The role of volatiles in the ecology of cactophilic *Drosophila*, in: *Ecological Genetics and Evolution: the Cactus-Yeast-Drosophila Model System* (J. S. F. Barker, and W. T. Starmer, eds), Academic Press Australia, Sydney, pp. 191-206.

Fogleman, J. C., Starmer, W. T., and Heed, W. B., 1981, Larval selectivity for yeast species by *Drosophila mojavensis* in natural substrates, *Proc. natn. Acad. Sci. USA* **78**:4435-4439.

Fogleman, J. C., Starmer, W. T., and Heed, W. B., 1982, Comparisons of yeast florae from natural substrates and larval guts of southwestern *Drosophila*, *Oecologia* **52**:187-191.

Fogleman, J. C., and Kircher, H. W., 1983, Senita alkaloids: No inhibition of sterol biosynthesis in yeasts or cacti, *J. Nat. Prod.* **46**:279-282.

Fogleman, J. C., and Kircher, H. W., 1986, Differential effects of fatty acid chain length on the viability of two species of cactophilic *Drosophila*, *Comp. Biochem. Physiol.* **83A**:761-764.

Fogleman J. C., and Heed, W. B., 1989, Columnar cacti and desert *Drosophila*: the chemistry of host plant specificity, in: *Special Biotic Relationships in the Arid Southwest* (J. O. Schmidt, ed.), University of New Mexico Press, Albuquerque, New Mexico, pp. 1-24.

Ganter, P. F., 1988, The vectoring of cactophilic yeasts by *Drosophila*, *Oecologia* **75**:400-404.

Gilbert, D. G., 1980, Dispersal of yeasts and bacteria by *Drosophila* in a temperate forest, *Oecologia* **46**:135-137.

Heed, W. B., 1977, Ecology and genetics of Sonoran Desert *Drosophila*, in:

Ecological Genetics: the Interface (P. F. Brussard, ed.), Springer-Verlag, New York, pp. 109-126.

Heed, W. B., 1982, The origin of *Drosophila* in the Sonoran Desert, in: *Ecological Genetics and Evolution: The Cactus-Yeast-Drosophila Model System* (J. S. F. Barker, and W. T. Starmer, eds), Academic Press Australia, Sydney, pp. 65-80.

Heed, W. B., Starmer, W. T., Miranda, M., Miller, M. W., and Phaff, H. J., 1976, An analysis of the yeast flora associated with cactiphilic *Drosophila* and their host plants in the Sonoran desert and its relation to temperate and tropical associations, *Ecology* 57:151-160.

Heed, W. B., and Mangan, R. L., 1986, Community ecology of the Sonoran Desert *Drosophila*, in: *The Genetics and Biology of Drosophila*, Vol. 3E (M. Ashburner, H. L. Carson, and J. N. Thompson, eds), Academic Press, New York, pp. 311-345.

Lachance, M. A., and Starmer W. T., 1986, The community concept and the problem of non-trivial characterization of yeast communities, *Coenoses* 1:21-28.

Lachance, M. A., Starmer, W. T., and Phaff, H.J., 1988, Identification of yeasts found in decaying cactus tissue, *Can. J. Microbiol.* 34:1025-1036.

LaRue, T. A., and Spencer, J. F. T., 1968, The utilization of purines and pyrimidines by yeasts, *Can. J. Microbiol.* 14:79-86.

Rodriquez-Felix, A., and Cantwell, M., 1988, Developmental changes in composition and quality of prickly pear cactus cladodes (nopalitos), *Pl. Fds hum. Nutr.* 38:83-93.

Sang, J. H., 1978, The nutritional requirements of Drosophila, in: *The Genetics and Biology of Drosophila*, Vol. 2 (M. Ashburner, and T. R. F. Wright, eds), Academic Press, New York, pp. 159-192.

Schultz, J. E., and Breznak, J. A., 1979, Cross-feeding of lactate between *Streptococcus lactis* and *Bacteroides* sp. isolated from termite hindguts, *Appl. & Environ. Microbiol.* 37:1206-1210.

Starmer, W. T., 1982, Associations and interactions among yeasts, *Drosophila* and their habitats, in: *Ecological Genetics and Evolution: The Cactus-Yeast-Drosophila Model System* (J. S. F. Barker, and W. T. Starmer, eds), Academic Press Australia, Sydney, pp. 159-174.

Starmer, W. T., and Gilbert, D. G., 1982, A quick and reliable method for sterilizing eggs, *Drosoph. Inf. Serv.* 58:170.

Starmer, W. T., Kircher, H. W., and Phaff, H. J., 1980, Evolution of host plant specific yeasts, *Evolution* 34:137-146.

Starmer, W. T., Lachance, M. A., Phaff, H. J., and Heed, W. B., 1989, The biogeography of yeasts associated with decaying cactus tissue in North America, the Caribbean and Northern Venezuela, *Evol. Biol.* (in press).

Starmer, W. T., Barker, J. S. F., Phaff, H. J., and Fogleman, J. C., 1986, Adaptations of *Drosophila* and Yeasts: their interactions with the volatile 2-propanol in the cactus-microorganism-*Drosophila* model system, *Aust. J. biol. Sci.* 39:69-77.

Starmer, W. T., and Barker, J. S. F., 1986, Ecological genetics of the *Adh-1* locus of *Drosophila buzzatii*, *Biol. J. Linn. Soc.* 28:373-385.

Starmer, W. T., and Fogleman, J. C., 1986, Coadaptation of *Drosophila* and yeasts in their natural habitat, *J. Chem. Ecol.* 12:1037-1055.

Starmer, W. T., Phaff, H. J., Bowles, J. M., and Lachance, M. A., 1988, Yeasts vectored by insects feeding on decaying saguaro cactus, *SWest. Nat.* 33:362-363.

Starmer, W. T., Peris, F., and Fontdevila, A., 1988, The transmission of yeasts by *Drosophila buzzatii* during courtship and mating, *Anim. Behav.* 36:1691-1695.

Vacek, D. C., 1982, Interactions between microorganisms and cactophilic *Drosophila* in Australia, in: *Ecological Genetics and Evolution: The*

Cactus-Yeast-Drosophila Model System (J. S. F. Barker, and W. T. Starmer, eds), Academic Press Australia, Sydney, pp. 175-190.

Wagner, R. P., 1944, The nutrition of *Drosophila mulleri* and *D. aldrichi*. Growth of the larvae on the extract and the microorganisms found in cactus, *Univ. Tex. Publs* **4445**:104-128.

Wagner, R. P., 1949, Nutritional differences in the Mulleri group, *Univ. Tex. Publs* **4920**:39-41.

Whiting, G. C., 1970, Sugars, in: *The Biochemistry of Fruits and their Products* (A. C. Hulme, ed.), Academic Press, New York, pp. 1-31.

CHAPTER 9

Experimental Analysis of Habitat Selection and Maintenance of Genetic Variation

J.S.F. BARKER

1. INTRODUCTION

A major goal of evolutionary genetics is to provide explanations for the maintenance of genetic variation within populations. Among numerous possibilities that have been considered, substantial attention has been devoted to effects of
heterogeneous environments (Hedrick, 1986), the basic premise
being that selection pressures will then be variable in space
and/or in time, and that such variable selection may play a
significant role in maintaining polymorphism. Nevo (1988, and
references therein) argues that there is sufficient evidence
from natural populations to implicate environmental heterogeneity as a significant factor in maintaining polymorphisms.
Nevertheless, he does point out that it is not at all clear
what proportion of polymorphic loci are maintained by environmental heterogeneity or by other mechanisms, and emphasises
the need for critical experimentation to substantiate causal
ecological-genetic relationships.

2. A STRATEGY FOR STUDY OF HETEROGENEOUS ENVIRONMENTS–GENETIC POLYMORPHISM RELATIONSHIPS

The habitats of most species certainly are heterogeneous,
but they are heterogeneous in many dimensions. Therefore
critical experimental analysis of the heterogeneous environ-
ments-genetic polymorphism relationship would be greatly
facilitated by identification of the relevant dimensions.
This would seem to be fundamental, yet it apparently has not
been seriously addressed.

The particular genetic variation of concern here is that at
structural protein-coding loci that can be assayed electrophoretically. Suppose polymorphism at a particular locus
is being maintained by variable selection due to a specific
heterogeneous component of the environment. Proof for that
direct association would be the endpoint of what may well be a

J.S.F. BARKER * Department of Animal Science, University of New England,
Armidale, N.S.W. 2351, Australia.

Ecological and Evolutionary Genetics of Drosophila, Edited by
J.S.F. Barker et al., Plenum Press, New York, 1990

series of diverse experimental studies. Now we may get there through serendipity, but excluding that fortunate possibility, there are three sources of information that could be used to give direction towards possibly relevant habitat dimensions:

1. Comparisons between species,
2. Comparisons among populations within species,
3. Analyses within populations.

To see how these might be used, take as a working hypothesis that polymorphism for at least some of these loci in the species of interest is being maintained by environmental heterogeneity.

2.1. Comparisons Between Species

The distribution of a species will be determined by climatic and ecological factors, and within this distribution individuals will utilize specific resources that allow reproduction and survival. While other factors such as genetic drift and historical events cannot be excluded, the distribution and the resources utilized must be partly the result of adaptation driven by natural selection. Thus closely related sympatric species may have been subject to similar selective forces, and selection affecting allele frequencies at specific loci might be detected by estimating correlations between allele frequencies at homologous loci. Significant geographic correlations (i.e., parallel patterns of variation) can arise only through the action of natural selection. Of course, the absence of parallel patterns of variation does not negate possible selection, and caution is necessary in that alleles in the two species with similar electrophoretic mobility may not be identical. Nevertheless, Clarke (1975) has characterized this correlation approach as "perhaps the most powerful method of all" for detecting selection, and although it has been applied in a few studies (e.g., Anderson and Oakeshott, 1984; Romano et al., 1987), it has been largely ignored.

The finding of significant geographic correlations does not implicate heterogeneous environments as a causative factor, but it would indicate loci for further study. It should lead to the study of these loci both between and within populations of each species in an attempt to identify environmental components that may be important.

2.2. Comparisons Among Populations Within Species

Different populations that are exposed to different physical environments or to differences in the relative frequency of resources may show differences in local adaptation and hence in allele frequencies. Environmental factors affecting such differentiation may be detected by multivariate statistical tests for association of allele frequencies with environmental variables (Mulley et al., 1979). Again lack of association does not negate possible selection, but significant associations for a particular locus provide clues to environmental variables for which spatial environmental variation is important. It might be argued that this approach is limited in that (i) commonly only physical environmental factors (based on climatological data) are included, and (ii) an association does not imply a causal relationship. Certainly cautious interpretation is necessary, as associations may not be due to direct effects of the environmental variables, and the climatological data may not truly reflect the environment as experienced by the organism. For example, Mulley *et al.* (1979) found significant associations for *Esterase-2* allele frequencies and heterozygosity in the cactophilic species, *D. buzzatii*, with seasonal extreme temperatures and variability in temperature. Such associations could be quite indirect, and mediated through effects of temperature on resource availability (frequency and types of cactus rots) or on the chemical and microbial composition of the rots. Other data on these variables could provide additional clues as to their likely significance. An alternative possibility worth pursuing is that extreme temperature stresses within populations impose environmental heterogeneity contributing to the maintenance of polymorphism at this locus. Within rot pockets, where relative humidity is near 100%, one might expect temperature to be less variable than ambient. However, rot pockets have been observed to exhibit temperatures much higher than ambient, e.g., as high as 44°C when ambient was 33°C, and overnight winter minimum temperatures less than 0°C have been recorded in localities where adults and larvae were present. Thus extreme temperature stress may be a significant environmental factor affecting the relative fitness of allozyme variants. Significant effects of heat shock have been found for *Esterase-2* allele frequencies (Watt, 1981), with some of the effects consistent with results of the genotype-environment association analyses, e.g., *Est-2^d* showed a positive correlation with mean summer maximum temperature, and increased frequency in survivors of heat shock. Further, biochemical studies (East, 1984) have shown dramatic differences among *Est-2* allozymes in thermostability (laboratory *in vitro* analyses), *Est-2^b* being much more stable, and *Est-2^c* much less stable than the other two alleles.

Apart from direct statistical tests of association, examination of the spatial patterns of genetic variation can give additional information. A method of directional spatial-autocorrelation analysis was introduced by Oden and Sokal (1986), and applied to geographic variation in allozyme frequencies in *D. buzzatii* in Australia by Sokal *et al.* (1987). These analyses do not reveal the nature of any selection, but they indicated that selection at different loci operates at different spatial scales ranging from a continental one to a strictly local one. However, the relevant variation determining selection effects must ultimately be at the level of individual rots. Thus, rot variables showing large-scale geographic variation, which may be mediated by and correlate with macroenvironmental factors such as temperature, could account for the continental patterns of allele frequency variation. On the other hand, those rot variables that on average are similar among localities, but which vary among rots within a locality, could determine micro-spatial patterns.

Comparisons between species and between populations within species thus could provide clues as to those environmental dimensions for which heterogeneity may be important in maintaining polymorphisms. But it is the polymorphism within populations that we wish to explain, and the final analysis must be within populations.

2.3. Analyses Within Populations

At this level, differences among individual organisms are of primary concern. Different individuals will be exposed to different physical environments, and will differ in their utilization of resources. By definition then, the environment is heterogeneous. If differences in response to physical environmental factors or in resource use affect survival and reproduction, and if these differences are associated with different genotypes, then selection is acting on those genotypes. Obviously this is neither original nor profound; but I consider it important in making the distinction between effects of physical environmental factors and of resource use.

Specifically in relation to *Drosophila* and to assessing the importance of environmental heterogeneity in maintaining polymorphisms, our strategy should take account of the following:

1. Effects of heterogeneity in physical environmental factors may be important, but will often be difficult to analyse,
2. Given information on the ecology of the species, resource use should be more amenable to analysis,
3. While theory shows that environmental heterogeneity in space or in time can maintain genetic polymorphisms, stable polymorphisms are more readily maintained if there is genotype specific habitat selection,
4. Habitat selection may involve the physical habitat, and behavioural responses to avoid environmental stresses, but except at a gross level, will be difficult to analyse in the field,
5. Specific resources - for mating, feeding or oviposition, can be defined and again given sufficient ecological information, could be studied in the field.

These observations lead me to one conclusion - if we are to maximise our chances of detecting significant effects of heterogeneous environments on the maintenance of polymorphism, the focus must be on habitat selection for resources.

Obviously this strategy is predicated on the assumption that genotype specific habitat selection for resources does occur, and evidence for this in the cactophilic *Drosophila* will be discussed later. Recall also that I am attempting to provide an experimental strategy; it is not suggested that physical environmental factors are not important, and that their study is unwarranted. As already noted, laboratory studies of environmental stresses can be informative, and can show what might be happening in nature. Further, studies in natural populations may detect cases where variability in physical environmental factors is important, such as in the study of Templeton *et al.* (Chapter 1, this volume) on abnormal abdomen in *D. mercatorum*. My point is simply that I at least find it difficult to envisage critical empirical study in natural populations, based on hypotheses relating specific genetic variation to specific physical environmental variation.

Studies of habitat fidelity may show that flies can discriminate among microhabitats, but such experiments are not very critical or informative. In these experiments, flies are captured from different microhabitats, marked according to origin and released from a common location. Recaptures show whether flies tend to return to their habitat of origin. Taylor (1987) claims a general finding from such experiments of consistent individual differences in habitat preferences. However, he does note exceptions (see also Hey and Houle, 1987) and variability in results due to weather conditions or even time of day - hardly the basis for a general finding. The difficulty with this approach is that habitats are defined only at a very gross level, e.g., forest versus grassland, or wet, dark woodland versus dry, light woodland. There is no information on the relevant environmental dimensions that are perceived by the flies, and that might be affecting any habitat choice. Further, even though genetic differences (in allozyme or inversion frequencies) have sometimes been found between microhabitats (see Taylor, 1987 for review), selection effects cannot be easily ruled out and no specific genotype-environment associations can be identified.

There is therefore a need to define ecological-genetic relationships, to determine the resources that are relevant to the organisms, if within population variability is to be analyzed and understood. The critical level of analysis is not the grossly defined habitats, as perceived by the human observers, but the finer level within this where flies might exhibit choice of resources that are relevant to the organism, i.e., for mating, feeding and oviposition. The distinction must be made between the broad habitat, as a place in which to live, and specific resources.

In addition to identifying relevant resources, the traits of the organism that are involved in the use of the resources need to be defined. As the aim is to elucidate the genetics of resource use and of habitat selection, two complementary approaches should be used - one ecological genetic analyzing resource-response interactions, the other behavioural and physiological genetic analyzing mechanisms of response and their genetic basis.

Finally as far as this discussion of strategy is concerned, although it should be the first consideration, there is the question of choice of the study organism. Not all species of *Drosophila* will be equally suitable, primarily because the ecology of many is not adequately known. Hey and Houle (1987) have recognized this problem in stating "The next step for those interested in population structure of *Drosophila* in relation to their ecology must be to identify the environmental stimuli flies perceive, and to quantify non-random behaviour in relation to these stimuli." and then going on rather plaintively "Unfortunately, this has proved very difficult in the *obscura* group."

3. THE CACTOPHILIC *DROSOPHILA*

In contrast, the cactophilic *Drosophila*, because of their relatively well-known ecology (Barker and Starmer, 1982), provide a model system for analysis. Two cactophilic species, *D. buzzatii* and *D. aldrichi* (both in the *mulleri* subgroup of the *repleta* group), have colonized in Australia in association with species of *Opuntia* (prickly pear) cactus. Both species feed and breed in necrotic rots in the cactus cladodes, and are specific to this cactus niche. In Australia, cactus rots primarily result from the mining of cladodes by the phytophagous larvae of the moth, *Cactoblastis cactorum*, which was introduced as a biological control agent. The decaying cactus tissue is a nutrient-rich environment for the growth of microorganisms, which in turn supply essential nutrients for the *Drosophila*. The microorganisms rely on insects for dispersal, and the *Drosophila* serve this vectoring role (Starmer *et al.*, 1988). This feeding and breeding habitat of the *Drosophila* can be studied in terms of the potentially important microenvironmental resources of the rot microflora and factors such as rot chemistry (Fogleman and Abril, Chapter 7, this volume), and it may be experimentally manipulated. In addition, both the *Drosophila* and the microorganisms can be grown and studied in the laboratory.

Although the strategy outlined earlier has not been followed in full detail for the cactophilic species in Australia, many facets have been completed, and in our search for habitat selection for resources, has led to a focus on yeast species as the resource and habitat selection for oviposition sites.

4. *D. BUZZATII* - YEAST SPECIES INTERACTIONS

A major stimulus to flies seeking a suitable feeding or breeding site is presumably olfactory, and volatiles produced by microorganisms most likely provide the cue (Fogleman, 1982). Thus environmental heterogeneity that should be perceived by the flies would result from variation in the microflora composition, both between and within rots. There is ample evidence for such heterogeneity. From one locality in the Hunter Valley, N.S.W. (locality 5 of Barker and Mulley, 1976) and from a number of localities in Queensland, 944 separate yeast isolates were identified. For the Hunter Valley locality, there were significant differences in yeast species distributions among seasons and among types of rot (Table I), i.e., significant temporal and microhabitat heterogeneity (Barker *et al.*, 1983). Analyses of the total yeast survey data (Barker *et al.*, 1984) showed that differences among yeast communities were greater between *Opuntia* species than between different localities within a single cactus species. Tempera-

Table I
Proportions of each yeast species within seasons and
within types of rot

Season	Yeast species[a]									
	Cs	Pc	Clo	Pb	Rhm	Po	Cm	Cra	Pa	Crc
Summer	0.27	0.45	0.14	0.04	0	0	0.06	0.02	0	0.02
Autumn	0.31	0.23	0	0.11	0.15	0.02	0.05	0.02	0.09	0.01
Winter	0.33	0.28	0.02	0.07	0.03	0.05	0.03	0.05	0	0.14
Spring	0.26	0.18	0.27	0.03	0	0.10	0.04	0.09	0.03	0.01

Type of rot										
Young cladode	0.27	0.16	0.27	0.02	0.04	0.07	0.02	0.08	0.05	0.01
Old cladode	0.35	0.29	0.07	0.08	0.05	0.05	0.06	0.02	0.02	0.02
Basal cladode	0.18	0.29	0.04	0.14	0.18	0.07	0	0	0.11	0
Basal rot	0.27	0.28	0.02	0.08	0.08	0.02	0.06	0.06	0.06	0.08

[a]Cs - *Candida sonorensis*, Pc - *Pichia cactophila*, Clo - *Clavispora opuntia*, Pb - *Pichia barkeri*, Rhm - *Rhodotorula minuta* var. *minuta*, Po - *Pichia opuntia*, Cm - *Candida mucilagina*, Cra - *Cryptococcus albidus* var. *albidus*, Pa - *Pichia amethionina* var. *amethionina*, Crc - *Cryptococcus cereanus* var.

ture, pH and age of rot all influenced the structure of the yeast community. Microenvironmental heterogeneity also has been found, in that multiple samples taken from individual rots at the one time showed substantial diversity in yeast species abundance and frequency, while repeated sampling over time for another set of rots showed significant spatial variability (among rots sampled at the same time) and temporal variability (within rots over time) in the effective number of yeast species and in yeast species diversity (Barker *et al.*, 1987). This heterogeneity in the environment of the flies could provide the basis for habitat selection.

However, given this heterogeneity, it is necessary to know whether it is important to the flies, i.e., do they perceive these differences and exhibit differential responses?

Laboratory and field experiments (Barker *et al.*, 1981 a,b) showed significant differences among yeast species in their attractiveness to adult flies. In the field experiments, different genotypes at the *Esterase-2* locus were differentially attracted to yeast species, providing the first indication of possible genotype specific choice of yeast species. However, at the time these experiments were done, our knowledge of the yeast flora was very limited and the field experiments measured only the numbers of flies attracted to and captured at each yeast bait. It might be expected that this measure would confound food preference and oviposition preference, but the sex x yeast interaction was not significant in any of the three experiments. Thus male and female food preference and female oviposition preference may be similar. However, the results of Jaenike (1986) suggest caution in interpreting just what is being measured in such experiments. He refers to this measure as settling behaviour, but for two strains of *D. tripunctata*, found that settling behaviour and oviposition preference were negatively correlated.

Because our field experiments provided no information on any differential use of the yeasts as resources, subsequent laboratory experiments measured actual feeding preference by identification of yeast species in the crops of experimental flies and oviposition preferences as the number of eggs laid on each yeast. Further, the natural environment was simulated more closely by growing each yeast species in association with a bacterial community on a cactus substrate. *Drosophila buzzatii* females showed strong discrimination among eight cactophilic yeast species, with discrimination increasing over three days, and there was a significant correlation ($r = 0.96$) between yeasts preferred for feeding and those preferred for oviposition (Vacek *et al.*, 1984). Clearly at the species level, *D. buzzatii* females discriminate among yeast species resources for feeding and for oviposition.

5. GENOTYPE-SPECIFIC HABITAT SELECTION

But are these expressed preferences simply a reflection of differential attractiveness of the yeasts to all *D. buzzatii* females, or do they derive, at least partly, from females of different genotypes exhibiting different preferences? Defining genotype in terms of seven polymorphic allozyme loci, Barker *et al.* (1986) found no significant differences in feeding preferences among adults of different genotypes. However, for oviposition preferences, there were significant microorganism-genotype associations for each of the seven loci. Analysis of the total electrophoretic genotype showed that the genotypes of eggs laid on the same microorganism species were genetically more similar than those laid on different species. That is, females of different genotypes show habitat selection for oviposition sites. This provided the first evidence, albeit indirect, for genotype-specific preferences for oviposition on these natural yeast resources.

In this case, there is no suggestion that the allozyme locus genotypes actually determined the oviposition preferences, and the genetic basis remains to be elucidated. A priori, it might seem unlikely that the preferences would be simply inherited, but if they were and if they segregated with an enzyme locus, the effect would be to contribute to the maintenance of polymorphism at that locus. Obviously, our result of significant microorganism-genotype interactions for all seven polymorphic loci argues against simply inherited preferences. Although there is a large literature on host plant utilization by herbivorous insects, there is surprisingly little information on genetic variation in resource use within insect populations (Futuyma and Peterson, 1985). For *Drosophila* species, Jaenike and Grimaldi (1983) demonstrated genetic variation in oviposition host preference in the polyphagous *D. tripunctata*, while Bird and Semeonoff (1986) successfully selected for oviposition preferences for different laboratory media in *D. melanogaster*. Genetic analysis of oviposition host preference in *D. tripunctata* has been initiated by Jaenike (1987). His results indicate genetic complexity, with autosomal genes with dominance and interaction effects, but with at least one locus linked to the *aconitase* locus.

Before pursuing genetic analysis of oviposition preferences for yeast species in *D. buzzatii*, we have sought (and obtained) direct evidence for genetic variation. However, these experiments have lead to further questions concerning factors affecting oviposition preferences in the laboratory.

Three experiments have been completed using 15 cm petri dishes as oviposition chambers, with females given a choice of five yeast species as oviposition sites, and the numbers of eggs laid on each yeast in a 30 hour test period were counted (for a similar procedure, see Jaenike and Grimaldi, 1983). Natural rots contain 1-5 yeast species with an average about 2-3 (Barker *et al.*, 1983, 1984), so that choice between rots likely depends on long distance attraction behaviour and responses to combinations of yeast species, bacteria species and other factors such as age of the rot. However, once within a rot where the yeast species distribution is patchy (Barker *et al.*, 1987), females will exhibit the short distance behaviour simulated in this experimental test system.

In all three experiments, four replicate oviposition chambers were set up for each of seven or eight isofemale lines, with these lines derived in total from seven different natural populations (Fig. 1), viz., four in experiment 1, three in experiment 2, and one in experiment 3 (new lines derived from one of the experiment 1 populations). The five yeast species used, viz., *Candida sonorensis*, *Pichia cactophila*, *Clavispora opuntiae*, *Candida mucilagina* and *Cryptococcus cereanus*, were chosen from those known to be most abundant in cactus rots in Australia (Barker *et al.*, 1984) to provide a wide range of expected preferences for oviposition by *D. buzzatii* females (Vacek *et al.*, 1985). Fifteen discs of cactus homogenate with growing yeasts (three discs for each yeast species) were placed around the periphery of each oviposition chamber, such that any disc of a particular yeast was located between two of the other species. Eggs per disc then were the primary data, and results were analyzed using analysis of variance of eggs per disc and of the proportions of eggs laid on each of the yeasts in each chamber.

Full details of the experiments will be published elsewhere (Barker, in prep.), but the main result is that ANOVA of the proportions of eggs per yeast showed a highly significant interaction for yeasts x isofemale lines within populations in experiments 1 and 2. These differences in preferences must be genetic, as all lines in each experiment were grown and tested together. From the repeatability over replicates, upper limits of the heritability of preferences were estimated as

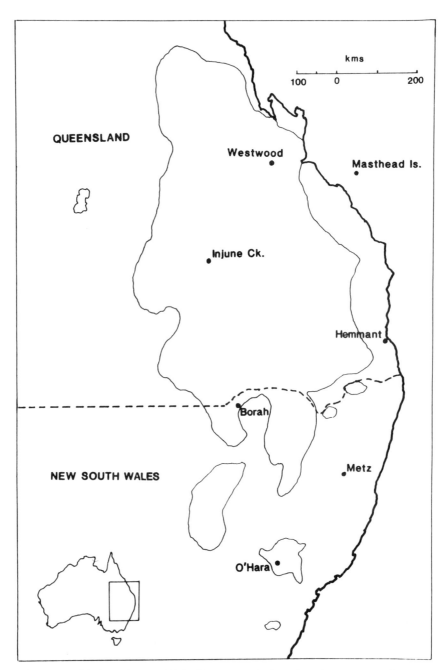

Figure 1. Localities from which lines used in the three experiments were derived.
Experiment 1 - Masthead Island, Hemmant, Metz, O'Hara,
Experiment 2 - Westwood, Injune Creek, Borah,
Experiment 3 - Metz.
The outlined areas indicate the major *Opuntia* infestations as at 1920, i.e., prior to initiation of biological control.

0.191 and 0.167 in the two experiments. Experiment 3, however, differed in that no genetic variation was detected among
seven lines from one population. This lack of variation most
probably is a sampling effect. Recall that these lines in
experiment 3 came from one of the populations (Metz) sampled
for experiment 1. Analysis of the experiment 1 data for each
population separately showed a non-significant isofemale line
x yeast interaction for one of these populations (O'Hara), but
that for Metz was significant. Another important feature of
the results is the marked difference between experiments in
the overall averages for yeast preferences (Fig. 2). The
average preferences in experiment 1 were generally similar to
those found by Vacek *et al.* (1985), and there were no differences among the four populations (non-significant yeast x population interaction). By comparison in experiment 2, the
preference for *C. sonorensis* was reduced and that for *C.
mucilagina* increased. This was particularly marked in two of
the three populations (Borah and Westwood), and the yeast x
population interaction term was significant (P<0.05). In
experiment 3, highest preferences were shown for *C. mucilagina*.

There are three aspects of these results:

1. Evidence for differences among populations in yeast
   preferences (in one experiment),
2. Marked differences in the yeast preferences of lines
   from the same (Metz) population (but collected at different times - viz. March 14, 1986 and April 27, 1988),
3. Increased preferences for *C. mucilagina* from experiment
   1 to 2 to 3.

The most obvious difference between the lines used in these
experiments is that they had been maintained in the laboratory
for different lengths of time - an average of 14 generations
(range of 10-20) for experiment 1, four generations for exper-

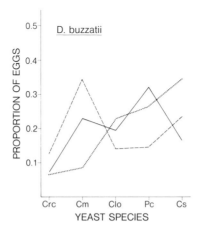

Figure 2. Average proportions of eggs laid on each yeast species over all
populations in each experiment.
· · · · ·Experiment 1, ———Experiment 2, -----Experiment 3.

iment 2 and one generation for experiment 3. The differences
in results between experiments then might result from adaptation to the laboratory environment causing changes in preferences and diluting any differences among populations. This
does not immediately explain all differences, however, as the
lines that had been in the laboratory for the most generations
(experiment 1) gave similar preferences to the average species
level preferences (Vacek *et al.*, 1985), where the experimental
flies were only one generation removed from the wild. Yet
even that similarity could be fortuitous, as the latter used
eight yeast species plus a mixed bacteria culture in population cages, and yeast preferences may be affected by what
other yeast species are included. Fifteen of the isofemale
lines used in experiment 1 were retested for their preferences
for *C. sonorensis* and *Cl. opuntiae* only, and the correlation
between the ratio of eggs on these two species in experiment 1
and in the retest was only 0.309.

Clearly we do not yet understand the full implications of
these results, even for the simplified laboratory test system.

6. HABITAT SELECTION IN NATURE

In the real world of cactus rots, the situation is more
complex, but we should at least begin to consider what the
situation might be. For models of polymorphism in heterogeneous environments, the factors increasing the likelihood of
polymorphism include habitat selection and mating within habitat patches rather than at random. If males and females
exhibit similar resource (yeast species) preferences, there
could be an increased probability of mating of genotypically
similar pairs. But with more than one yeast species per rot, a
rot is not a patch in this sense unless it is not a single
yeast species that determines patch type, but some particular
combinations of yeast species. The primary habitat selection
then would be between rots rather than within rots. Further,
some models for maintaining genetic variation in habitat preference (e.g., Rausher, 1984) assume that population size is
regulated independently in the different habitats. This again
could be relevant to choice between rots. But it is not appropriate for yeast species oviposition preferences expressed
within a rot because larval movement during growth and development must mean that population size regulation is a function
of the total rot resource.

Assortative mating for preference alleles, i.e., mating
within niche after habitat selection, would increase the
effectiveness of oviposition habitat choice, but a similar
effect may be achieved by conditioning (habituation or learning), whereby adults exposed to a particular yeast on emergence are more likely to choose that yeast for feeding and/or
oviposition (Hoffmann, 1988; Jaenike, 1988). Multiple yeast
species per rot and the changing spectrum of yeast species

during the life of a rot pose problems for this mechanism, i.e., if eggs are laid on a particular yeast, will it still be common when these progeny emerge?

The changing spectrum of yeast species during the life of a rot also has implications for any association between female oviposition preferences and larval performance on the yeast species. Although polymorphism may be maintained for habitat preference without genetic variation affecting fitness in the different habitats (Rausher, 1984), we still need to know if adult choice and larval performance are genetically correlated (Thompson, 1988).

Extrapolation to the field and to between rot habitat choice will depend on knowledge of the ecological genetic relationships of fly genotypes to combinations of yeast species. Before that can be attempted, more study of the simplified laboratory system is needed. For the present, that study will remain as our major focus.

7. SUMMARY

Assuming that polymorphism at some proportion of structural protein-coding loci is due to environmental heterogeneity, a strategy is presented that could be used to identify possibly relevant environmental (habitat) dimensions. This involves study of (1) comparisons between species - seeking parallel patterns of variation for individual loci, (2) comparisons among populations within species - multivariate statistical tests for association of allele frequencies with environmental variables and directional spatial auto-correlation analysis, and (3) analyses within populations. At this last level, it is argued that ecological-genetic relationships need to be defined, and that emphasis should be placed on habitat selection for resources. Advantages of the cactophilic *Drosophila* for such studies are noted. Interactions between *D. buzzatii* and yeast species indicate that spatial and temporal heterogeneity in the yeast species found in cactus rots could provide the basis for habitat selection. There is indirect evidence for genotype-specific preferences for oviposition on these natural yeast resources. Laboratory experiments have shown significant genetic variation in oviposition preferences (upper limits to heritability of 0.191 and 0.167 in two experiments). However, differences in average preferences between experiments remain unresolved. Problems in extrapolating these results to habitat selection in nature, and additional factors that might operate there, are briefly discussed.

ACKNOWLEDGEMENTS.   I thank P.W. Hedrick for discussions during preparation of this paper.   Tim Armstrong, Annette Edmonds, Don Fredline, Chris Leger, Margaret Low, Frances McDonald, Richard Thomas and Gillian Whitington provided valued technical assistance.   The research was supported by an Australian Research Grants Committee grant.

References

Anderson, P. R., and Oakeshott, J. G., 1984, Parallel geographical patterns of allozyme variation in two sibling *Drosophila* species, *Nature, Lond.* **308**:729-731.

Barker, J. S. F., and Mulley, J. C., 1976, Isozyme variation in natural populations of *Drosophila buzzatii*, *Evolution* **30**:213-233.

Barker, J. S. F., and Starmer, W. T., eds, 1982, *Ecological Genetics and Evolution: The Cactus-Yeast-Drosophila Model System*, Academic Press Australia, Sydney.

Barker, J. S. F., Starmer, W. T., and Vacek, D. C., 1987, Analysis of spatial and temporal variation in the community structure of yeasts associated with decaying *Opuntia* cactus, *Microb. Ecol.* **14**:267-276.

Barker, J. S. F., East, P. D., Phaff, H. J., and Miranda, M., 1984, The ecology of the yeast flora in necrotic *Opuntia* cacti and of associated *Drosophila* in Australia, *Microb. Ecol.* **10**:379-399.

Barker, J. S. F., Parker, G. J., Toll, G. L., and Widders, P. R., 1981a, Attraction of *Drosophila buzzatii* and *D. aldrichi* to species of yeasts isolated from their natural environment. I. Laboratory experiments, *Aust. J. biol. Sci.* **34**:593-612.

Barker, J. S. F., Toll, G. L., East, P. D., and Widders, P. R., 1981b, Attraction of *Drosophila buzzatii* and *D. aldrichi* to species of yeasts isolated from their natural environment. II. Field experiments, *Aust. J. biol. Sci.* **34**:613-624.

Barker, J. S. F., Vacek, D. C., East, P. D., and Starmer, W. T., 1986, Allozyme genotypes of *Drosophila buzzatii*: feeding and oviposition preferences for microbial species, and habitat selection, *Aust. J. biol. Sci.* **39**:47-58.

Barker, J. S. F., Toll, G. L., East, P. D., Miranda, M., and Phaff, H. J., 1983, Heterogeneity of the yeast flora in the breeding sites of cactophilic *Drosophila*, *Can. J. Microbiol.* **29**:6-14.

Bird, S. R., and Semeonoff, R., 1986, Selection for oviposition preference in *Drosophila melanogaster*, *Genet. Res.* **48**:151-160.

Clarke, B., 1975, The contribution of ecological genetics to evolutionary theory: detecting the direct effects of natural selection on particular polymorphic loci, *Genetics* **79**:101-113.

East, P. D., Biochemical genetics of two highly polymorphic esterases in *Drosophila buzzatii*, Ph.D. thesis, University of New England, Armidale.

Fogleman, J. C., 1982, The role of volatiles in the ecology of cactophilic *Drosophila*, in: *Ecological Genetics and Evolution. The Cactus-Yeast-Drosophila Model System.* (J. S. F. Barker, and W. T. Starmer, eds), Academic Press Australia, Sydney, pp. 191-206.

Futuyma, D. J., and Peterson, S. C., 1985, Genetic variation in the use of resources by insects, *A. Rev. Ent.* **30**:217-238.

Hedrick, P. W., 1986, Genetic polymorphism in heterogeneous environments: A decade later, *Annu. Rev. Ecol. & Syst.* **17**:535-566.

Hey, J., and Houle, D., 1987, Habitat choice in the *Drosophila affinis* subgroup, *Heredity* **58**:463-471.

Hoffmann, A. A., 1988, Early adult experience in *Drosophila melanogaster*, *J. Insect Physiol.* **34**:197-204.

Jaenike, J., 1986, Genetic complexity of host-selection behavior in *Drosophila*, *Proc. natn. Acad. Sci. USA* **83**:2148-2151.

Jaenike, J., 1987, Genetics of oviposition-site preference in *Drosophila tripunctata*, *Heredity* **59**:363-369.

Jaenike, J., 1988, Effects of early adult experience on host selection in insects: some experimental and theoretical results, *J. Insect Behav.* **1**:3-15.

Jaenike, J., and Grimaldi, D., 1983, Genetic variation for host preference within and among populations of *Drosophila tripunctata*, *Evolution* 37:1023-1033.

Mulley, J. C., James, J. W., and Barker, J. S. F., 1979, Allozyme genotype-environment relationships in natural populations of *Drosophila buzzatii*, *Biochem. Genet.* 17:105-126.

Nevo, E., 1988, Genetic differentiation in evolution, *ISI Atlas of Science: Animal and Plant Sciences* 195-202.

Oden, N. L., and Sokal, R. R., 1986, Directional autocorrelation: an extension of spatial correlograms to two dimensions, *Syst. Zool.* 35:608-617.

Rausher, M. D., 1984, The evolution of habitat preference in subdivided populations, *Evolution* 38:596-608.

Romano, M. A., Ralin, D. B., Guttman, S. I., and Skillings, J. H., 1987, Parallel electromorph variation in the diploid-tetraploid gray treefrog complex, *Am. Nat.* 130:864-878.

Sokal, R. R., Oden, N. L., and Barker, J. S. F., 1987, Spatial structure in *Drosophila buzzatii* populations: simple and directional spatial auto-correlation, *Am. Nat.* 129:122-142.

Starmer, W. T., Phaff, H. J., Bowles, J. M., and Lachance, M.-A., 1988, Yeasts vectored by insects feeding on decaying saguaro cactus. *SWest. Nat.* 33:362-363.

Taylor, C. E., 1987, Habitat selection within species of *Drosophila*: a review of experimental findings, *Evol. Ecol.* 1:389-400.

Thompson, J. N., 1988, Evolutionary ecology of the relationship between oviposition preference and performance of offspring in phytophagous insects, *Entomol. exp. Appl.* 47:3-14.

Vacek, D. C., East, P. D., Barker, J. S. F., and Soliman, M. H., 1985, Feeding and oviposition preferences of *Drosophila buzzatii* for microbial species isolated from its natural environment, *Biol. J. Linn. Soc.* 24:175-187.

Watt, A. W., 1981, The genetics of temperature tolerance in *Drosophila buzzatii*, in: *Genetic Studies of Drosophila Populations*. (J. B. Gibson, and J. G. Oakeshott, eds), Proceedings of the Kioloa Conference, Australian National University, pp. 139-146.

CHAPTER 10

Heritable Variation in Resource Use in Drosophila in the Field

ARY A. HOFFMANN, AND SHARON O'DONNELL

1. INTRODUCTION

Some recent work on the ecological genetics of resource use in insects has considered the question of whether genetic variation in behavior (resource response) or performance is associated with resource heterogeneity in the wild. In other words, do individuals in natural populations tend to be associated with particular resources because of their genetic constitution?

One approach to this question is to test for associations between environmental heterogeneity and specific genotypes (usually protein polymorphisms) that may be involved in resource use. A few associations have been detected (reviewed in Futuyma and Peterson, 1985; Hedrick, 1986), including a recent demonstration of electrophoretic differences between *Rhagoletis pomonella* from apples and hawthorn (Feder *et al.*, 1988) although there are some published (and probably many unpublished) unsuccessful attempts (e.g., Mitter and Futuyma, 1979; Jaenike and Selander, 1979; Nielsen *et al.*, 1985). A limitation of this approach is that a negative result does not necessarily imply the absence of an association between genetic variation and resource heterogeneity, but only that any association is not related to the loci under study.

Another approach is to test for a genetic contribution to phenotypic variation in resource use. A number of experimental designs have been used, particularly for behavioral traits. In a common design (Fig. 1a), individuals are collected from different natural resources and lines derived from them are bred under the same laboratory conditions for one or more generations. Lines are tested for performance or response in laboratory assays, using resources that can be related to those from which the insects were collected. Examples of studies using this design are McKenzie and Parsons (1974), Tavormina (1982) and Smith (1988). A limitation of this design

ARY A. HOFFMANN, AND SHARON O'DONNELL ⋆ Department of Genetics and Human Variation, La Trobe University, Vic. 3083, Australia.

Ecological and Evolutionary Genetics of Drosophila, Edited by
J.S.F. Barker *et al.*, Plenum Press, New York, 1990

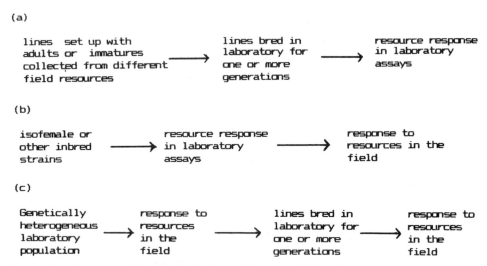

Figure 1. Three experimental designs for examining the association between resource heterogeneity in the field and genetic variation in resource response.

is that genotypic differences expressed in the field may not be expressed under laboratory conditions because of genotype-environment interactions. Another limitation is that laboratory behavior and performance tests may not be realistic measures of resource use in the field (see Singer, 1986). The potential importance of this problem can be illustrated by experiments on experience effects in *Drosophila*: Laboratory experiments often provide evidence for strong positive effects of early adult experience (Jaenike, 1982; Hoffmann, 1985), but they are not evident in field tests with similar resources (Hoffmann, 1988; Turelli and Hoffmann, 1988). These limitations can be at least partly overcome by making detailed observations on resource response in the field, which can be used to generate meaningful laboratory assays (e.g., Singer, 1986). In addition, stocks can be raised in environments that reflect field conditions to minimize the problem of genotype-environment interaction.

Several modifications of this design have been used, although these tend to have similar limitations. Resource response variation in some studies is tested in field-collected individuals without distinguishing the natural resources from which individuals originated. For example, Singer *et al.* (1988) collected adults of the butterfly *Euphydryas editha* from a natural population and tested the oviposition response of females to hosts in the field before breeding their offspring in a laboratory environment. Another modification is to characterize inbred strains in the laboratory before examining their field response (Fig. 1b). For example, Jaenike (1986) tested the field response of two strains of *D. tripunctata* which behaved differently in a laboratory assay.

Two other designs that have not been widely used may overcome the problem of relating laboratory assays to field behaviors. Resource use could be tested in the field rather than in
laboratory assays after stocks have been cultured in a laboratory environment. This method was used in one of the releases
with *Drosophila* carried out by Taylor and Powell (1978).
Another possibility (Fig. 1c) is to start with a laboratory
population and release it in the field. Individuals are captured on different field resources and lines are cultured for
a generation in the laboratory before their progeny are
released in the field. This design also overcomes the problem
of genotype-environment interaction because both generations
are bred in the same environment although the environmental
component of variance for insects cultured in the laboratory
may be smaller than for insects collected in the more heterogeneous field situation, inflating the heritability estimate.

In *Drosophila*, several attempts have now been made to
extend the numerous studies on resource response carried out
in the laboratory to field conditions. Early work on field
responses involved capture-mark-release-recapture experiments
(reviewed in Turelli *et al.*, 1984; Hedrick, 1986). In these
experiments, flies were captured from two or more ecologically
distinct habitats and usually released at a point between the
habitats. A tendency of flies to return to their original
habitat was interpreted as evidence for habitat fidelity.
Unfortunately, these experiments produced ambiguous results
and failed to consider the possibility that flies from different habitats varied in their physiological state, which can
lead to apparent habitat fidelity (Hoffmann and Turelli,
1985).

Some of the more recent *Drosophila* studies have used the
designs discussed above, and cases of resource-associated heritable variation in response within natural populations have
been described (e.g., Hoffmann *et al.*, 1984; Jaenike, 1986).
These studies have not considered variation in performance as
well as response, although performance variation has been
examined in winery populations of *D. melanogaster* in relation
to environmental ethanol. Differences in larval and adult
ethanol resistance between lines originating inside and outside winery cellars have been found (McKenzie and Parsons,
1974; Gibson and Wilks, 1988) but these differences have not
always been related to ethanol levels in natural breeding or
feeding sites. Performance tests using natural resources from
wineries have not been carried out.

In this paper we present the results of experiments with
design (a) in Figure 1 to test for resource-associated variation in response and performance in *Drosophila melanogaster*
and *D. simulans* populations. We considered several field situations where alternative breeding or adult feeding resources
occurred in close proximity. We also present some results from
experiments with design (c) in Figure 1 to test for heritable
variation in the response of *D. melanogaster* to two pairs of
fruit types.

Table I
Description of collections

Collection site	Date	Species	Resources at site
California			
Gilroy	8/84	*D. melanogaster*	pear, persimmon
Asti	8/84	*D. melanogaster*	apple, pear
Napa	11/84	*D. melanogaster*	pomace, seepage
Victoria			
Silvan	2/88	*D. melanogaster*	apple, peach
Donvale	3/87	*D. simulans*	apple, lemon, pear
Donvale	3/88	*D. simulans*	apple, pear
Tahbilk	4/86,4/87, 4/88	*D. melanogaster*	seepage,orange/apple, traps, grape residue

2. FIELD COLLECTIONS

2.1. Flies and Field Sites

Three collections were made in California and six in Victoria (Table I). Flies were collected from orchards by rearing adults from fallen fruit. Only one female was sampled from each fruit item. Adults were collected directly from field resources at the winery sites. Except for the Gilroy collection, two replicate lines were set up for each type of resource. Each line was started with progeny from 25-30 females. Experiments were carried out after lines had been cultured for 1-3 generations on laboratory medium. Some details about each collection follow.

Gilroy: Fruit was collected from 10 persimmon trees and the adjacent three rows of pear trees.

Asti: Fruit was collected from about 20 adjacent apple and pear trees.

Napa: Adults were aspirated from the seepage of oak storage casks in a cellar of the Krug winery. The seepage served as breeding sites for *D. melanogaster* as indicated by the presence of pupae and larvae. A second collection was made from a grape pomace area 5 km away. There were numerous *D. melanogaster* adults at this site, but immature stages were not found. The pomace probably attracted adults from the surrounding area that included apple orchards as well as wineries, so we used apple as an alternative resource in the laboratory tests.

Silvan: This orchard is described in Nielsen and Hoffmann (1985). Fruit was collected from adjacent rows of apple and peach trees.

Donvale: This orchard is in the Melbourne metropolitan area. In 1987, lemons were collected from three rows of trees, and apples and pears were collected from four rows adjacent to the lemon trees. Flies were only collected from apples and pears in 1988.

Tahbilk: This winery is described in McKenzie and Parsons (1972). Adults were collected by aspiration during three vintage periods from cask seepage in the cellar and from traps with fruit (apples and oranges) placed in a nearby lemon orchard. Adults were also collected from a grape residue pile described in McKenzie and McKechnie (1979) which is close to the lemon orchard.

2.2. Performance Tests

We used two of the performance tests described in Hoffmann and Turelli (1985). The ability of adults to utilize a resource as a feeding site was tested by their starvation resistance under 100% humidity after flies had been exposed to a resource for three days. Resistance was measured by scoring adult mortality at 12-hour intervals and calculating the time

Table II
Probabilities for terms from ANOVAs on data from starvation tests

		Origin	Resource	Line within Origin	Resource x Origin	Resource x Line
Gilroy	females	**	***	–	NS	–
	males	NS	NS	–	NS	–
Asti	females	NS	*	*	NS	NS
	males	NS	NS	NS	NS	NS
Napa	females	NS	NS	NS	*	NS
	males	NS	NS	*	**	NS
Donvale	females	NS	NS	NS	NS	NS
(1987)	males	NS	**	*	NS	NS
Donvale	females	NS	NS	NS	NS	NS
(1988)	males	NS	NS	NS	NS	NS
Silvan	females	NS	NS	NS	NS	NS
	males	NS	*	NS	NS	NS
Tahbilk	females	*	***	NS	*	NS
(1986)	males	NS	***	**	NS	NS
Tahbilk	females	NS	***	NS	*	NS
(1987)	males	NS	***	NS	*	NS
Tahbilk	females	NS	***	NS	NS	NS
(1988)	males	NS	***	NS	*	NS

* $P < 0.05$, ** $P < 0.01$, *** $P < 0.001$.

Table III
Mean LT50s (in hours) for the Napa lines
after exposure to wine or apple

Origin		Test Resource			
		Apple		Wine	
		Mean	SD	Mean	SD
Males					
Pomace	line 1	30.7	6.9	15.4	8.2
	line 2	32.6	4.4	21.9	10.8
Cellar	line 1	23.6	7.0	22.4	11.4
	line 2	31.4	8.1	32.5	2.9
Females					
Pomace	line 1	39.7	18.1	24.6	11.3
	line 2	55.7	10.1	16.7	5.3
Cellar	line 1	33.7	18.1	47.3	7.8
	line 2	35.1	18.9	47.1	14.1

Means and standard deviations are based on six replicates.

taken for half the flies to die by linear interpolation. Males and females were starved separately in groups of 10. We tested 4-6 replicate vials per line. The ability of lines to utilize a resource as a breeding site was tested by holding flies on a resource at a low density (three flies of each sex on 30 ml of pulp or wine) for one day and counting the number of emerging progeny.

Fruit resources for these tests consisted of ripe fruit that had been purchased or obtained from the collection sites. The fruit was cut into segments and left to rot for three days before it was pulped in a food processor. Fruit was not seeded with bacteria or yeasts. To mimic seepage from wine casks, we used red wine from the wineries where flies were collected. For experiments with Tahbilk lines, wine was seeded with residue that had been scraped from seepage areas in the cellar and placed in open containers overnight to simulate cask seepage. Vermiculite was added to the wine or "simulated seepage" to prevent flies from drowning.

Data were analyzed by nested analyses of variance (ANOVAs) with the replicate line term nested within the resource of origin term. Resource-associated differentiation was indicated by a significant resource-by-origin interaction. The significance of this term was determined by placing it over the resource by replicate line interaction term. We did not expect the resource by replicate line term to be significant because genetic differences between replicate lines were expected to be small and not resource-specific. We therefore pooled this term with the error whenever its probability

exceeded 0.25 (see Sokal and Rohlf, 1981) to increase the number of degrees of freedom.

A summary of the ANOVA results for the starvation data is
given in Table II. There was evidence for resource-associated
variation at the Napa and Tahbilk winery sites as indicated by
the significant origin-by-resource interactions. The only suggestion of heritable variation in the orchard collections was
the significant origin term in the Gilroy collection, although
this result is difficult to interpret because replicate lines
for the two types of fruit were not set up in this collection.

For the Napa flies, line means were similar after feeding
on apple, but lines from the cellar lived longer than pomace
lines after exposure to wine (Table III). The starvation test
was repeated on the next generation of flies to emerge from
these lines with similar results (not presented).

A summary of the Tahbilk results with *D. melanogaster* is
given in Figure 2. Details will be presented elsewhere
(Hoffmann and McKechnie, in preparation). For the 1986 collection, females from cellar lines were more resistant to starvation than those from lemon orchard lines after exposure to
seepage, but not after exposure to apple or orange pulp. For
the 1987 collection, males and females from both the cellar

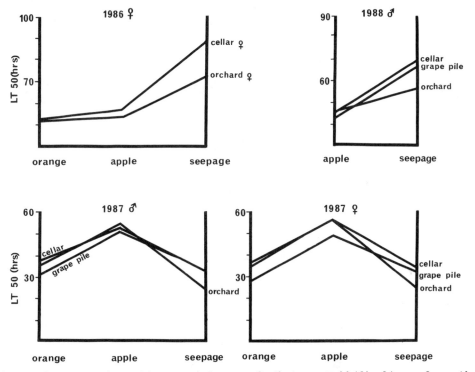

Figure 2. Mean starvation resistance of Chateau Tahbilk lines from the
cellar, lemon orchard or grape residue pile after exposure to orange, apple
or seepage.

lines and the grape residue pile lines were more resistant to starvation than those from the lemon orchard lines after seepage exposure. The residue pile lines performed more poorly than the other lines after exposure to fruit. In the 1988 collection, males from the cellar and residue pile lines performed better than those from the lemon orchard lines after seepage exposure, but there were no differences after exposure to apple. Figure 2 also indicates that resource type had a marked effect on starvation resistance, and that resource ranking changed from year to year. Flies exposed to seepage lived longer than flies exposed to apple in 1986 and 1988, while the reverse occurred in 1987. These changes probably reflect differences in the seepage batches and/or fruit pulp. In summary, flies were more starvation resistant after exposure to resources that resembled those from which they originated.

Productivity tests were carried out for some of the collections (Table IV). We only found evidence for heritable variation in the 1987 *D. simulans* Donvale lines, as indicated by the significant origin term and the significant origin-by-resource interaction. Line means (Table V) suggest that differences between fruit types were in the predicted direction. Lemon lines performed poorly on apple but performed well on lemon pulp compared to the other lines, while pear lines performed better on pear than the other lines. No resource-associated differences were evident in the 1988 collection from the Donvale site when we did not obtain flies from lemons.

2.3. Resource Response

We only tested behaviors involved in resource finding by *Drosophila* adults, and not those involved in resource acceptance. In one assay, adults were released in a large population cage (400 x 400 x 800 mm) covered with gauze at two ends as described in Hoffmann (1988). Attraction to two resources was tested by placing 10 ml of pulp in trap vials. Each vial was covered with gauze and entry was through a funnel suspended in the gauze. Five vials of each fruit type were placed on the bottom of the cage. In a second assay, flies were

Table IV
Probabilities for terms from ANOVAs on data
from breeding site tests

	Origin	Resource	Line within Origin	Resource x Origin	Resource x Line
Donvale (1987)	**	***	NS	*	NS
Donvale (1988)	NS	NS	NS	NS	NS
Silvan	NS	NS	**	NS	NS
Tahbilk (1986)	NS	NS	NS	NS	NS
Tahbilk (1988)	NS	***	NS	NS	NS

* $P < 0.05$, ** $P < 0.01$, *** $P < 0.001$.

Table V
Mean productivities for lines from the 1987 Donvale collection
in the breeding site test

Origin		Test Resource					
		Apple		Lemon		Pear	
		Mean	SD	Mean	SD	Mean	SD
Apple	line 1	34	26	136	26	151	11
	line 2	61	36	96	55	161	29
Lemon	line 1	13	11	155	12	137	40
	line 2	9	6	158	34	152	15
Pear	line 1	36	57	148	19	174	63
	line 2	50	26	132	11	190	9

Means and standard deviations are based on three replicates.

released in a vertical wind tunnel described in Hoffmann *et al.* (1984). This wind tunnel tests for optomotor anemotactic behavior (upwind responses to odors using visual cues for orientation) which is important for the long distance location of food by insects. Flies were trapped after they had responded by flying upwind and following scented columns in an airflow. The resources used as attractants were prepared as described for the performance tests. In both assays, 200-300 flies (3-4 days old) from each of the two lines being compared were released. More than 80% of the flies were retrieved from traps in both assays after they were left overnight. Five replicate trials were carried out for each comparison.

Counts from each trial formed a 2x2 table (origin versus attractant) and two probability estimates ($P_{1|1}$, $P_{1|2}$) were obtained from each table to estimate the attraction index (Turelli *et al.*, 1984)

$$\Delta = P_{1|1} - P_{1|2}$$

where $P_{1|1}$ is the probability of flies originating from resource 1 being captured on that resource and $P_{1|2}$ is the probability of flies originating from resource 2 being captured on resource 1. This index varies from -1 to +1, and positive values indicate that lines are relatively more attracted to the resource from which they originated. The significance of the association between resource of origin and attractant was determined by combining the χ^2 values from chi-square contingency analyses on the replicate trials, and calculating the normal deviates as described in Hoffmann (1988).

Attraction indices from the wind tunnel experiments were positive for the two Californian collections with *D. melanogaster* and these were significant in three out of four cases (Table VI). Flies in the Asti and Gilroy populations

were therefore relatively more attracted to the type of
resource from which they originated, although the small
Δ values indicate this tendency was weak (i.e., flies were
only 5-11% more likely to be attracted to their resource of
origin than those from the alternative resource). Results from
the other orchard collection with D. *simulans* (Donvale) were
not significant even though flies were collected from similar
types of fruit.   For the Tahbilk experiments with D. *melano-*

Table VI
Resource response experiments in the wind tunnel

Collection site	Resource combination	Females		Males	
		Δ	Normal deviate	Δ	Normal deviate
Asti	apple/pear	0.05	1.84	0.08	2.55*
Gilroy	persimmon/pear	0.11	3.89***	0.08	2.25*
Donvale 1987	apple/lemon	0.01	0.01	-0.01	-0.20
	apple/pear	-0.01	-0.37	-0.07	-1.49
	lemon/pear	-0.06	-1.19	0.02	0.35
Tahbilk 1986	apple/seepage	0.07	1.92	0.03	1.51
	orange/seepage	-0.01	-0.58	0.06	1.98*
Tahbilk 1987	apple/seepage	0.01	0.14	-0.05	-0.85
	orange/seepage	-0.08	-1.98*	-0.08	-1.44

The association index (Δ) is defined in the text, and Δ values are based
on five replicate trials.
* P < 0.05, *** P < 0.001.

Table VII
Resource response experiments in population cages

Collection site	Resource combination	Females		Males	
		Δ	Normal deviate	Δ	Normal deviate
Donvale 1987	apple/lemon	0.05	1.36	0.03	0.99
	apple/pear	0.03	0.84	0.02	0.30
	lemon/pear	0.02	0.61	0.02	0.38
Tahbilk 1986	apple/seepage	0.15	3.67***	0.16	3.42***
	orange/seepage	0.11	3.28***	0.07	1.40
Tahbilk 1987	apple/seepage	0.02	0.38	0.05	1.05
Tahbilk 1988	apple/seepage	-0.04	-0.69	-0.02	-0.35

The association index (Δ) is defined in the text, and Δ values are based
on five replicate trials.
*** P < 0.001.

gaster, only cellar and lemon orchard flies were compared. The results suggest that flies were attracted to their resource of origin in the 1986 collections but not in the 1987 collections when there was a significant negative Δ value for one of the resource combinations.

Response tests in population cages were only carried out with the Victorian lines. There was no evidence for resource response variation in the 1987 Donvale collection or the 1987 and 1988 Tahbilk collections, but significant positive Δ values were obtained with the 1986 Tahbilk lines (Table VII). Cellar flies were relatively more attracted to seepage than lemon orchard flies regardless of whether seepage was paired with apple or orange.

3. FIELD RELEASES

Experiments were carried out with a mass-bred *D. melanogaster* laboratory stock initiated with the progeny of 30 inseminated females collected from an orchard with cherries, plums, peaches, apples and lemons. Flies were released at two places between paired capture sites (Fig. 3) at times when native *D. melanogaster* and *D. simulans* were rare. A capture

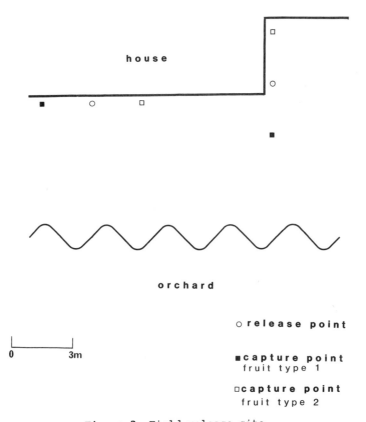

Figure 3. Field release site.

site consisted of five buckets each containing 500 ml of one
type of fruit (lemon, orange or apple) prepared as for the
performance tests, except that orange was only rotted for two
days. Two types of fruit were compared in an experiment and
two fruit combinations (lemon/apple, lemon/orange) were
tested. We captured between 15% and 35% of the released flies
at the capture sites.

An initial release of 11000-15000 adults (3-5 days old) was
made in the afternoon. Flies were netted from buckets the next
morning and held on laboratory medium until the afternoon.
Flies were then released again after they had been marked with
fluorescent dust to distinguish the type of fruit on which
they had been captured. The same buckets were used to recap-
ture the flies, although the position of the two fruit types
at the capture sites was reversed. Flies were retrieved the
next morning. Adults attracted to the same resource in both
releases were kept and placed on laboratory medium for oviposition. At least 30 females were set up for each fruit type.
Progeny were aged for 3-5 days and were marked according to
the fruit type to which their parents were attracted before
they were released in the afternoon and retrieved from capture
sites the next morning. A new batch of fruit pulp was used in
the progeny release.

Results from two trials with the two fruit combinations are
presented in Table VIII. The Δ values in this experiment represent the tendency of progeny to be attracted to the same
resource as their parents. There was a significant association
between the type of fruit to which adults were attracted and
the type of fruit to which their progeny were attracted for
one sex in all four releases. The Δ values were positive but
(with two exceptions) tended to be small. These data indicate
some heritable variation for field responses to fruit.

4. DISCUSSION

The experiments provide evidence for heritable variation in
performance at some field sites. The winery results are particularly convincing because differences in starvation resistance were found for at least one sex in each of the four collections. However, there were several cases where performance
differences were not found, including all but one of the
orchard collections. In general, results were inconsistent
between the sexes and between collections within the same
site.

A limitation of this study is that the resources we used in
the laboratory tests may not accurately reflect the resources
where we sampled the flies. We used the same types of fruit in
the orchard collections, but it is possible that rotting fruit
in the field differs from fruit rotted in the laboratory
because different microorganisms may be associated with decom-
position. The effect of microorganisms on the performance of
cosmopolitan *Drosophila* species is illustrated by the effect
of *Penicillium* molds on species distributions in citrus

Table VIII
Response of progeny in the field releases

Fruit combination		Females			Males		
		Δ^a	χ^2	N	Δ	χ^2	N
lemon/apple	trial 1	0.04	3.01+	510	0.15	13.77***	472
	trial 2	0.00	0.08	1223	0.03	6.50**	1707
lemon/orange	trial 1	0.04	3.91*	548	0.06	3.71+	304
	trial 2	0.14	14.10***	371	0.00	0.01	198

[a]Δ is defined in the text. N is the total number of marked flies captured.
+ $P < 0.10$, * $P < 0.05$, ** $P < 0.01$, *** $P < 0.001$.

(Atkinson, 1981): Both *D. immigrans* and *D. melanogaster* emerge from uninfected fruit, but *D. immigrans* predominates and survives better on infected citrus. In addition, simulating cask seepage in the winery tests was problematical. The wine we used for the Napa collection had an initially high alcohol content that is not found in seepage (Gibson *et al.*, 1981). The simulated seepage used in experiments with the Tahbilk flies is probably more similar than wine to cellar resources because most of the alcohol has evaporated (S.W. McKechnie, pers. comm.). The extent to which seepage reflects resources at the grape residue pile is unknown, although ethanol levels at the surface of this pile are low (McKenzie and McKechnie, 1979). Despite these difficulties, significant differences in performance were always in the "right" direction: Flies performed relatively better on the resource type which appeared to be more closely related to their resource of origin. The large number of cases where performance differences were not found may reflect an absence of heritable variation or our inability to simulate performance on the natural resources.

Do performance differences have a genetic basis? Differences are heritable because lines were cultured under the same environmental conditions and replicate lines originating from a resource showed similar results. However, they may not be genetic if flies from different resources carry microorganisms that persist during laboratory culturing. Transmitted microorganisms may colonize fruit or seepage once flies are transferred onto resources, and they may provide a different feeding or breeding environment. Transmission of yeasts on laboratory medium by *D. melanogaster* has been demonstrated by Begon (1974). An effect of microorganisms on the quality of resources as breeding sites has been demonstrated in cactophilic *Drosophila* (e.g., Vacek, 1982).

We have evidence against this possibility for the Tahbilk lines (Hoffmann and McKechnie, in preparation). In one experiment with the 1987 lines, adults from cellar and lemon orchard lines held in bottles with seepage were swapped after 1.5 days, so that lemon flies were held for 1.5 days on seepage that had been exposed to cellar flies (and vice versa). This did not alter performance differences between the lines. In another experiment with the 1988 Tahbilk lines, F_1 males were obtained from reciprocal crosses between one of the grape residue pile and lemon orchard lines. Performance of the F_1's was intermediate regardless of the maternal parent. Finally, we have shown that performance differences persist in adults bred from eggs that have been dechorionated and that differences do not arise after lines have been cultured on seepage and fruit medium under conditions that minimize selection. Nevertheless it is possible that the transmission of microorganisms accounts for productivity differences between the 1987 Donvale lines.

Genetic differences in performance may have arisen as a consequence of natural selection at the adult or larval stages. The starvation results suggest that selection may have occurred at the adult stage although there could be a correlation between larval and adult performance. Another possibility is that flies were differentially attracted to a resource and that there was an association between performance and behavioral variation (see below). Such an association could lead to performance differences even in the absence of strong selection.

Performance differences seemed to be more common in winery collections than in the orchard collections and this could be related to several factors. Differences between the composition of seepage and fruit may have been larger than composition differences between types of fruit, and this could affect the intensity of divergent selection. Resource-associated divergence may depend on relatively permanent resources because repeated episodes of selection will increase genetic divergence if individuals return to the resource types from which they eclosed. Wineries with large storage casks such as found at Chateau Tahbilk and Krug probably provide permanent resources, but most types of fruit are only available ephemerally in orchards. Another possibility is that gene flow was lower in winery sites than in orchard sites so that genetic differences between resources were more likely to persist. However, this seems unlikely at Chateau Tahbilk because there is movement of flies into the cellar during vintage (McKenzie, 1975) and we found differences between the grape residue pile and lemon orchard even though these sites were a few meters apart.

We found evidence for resource-associated variation in olfactory response in two Californian populations and one of the Tahbilk collections. The low Δ values indicate that these are weak effects, although differences in resource attraction were always in the predicted direction. This supports previous work with *D. melanogaster* from the Silvan orchard site which indicated that lines derived from peach, plum and apple showed a weak tendency to be captured with their resource of origin (Hoffmann *et al.*, 1984). The inconsistent results from the Tahbilk collection may reflect an absence of heritable variation in 1987-1988, although there may also have been variation in odors associated with different batches of seepage or fruit pulp.

The absence of response differences in some of the collections does not imply the absence of heritable variation for other behaviors affecting resource response, particularly as behaviors involved in oviposition and attraction to food traps may not be genetically correlated in *Drosophila* (Jaenike, 1986). The field releases suggest that heritable variation may often contribute to phenotypic variation in resource response, although this contribution may be small and only detected when large sample sizes are used. We should emphasize that flies for the field releases were bred on laboratory medium and the environmental component of variance may be larger when flies are bred on natural resources. Consequently, the field heritability may be lower than indicated in these experiments, although it is difficult to predict if the genetic component of variance would be larger or smaller in field-cultured flies.

We have not directly addressed the question of whether heritable variation in resource performance and field responses are correlated. Obviously such a correlation cannot occur when there is no resource-associated variation in response or performance as may be the case for some of our field sites. A correlation between performance and response could also arise as a consequence of adult experience rather than behavior-genetic variation if adults become attracted to the same resource from which they eclose or feed. Laboratory studies have shown that oviposition and attraction behaviors in *Drosophila* can be modified by the resource that adults encounter (Jaenike, 1982; Hoffmann, 1985; 1988). However, experience with fruit pulp does not seem to affect the field behavior of *D. melanogaster* in a consistent manner (Hoffmann, 1988; Turelli and Hoffmann, 1988).

In conclusion, we have evidence for some resource-related heritable variation in performance and response in some field sites. This suggests that genetic variation may be associated with environmental heterogeneity in natural populations of *D. melanogaster* and *D. simulans* even in cases where there is substantial gene flow between sites. These observations may provide a starting point for answering questions about the maintenance of genetic variation by environmental heterogeneity and host range expansion.

5. SUMMARY

Methods for testing whether heritable variation in resource use (response and performance) is associated with resource heterogeneity in nature are briefly reviewed. Two methods were applied to *Drosophila melanogaster* and *D. simulans* populations. In the first method, lines originating from different pairs of food resources in the field were bred under the same laboratory environment before being tested for resource use in laboratory assays. Heritable variation in performance was found in five out of nine collections. There were heritable differences in resource attraction in three out of six collections. Differences between lines were always in the predicted direction: Lines performed relatively better and were relatively more attracted to the type of resource from which they originated. In the second method, flies from a laboratory stock were released in the field and captured on two different pairs of fruit types. Progeny from the attracted flies were raised on laboratory medium and their responses to the same fruit types were tested in the field. The flies showed a weak tendency to be attracted to the same resources as their parents, indicating some heritable variation for resource response in the field.

ACKNOWLEDGEMENTS.   We are grateful to Larry Harshman and Steve McKechnie for help with collecting flies, and to David Edwards for technical assistance. We thank Phil Hedrick, Michael Turelli and an anonymous referee for useful comments on an earlier draft of this paper. The Hasan family and the owners of the Chateau Tahbilk and Krug wineries kindly provided access to their fly populations. This work was supported by a grant from the Australian Research Committee.

References

Atkinson, W. D., 1981, An ecological interaction between citrus fruit, *Penicillium* moulds and *Drosophila immigrans* Sturtevant (Diptera: Drosophilidae), *Ecol. Entomol.* **6:** 339-344.

Begon, M., 1974, *Drosophila* and dead laboratory medium, *Drosoph. Inf. Serv.* **51:**106.

Feder, J. L., Chilcote, C. A., and Bush, G. L., 1988, Genetic differentiation between sympatric host races of the apple maggot fly *Rhagoletis pomonella*, *Nature, Lond.* **336:** 61-64.

Futuyma, D. J., and Peterson, S., 1985. Genetic variation in the use of resources by insects, *A. Rev. Ent.* **30:** 217-238.

Gibson, J. B., May, T. W., and Wilks, A. V., 1981, Genetic variation at the alcohol dehydrogenase locus in *Drosophila melanogaster* in relation to environmental variation: ethanol levels in breeding sites and allozyme frequencies, *Oecologia* **51:** 191-198.

Gibson, J. B., and Wilks, A. V., 1988, The alcohol dehydrogenase polymorphism of *Drosophila melanogaster* in relation to environmental ethanol, ethanol tolerance and alcohol dehydrogenase activity, *Heredity* **60:** 403-414.

Hedrick, P. W., 1986, Genetic polymorphism in heterogeneous environments: A decade later, *Annu. Rev. Ecol. & Syst.* **17:** 535-566.

Hoffmann, A. A., 1985, Effects of experience on oviposition and attraction in *Drosophila*: Comparing apples and oranges, *Am. Nat.* **126:** 41-51.

Hoffmann, A. A., 1988, Early adult experience in *Drosophila melanogaster*, *J. Insect Physiol.* **34:** 197-204.

Hoffmann, A. A., Parsons, P. A., and Nielsen, K. M., 1984, Habitat selection: olfactory response of *Drosophila melanogaster* depends on resources, *Heredity* **53:** 139-143.

Hoffmann, A. A., and Turelli, M., 1985, Distribution of *Drosophila melanogaster* on alternative resources: Effects of experience and starvation, *Am. Nat.* **126:** 662-679.

Jaenike, J., 1982, Environmental modification of oviposition behavior in *Drosophila*, *Am. Nat.* **119:** 784-802.

Jaenike, J., 1986, Genetic complexity of host-selection behavior in *Drosophila*, *Proc. natn. Acad. Sci. USA* **83:** 2148-2151.

Jaenike, J., and Selander, R. K., 1979, Ecological generalism in *Drosophila falleni:* genetic evidence, *Evolution* **33**: 741-748.

McKenzie, J. A., 1975, Gene flow and selection in a natural population of *Drosophila melanogaster, Genetics* **80**: 349-361.

McKenzie, J. A., and McKechnie, S. W., 1979, A comparative study of resource utilization in natural populations of *Drosophila melanogaster* and *D. simulans, Oecologia* **40**: 299-309.

McKenzie, J. A., and Parsons, P. A., 1972, Alcohol tolerance: an ecological parameter in the relative success of *Drosophila melanogaster* and *D. simulans, Oecologia* **10**: 373-388.

McKenzie, J. A., and Parsons, P. A., 1974, Genetic microdifferentiation in a natural population of *Drosophila melanogaster* in relation to alcohol in the environment, *Genetics* **77**: 385-394.

Mitter, C., and Futuyma, D. J., 1979, Population genetic consequences of feeding habits in some forest Lepidoptera, *Genetics* **92**: 1005-1021.

Nielsen, K. M., and Hoffmann, A. A., 1985, Numerical changes and resource utilization in orchard *Drosophila* populations, *Aust. J. Zool.* **33**: 875-884.

Nielsen, K. M., Hoffmann, A. A., and McKechnie, S. W., 1985, Population genetics of the metabolically related *Adh, Gpdh* and *Tpi* polymorphisms in *Drosophila melanogaster*: II. Temporal and spatial variation in an orchard population, *Génét. Sél. Evol.* **17**: 41-58.

Singer, M. C., 1986, The definition and measurement of oviposition preference in plant-feeding insects, in: *Insect- Plant Interactions* (T. A. Miller, and J. Miller, eds), Springer-Verlag, New York, pp. 65-94.

Singer, M. C., Ng, D., Thomas, C. D., 1988, Heritability of oviposition preference and its relationship to offspring performance within a single insect population, *Evolution* **42**: 977-985.

Smith, D. C., 1988, Heritable divergence of *Rhagoletis pomonella* host races by seasonal asynchrony, *Nature, Lond.* **336**: 66-67.

Sokal, R. R., and Rohlf, F. J., 1981, *Biometry*, Freeman, New York.

Taylor, C. E., and Powell, J. R., 1978, Habitat choice in natural populations of *Drosophila, Oecologia* **37**: 69-75.

Tavormina, S. J., 1982, Sympatric genetic divergence in the leaf-mining insect *Liriomyza brassicae* (Diptera: Agromyzidae), *Evolution* **36**: 523-534.

Turelli, M., Coyne, J. A., and Prout, T., 1984, Resource choice in orchard populations of *Drosophila, Biol. J. Linn. Soc.* **22**: 95-106.

Turelli, M., and Hoffmann, A. A., 1988, Effects of starvation and experience on the response of *Drosophila* to alternative resources, *Oecologia* **77**: 497-505.

Vacek, D. C., 1982, Interactions between microorganisms and cactophilic *Drosophila* in Australia, in: *Ecological Genetics and Evolution: The Cactus-Yeast-Drosophila Model System* (J. S. F. Barker, and W. T. Starmer, eds), Academic Press Australia, Sydney, pp. 175-190.

Factors Maintaining Genetic Variation For Host Preference in Drosophila

JOHN JAENIKE

1. INTRODUCTION

With the recognition in recent years that behavior may be the most important proximate determinant of patterns of host utilization in insects (Futuyma, 1983; Jermy, 1984), increasing efforts are being made to determine if populations harbor detectable levels of genetic variation for host-selection behavior. Such variation, it is believed, may be the raw material on which selection could bring about evolutionary changes in host usage. Because of the diversity of resources they use and the facility with which genetic analyses can be carried out, a substantial number of recent studies have focused on various species of *Drosophila*.

Among the species in which genetic variation for host preference has been demonstrated are *D. melanogaster* (Hoffmann *et al.*, 1984), *D. pseudoobscura* (Klaczko *et al.*, 1986), *D. mojavensis* (Lofdahl, 1986), *D. buzzatii* (Barker *et al.*, 1981, 1986), *D. tripunctata* (Jaenike, 1985), *D. grimshawi* (Carson and Ohta, 1981), and D. *suboccidentalis* (Courtney and Chen, 1988). Further reports of such variation in other species are unlikely to come as a great surprise. The questions now become: what is this variation good for, and how is it maintained within populations? I focus here on the latter question, although knowing how this variation is maintained may well shed light on its adaptive significance.

Most of my discussion concerns *D. tripunctata*, in which a good deal of genetic variation for host preference is found within natural populations. After briefly reviewing the evidence for this variation, I turn to possible mechanisms that may be responsible for its maintenance. I conclude that the adaptive significance of this variation is far from clear.

JOHN JAENIKE * Department of Biology, University of Rochester, Rochester, New York 14627, USA.

Ecological and Evolutionary Genetics of Drosophila, Edited by
J.S.F. Barker *et al.*, Plenum Press, New York, 1990

2. GENETIC VARIATION FOR ADULT PREFERENCE AND LARVAL PERFORMANCE ON DIFFERENT HOSTS IN *DROSOPHILA TRIPUNCTATA*

Drosophila tripunctata is among the more polyphagous drosophilids in North America, breeding on a variety of fruits and mushrooms in nature (Sturtevant, 1921; Carson and Stalker, 1951; Lacy, 1984). Most of the assays of host preference I have conducted involved tomatoes (*Lycopersicum esculentum*) and commercial mushrooms (*Agaricus bisporus*) as alternative breeding sites. These were chosen because mushroom- and fruit-breeding species of *Drosophila* differ substantially in their preference for and ability to develop on these foods (Jaenike, 1989). Thus, within *D. tripunctata*, a fly's propensity to use tomatoes or mushrooms as breeding sites is one measure of the degree to which it resembles a fruit- or mushroom-breeding species of *Drosophila*. Additionally, tomatoes have been cultivated for the past 200 years or so in the United States (Rick, 1978) and it is not uncommon to find *D. tripunctata* feeding on its decomposing fruits (personal observations).

Within a given habitat, the first stage in the host-selection process entails movement to and settling at a potential breeding site. This is followed by acceptance or rejection of this site for oviposition. Genetic variation for the initial stage of host selection, which I refer to as settling behavior and assay by determining the fraction of flies captured at different breeding sites, has been found between two isofemale strains of *D. tripunctata* that were derived from flies collected in 1981 from a single population in the Smoky Mountains of Tennessee. When flies of these strains are released in the woods, they show consistent and large differences in their probabilities of settling at mushrooms versus tomatoes. Over three replicate experiments, an average of 62% of the captured individuals of strain S64 were collected at mushrooms, whereas 81% of strain S74 flies were found at tomato baits (Jaenike, 1985). Thus, in terms of settling behavior, these two strains differ in their rank order preferences.

Oviposition-site preference (OSP) has been assayed in this species by placing flies, individually or in groups, in petri dishes containing slices of media prepared from tomatoes and mushrooms. The fraction of eggs laid on each food indicates the tendency of a fly to accept a potential breeding site for oviposition, once it has come into close proximity to it. Such assays have revealed great differences among 175 isofemale strains collected in 1984 from seven populations of *D. tripunctata* (Fig. 1) with the fraction of eggs laid on tomatoes varying from 23% to 93%. These strains vary not only in their specificity but also their rank order preference for these different hosts (see Singer (1982) for a discussion of these terms). It is worth noting that each of the seven populations studied harbored significant genetic variation for OSP. Crosses among strains have shown that this variation has a heritable genetic basis, with no detectable maternal effects (Jaenike, 1987).

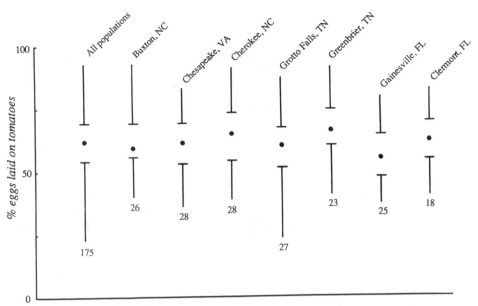

Figure 1. Box plots of variation among isofemale strains in oviposition-site preference. Range, median, and 25% and 75% quartiles are indicated for each population studied and for strains of all populations pooled. Numbers of strains studied are shown below each bar. The average within-strain standard error of the mean is 5.2%.

Strain variation has also been found in the ability of larvae to develop on different substrates (Jaenike, 1989). Because strains may differ in general vigor, evidence for genetic variation in relative performance is revealed by strain-by-breeding site interactions in analyses of variance of various components of fitness. The breeding sites considered were tomatoes versus mushrooms, and instant *Drosophila* medium prepared with or without alpha-amanitin, a potent toxin found in some mushrooms of the genus *Amanita*. (Mycophagous species of *Drosophila* are far more tolerant of amanitin than are non-mycophagous species (Jaenike *et al.*, 1983).) Genetic variation in relative performance was found for both pairs of breeding sites tested, and it is manifested by differences in development time. The strain-by-breeding site interaction was significant at the 0.05 level in the experiment using tomatoes and mushrooms and at the 0.01 level in the test of amanitin tolerance (Jaenike, 1989). As with OSP, this variation in larval performance was significant at the within-population level.

In summary, *D. tripunctata* harbors substantial genetic variation within populations both for host preference, including settling behavior and oviposition-site preference, and for relative larval performance on different breeding sites. Possible mechanisms for the maintenance of this variation are explored in the following sections.

3. GENE FLOW AMONG DIFFERENTIATED POPULATIONS

Many species of insects exhibit geographic variation in their patterns of host utlization (Fox and Morrow, 1981). In a few cases this has been shown to have a genetic basis: for instance, differences among populations of the butterfly *Euphydryas editha* in oviposition behavior and larval performance on different host plants is probably based to some extent on genetic differentiation (Singer, 1971; Rausher, 1982). Moderate levels of gene flow among such differentiated populations could maintain genetic variation for resource use within local populations.

This mechanism does not seem to be a likely explanation for the maintenance of variation within populations of *D. tripunctata*. Although there is some variation among populations in their mean OSPs (Fig. 1), this variation is not statistically significant (Jaenike, 1989); even if it were, it is not nearly great enough to account for the wide range in OSP among individual strains within populations. Similarly, for no measure of resource-specific larval performance were significant differences among populations found (Jaenike, 1989). Therefore, in order to explain the maintenance of this genetic variation for resource use, it would seem necessary to invoke processes that operate within populations.

4. GENETIC CORRELATIONS BETWEEN PREFERENCE AND PERFORMANCE

Rausher (1983) has argued that, in terms of preferences manifested after potential breeding sites are encountered, female insects should prefer to oviposit on the hosts that are most suitable for the development of their offspring. Indeed, phenotypic correlations between female oviposition preference and offspring performance on different host plants have been reported in two species of phytophagous insects, *Liriomyza sativae* (Via, 1986) and *Euphydryas editha* (Singer *et al.*, 1988; Ng, 1988). However, a genetic basis for these correlations has not yet been established.

Because genetic variation for oviposition-site preference and substrate-specific larval performance has been demonstrated in *D. tripunctata*, genetic correlations between preference and performance are possible in this species. Because no significant variation among populations in either OSP or larval performance has been found, the following correlations are based on strains of all populations pooled. In order to correct for differences among strains in general vigor, measures of larval performance were based on the difference in development time on alternative substrates: DTMT is a strain's mean development time on mushrooms minus its mean development time on tomatoes, and DTCA is the mean development time on amanitin-free medium minus that on medium containing amanitin. The correlation among strains between OSP (the fraction of eggs laid by a strain on tomatoes) and DTMT was only 0.06 (n = 148 strains), while the correlation between OSP and DTCA was

Table I

Analyses of variance of larval performance of strains as a function of the bait at which the founding females were caught

Performance measure[a]	Source of variation	df	SS	F	P
DTMT	bait	2	0.222	0.80	0.45
	error	145	20.234		
	total	147	20.456		
DTCA	bait	2	6.110	1.57	0.21
	error	130	252.281		
	total	132	258.391		

[a]See text for definitions of performance measures.

−0.001 (n = 133) (Jaenike, 1989). Since neither correlation approached significance, this survey of isofemale strains provides no evidence for a genetic correlation between oviposition-site preference and larval performance in *D. tripunctata*.

Elsewhere, I have shown that two important behavioral components of host selection - settling behavior and oviposition-site preference - are genetically independent (Jaenike, 1986). Perhaps larval performance is correlated with settling bevhavior. The following is an indirect test of this possibility. At each site where *D. tripunctata* was collected in 1984, tomato, mushroom, and banana baits were set out, and I recorded the bait type at which the founder of each isofemale strain was collected. Because such strains may differ substantially in settling behavior (see above), the bait at which an individual fly was caught is probably partly a function of its genotype. Thus, analyses of variance of larval performance were carried out with bait of initial capture as the independent variable. As can be seen in Table I, neither DTMT nor DTCA varied significantly among strains descended from flies caught at different bait types. (ANOVAs in which tomato baits are lumped with bananas or in which baits are nested within collection sites similarly fail to show any correlation between bait of capture and larval performance). Thus, these tests, which admittedly are weak, provide no evidence for the idea that females tend to settle at breeding sites that are particularly suitable for the development of their offspring.

As a final note, I would point out that, contrary to what may be one's intuitive expectation, there are theoretical problems with the maintenance of genetically-based preference-performance correlations within populations, if performance on a given substrate is a density-independent property of an individual's genotype (Jaenike and Holt, 1990). In general, one would expect fixation of alleles whose carriers prefer the most suitable breeding sites.

5. DENSITY-DEPENDENT PROCESSES

If the suitability of a given breeding site decreases as the number of individuals utilizing it increases, then there can be frequency-dependent selection favoring alleles whose carriers prefer relatively under-utilized resources (Rausher, 1984). Hedrick (Chapter 12, this volume) has modelled mixtures of hard- and soft-selection and found that even a small amount of soft-selection can be very important for maintaining genetic polymorphisms affecting habitat use. Here I consider two density-dependent processes, competition and parasitism, that could bring about soft-selection on host preference.

Beaver (1984) has argued that insects which breed in discrete patchy resources frequently experience intense larval competition for food. Such competition has been found in natural populations of both fruit- and mushroom-breeding species of *Drosophila* (Atkinson, 1979; Grimaldi and Jaenike, 1984). Although *D. tripunctata* has not yet been studied in this regard, my hunch is that those larvae feeding in mushrooms may frequently experience food shortage as a result of intra-specific competition. This is based on several observations. (1) Among mycophagous species studied in New York, not only were all four species of *Drosophila* studied found to experience food limitation in nature, but so were a cranefly (*Limonia triocellata*) and a wood gnat (*Silvicola alternatus*) that were feeding as larvae in the same fungi as the *Drosophila* species (Grimaldi and Jaenike, 1984). Thus, larval food limitation may be a general feature of Diptera that feed in fleshy fungi. (2) Shorrocks and Rosewell (1987) have shown that within species of fruit- and mushroom-feeding *Drosophila*, the distributions of eggs tend to be aggregated across breeding sites, and that these distributions are generally independent among species. As a result, most competition among larvae is expected to be within species. Data from Grimaldi (1983) confirm the aggregated distribution of larvae for North American species of mycophagous *Drosophila*. Larvae feeding in naturally occurring mushrooms were provided with supplemental food in the field, in order to reduce larval competition and get a better estimate of the numbers of eggs which had been laid on these mushrooms. Using numbers of flies bred from these supplemented mushrooms, the following values of k of the negative binomial distribution were obtained: *D. falleni*, 0.57; *D. recens*, 0.33; *D. putrida*, 0.36; and *D. testacea*, 0.27. Such values of k indicate highly aggregated distributions. (3) Finally, because *D. tripunctata* is an abundant species, at least in the southern half of its range, I would guess that it frequently experiences significant intraspecific competition for larval food.

For these reasons, it seems at least possible that such competition could bring about selection for an allele in *D. tripunctata* whose carriers prefer breeding sites where competition is less severe. If the level of competition on such a breeding site increases with the frequency of this allele, then frequency-dependent selection could maintain genetic variation for host preference by females seeking oviposition sites.   This possibility remains to be tested in the field.

Another potential means by which density-dependent population regulation may occur in a resource-specific manner is parasitism by the nematode *Howardula aoronymphium*.   Inseminated females of these worms infect larvae of mycophagous *Drosophila* by penetrating through their cuticle (Welch, 1959). These nematodes then produce hundreds of offspring within an adult fly, which escape from it via the anus or ovipositor when the fly visits a mushroom.   The net effects of these nematodes on *Drosophila* populations can be quite substantial. In *D. testacea*, for instance, it is not uncommon for 50% of the flies to be parasitized at any one time, and virtually none of the parasitized females are fertile (Jaenike, unpublished data).

Because nematodes are passed from one generation of flies to the next via the mushroom flies visit, it seems likely that the probability of infection of a larval *Drosophila* would be an increasing function of the number of flies that visit a mushroom.   As a result, there could be frequency-dependent selection in favor of flies that prefer to oviposit on resources that are relatively unattractive to the rest of the population.

This mechanism for maintaining variation in resource preference may work for mycophagous members of the *Drosophila quinaria* and *testacea* species groups that occur in the eastern United States, because all of them are parasitized by *H. aoronymphium* in nature (Montague and Jaenike, 1985).   However, it cannot work for *D. tripunctata*, as this species appears to be virtually immune to parasitism by nematodes.   Among flies collected from areas where other mycophagous *Drosophila* are commonly parasitized, not one out of 316 *D. tripunctata* was infected.   Furthermore, in three experiments attempting to infect *D. tripunctata* in the laboratory, only two flies out of 1038 became parasitized.   Similar experiments on other mycophagous *Drosophila* yield parasitism frequencies from 20% to 60%.   *D. tripunctata*'s high degree of resistance to parasitism means that this potentially resource-specific density-dependent factor cannot be invoked to explain the maintenance of genetic variation in host preference.

6. MUTATION-SELECTION BALANCE

With the exception of larval competition, it appears that none of the mechanisms outlined above can explain the maintenance of variation in host-selection behavior within populations of *D. tripunctata*.   An alternative hypothesis, which ought to be considered seriously, is that genetic variation in host preference is essentially neutral, maintained by mutation-selection balance.

In modelling the dynamics of host-selection behavior by insects, Courtney *et al.* (1989) have suggested that as an individual's motivation to oviposit increases, it will accept more and more of the potential hosts it encounters. Thus, it would seem that even if two individuals differ substantially in their genetically programmed ranking of hosts, their actual rates of oviposition and fractions of eggs laid on various hosts will be similar if each is highly motivated to oviposit much of the time. Under such circumstances, the genetic differences in preference are not manifested in differential patterns of host use. Although there may be an optimal behavioral phenotype and a single genotype corresponding to it, the patterns of host use of other genotypes may be sufficiently similar to the optimal type to render selection against them quite weak. Thus, genetic variation in preference could be maintained by a balance between mutation and weak stabilizing selection.

Courtney *et al.* (1989) reason that an insect's motivation to oviposit should be an increasing function of its current eggload. This then suggests that the degree to which genetic variation in host preference may be neutral will depend on the potential rate at which eggs can be produced relative to the rate at which they are laid. The former should reflect the rate at which suitable feeding sites are discovered, while the latter should be a function of the rate of finding acceptable oviposition sites.

Carson (1971) states that within most species of *Drosophila*, the variety of substrates on which adults feed is significantly greater than the range on which they will oviposit. Because *D. tripunctata* breeds to a large extent on fleshy fungi, which are unpredictable and ephemeral in occurrence, it may be that they rarely find acceptable oviposition sites. A mushroom that is already crowded with larvae or that is drying out may serve as food for an adult, but it offers no chance of survival for any eggs that would be laid there. If these speculations turn out to be true, then mycophagous species of *Drosophila*, including *D. tripunctata*, may frequently be highly motivated to oviposit, thus reducing the difference in host usage among flies that differ genetically in host preference.

Similar arguments could be applied to the genetic variation in settling behavior that has been found in *D. tripunctata*. If odors emanating from a potential breeding site are detectable over only short distances, then a fly highly motivated to oviposit will often be attracted to any such breeding site it detects. In this manner, flies that differ in their genetic predisposition to settle at different breeding sites may in fact differ rather little in their actual patterns of resource use, if breeding sites are encountered only rarely.

These ideas suggest that variation in host preference is most likely to be neutral when breeding sites are scarce. When they are more abundant, higher rates of oviposition lead to lower eggs loads, which then make it more likely that a fly will reject low ranking hosts. Under such conditions, flies that differ in host preference will also differ in their patterns of host utilization.

Occasionally, mushrooms of many species are simultaneously very abundant. At such times, when flies are most able to manifest genetically-conferred differences in host preference, there will be strong selection to oviposit only on the most suitable breeding sites. Therefore, if there is stabilizing selection on host preference, that selection is likely to be greatest when resources are most abundant. Under the motivational model, then, host preference in mycophagous drosophilids is expected to be nearly neutral under most circumstances but strongly selected at times of mushroom abundance.

7. LACK OF GENETIC VARIATION FOR HOST PREFERENCE IN *DROSOPHILA QUINARIA*

The thoughts in the previous section stem from a selection experiment I carried out on *Drosophila quinaria*, a specialist that breeds almost exclusively on skunk cabbages, *Symplocarpus foetidus* (Araceae). Because larvae of this species develop very well on mushrooms (James *et al.*, 1988), it seemed that host-selection behavior of females was the proximate determinant of its restriction to skunk cabbage. In order to test this, I had hoped to develop a population of this species that preferred mushrooms as oviposition sites.

Sixty-seven isofemale strains of *D. quinaria* were established from flies bred from skunk cabbages collected on Deer Isle, Maine in 1986. A control population set up from these strains was maintained in large cages with skunk cabbage inflorescenses, fruits, and spathes as the sole food resource. A selected population was maintained in a large (3m x 3m x 2.5m) screened enclosure designed to simulate the natural habitat of these flies. This enclosure, which was located in a greenhouse, contained three 1m to 2m tall white spruce trees (*Picea glauca*) and eight to ten skunk cabbage plants, next to which were placed additional fruits and spathes. In addition, various common species of field-collected mushrooms (e.g., *Amanita rubescens*, *A. muscaria*, *Russula subfoetans*, *Boletus edulis*) were placed in the enclosure twice each week. The population was maintained by collecting the mushrooms after they had been in the cage for three to four days, breeding out any *D. quinaria* from them, and releasing these flies back into the cage. The skunk cabbage breeding sites were replaced at regular intervals, preventing emergence of any adults from them. Thus, the selected population was maintained solely from eggs that were laid on mushrooms. At any one time, the selected population comprised several hundred individuals.

After two years of continuous selection, the control and selected populations were assayed for use of mushrooms as breeding sites. For each population, five replicate groups of flies were reared independently. When they had matured as adults, they were released into large population cages containing mushrooms (*Agaricus campestris* and *Amanita rubescens*) and skunk cabbage fruits as oviposition sites. Thirty males and thirty females were released into each cage, with five replicate cages for control and selected flies. Flies were removed after 48 hours and the number of flies emerging from these breeding sites subsequently counted. The number of flies emerging is a function of the number of eggs laid and the probability of egg to adult survival on a given breeding site.

Despite two years of continuous selection, the control and selected flies were virtually identical in this assay. The fractions of control and selected flies emerging from mushrooms were 42.4% ± 6.4% and 43.4% ± 4.9%, respectively. This experiment suggests that the Deer Isle population of *D. quinaria* harbored little genetic variation for oviposition-site preference. (It is conceivable, but very implausible, that the similarity between the control and selected flies was due to an increase in oviposition preference for mushrooms in the selected flies that was exactly balanced by a decrease in their ability to survive on them.)

Suitable oviposition sites would seem to be abundant and predictable for populations of *D. quinaria*. Skunk cabbages are long-lived perennials that appear, flower, and set fruit in the same sites every year. Thus, with such an abundance of oviposition sites, individual flies may generally have only low levels of motivation to oviposit. If so, then genetic differences among them in host preference may well lead to differences in host utilization. And if alternative hosts are inferior resources, this will select against alleles whose carriers prefer breeding sites other than skunk cabbage for oviposition. In particular, because *D. quinaria* is a large fly with fairly slow development, larvae developing in mushrooms may lose out in competition with the smaller, rapidly developing mycophagous drosophilids.

8. SUMMARY AND CONCLUSIONS

Genetic variation for host preference appears to be common in the genus *Drosophila*. In *D. tripunctata*, significant genetic variation has been found within populations for both long distance movement to potential breeding sites and for oviposition-site preference once flies have encountered such resources. This paper addresses the question of how such variation can be maintained within populations.

The lack of significant genetic differentiation among populations of *D. tripunctata* in host preference indicates that gene flow among them cannot explain the substantial degree of variation found within individual populations. This suggests that explanations for the maintenance of such variation should be sought among processes that occur within populations. There is no empirical support for the idea that variation in host preference is maintained by a genetic correlation with larval performance on different hosts, nor for the possibility that parasitism by the nematode *Howardula aoronymphium* can bring about soft-selection on host preference in *D. tripunctata*, as this species, unlike many other mycophagous drosophilids, is resistant to infection by this parasite. The most likely adaptive mechanism for the maintenance of genetic variation in host preference is larval competition for food, which is known to occur in other mycophagous species of *Drosophila*, as it could bring about soft-selection favoring the use of a variety of resources by a population.

As an alternative to the possibility of soft-selection mediated via larval competition, I have proposed a neutral model, in which variation is maintained as a balance between mutation and weak stabilizing selection. The strength of stabilizing selection is hypothesized to be related to potential rates of egg production and oviposition. When potential rates of oviposition on suitable substrates are low, as they may frequently be in *D. tripunctata*, this selection is likely to be weak, because flies will use most breeding sites encountered, regardless of their genetically-conferred preferences. Under these conditions, genetic variation for host preference is postulated to be selectively neutral. Where potential rates of oviposition are higher, as in the skunk-cabbage specialist *D. quinaria*, selection should weed out genetic variants that do not rank skunk cabbage as the most acceptable breeding site. In fact, a long-term selection experiment failed to reveal any genetic variation for host preference in this species. Among mycophagous species of *Drosophila*, selection on host preference is likely to be strongest during occasional periods when a variety of mushroom species are abundant.

While much of the foregoing is speculative, it does suggest that some insight into the patterns of host utilization and the mechanisms maintaining genetic variation for host preference could be gained by determining rates of egg production and oviposition under natural conditions.

ACKNOWLEDGEMENTS. I wish to thank Alan Templeton for insightful comments on the ideas developed here. I am grateful to Avis James and Steve Iveson for technical help with the selection experiment on *D. quinaria*. The work reported here was supported by NSF grants BSR 83-10141 and BSR 86-05196.

References

Atkinson, W. D , 1979, A field investigation of larval competition in domestic *Drosophila*, *J. Anim. Ecol.* **48**:91-102.

Barker, J. S. F., Toll, G. L., East, P. D., and Widders, P. R., 1981, Attraction of *Drosophila buzzatii* and *D. aldrichi* to species of yeasts isolated from their natural environment, II. Field experiments, *Aust. J. biol. Sci.* **34**:613-624.

Barker, J. S. F., Vacek, D.C., East, P. D., and Starmer, W.T., 1986, Allozyme genotypes of *Drosophila buzzatii*: feeding and oviposition preferences for microbial species, and habitat selection, *Aust. J. biol. Sci.* **39**:47-58.

Beaver, R. A., 1984, Insect exploitation of ephemeral habitats, *S. Pac. J. Nat. Sci.* **6**:3-47.

Carson, H. L., 1971, The ecology of *Drosophila* breeding sites, University of Hawaii, Harold L. Lyon Arboretum Lecture Number Two.

Carson, H. L., and Ohta, A. T., 1981, Origin of the genetic basis of colonizing ability, in: *Evolution Today* (G. G. E. Scudder, and J. L.

Reveal, eds), Carnegie-Mellon University, Pittsburgh, pp. 365-370.

Carson, H. L., and Stalker, H. D., 1951, Natural breeding sites for some wild species of *Drosophila* in the eastern United States, *Ecology* 32:317-330.

Courtney, S. P., and Chen, G. K., 1988, Genetic and environmental variation in oviposition behaviour in the mycophagous *Drosophila suboccidentalis* Sper., *Functional Ecol.*, 2:521-528.

Courtney, S. P., Chen, G. K., and Gardner, A., 1989, A general model for individual host selection, *Oikos* 55:55-65.

Fox, L. R., and Morrow, P. A., 1981, Specialization: species property or local phenomenon? *Science* 21:887-893.

Futuyma, D. J., 1983, Selective factors in the evolution of host choice by phytophagous insects, in: *Herbivorous Insects: Host-seeking Behavior and Mechanisms* (S. Ahmad, ed.), Academic Press, New York, pp. 227-244.

Grimaldi, D. A., 1983, Ecology and competitive interactions of four coexisting species of mycophagous *Drosophila*, Masters Thesis, State University of New York at Binghamton.

Grimaldi, D., and Jaenike, J., 1984, Competition in natural populations of mycophagous *Drosophila*, *Ecology* 65:1113-1120.

Hoffmann, A. A., Parsons, P. A., and Nielsen, K. M., 1984, Habitat selection: olfactory response of *Drosophila* melanogaster depends on resources, *Heredity* 53:139-143.

Jaenike, J., 1985, Genetic and environmental determinants of food preference in *Drosophila tripunctata*, *Evolution* 39:362-369.

Jaenike, J., 1986, Genetic complexity of host-selection behavior in *Drosophila*, *Proc. natn. Acad. Sci. USA* 83:2148-2151.

Jaenike, J., 1987, Genetics of oviposition-site preference in *Drosophila tripunctata*, *Heredity* 59:363-369.

Jaenike, J., 1989, Genetic population structure of *Drosophila tripunctata*: patterns of variation and covariation of traits affecting resource use, *Evolution* 43:1467-1482.

Jaenike, J., and Holt, R. D., 1990, Genetic variation for habitat preference: evidence and explanations, *Am. Nat.*, in press.

Jaenike, J., Grimaldi, D. A., Sluder, A. E., and Greenleaf, A. L., 1983, Alpha-amanitin tolerance in mycophagous *Drosophila*, *Science* 221:165-167.

James, A. C., Jackubczak, J., Riley, M. P., and Jaenike, J., 1988, On the causes of monophagy in *Drosophila quinaria*, *Evolution* 42:626-630.

Jermy, T., 1984, Evolution of insect/host plant relationships, *Am. Nat.* 124:609-630.

Klaczko, L. B., Taylor, C. E., and Powell, J. R., 1986, Genetic variation for dispersal by *Drosophila pseudoobscura* and *Drosophila persimlis*, *Genetics* 112:229-235.

Lacy, R. C., 1984, Ecological and genetic responses to mycophagy in Drosophilidae (Diptera), in: *Fungus-Insect Relationships* (Q. Wheeler, and M. Blackwell, eds), Columbia Univ. Press, New York, pp. 286-301.

Lofdahl, K. L., 1986, A genetic analysis of habitat selection in the cactophilic species, *Drosophila mojavensis*, in: *Evolutionary genetics of Invertebrate Behavior* (M. D. Huettel, ed.), Plenum Press, New York, pp. 153-162.

Montague, J. R., and Jaenike, J., 1985, Nematode parasitism in natural populations of mycophagous drosophilids, *Ecology* 66:624-626.

Ng, D., 1988, A novel level of interactions in plant-insect systems, *Nature, Lond.* 334:611-613.

Rausher, M. D., 1982, Population differentiation in *Euphydryas editha* butterflies: larval adaptation to different hosts, *Evolution* 36:581-590.

Rausher, M. D., 1983, Ecology of host-selection behavior in phytophagous insects, in: *Variable Plants and Herbivores in Natural and Managed Systems* (R. F. Denno, and M. S. McClure, eds), Academic Press, New York, pp. 223-257.

Rausher, M. D., 1984, The evolution of habitat preference in a subdivided population, *Evolution* **38**:596-608.

Rick, C. M., 1978, The tomato, *Scient. Am.* **239**(2):76-87.

Shorrocks, B., and Rosewell, J., 1987, Spatial patchiness and community structure: coexistence and guild size of drosophilids on ephemeral resources, in: *Organization of Communities: Past and Present* (J. H. R. Gee, and P. S. Giller, eds), Blackwell, Oxford, pp. 29-51.

Singer, M. C., 1971, Evolution of food-plant preferences in the butterfly *Euphydryas editha*, *Evolution* **25**:383-389.

Singer, M. C., 1982, Quantification of host preference by manipulation of oviposition behavior in the butterfly *Euphydryas editha*, *Oecologia* **52**:230-235.

Singer, M. C., Ng, D., and Thomas, C. D., 1988, Heritability of oviposition preference and its relationship to offspring performance within a single insect population, *Evolution* **42**:977-985.

Sturtevant, A. H., 1921, The North American Species of Drosophila, Carnegie Institution of Washington, Publ. **301**:1-150.

Via, S., 1986, Genetic covariance between oviposition preference and larval performance in an insect herbivore, *Evolution* **40**:778-785.

Welch, H. E., 1959, Taxonomy, life cycle, development, and habits of two new species of Allantonematidae (Nematoda) parasitic in drosophilid flies, *Parasitology* **49**:83-103.

CHAPTER 12

Theoretical Analysis of Habitat Selection and the Maintenance of Genetic Variation

PHILIP W. HEDRICK

1. INTRODUCTION

Habitat selection is often suggested as a factor that may result in easier maintenance of genetic polymorphisms (e.g., Hedrick, 1986 and references therein). However, habitat selection may take several different conceptual forms, some of which appear to greatly broaden the conditions for a polymorphism while others have a much smaller impact. In particular, habitat selection based on homing that is independent of genotype has only a marginal influence on the conditions for polymorphism (e.g., Hoekstra *et al.*, 1985) while models that are based on genotypic-specific habitat selection may greatly broaden the conditions for a genetic polymorphism.

Here I will discuss only genotypic-specific habitat selection and will consider four different aspects of this type of habitat selection: (1) the approach used to parameterize the extent of habitat selection, i.e., whether habitat selection is a function of the frequency of the habitats or not, (2) which individuals are selecting habitats, progeny of both sexes or mated females, (3) the type of selection occuring, whether it is soft selection, hard selection, or some combination of the two, and (4) the number of genes involved, i.e., whether a single gene pleiotropically influences both habitat and viability selection or two loci are involved, one affecting habitat selection and the other affecting viability selection. Finally, I will discuss some experimental work and evaluate it as much as possible in view of these theoretical findings (see also Barker, Chapter 9, this volume).

2. GENOTYPIC-SPECIFIC HABITAT SELECTION

2.1. The General Model

Let us assume initially that a single locus with two alleles pleiotropically influences both habitat selection and

PHILIP W. HEDRICK * Department of Biology, Pennsylvania State University, University Park, PA 16802, USA.

Ecological and Evolutionary Genetics of Drosophila, Edited by J.S.F. Barker *et al.*, Plenum Press, New York, 1990

relative viability. In this case, the probabilities that
genotypes A_1A_1, A_1A_2, and A_2A_2 select habitat (or niche) i are
$h_{11.i}$, $h_{12.i}$, and $h_{22.i}$, respectively, and the relative survival of
genotypes A_1A_1, A_1A_2, and A_2A_2 in habitat i are $w_{11.i}$, $w_{12.i}$, and
$w_{22.i}$, respectively. Assuming that young of both sexes choose
habitats and then viability selection occurs (see Fig. 1a),
then the frequency of allele A_1 after habitat and viability
selection in habitat i is

$$p'_{1.i} = (p_1/w_i)(p_1 h_{11.i} w_{11.i} + p_2 h_{12.i} w_{12.i}) \tag{1}$$

where p_2 is the frequency of A_2 and

$$w_i = p_1^2 h_{11.i} w_{11.i} + 2p_1 p_2 h_{12.i} w_{12.i} + p_2^2 h_{22.i} w_{22.i}$$

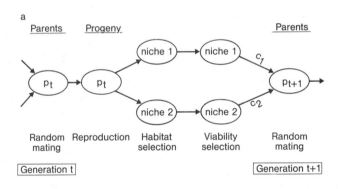

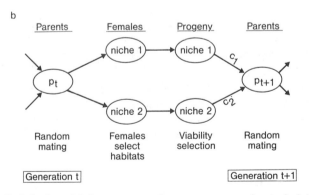

Figure 1. (a) Model in which progeny in sequence select habitats, undergo
viability selection within those habitats, and then the survivors mate at
random. (b) Model in which random-mated females select habitats, then
progeny undergo viability selection within these habitats and then the survivors mate at random.

The frequency of allele A_1 over habitats is then

$$p_1' = \sum_i c_i p_{1.i} \tag{2}$$

where c_i is the proportion of individuals from habitat i.

In this general situation, Templeton and Rothman (1981) and Garcia-Dorado (1987) have shown that the conditions for a stable polymorphism are

$$\sum_i c_i \frac{h_{12.i} w_{12.i}}{h_{11.i} w_{11.i}} > 1$$

and

$$\sum_i c_i \frac{h_{12.i} w_{12.i}}{h_{22.i} w_{22.i}} > 1 \tag{3}$$

When there is no viability selection and only two habitats, then the conditions for a stable polymorphism are reduced to $h_{11.1} > c_1 > h_{22.1}$.

2.2. Parameterization of Habitat Selection

Genotypic-specific habitat selection can be parameterized in several different ways. For example, Templeton and Rothman (1981) assumed that the probability that a genotype selects a given habitat is independent of the proportion of that habitat present. The simplest representation of this model is given in Table Ia where there are two niches, the two homozygotes have the same preference h for different niches (generally $\frac{1}{2}$ <h<1), and the heterozygotes are exactly intermediate. However, these values do not change with the proportion of the two niches so that one can have the improbable situation as one niche becomes very rare that a proportion h of A_1A_1 still select this habitat. In another way of parameterization, Garcia-Dorado (1987) used a genotypic adaptation of Maynard Smith's (1966, 1970) habitat selection model that incorporates niche frequency into habitat selection (Table Ib, the frequency of niche 1 is c_1, niche 2 is c_2, and $c_1 + c_2 = 1$). This model can be viewed in two equivalant ways: the top one in Table I is as written by Garcia-Dorado, the bottom one is in the form used by Maynard Smith. Again the simplest case is given, equal but opposite habitat selection of the two homozygotes, with the heterozygotes intermediate. In this model, it is assumed that a fraction h ($0 \leq h \leq 1$) of the homozygotes choose a given niche and the remaining fraction (1-h) settles at random with a proportion c_1 going to niche 1 and a proportion c_2 going to niche 2.

Here I propose another approach of specifying habitat selection (Table Ic) that may be more appropriate for reasons I

Table I
Models of genotypic habitat selection of (a) Templeton and Rothman (1981),
(b) Garcia-Dorado (1987), and (c) this paper

	Niche	A_1A_1	A_1A_2	A_2A_2
(a)	1	h	$\frac{1}{2}$	$1-h$
	2	$1-h$	$\frac{1}{2}$	h
(b)	1	$h+(1-h)c_1$	$\frac{1}{2}h+(1-h)c_1$	$(1-h)c_1$
	2	$(1-h)c_2$	$\frac{1}{2}h+(1-h)c_2$	$h+(1-h)c_2$
	or			
	1	c_1+hc_2	$c_1+h(\frac{1}{2}-c_1)$	c_1-hc_1
	2	c_2-hc_2	$c_2-h(\frac{1}{2}-c_1)$	c_2+hc_1
(c)	1	$\dfrac{hc_1}{1-h-c_1+2hc_1}$	c_1	$\dfrac{(1-h)c_1}{c_1+h-2hc_1}$
	2	$\dfrac{(1-h)c_2}{1-h-c_1+2hc_1}$	c_2	$\dfrac{hc_2}{c_1+h-2hc_1}$

will discuss below. In this model, the probability of A_1A_1 choosing niche 1 is given by the product of the preference for that niche h and the frequency of that niche c_1 standardized by the sum of this product and the analogous product for genotype A_1A_1 in niche 2. In this simplest case, heterozygotes are assumed to be intermediate in preference (to have no preference) so that a proportion c_1 go to niche 1 and a proportion c_2 go to niche 2.

Let us compare the model of Garcia-Dorado and my model at two extreme values, c_1 = 0 and h = 1 (Table II). First, when c_1 = 0 in the model of Garcia-Dorado the probability of A_1A_1 choosing niche 1 is still h. Obviously this is not possible because niche 1 does not exist. On the other hand, in my model a proportion 0 of all genotypes choose niche 1 when c_1 = 0. Second, if h = 1, then in Garcia-Dorado's model 1/2 the heterozygotes choose each niche, independent of the frequency of the niche. Again this is not realistic if the frequencies of the two habitats are not equal. On the other hand, in my model a proportion c_1 of the heterozygotes choose habitat 1, as expected if the heterozygotes show no preference. I should note that dominance within a niche is actually a function of the frequency of the two niches in my model.

Table II
Models (b) and (c) from Table I of genotypic habitat selection where
$c_1 = 0$ and $h = 1$

	Niche	A_1A_1	A_1A_2	A_2A_2
$c_1 = 0$				
(b)	1	h	$\frac{1}{2}$h	0
	2	1-h	$1-\frac{1}{2}$h	1
(c)	1	0	0	0
	2	1	1	1
$h = 1$				
(b)	1	1	$\frac{1}{2}$	0
	2	0	$\frac{1}{2}$	1
(c)	1	1	c_1	0
	2	0	c_2	1

2.3. Habitat Selection by Females

Let us assume that only females select different habitats and that they have previously mated at random as in Figure 1b. In this case, viability selection occurs on the progeny of these females and Table III gives the frequency of progeny expected from the three female genotypes. As a result, the frequency of the three progeny genotypes A_1A_1, A_1A_2, and A_2A_2 are $P'_{11.i}$, $P'_{12.i}$, and $P'_{22.i}$, respectively, as given by

$$P'_{11.i}\,\bar{w}_i = w_{11.i}p_1^2\,(p_1h_{11.i} + p_2h_{12.i})$$

$$P'_{12.i}\,\bar{w}_i = w_{12.i}p_1p_2\,(p_1h_{11.i} + h_{12.i} + p_2h_{22.i}) \qquad (4)$$

$$P'_{22.i}\,\bar{w}_i = w_{22.i}p_2^2\,(p_1h_{12.i} + p_2h_{22.i})$$

where $\bar{w}_i$ is the sum of the right hand side of these equations.

The frequency of A_1 in habitat i is

$$p'_{1.i} = P'_{11.i} + \frac{1}{2}\,P'_{12.i} \qquad (5)$$

When there is no viability selection and two habitats, Rausher (1984) showed that sufficient conditions for maintenance of a polymorphism when habitat selection occurs only in females are $h_{11.1} > c_1 > h_{22.1}$, the same as when both sexes select habitats.

Table III
The frequency of progeny produced when only random-mated females select habitats

Genotype	Frequency	Female Progeny		
		A_1A_1	A_1A_2	A_2A_2
A_1A_1	p_1^2	$p_1^3 h_{11.i}$	$p_1^2 p_2 h_{11.i}$	–
A_1A_2	$2p_1p_2$	$p_1^2 p_2 h_{12.i}$	$p_1 p_2 h_{12.i}$	$p_1 p_2^2 h_{12.i}$
A_2A_2	p_2^2	–	$p_1 p_2^2 h_{22.i}$	$p_2^3 h_{22.i}$

2.4. Hard or Soft Selection

The conditions for a stable polymorphism when there is hard selection are much more restrictive than when there is soft selection. For example, with habitat and viability selection, Templeton and Rothman (1981) found that to maintain a polymorphism with hard selection

$$\sum_i h_{11.i} w_{11.i} < \sum_i h_{12.i} w_{12.i} > \sum_i h_{22.i} w_{22.i} \tag{6}$$

(see also Jaenike and Holt (1990) for a graphical model of hard selection). With hard selection, the contribution from a given niche is a function of the mean fitness within that niche or

$$c_i' = \frac{\bar{w}_i}{\sum_i \bar{w}_i} \tag{7}$$

If we assume that both hard and soft selection are present, then the contribution from a niche is a composite of the two types of selection. A straightforward approach to incorporate the two types of selection into a single population is to assume that proportions x and 1-x are soft and hard selection, respectively, so that the contribution from niche i is

$$\bar{c}_i = xc_i + (1-x)c_i' \tag{8}$$

2.5. Two Loci

Let us assume that locus A influences habitat selection as given in Table I and that a second locus B determines relative viabilities so that the fitnesses of genotypes B_1B_1, B_1B_2, and B_2B_2 in habitat i are $w_{11.i}$, $w_{12.i}$, and $w_{22.i}$, respectively. If we let the frequencies of gametes A_1B_1, A_1B_2, A_2B_1, and A_2B_2 be x_1, x_2, x_3, and x_4, respectively, then after habitat and viability selection, they become in habitat i

$$x_{1.i}' \bar{w}_i = x_1(x_1 h_{11.i} w_{11.i} + x_2 h_{11.i} w_{12.i} + x_3 h_{12.i} w_{11.i} + x_4 h_{12.i} w_{12.i}) - h_{12.i} w_{12.i} CD$$

$$x_{2.i}' \bar{w}_i = x_2(x_1 h_{11.i} w_{12.i} + x_2 h_{11.i} w_{22.i} + x_3 h_{12.i} w_{12.i} + x_4 h_{12.i} w_{22.i}) + h_{12.i} w_{12.i} CD$$

$$x'_{3.i}\bar{w}_i = x_3(x_1 h_{12.i}w_{11.i} + x_2 h_{12.i}w_{12.i} + x_3 h_{22.i}w_{11.i} + x_4 h_{22.i}w_{12.i}) + h_{12.i}w_{12.i}cD$$

$$x'_{4.i}\bar{w}_i = x_4(x_1 h_{12.i}w_{12.i} + x_2 h_{12.i}w_{22.i} + x_3 h_{22.i}w_{12.i} + x_4 h_{22.i}w_{22.i}) - h_{12.i}w_{12.i}cD$$

$$(9)$$

where c is the proportion of recombination between the loci, $D = x_1 x_4 - x_2 x_3$ and $\bar{w}_i$ is the sum of the righthand side of the four equations.

For this situation, Garcia-Dorado (1986) gave the "mimimal" conditions for a polymorphism as

$$(1-c)\sum_i c_i \frac{h_{12.i}w_{12.i}}{h_{11.i}w_{11.i}} > 1$$

and $$(10)$$

$$(1-c)\sum_i c_i \frac{h_{12.i}w_{12.i}}{h_{22.i}w_{22.i}} > 1$$

3. RESULTS

3.1. Parameterization of Habitat Selection

Let us compare the conditions for a polymorphism for the three habitat selection models discussed above. Figure 2 gives between the symmetrical lines the limit of c_1 for the full range of s when there is 1/4 of the maximum habitat selection (h = 0.625 for Templeton and Rothman and Hedrick and h = 0.25 for Garcia-Dorado). It is assumed here that the relative fitness of genotypes A_1A_1, A_1A_2, and A_2A_2 are 1, 1 − (1/2)s, and 1 − s in niche 1 and the fitness of the two homozygotes are reversed in niche 2. For comparison, the limits under the Levene (1953) model, which has no habitat selection, is also given. Obviously all of the habitat selection models broaden the conditions for a polymorphism, with greatest effect under the Garcia-Dorado model and the least under Templeton and Rothman. When there is no viability selection (s = 0), there is a stable polymorphism for all proportions of the two environments for the Garcia-Dorado model and the present model, while for the Templeton and Rothman model, the limits are between c_1 = h and 1 − h. Most importantly, all of the models significantly broaden the conditions for a polymorphism when s is low.

Initially it is not apparent why the region of stability should narrow for intermediate values of s under the Garcia-Dorado and Hedrick models. To explain this phenomenon, I have calculated the expected change in A_1 when c_1 = 0.85 for habitat selection alone (s = 0, h = 0.625), viability selection alone (s = 0.25, h = 0.5) and the presence of both types of selection (s = 0.25, h = 0.625) (Fig. 3). This combination of se-

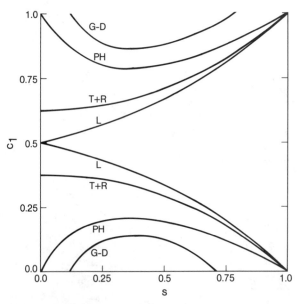

Figure 2. Comparison of the regions of stability (between pairs of curved lines) for the Levene model (L) and the three models of habitat selection (h = 0.625 for T + R and PH, h = 0.25 for G-D) where c_1 is the proportion of niche 1 and s is selective difference between the two homozygotes.

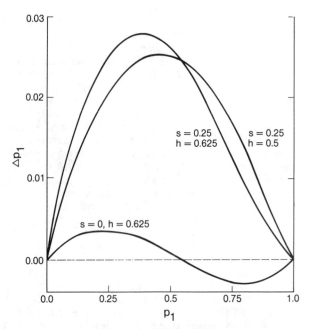

Figure 3. The expected change in the frequency of A_1, when there is only habitat selection, only viability selection, and both forms of selection given that c_1 = 0.85.

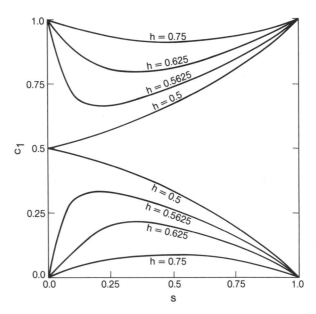

Figure 4. The region of stability for different levels of habitat selection for the PH model.

lection factors lies just out of the region of stability. As shown in Figure 3, habitat selection alone results in a stable equilibrium at 0.54 although the resulting Δp_1 values are relatively small. On the other hand, because $c_1 = 0.85$ viability selection results in strong selection favoring A_1, even at high values of p_1. As a result, the net effect of both factors results in a positive Δp_1 for all p_1 and no genetic polymorphism. In other words, at high (or low) c_1 values, the directional change in allelic frequencies from viability selection destabilizes the balancing influence of habitat selection.

Next let us examine the region of stability for different levels of habitat selection in my model (Fig. 4). Two important observations should be noted here. First, even when h is only 0.5625, habitat selection has a substantial impact on maintenance of polymorphism when s is low. Second, if h is 0.75 or greater, then genetic polymorphism is maintained for virtually all levels of s except when c_1 is very near 0 or 1.

3.2. Habitat Selection by Females

How does the limitation of habitat selection to only females influence the possibility of genetic polymorphism. As an example, Figure 5 gives the region of stability when h = 0.625 for both sexes (as in Fig. 1a) or only females (as in Fig. 1b). As expected, the region of stability is smaller when only females select habitats. However, when viability selection is low, the region of stability is nearly as large for habitat selection by only females as when habitat selection occurs in both sexes. As s -> 0, the region of stability

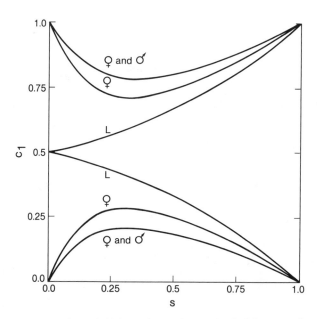

Figure 5. The regions of stability when there is habitat selection for both sexes and when only females select habitats (h = 0.625).

encompasses the whole range of niche frequencies, as was also found by Rausher (1984) for female habitat selection.

3.3. Hard or Soft Selection

If we allow niche contributions to the mating pool to be determined by a proportion x of soft selection and 1 - x of hard selection, what is the impact on genetic polymorphism? Remember that if there is only hard selection, then for all values of viability and habitat selection we have been considering, no stable polymorphism is possible. Figure 6 gives the regions of stability with only soft selection and two proportions of hard selection. As expected, hard selection greatly reduces the region of stability over most of the range of s. However, when s is low, even a relatively small proportion of soft selection has a substantial impact on the region, allowing the existence of a polymorphism.

3.4. Two Loci

Let us now examine the situation in which different loci influence habitat selection and viability selection. Because the habitat selection values we have been using always give a stable polymorphism for the locus when there is only habitat selection, we will examine the region of stability for the viability locus as it is influenced by the habitat selection locus. As an example, Figure 7 gives the region of stability

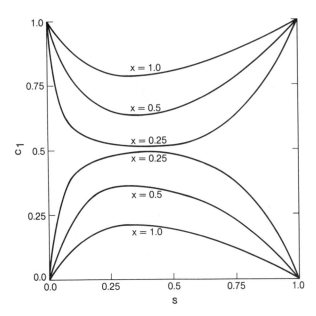

Figure 6. The regions of stability when there is only soft selection (x = 1.0) and mixed soft and hard selection (x = 0.5 and x = 0.25) when h = 0.625.

for h = 0.75 and different levels of recombination. The c = 0.0 and 0.5 values are the single locus and Levene limits. Notice that the two other levels of recombination given, c = 0.01 and 0.05, give regions of stability quite close to those in which there is no habitat selection. Furthermore as s -> 0, the limits approach that of separate loci (c = 0.5), not that for a single locus. In other words, when two loci are involved, the effect of linked association on the region of stability is quite small except for very tight linkage.

How does the presence of the linked habitat selection locus influence the frequency of the viability alleles? Figure 8 illustrates this for selection values of the viability locus (s = 0.25 and c_1 = 0.4) that by themselves do not give stability. Notice that when habitat selection is low (h = 0.625) only very tight linkage (c < 0.027) will maintain the polymorphism. Furthermore, habitat selection will only increase q, say above 0.1, at quite tight linkage.

The linkage necessary as given by Garcia-Dorado (1986) (given here in equation 10) is shown below the values found here (Fig. 8). For example, when h = 0.625, her minimal conditions suggest that if c < 0.086, a stable polymorphism could be maintained. However, the actual limits using numerical iteration are much more restrictive. In other words, the potential for a linked habitat selection gene to maintain polymorphism at a viability locus are much less likely than suggested by Garcia-Dorado.

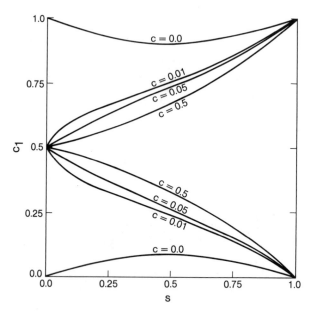

Figure 7. The regions of stability when locus A causes habitat selection ($h = 0.75$) and locus B results in viability selection where c is the proportion of recombination between the two loci.

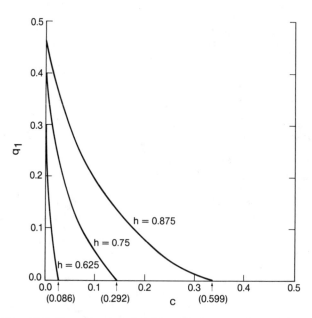

Figure 8. The equilibrium frequency of allele B_1, for different levels of recombination and habitat selection when $s = 0.25$ and $c_1 = 0.4$. The numbers in parentheses are limits of c given by Garcia-Dorado for the three h values.

4. DISCUSSION

From this examination of models of genotypic-specific habitat selection, a number of conclusions are possible. First, the parameterization of habitat selection in a logical way as proposed here, based on the frequences of the niches, results in broad conditions for a stable polymorphism. In particular, the robustness is greatly expanded when viability selection is weak, a problem emphasized by Hoekstra *et al.* (1985) in their discussion of non-genotypic-specific habitat selection models. Furthermore, previous approaches to parameterization of habitat selection, e.g., that of Templeton and Rothman (1981) and Garcia-Dorado (1987) appear biologically unrealistic.

Second, habitat selection by only females restricts the conditions for a polymorphism somewhat, but when viability selection is low, the region of stability is nearly as large as when both sexes select habitats. Third, although pure hard selection greatly restricts the conditions for polymorphism, a small proportion of soft selection substantially broadens, particularly at low viability selection levels, the conditions for a polymorphism. Overall then, genotypic-specific habitat selection greatly broadens the condition for a polymorphism even when viability selection is low, only females select habitats, or there is mostly hard selection.

The above conclusions are for a single gene that pleiotropically influences habitat and viability selection. When different genes influence these two traits, then the conditions become somewhat more restrictive. For example, for an association between a habitat selection allele at locus A and a viability allele at linked locus B to be maintained, linkage must be rather tight. I should note that the minimal conditions given by Garcia-Dorado are not nearly as restrictive as I have found using numerical iteration.

There are several other factors that may influence the conditions for a polymorphism that I have not discussed here. First, there may not be random mating between individuals from different niches. In many insects, mating may take place within the niche, resulting in a type of positive assortative mating. When this occurs, it will greatly broaden the conditions for a polymorphism (e.g., Jaenike, 1988; Diehl and Bush, 1989). Second, not all individuals may actively choose habitats. In other words there may be a high proportion of individuals that remain in their natal habitat and thereby perpetuate between niche differences. This effect may be much like limited gene flow and again broadens the conditions for a polymorphism (e.g., Christiansen, 1975). Finally, early adult experience may accentuate habitat selection and broaden the conditions for a polymorphism (Jaenike, 1988). However, this may not occur for all types of learning, e.g., aversion learning may result in the opposite effect.

In light of these theoretical findings, I would like to discuss a series of experiments in *Drosophila* that have attempted to examine the relationship between environmental heterogeneity and genetic variation (electrophoretic variation)

by varying environmental parameters. Before I discuss the results of these experiments, let us examine the different experimental designs that have been used in order to better compare and evaluate these experiments (Table IV). First, notice
that the characteristics of the populations differ at the time
of their initiation characteristics. For example, the number
of initial isofemale lines was highest in the experiments of
McDonald and Ayala (1974) and Yamazaki *et al.* (1983). The
stock used by Yamazaki *et al.* was in culture for six years,
much longer than that of McDonald and Ayala, Powell and Wistrand (1978), or Haley and Birley (1983). One might expect
from these differences that the disequilibrium between loci
(say between electrophoretic loci and other loci undergoing
either habitat or viability selection) would be highest in
Powell and Wistrand's or in Haley and Birley's experiments,
and the least in that of Yamazaki *et al.*

Second, the environmental regime used differs among the experiments (Table IV). If we use the general conclusions from
theoretical models (see discussion in Hedrick *et al.*, 1976;
Hedrick, 1986 and above), and if the fitness of a locus is affected by environmental differences, the locus would most
likely be polymorphic when the environment varies spatially in
a coarse-grained fashion and there is the potential for habitat selection and least likely when the environment varies in
a fine-grained manner. However, in all these experiments except that of Haley and Birley, the environments appear to be
varied either in a spatial, fine-grained manner or a temporal,
coarse-grained manner. From a theoretical perspective then,
one might expect the most maintenance of genetic variation in
the experiments of Haley and Birley, McDonald and Ayala, or
Powell and Wistrand and the least in Yamazaki *et al.*

The mean heterozygosities for these experiments are given
in Table V. Note that in all of the experiments, there is a
small decline in heterozygosity over time, or there may be
virtually no change, as in Yamazaki *et al.*, Haley and Birley,
and Powell and Wistrand. However, note that there is no initial estimate of genetic variation in the experiment of Powell
and Wistrand (none until after 12 months); this leaves open
the possibility that the differences observed were the result
of unknown factors in the early generations and unrelated to
the environmental regime. Thus, the general impression is
that environmental heterogeneity may slightly reduce the rate
of decay of genetic variation but that it does not seem to increase the amount of heterozygosity. The small difference between the experiments could be accounted for by the influence
of the experimental design on the extent of initial disequilibrium and the potential for effective variable selection and
habitat selection, as discussed above.

A number of other possible considerations are important to
the evaluation of these results. First, maybe only a few
loci, marking only a small proportion of the genome, are affected by these environmental factors, and the effects of
these loci are diluted by that of others at which there may be
little or no environmental effect. Second, the time scale may

Table IV

Characteristics of the six experiments discussed in Table V

| Reference | Species | Initial population | | Environmental factors[a] |
		Number	Time in culture	
McDonald and Ayala (1974)	D. pseudoobscura	141 isofemale lines	3 months	food (S) light (T?) temperature (T) yeast
Powell and Wistrand (1978)	D. pseudoobscura	23 isofemale lines	2-4 generations	food (S) D.persimilis (S) temperature (T)
Minawa and Birley (1978)	D. melanogaster	30 isofemale lines	130 generations	food (S) temperature (T)
Oakeshott (1979)	D. melanogaster	?	?	alcohol (T,S) temperature (T) yeast (T,S)
Yamazaki et al. (1983)	D. melanogaster	400 isofemale lines	6 years	food (T*) light (T*) temperature (T*)
Haley and Birley (1983)	D. melanogaster	100 isofemale lines	3 generations	food (S)

[a]S=spatial variation
T=temporal variation
T*=coarse-grained temporal

Table V

The mean heterozygosity in six different experiments in which the environment was varied

Reference	Time	Number of heterogeneous environmental factors				
		0	1	2	3	4
McDonald and Ayala (1974)	0	0.212				
	12 mo	0.146	0.186	0.201	0.208	0.186
Powell and Wistrand (1978)	12 mo	0.254	0.297	0.326	0.338	-
	18 mo	0.236	0.291	0.322	0.333	-
	24 mo	0.220	0.288	0.318	0.337	-
Minawa and Birley (1978)	0	0.293				
	6 mo	0.271	-[a]	0.284	-	-
	18 mo	0.284	-	0.062	-	-
Oakeshott (1979)	0	0.338				
	20 gns	0.288	0.297	0.308	0.274	-
Yamazaki et al. (1983)	0	0.333				
	12 mo	0.332	0.343	0.343	0.331	-
	24 mo	0.346	0.355	0.353	0.343	-
Haley and Birley (1983)	0	0.333				
	11 mo	0.345	0.332	-	-	-
	22 mo	0.342	0.339	-	-	-

[a]not done

be too short to pick up selective differences, i.e., if we are comparing the relative effect of genetic drift on neutral versus marginally selected loci in relatively large populations such as these, then many generations may be necessary to tell the difference. Third, the optimum environmental regime for maintaining genetic variation - a spatial, coarse-grained environment with opportunity for habitat selection - was not present. In fact, generally the environmental regime used was one in which retention of genetic variation for selected loci should be only slightly higher than neutrality.

In a different vein, there has been a series of field experiments, apparently stimulated by the initial positive findings of Taylor and Powell (1978), to determine whether there is habitat fidelity in Drosophila. I will not discuss these studies here (see discussions by Hedrick, 1986; Hoffmann and O'Donnell, Chapter 10, this volume) but only say that a number of factors such as experience and the physiological state of the flies as well as the assay used may greatly influence these experiments. Furthermore, in most of these experiments particular genotypes are not used so that the results are for groups of flies that may only share some genotypes in common.

What is the evidence that genotypes select specific habitats? A good demonstration of the potential for environmental heterogeneity to maintain a polymorphism in a coarse-grained, spatial environment with opportunity for habitat selection is the experiment conducted by Jones and Probert (1980) using the white-eye (w) mutant in *D. simulans*. In a uniform habitat of either normal or dim red light, the w mutant is lost; it went from 0.5 to 0.01 in normal light and 0.5 to 0.06 in dim red light in 30 weeks. However, in a population cage split so that half had normal light and half had dim red light, the frequency of w after 30 weeks was still 0.32 (average of 5 replicates). In addition, the frequency of w was quite different between the two halves of the cage - much higher in the dim-red-light sector (Fig. 9). Because in the control experiments w was always lost, this experiment indicates that heterogeneity in the enviroment is necessary for the maintenance of a polymorphism. The apparent basis for the polymorphism is habitat preference (and perhaps limited gene flow); wild type flies are positively phototactic, and the white-eyed flies prefer (or perhaps are neutral to) the dim red light. The habitat selection model of Templeton and Rothman (1981) predicts that a polymorphism can be maintained even in the face of a fitness disavantage; this seems an appropriate theoretical explanation here.

Do phenotypes or genotypes prefer the habitat in which they have the highest fitness? Two studies show clearly that they may for larval behavior. First, de Souza *et al.* (1970) observed a genetic polymorphism in *D. willistoni* larvae for pupation inside or outside of cups in a population cage. Under moist conditions the genetically outside pupae had higher survival outside the cups while the genetically inside pupae had higher survival inside the cups. Second, Sokolowski *et al.* (1986) found in *D. melanogaster* an association between the ge-

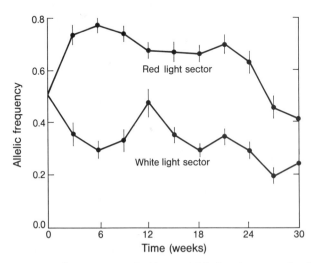

Figure 9. The average frequency of the w allele in *D. simulans* over five replicates in the red light and white light sectors (Jones and Probert, 1980).

netically different "rover" and "sitter" larvae and traits
such as pupation site selection and survival at different soil
moisture levels.   They presented evidence that suggests that
sitter larvae have a greater tendency to pupate on fruit and
that there is higher pupal emergence on fruit when the soil
moisture is low.   Rover larvae have a tendency to pupate in
the soil, and there is a higher emergence in soil when soil
moisture is high.   Overall, it appears from these studies that
habitat selection at specific loci may be important in
Drosophila.   Further examination both experimentally and theoretically of the examples provided by Jones and Probert
(1980), Sokolowski *et al.* (1986), and de Souza *et al.* (1980)
should provide insight into how habitat selection maintains
genetic polymorphism in these instances and be useful as background for understanding other cases of habitat selection in
Drosophila.

5. SUMMARY

Theory discussed or developed here suggests that maintenance of polymorphism may be greatly broadened by genotypic-
specific habitat selection.   In fact, even when only females
select habitats or there is mostly hard selection, the conditions are expanded, particularly at low values of viability
selection.   However, when different genes influence viability
selection and habitat selection, the impact is much less.   One
possible reason why the experiments examining the relationship
of environmental heterogeneity and genetic variation have
shown only slight effects is that most of these experiments
have not allowed habitat selection.   The studies by Jones and
Probert (1980) and Sokolowski *et al.* (1986) on specific genes
suggest that strong habitat selection can occur and can maintain genetic variation in *Drosophila*.

ACKNOWLEDGEMENTS. I appreciate the comments of J.S.F. Barker and
J. Jaenike on the manuscript.   The paper was prepared while I
held a University of New England Visiting Research Fellowship.
T. Prout encouraged me to investigate this general problem.

References

Christiansen, F. B., 1975, Selection and population regulation with habitat
     variation, *Am. Nat.* **126**:418-429.
de Souza, H. M. L., da Cunha, A. B., and dos Santos, E. P., 1970, Adaptive
     polymorphism of behavior evolved in laboratory populations of
     Drosophila willistoni, *Am. Nat.* **124**:175-189.
Diehl, S. R., and Bush, G. L., 1989, The role of habitat preference in
     adaptation and speciation, in: *Speciation and its Consequences* (D.
     Otte, and J. Endler, eds), Sinauer, Sunderland, Massachusetts, pp. 345-
     365.
Garcia-Dorado, A., 1986, The effect of niche preference on polymorphism
     protection in a heterogeneous environment, *Evolution* **40**:936-945.
Garcia-Dorado, A., 1987, Polymorphism from environmental heterogeneity:
     some features of genetically induced niche preference, *Theor. Pop.
     Biol.* **32**:66-75.

Haley, C. S., and Birley, A. J., 1983, The genetical response to natural selection by varied environments. II. Observations on replicate populations in spatially varied laboratory environments, *Heredity* **51**:581-606.

Hedrick, P. W., Ginevan, M. E., Ewing, E. P., 1976, Genetic polymorphism in heterogeneous environments, *Annu. Rev. Ecol. & Syst.* **7**:1-32.

Hedrick, P. W., 1986, Genetic polymorphism in heterogeneous environments: A decade later, *Annu. Rev. Ecol. & Syst.* **17**:535-566.

Hoekstra, R. F., Bijlsma, R., and Dolman, A. J., 1985, Polymorphism from environmental heterogeneity: Models are only robust if the heterozygote is close in fitness to the favored homozygote in each environment, *Genet. Res.* **45**:299-314.

Jaenike, J., 1988, Effects of early adult experience on host selection in insects: Some experimental and theoretical results, *J. Insect Behav.* **1**: 3-15.

Jaenike, J., and Holt, R. D., 1990, Genetic variation for habitat preference: evidence and explanations, *Am. Nat.* (in press).

Jones, J. S., and Probert, R. F., 1980, Habitat selection maintains a deleterious allele in a heterogeneous environment, *Nature, Lond.* **287**:632-633.

Levene, H., 1953, Genetic equilibrium when more than one ecological niche is available, *Am. Nat.* **87**:331-333.

Maynard-Smith, J., 1966, Sympatric speciation, *Am. Nat.* **100**: 637-650.

Maynard-Smith, J., 1970, Genetic polymorphism in a varied environment, *Am. Nat.* **104**:487-490.

McDonald, J. F., and Ayala, F. J., 1974, Genetic response to environmental heterogeneity, *Nature, Lond.* **250**:572-574.

Minawa, A., and Birley, A. J., 1978, The genetical response to natural selection by varied environments. I. Short-term observations, *Heredity* **40**: 39-50.

Oakeshott, J. G., 1979, Selection affecting enzyme polymorphisms in laboratory populations of *Drosophila melanogaster*, *Oecologia* **43**: 341-354.

Powell, J. R., and Wistrand, H., 1978, The effect of heterogeneous environments and a competitor on genetic variation in *Drosophila*, *Am. Nat.* **112**:935-947.

Rausher, M. D., 1984, The evolution of habitat preference in subdivided populations, *Evolution* **38**:596-608.

Sokolowski, M. B., Bauer, S. J., Wai-Ping, V., Rodriquez, L., Wang, J. L., and Kent, C., 1986, Ecological genetics and behavior of *Drosophila melanogaster* larvae in nature, *Anim. Behav.* **34**:403-408.

Taylor, C. E., and Powell, J. R., 1978, Habitat choice in natural populations of *Drosophila*, *Oecologia* **37**:69-75.

Templeton, A. R., and Rothman, L. D., 1981, Evolution in fine-grained environments. II. Habitat selection as a homeostatic mechanism, *Theor. Pop. Biol.* **19**:326-340.

Yamazaki, T., Kusakabe, S., Tachida, H., Ichinose, M., Yoshimaru, H., Matsuo, Y., and Mukai, T., 1983, Reexamination of diversifying selection of polymorphic allozyme genes by using population cages in Drosophila melanogaster, *Proc. natn. Acad. Sci. USA* **80**:5789-5792.

Biochemical Genetics: Introduction

Biochemical genetics examines the relationship between nucleotide sequence variation and physiological differences among members of the same or different species. If these differences are adaptive, the findings are central to a complete understanding of the ecological and evolutionary significance of the genetic variation. Historically, two approaches have been taken. In the first, genetic variation at a locus is initially detected, often as allozymes or null mutants, and the effect of this variation on the properties of the protein product and/or the fitness of the individuals with the different genotypes is assessed. In other words, the investigator attempts to demonstrate that the genetic variants have meaningful physiological effects. The second approach begins with the detection of adaptively important phenotypic differences, and proceeds toward an analysis of the underlying, often cryptic, genetic variation.

In either approach, a complete analysis of the variation at the metabolic level is not a trivial task. Differences in both the structure of the gene product and the transcriptional, translational and post translational regulation of the gene product must be understood in the context of its biochemical pathway or pathways. For example, do the variants of a single enzyme affect catalysis? If so, are the amino acid sequence differences or differences in the level of the enzyme's activity meaningful with regard to the metabolic flux of intermediates through the pathway? And do such effects on the pathway reflect on the fitness of the individual? It is important to realize that no enzyme or metabolic pathway operates in a cellular vacuum. Thus, the interactions with a specific gene product may involve enzymatic or even non-enzymatic components (e.g. receptors or membrane proteins involved in the active transport of pathway precursors) which are not obviously related to the apparent metabolic role of that gene. In this regard it is difficult to unravel a network of pleiotropic effects resulting from even a single mutant gene, even when the effects may be seen at the enzymatic, the metabolic and the morphological levels.

Despite the complexity of the task, there have been some notable successes in unraveling the phenotypic effects of single gene mutations. Several of these are documented in the following chapters. Rollin Richmond and colleagues, in pursuing the first approach, i.e. working upward from naturally occurring *Esterase-6* alleles in *D. melanogaster*, have shown how these variants can significantly affect male reproductive physiology. Robyn Russell and her collaborators, beginning with insecticide resistant individuals of *Lucilia cuprina*,

Ecological and Evolutionary Genetics of Drosophila, Edited by
J.S.F. Barker *et al.*, Plenum Press, New York, 1990

have traced the resistance back to genes coding for carboxylesterases. In another study which concentrates on differences in an obvious fitness component, i.e. premating behavior in *D. mojavensis*, Teri Markow and Eric Toolson have defined associated differences in the hydrocarbons of the epicuticle. Differences in the enzymes involved in hydrocarbon biosynthesis coupled with environmental factors such as hot, dry conditions may select for epicuticle differences in flies living in different regions. These differences may then result in differential mate recognition responses for males and females of the different regions. Bill Geer and Steve McKechnie and their respective collaborators have made impressive progress in understanding the roles of several glycolytic enzymes in determining the fitness of *D. melanogaster* lines reared on diets containing different levels of ethanol and/or sucrose. Geer clearly documents the pleiotropic effects of *Adh* variants on the levels of other enzymes, cofactors and triglycerides. McKechnie finds that dietary components can affect the transcriptional regulation of the mitochondrial enzyme, αglycerophosphate oxidase. There could well be feedback effects from these diets on the levels of αglycerophosphate and acetyl Co-A, both precursors of lipids. Lipid content, in turn, could affect the ability of the flies to react to stress.

It is clear from discussions at the meeting, that future efforts in the area of biochemical genetics might profit more by following the second approach, i.e. first selecting for study phenotypes and "candidate" loci with obvious effects on fitness. Researchers would be advised, whenever possible, to utilize genetic differences between closely related species adapted to different environments. In addition, during investigations of the metabolic interactions between the products of "candidate loci", it is well to keep in mind that extrapolation from findings in vertebrate systems to Dipteran insects may be fraught with danger. Studies on lysosomal enzymes and cytoplasmic-mitochondrial shunts have shown insects and vertebrates to be fundamentally different in their basic metabolic processes. Despite these caveats, continued research in biochemical genetics should lead to significant insights into the relationship between genetic variation and fitness, and should provide examples of co-adaptation at the genetic and molecular levels.

and the characteristics that were monitored in the various tolerance tests may not be affected by ethanol to the same degree.

Thus, it is important to test alcohol tolerance under controlled environmental conditions and to compare the same life stage. Canton-S and Adh^{n2} larvae were cultured on a defined 1% sucrose diet for 4 days, then the larvae were transferred to an intervening test diet for two days, and finally the larvae were transferred to an ethanol-free control diet or a diet containing a growth-limiting concentration of ethanol. We assessed the effects of test nutrients in the intervening diet on larval growth rate, survival to pupation and eclosion, and adult male weight by comparing the performance of larvae on the final control and growth-limiting diets. We also examined the ultrastructure of larval fat body cells to judge the effects of alcohols at the cellular level (Geer, Dybas and Shanner, 1989).

Exposure of early third instar larvae of *D. melanogaster* to a non-lethal dose of ethanol was detrimental to larvae lacking alcohol dehydrogenase (ADH, EC 1.1.1.1), but beneficial to wild-type larvae in terms of surviving a later ethanol tolerance test (Geer, Dybas and Shanner, 1989). This implies that one of the important functions of the ADH system is to supply derivatives of ethanol to larvae that in turn promote ethanol tolerance. High intracellular concentrations of ethanol in Adh^{n2} larvae fed ethanol were accompanied by a decrease in the cell membrane infoldings of fat body cells, suggesting that the capacities to absorb and release molecules were reduced. Marked effects of ethanol on the endoplasmic reticulum and mitochondria of ADH-null larvae were also evident. The absence of similar changes in wild-type larvae that were fed moderate levels of ethanol showed that the ADH system kept the intracellular level of ethanol at a concentration low enough to avoid cell damage. A cytometric analysis of electron micrographs showed that there were ethanol-induced reductions in glycogen, lipid and protein stores in the fat body cells of ADH-null larvae fed 1.25% ethanol (v/v) compared to null larvae fed an ethanol-free diet (Fig. 1). Apparently, the capacities to synthesize or store these compounds are limited by high intracellular concentrations of ethanol. The consumption of diets containing 2.5% and 4.5% ethanol by Canton-S wild-type larvae for three days resulted in decreases in glycogen and protein deposits in fat body cells, but increased the amount of lipid deposits compared to larvae fed an ethanol-free diet. These findings coupled with the greater weight of wild-type adults that were fed a growth-limiting concentration of ethanol compared to control adults suggested that one defense mechanism is to convert toxic ethanol to non-toxic storage products (see also below). A low dietary level of propan-2-ol (1%) completely blocked glycogen deposition in wild-type larvae, whereas ethanol did not. Thus, ethanol and propan-2-ol exert some different toxic effects on larval fat bodies. Propan-2-ol has been shown to be converted to acetone, a metabolic endproduct and an inhibitor of ADH (Heinstra *et al.*, 1986).

1.2. Changes in Lipid Composition Induced by Low Levels of Ethanol

Ethanol is used efficiently as a substrate for lipid synthesis by *D. melanogaster* wild-type larvae (Geer *et al.*,
1985). The activities of the lipogenic enzymes; ADH, fatty
acid synthetase and sn-glycerol-3-phosphate dehydrogenase
(GPDH, EC 1.1.1.8) and the triacylglycerol (TG) content of
larvae are all correlated with the dietary ethanol concentration (Geer *et al.*, 1983; McKechnie and Geer, 1984). Lipid
that is stored during the larval period is a primary energy
source for pupal development (Green and Geer, 1979); consequently, differences in the abilities of larvae to convert
ethanol to lipid could be of adaptive significance.

At a moderate concentration (2.5%, v/v), dietary ethanol
reduced the chain length of total fatty acids and increased
the desaturation of short-chain fatty acids in *D. melanogaster*
larvae with a functional ADH (Geer *et al.*, 1986). The changes
in length in total fatty acids were postulated to be due to
the modulation of the termination specificity of fatty acid
synthetase. Because the ethanol-stimulated reduction in the
length of unsaturated fatty acid was blocked by linoleic acid,
the fatty acid modification by ethanol was thought to reflect
the properties of fatty acid Δ9-desaturase. Although the
ethanol-stimulated reduction in chain length of unsaturated
fatty acid was also observed in ADH-null larvae, ethanol pro-
moted an increase in the length of total fatty acid of the mu-

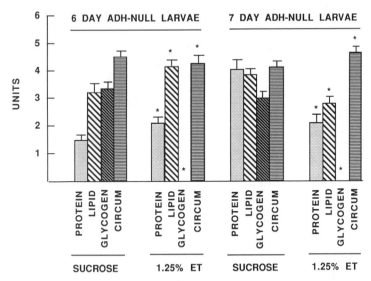

Figure 1. The relative areas of the fat body cells of day 4 ADH-null larvae
that were transferred to 0.3% sucrose (sucrose) and 0.3% sucrose with 1.25%
ethanol (1.25% ET) for two days (6 day larvae) and for three days (7 days
larvae) before observation. The protein units are 10^{-1}, the lipid units 2 X
10^{-2}, the glycogen units 10^{-2} and the circumference of the cell X1. Cell
traits that are different from the 0.3% sucrose value of the same age group
are indicated by *P<0.01 (from Geer *et al.*, 1989).

tant larvae. Thus, the ethanol-stimulated change in fatty acid length was ADH dependent but the ethanol effect on fatty acid desaturation was not (Geer *et al.*, 1986).

D. melanogaster is among the species of insects in the order Diptera which require dietary choline for optimal growth (Geer and Vovis, 1965; Geer *et al.*, 1968) and reproduction (Geer and Dolph, 1970). Choline functions in *D. melanogaster* primarily as a component of phospholipid. Phosphatidylcholine constitutes about 25% of the cell membrane phospholipids of *D. melanogaster* (Geer *et al.*, 1971; Tilghman and Geer, 1988) and is also a major component of the phospholipids of other Dipterans (Fast, 1966). Phosphatidylethanolamine is the prominent phospholipid in membranes (Fast, 1966; Geer *et al.*, 1971). A reduction in the dietary concentration of choline to 0.8 µg/ml from the optimal concentration of 80 µg/ml diminished the level of tissue phosphatidylcholine to less than one-third the normal level in third instar larvae without significantly altering the amount of phosphatidylethanolamine (Tilghman and Geer, 1988). The rates of synthesis of phospholipids, TGs, diglycerides and monoglycerides were reduced by the choline-deficiency, and the chain length of fatty acids in lipids was shortened. The activity of succinic dehydrogenase, a mitochondrial enzyme, was decreased by the deficiency, but the activities of fumarase, ADH and some other enzymes associated with lipogenesis were unaffected. A choline-deficiency did not alter the ultrastructure of mitochondria of larval fat body cells. Choline-deficient individuals were more susceptible to the toxic effects of ethanol during larval and pupal development, and less adept at utilizing ethanol as a substrate for adult tissue synthesis (Tilghman and Geer, 1988). Because similar ethanol-induced changes have been found in membrane lipids of other animals, ethanol may alter the properties of membranes in larvae. Ethanol tolerance in *D. melanogaster* may also be dependent on genes that specify lipids that are resistant to the detrimental effects of ethanol.

2. THE METABOLISM OF ALCOHOLS

2.1. Metabolic Pathways that Degrade Ethanol

A number of short chain alcohols including methanol, ethanol, n-propanol, propan-2-ol and n-butanol are found in the natural feeding sites of *D. melanogaster* (McKechnie and Morgan, 1982). At least three metabolic systems for degrading ethanol and other short-chain alcohols have been proposed. One pathway is initiated by ADH (Fig. 2). This ethanol degradation pathway generates first acetaldehyde. Both of the first two steps of the pathway form NADH from NAD^+. The acetic acid in turn is converted into acetyl-CoA which is degraded by the Citric Acid Cycle or is diverted into the synthesis of fatty acids. The products of ethanol degradation are coupled together to form TG. The TG content of wild-type larvae has been found to be proportional to the dietary ethanol concentration to about 5% ethanol (v/v). Consequently, the ADH pathway represents a mechanism for converting a potentially toxic compound, ethanol, into a non-toxic energy-storage compound, TG. By comparing the capacities of Canton-S wild-type larvae and ADH-deficient larvae to incorporate label from ^{14}C-ethanol into lipid, we have found that more than 90% of the capacity of *D. melanogaster* larvae to degrade ethanol resides in the ADH-pathway (Geer *et al.*, 1985; Heinstra *et al.*, 1987).

Another pathway initiated by catalase (EC 1.11.1.6) has been proposed to degrade alcohols in *D. melanogaster* (Deltombe-Lietaert *et al.*, 1979). Catalase protects cells from toxic effects of peroxides by degrading them in a two-step process. However, compound I in scheme A is postulated to have the capacity to convert ethanol into acetaldehyde. Other alcohols may be converted to the corresponding aldehydes by catalase.

(A) catalase + H_2O_2 ----->catalase-H_2O_2 (compound I)

catalase-H_2O_2 + C_2H_5OH ----->Catalase + CH_3CHO + 2 H_2O

There is some question as to whether ethanol reaches an intracellular concentration high enough for this reaction to be of importance. The catalase pathway may be valuable to the individual because it has a wide tissue distribution (Bewley *et al.*, 1983), whereas ADH is restricted to the fat body, gut and Malpighian tubules in larvae and fat body and gut in adult *D. melanogaster* (Maroni and Stamey, 1983; Heinstra *et al.*, 1989). Moreover, catalase activity increases in the pupal stage while ADH activity declines (Bewley *et al.*, 1983), indicating that catalase activity may be important in ethanol degradation during the pupal stage.

The function of catalase in the alcohol metabolism of larvae of *D. melanogaster* was studied recently by using two different experimental approaches (Geer, van der Zel and Heinstra, 1989). In one experimental regime catalase was inhibited in Adh^{n2} third instar larvae with 3-amino-1,2,4-triazole (3-AT). 3-AT is an irreversible catalase inhibitor. ADH-null larvae were cultured for 4 days on a 1% sucrose defined fat-

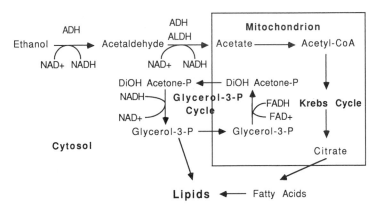

Figure 2. The alcohol dehydrogenase pathway for ethanol degradation.

free diet, then exposed to the same diet with varying concentrations of 3-AT for one day to inhibit catalase, and finally administered a diet with 3-AT and ^{14}C-ethanol (1% ethanol v/v) for 17 hours. When the catalase activities and fluxes of label from ^{14}C-ethanol into lipid were analyzed, a simple linear relationship was observed (Fig. 3). Thus, catalase participates in the conversion of dietary ethanol into lipid at least in larvae lacking a functional ADH pathway.

The other experimental approach used to show the interaction of alcohols with catalase in wild-type larvae was as follows. Canton-S larvae were cultured on a 1% sucrose fat-free diet to 5 days of age, then transferred for 17 hours to a series of diets containing no 3-AT or 4 mM 3-AT with one of several different alcohols before the larvae were tested for catalase activity. The rationale was that 3-AT and the alcohols may compete for compound I (scheme A), and that compound I may be protected from 3-AT inhibition when associated with an alcohol molecule. The rank order of protection against 3-AT inhibition was methanol>ethanol> propan-2-ol>n-propanol (Fig. 4). Thus, catalase is able to bind ethanol at low dietary concentrations when the ADH-pathway is functional. Catalase may be more active in the degradation of methanol, which is not a substrate for *Drosophila* ADH, than in the degradation of ethanol.

In summary the ADH-pathway represents about 91-93% and the catalase-pathway 3-5% of the total capacity of wild-type larvae to degrade ethanol. Theoretically, this leaves room for

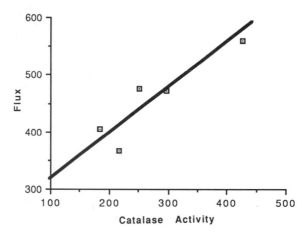

Figure 3. A linear regression of catalase activity versus the flux of label from ^{14}C-ethanol into lipid in Adh^{f6n}; ry+ larvae. The catalase activity was inhibited to different degrees *in vivo* by 3-amino-1,2,4-triazole. Each point represents a different experiment. The mean for six replicate catalase assays and the mean for six flux determinations are plotted for each experiment. The activity is nmoles of H_2O_2 metabolized/mg protein/min (from Geer *et al.*,1989).

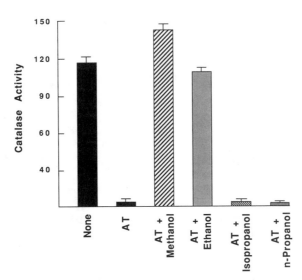

Figure 4. The *in vitro* catalase activity in homogenates of Canton-S wild-type larvae that were cultured on a 1% sucrose diet to day 5 post-hatch, then given a diet with 4 mM 3-amino-1,2,4-triazole (AT) for 17 hr with no alcohol, 1% methanol (v/v), 1% ethanol, 1% isopropanol (propan-2-ol) or 1% n-propanol. The mean+S.D. of six determinations are shown. The activity is nmoles of H_2O_2 metabolized/mg protein/min (from Geer *et al.*, 1989).

the participation of other minor pathways in *D. melanogaster* such as the microsomal ethanol oxidizing system.

2.2. Metabolic Pathways that Degrade Ethanol-Derived Acetaldehyde

Recently, it has been demonstrated that *Drosophila* ADH is capable of catalyzing the oxidation of acetaldehyde into acetic acid concomitant with the reduction of NAD^+ to NADH (Heinstra *et al.*, 1983; Eisses *et al.*, 1985; Geer *et al.*, 1985; Moxom *et al.*, 1985). To monitor the conversion of acetaldehyde to acetic acid (the second-half reaction of ADH) spectrophotometrically, the reaction mixture must be at a high pH (8.5 to 10) or include thiazolyl blue and phenazine methosulfate to drive the reaction in the oxidative direction (Heinstra *et al.*, 1983, 1986). The high pH "optimum", it has been argued makes the second-half reaction physiologically unimportant (e.g.,Benner *et al.*, 1985; Chambers, 1988). Another enzyme, aldehyde dehydrogenase (ALDH, EC 1.2.1.3), has been postulated to catalyze the same reaction (Garcin *et al.*, 1983, 1985), but has a similar high pH optimum. Nonetheless, the equilibria of the reactions of the ADH-pathway favor ethanol degradation *in vivo* (Middleton and Kacser, 1983; Geer *et al.*, 1985; Heinstra *et al.*, 1987; Heinstra *et al.*, 1989). By using cyanamide to inhibit ALDH in third instar larvae, we have determined that about 80% of the capacity for flux from ethanol to lipid is retained when more than 90% of the ALDH activity is inhibited. These and other data strongly suggest that ADH is the major enzyme for the conversion of acetaldehyde in *D. melanogaster* larvae (Heinstra *et al.*, 1989).

Non-insect ADHs, which lack the ability to catalyze the conversion of acetaldehyde to acetate, have a very low reactivity with secondary alcohols (Dalziel, 1975; Branden *et al.*, 1975); whereas, Drosophila ADH catalyzes the oxidation of secondary alcohols more rapidly than the oxidation of primary alcohols (Winberg *et al.*, 1982). This may reflect the ability of the Drosophila enzyme to catalyze the conversion of acetaldehyde to acetic acid. In aqueous solutions, aldehydes are present in two distinct forms, the aldehyde- and its hydrated *gem*-diol form:

$$
\underset{\textbf{aldehyde}}{R-\overset{\overset{\displaystyle H}{|}}{C}=O} \quad \underset{\longleftarrow}{\xrightarrow{\hspace{2cm}}} \quad \underset{\textbf{gem-diol}}{R-\overset{\overset{\displaystyle H}{|}}{\underset{\underset{\displaystyle OH}{|}}{C}}-OH}
$$

(see Eisses, 1989). The equilibrium of this hydration (non-enzymatic) reaction depends on the chemical nature of the R-group. For instance, acetaldehyde is about 55% hydrated, trichloroacetaldehyde nearly 100%, but benzaldehyde is not hydrated at all (Bodley and Blair, 1971; Eisses, 1989). The substrate specificity of the second-half reaction of *Drosophila* ADH suggests that the *gem*-diol form of acetaldehyde is the most reactive substrate for that enzyme (Eisses, 1989). *Drosophila* ADH may have high activities for the oxidation of secondary alcohols because the structures of secondary alcohols resemble the hydrated forms of aldehydes.

2.3. Catalytic Properties of *D. melanogaster* ADH

The ADH of *D. melanogaster* is encoded in a single structural gene, *Adh*, located at 50.1 on chromosome 2 (Lindsley and Grell, 1968). *Adh* is transcribed by use of a proximal promoter during the larval period and a distal promoter late in the third larval instar and during the adult stage (Savakis *et al.*, 1986). Most natural populations of *D. melanogaster* are polymorphic for electromorphs encoded in Adh^F and Adh^S alleles. A latitudinal cline in allele frequency is evident on at least three continents, with Adh^S being more frequent in tropical regions (Oakeshott *et al.*, 1982). There are only four amino acid differences out of a total of 255 residues between the allozymes of *D. melanogaster* and the ADH of *D. simulans* (Table I). The allozyme encoded by the rare allele, Adh^{71k}, apparently originated from a mutation of Adh^f; and the ADH of the sibling species *D. simulans* is most closely related to ADH-S.

The mammalian ADHs are a dimer of a polypeptide of about 350 amino acid residues and zinc; whereas *Drosophila* ADH is also a dimer, but is a member of an evolutionary branch characterized by a shorter, nonmetallic enzyme (Jörnvall *et al.*,

Table I
Differences in primary structure between
Drosophila ADH variants

Variant	Amino acid position NH_2————————————————COOH			
	1	82	192	214
D. simulans ADH	Ala	Lys	Lys	Pro
D. melanogaster ADH-S	Ser	Gln	Lys	Pro
D. melanogaster ADH-F	Ser	Gln	Thr	Pro
D. melanogaster ADH-71k	Ser	Gln	Thr	Ser

Data are from Heinstra *et al.* (1988b and original references therein).

1981, 1984; Place *et al.*, 1980). Yeast ADH, in contrast, is a tetramer. The homology of the primary sequences of yeast and mammals is significant, but low compared with the homology of each with Drosophila ADH (Jörnvall *et al.*, 1981). The effects of evolutionary divergence on the catalytic functions of the ADHs are difficult to judge. The different forms of ADH have different efficiencies for the oxidation of ethanol, and the k_{cat}/K_m^{eth} specificity constants of the different ADHs are markedly different (Table II). Considering all of the kinetic characteristics, Drosophila ADH is more like the mammalian ADH than the yeast ADH. However, the comparison of the absolute values of the kinetic constants can be misleading for a number of reasons. Acquisition of comparative values for the kinetic constants depends on the catalytic mechanisms of the enzymes (Hall and Koehn, 1983). Both the yeast and the mammalian ADHs obey an "ordered" mechanism in the oxidation of ethanol (Dalziel, 1975), whereas Drosophila ADH seems to function by a "rapid equilibrium random" mechanism (Heinstra *et al.*, 1988a, b). For this reason the most meaningful comparisons of kinetic values can be made for the ADH variants of Drosophila (Table II). The effects of the amino acid substitutions on the Michaelis constants for ethanol and NAD$^+$ are small, but the effects on the inhibition values for acetaldehyde and NADH are relatively large. The observed differences suggest that natural selection favored ethanol degradation rather than fermentation during the evolution of *D. simulans* and *D. melanogaster*. Moreover, catalytic properties for the turnover number, the specificity constants, and in various thermodynamic parameters apparently became more favorable for the oxidation of ethanol during the proposed evolution of *D. simulans* ADH to *D. melanogaster* ADH-S and to ADH-F (Heinstra *et al.*, 1988a, b). Thus, the mutations that produced the ADH allozymes of *D. melanogaster* and the ADH of *D. simulans* did not cause random effects, but were directional. This is consistent with the notion that strong purifying selection acts on base pair substitutions in the exons of *Adh* (Kreitman, 1983).

Many efforts have been made to link the *Adh* polymorphism to differential selection for alleles in an ethanol environment

Table II
Kinetic parameters for alcohol dehydrogenases from various sources

Source/variants	k_{cat} (s^{-1})	K_m^{eth} (mM)	k_{cat}/K_m^{eth} $(s^{-1}.M^{-1}.10^{-3})$	K_m^{NAD} (μM)	K_i^{NADH} (μM)	$K_i^{Acetald}$ (μM)
Yeast	60.0	13.0	5.0	70	18.0	670
Horse liver	3.0	0.5	60.0	20	1.5	90
Rat liver	2.6	0.5	50.0	30	1.0	12
D. simulans	2.4	6.1	0.4	200	1.0	310
D. melanogaster/Slow	3.2	3.0	1.0	135	3.0	350
D. melanogaster/Fast	4.9	4.0	1.2	160	12.0	110
D. melanogaster/71k	4.1	4.0	1.0	140	5.0	50

The data were extracted from: Yeast ADH (pH 7.2), Bloomfield *et al.*, 1962, Mahler and Douglas, 1957, and Wratten and Cleland, 1963; horse liver ADH (pH 7.2), the same references as for yeast ADH; rat liver ADH (pH 7.3), Crabb *et al.*, 1983, and *Drosophila* ADHs (pH 7.4), Heinstra *et al.*, 1988a and 1988b.

(van Delden *et al.*, 1978; Ziola and Parsons, 1982; Zera *et al.*, 1985). Lines homozygous for Adh^F-alleles have two to three times as much ADH activity as lines homozygous for Adh^S alleles. This is due to a higher *in vivo* ADH-F protein content than ADH-S and to differences in the turnover number (Gibson, 1972; Lewis and Gibson, 1978; Winberg *et al.*, 1985) Recently, it has been shown that Adh mRNA levels do not differ significantly (Geer *et al.*, 1988; Laurie and Stam, 1988). The translation rates of the Adh mRNAs or the stabilities of the allozymes in the two *Adh* genotypes may differ (Laurie and Stam, 1988). Attempts to relate the ADH allozyme variation to differences in ethanol degradation rates have yielded mixed results. Middleton and Kacser (1983) found no difference between the flux from ethanol to lipid and CO_2. However, third instar larvae with either ADH-F or ADH-S exhibited differences of about 20% in the rates of flux through the ADH pathway (Heinstra *et al.*, 1987; Geer *et al.*, 1988a). The flux from ethanol into lipids differed by a factor of two between *D. simulans* and *D. melanogaster* (Heinstra and Geer, unpublished). The flux control coefficient of Kacser and Porteous (1987), which relates ADH activity to flux through the ADH pathway, was close to 1.0. This indicates that ADH exerts complete control over this catabolic pathway.

3. THE GENETIC BASIS FOR SHORT TERM ADAPTATION TO ETHANOL

3.1. Regulation of the Expression of the *Adh* Gene

The regulation of the *Adh* structural gene of *D. melanogaster* is increasingly gaining the attention of geneticists. Dickinson *et al.* (1984) observed that there are cis-acting factors near *Adh* genes of *D. melanogaster* and *D. simulans* that control the pattern of expression of the gene in the hybrid. Heberlein *et al.* (1985), Posakony *et al.* (1986), Heberlein and Tjian (1988), and Corbin and Maniatis (1989)

have identified DNA sequences upstream from the promoters of
Adh that are important to the level of expression and the tissue specificity for the expression of the gene. Further evidence of the importance of the nontranscribed DNA of the *Adh*
region was gained when Aquadro *et al.* (1986) found a nonrandom
association among ADH activity levels, ADH allozyme and several restriction site polymorphisms.

Because the larvae of *D. melanogaster* are bound to a single
feeding site and adults are mobile and may move to feeding
sites that match their level of tolerance to the toxic effects
of alcohol, larvae were examined for short term responses to
dietary ethanol (McKechnie and Geer, 1984). Canton-S larvae
were cultured for four days on a defined, fat-free diet, then
transferred to test diets with no ethanol or with varied concentrations of ethanol. The ADH activity and ADH CRM were proportional to the dietary concentration of ethanol to about 5%
(v/v) (McKechnie and Geer, 1984).

A time course study using specific riboprobes for the distal and proximal transcripts of *Adh* indicated that the proxi-

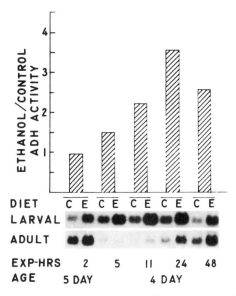

Figure 5. Northern blots of total RNA (10 µg of total RNA/lane) from
Canton-S adults and third instar larvae. After being fed a 1% sucrose diet
until day 4 or 5 of the larval period, test individuals were fed a 0.3%
sucrose diet (C) or a 0.3% sucrose, 2.5% ethanol diet (E) for the number of
hours indicated by Exp-hrs. The proximal transcript (larval) was visualized
by hybridization with ^{32}P-labeled probe prepared from pSP65L and the distal
transcript (adult) was visualized by hybridization with probe prepared from
pSP65A. The ratio of the ADH activity in homogenates of larvae fed the
ethanol diet to the ADH activity in homogenates of larvae fed the control
diet is given as an indicator of the degree of the induction of ADH by
ethanol. Canton-S larvae pupate in about 9 days when fed the defined test
diet. This figure is from Geer *et al.* (1988).

mal transcript is increased to a much greater extent by a low level of dietary ethanol than the distal transcript in third instar larvae (Fig. 5). Increases in the transcripts were evident within 2 hours after exposure to ethanol, but the increase in ADH activity was not detectable until 5 hours of exposure. The induction of ADH activity in larvae increased the flux from ethanol into lipids (Geer et al., 1988), consequently the modulation of ADH activity by dietary ethanol appears to be a mechanism by which larvae increase their abilities to degrade ethanol.

3.2. DNA Sequences that Affect the Induction of *Adh* mRNA

P-element transformed strains of *D. melanogaster* with deletions of part of the 5' regulatory DNA of *Adh* were examined to determine if any DNA sequences were required for the induction of *Adh* by ethanol (Kapoun et al., 1990). A Northern blot analysis of strains with deletions between -660 from the start of the distal transcript to -80 from the start of the proximal transcript (+686 from the distal transcript start site) revealed that the strains responded differently to dietary ethanol (Fig. 6). The proximal *Adh* mRNA levels in both the M and N lines increased in response to 1.5% dietary ethanol. However, the proximal transcript levels of four P lines and

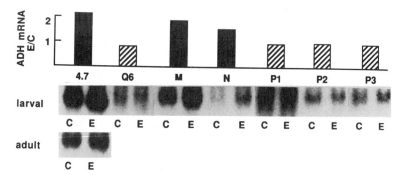

Figure 6. Northern blots of total RNA (12 µg of total RNA/lane) from P-element transformant larvae. Four day old larvae were transferred to one of two types of media: 1) 0.3% sucrose (C) and 2) 0.3% sucrose with 1.5% ethanol (E) for about 24 hours. Proximal and distal transcripts of *Adh* were visualized with specific probes. The ratio of the *Adh* mRNA of larvae fed the ethanol diet to those fed the control diet in each pair as determined by densiometric scanning is given as an indicator of the degree of the induction of *Adh* mRNA by ethanol. An *Adh* mRNA E/C ratio of one indicates that induction did not occur. The host Adh^{f6n} ; ry^{506} strain was transformed with Adh^f and about 5,000 bp of regulatory DNA upstream from *Adh* (line 4.7) or a regulatory region with a deletion. The deletion left endpoint for all the deletion lines is -660 from the transcription start site of the distal promoter. The right endpoints for the deletions relative to the start site of the transcription of the proximal promoter were -80 for Q6 (+686 from the distal transcript start site), -110 for the P lines, -242 for the M line and -187 for the N line. Thus, each of the deletion lines lacked the distal promoter and part of the regulatory DNA between the distal and proximal promoters of *Adh*. This figure is from Kapoun et al. (1990).

Q6 were not significantly altered by 1.5% ethanol. A comparison of the deletion endpoints for the test lines suggest that the region extending from -110 to -187 from the proximal transcript start site is necessary for the induction process.

3.3. Induction of mRNA in Adh^{n2}

It is of interest to know which compound(s) is involved in the induction of Adh. The ADH-null strain, Adh^{n2}, was examined to determine if ethanol itself or a derivative of ethanol is the inducer of the Adh mRNAs and ADH protein. Adh mRNAs are formed in Adh^{n2} individuals, but the ADH protein is unstable and larvae have only about 0.3% of the wild-type level of ADH activity. A chromatographic analysis showed that Adh^{n2} larvae fed 1.25% ethanol contained more than 15 times as much ethanol as Canton-S larvae fed the same diet (Kapoun et al., 1990), confirming that the ADH pathway for degrading ethanol is blocked in Adh^{n2} larvae. The proximal transcript was about five times greater in quantity in Adh^{n2} larvae than in Canton-S wild-type larvae, whereas the distal message was not significantly altered in the mutant larvae (Fig. 7). This shows that the degradation of ethanol by the ADH pathway is not necessary for Adh induction.

As mentioned, the ADH enzyme apparently obeys a "rapid equilibrium random" mechanism for the oxidation of ethanol. Compared to ethanol, the maximum velocity of ADH was significantly less with d6-ethanol as the substrate (Hovik et al., 1984; Heinstra et al., 1988a). "Rapid equilibrium" indicates that the transfer of hydrogen from ethanol to NAD^+ is the rate-limiting step. In contrast, the rate-limiting step for mamma-

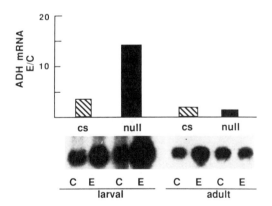

Figure 7. Northern blots of total RNA (12 μg of total RNA/lane) from Canton-S and Adh^{n2} four day old larvae were transferred to one of two types of media: 1) 0.3% sucrose and 2) 0.3% sucrose with 1.25% ethanol for about 48 hours. Proximal and distal transcripts were visualized with ^{32}P-labelled pSP65L and pSP65A respectively. The ratio of the Adh mRNA of larvae fed the ethanol diet as determined by densiometric scanning to those fed the control diet in each pair is given as an indicator of the degree of the induction of Adh mRNA by ethanol. An Adh mRNA E/C ratio of one indicates that induction did not occur (from Kapoun et al., 1990).

lian ADH is the dissociation of the ADH:NADH binary complex
(Dalziel, 1975). Studies were performed to test whether a similar catalytic mechanism prevails *in vivo*, and whether d6-
ethanol accumulated in larvae compared to protonated-ethanol
(Kapoun *et al.*, 1990). Apparently, a similar slower turnover
of d6-ethanol occurred in larvae. Northern blot analysis
showed once again a strong correlation with proximal Adh mRNA
accumulation and *in vivo* alcohol levels. This again suggested
that ethanol, and not a derivative, was the inducing compound
and that the two promoters are sensitive to the inducing action of ethanol to different degrees.

4. THE GENETIC BASIS FOR LONG TERM ADAPTATION

4.1. The Epistatic Interaction of *Adh* and *Gpdh*

GPDH, a cytoplasmic enzyme, provides sn-glycerol-3-phos-
phate for the synthesis of phospholipid (PL) and TG (O'Brien
and MacIntyre, 1972; Geer *et al.*, 1983; MacIntyre and Davis,
1987) and operates in concert with sn-glycerol-3-phosphate
oxidase (GPO, EC 1.1.99.5), a mitochondrial enzyme, to form
the sn-glycerol-3-phosphate cycle that generates NAD^+ to drive
glycolysis and ethanol degradation in the fat body (Geer *et
al.*, 1983). Equally important, the cycle provides ATP in
flight muscle (Sacktor, 1970). The gene loci for GPDH (*Gpdh*)
and GPO (*Gpo*) are located on chromosome 2 at 20.5 and 75, respectively, and modifiers of the activities of both enzymes
have been localized to chromosomes 2 and 3 of *D. melanogaster*
(Laurie-Ahlberg *et al.*, 1980). GPDH has been implicated in
ethanol tolerance. GPDH is induced by ethanol to higher tissue activity levels in larvae (Geer *et al.*, 1983), as is ADH
(McKechnie and Geer, 1984). This suggests that GPDH and ADH
may be parts of the same metabolic process that degrades
ethanol. Population cage studies by Cavener and Clegg (1978,
1981) suggested that epistatic interaction between the *Adh* and
Gpdh loci may be important in flies under ethanol stress.

The involvement of epistatic interaction of allozymes in
larval ethanol tolerance variation of *D. melanogaster* was examined (McKechnie and Geer, 1988). A set of 42 isofemale
lines derived from flies collected at the Chateau Tahbilk Winery and Wandin North Orchard of Victoria, Australia were made
homozygous for the common alleles of *Adh* and *Gpdh*. When fed
6% ethanol, all of the test lines had reduced survival levels
and, for the survivors, body weight was reduced and developmental time to eclosion was lengthened. Alleles responsible
for slowing development appeared to contribute to ethanol tolerance. Factorial ANOVA of larval ethanol tolerance measured
at each of four consecutive generations on all 42 test lines
revealed that larvae homozygous for Adh^F and $Gpdh^S$, and Adh^S and
$Gpdh^S$ were more tolerant than larvae with the other allelic
combinations. However, these genotypes were not associated
with the slowing of development nor the weight loss on
ethanol. The epistatic interaction of the two loci implies
that certain combinations of ADH and GPDH allozymes are most
effective in metabolizing ethanol in larvae in ways that promote tolerance.

GPO was induced several-fold by a moderate concentration of
dietary ethanol in larvae of all the test lines (McKechnie and
Geer, 1986) and appears to have an important role in ethanol
tolerance in *D. melanogaster* (McKechnie and Ross, Chapter 14,
this volume).

4.2. Metabolic Flux, Lipid Composition and Ethanol Tolerance

The associations of tolerance with physiological traits
were recently examined in seven isochromosomal lines derived
from *D. melanogaster* females caught at the Tahbilk winery
(Geer *et al.*, 1990). The test lines were cultured under axenic
conditions throughout the experiment. Larvae were grown on
ethanol-free, 1% sucrose defined medium for four days, then
transferred to an ethanol-free medium or a medium with 4.5%
ethanol (v/v). The percent of larvae to survive to pupation
(pupation score), the percent of larvae to survive to eclosion
(eclosion score), the length of the larval growth period
(development time), and the mean adult male weight were monitored for the ethanol-free control diet and the ethanol diet.
Six cultures of about fifty larvae of each line for the
ethanol-free and ethanol-supplemented diets were examined in
four separate runs. The flux from ^{14}C-ethanol into lipid was
determined in two separate experiments for larvae fed a diet
with 2.5% (v/v) total ethanol. The activities of ADH, GPDH and
GPO and the fatty acid compositions were monitored for the
seven lines in larvae fed the ethanol-free control diet. Correlations between the tolerance test parameters and the physiological traits were then assessed.

The pupation score was very closely correlated with eclosion score (Table III). This implied that survival of the
individual during the pupal period was very dependent upon its
growth during the larval period. The larval development time
was not significantly correlated with either pupation score or
eclosion score, further establishing that the rate of larval
growth is not critical to survival of the individual in an
ethanol environment. The ADH and GPDH activities were not significantly related to pupation score, to eclosion score or to
development time. There was a modest correlation between ADH
activity and GPDH activity (r=0.73, p=0.08). GPO activity was
significantly related to both pupation score and eclosion
score, further evidencing that the activity of this enzyme is
important to survival in an ethanol environment.

The flux from ethanol into lipid appeared to be very important to growth and survival of individuals in an ethanol environment (Table III). The larval development time was strongly
correlated with flux (r=0.9, p=0.02). Flux through the ADH
pathway was moderately correlated with the pupation score
(r=0.76, p=0.07) and eclosion score (r=0.76, p=0.07). The rate
of flux from ethanol to lipid was also positively correlated
with GPO activity (r=0.81, p=0.05).

The strongest correlations were observed between the fatty
acid compositions of larval lipids and the survival growth
traits (Table III). The percentage unsaturated fatty acids was

Table III

A Pearson correlation coefficient analysis of the growth and biochemical traits of larvae of the Tahbilk lines (from Geer et al., 1990)

	Pupation score	Development time	% 18-C fatty acid	% unsaturated fatty acid	Flux	GPO activity	ADH activity	GPDH activity
Adult score	0.98 (6) P=.00**	0.30 (6) P=0.28	0.87 (6) P=0.01**	0.87 (6) P=0.01**	0.76 (5) P=0.07	0.94 (5) P=0.01**	0.02 (5) P=0.48	0.21 (5) P=0.36
Pupation score		0.24 (6) P=0.32	0.92 (6) P=0.005**	0.93 (6) P=0.003**	0.76 (5) P=0.07	0.88 (5) P=0.02*	-00.01 (5) P=0.49	0.09 (5) P=0.44
Development time			0.42 (6) P=0.21	0.07 (6) P=0.45	0.90 (5) P=0.02*	0.56 (5) P=0.16	-0.04 (5) P=0.47	0.53 (5) P=0.18
% 18-carbon fatty acids				0.89 (6) P=0.009**	0.85 (5) P=0.033**	0.82 (5) P=0.047*	0.12 (5) P=0.42	0.16 (5) P=0.40
% unsaturated fatty acids					0.71 (5) P=0.09	0.76 (5) P=0.07	0.12 (5) P=0.42	0.01 (5) P=0.50
Flux						0.81 (5) P=0.047*	0.14 (5) =0.42	0.49 (5) P=0.20
GPO activity							0.34 (5) P=0.29	0.53 (5) P=0.18
ADH activity								0.73 (5) P=0.08

$*P<0.05$, $**P<0.01$.

most strongly correlated with pupation score (r=0.93, p=0.003) but was also closely related to eclosion score (r=0.87, p=0.01). The percentage 18-carbon fatty acid content was also strongly correlated to pupation score (r=0.92, p=0.005) and eclosion score (r=0.87, p=0.01). Thus, both the degree of desaturation and the chain length of fatty acids appeared to be prominent factors for ethanol tolerance.

5. SUMMARY

Alcohol tolerance is the sum of many traits whose total effect is to allow the individual to grow and reproduce in the presence of alcohol. Wild-type *D. melanogaster* is relatively tolerant to the toxic effects of ethanol and can use low dietary concentrations as a nutrient. These characteristics are dependent on the alcohol dehydrogenase (ADH) system for the degradation of ethanol. Some of the toxic effects of ethanol were evident upon the examination of the ultrastructure of fat body cells of ADH-null third-star larvae fed a near lethal concentration of ethanol. Membrane transport and the depositions of protein, glycogen and lipid were seen to be influenced. Low doses of dietary ethanol were detrimental to ADH-null larvae, but advantageous to wild-type larvae, when a later, growth-limiting dose of ethanol was administered. At low dietary concentrations ethanol stimulated an increase in phosphatidylethanolamine and a decrease in phosphatidylcholine in wild-type larvae, and there was a shift to shorter chain, unsaturated fatty acids. Ethanol may be toxic because of its effects on the lipid composition of larvae.

The ADH, catalase and the microsomal ethanol oxidizing systems have been suggested to be metabolic pathways for the degradation of alcohols in *D. melanogaster*. The ADH pathway accounts for 90-93% and the catalase system 3-5% of the total ethanol-degrading capacity of *D. melanogaster*. Drosophila ADH is smaller than the mammalian ADH and it is not a metalloenzyme. *D. melanogaster* ADH catalyzes both the conversion of ethanol to acetaldehyde, and acetaldehyde to acetic acid. The mechanism for the first half reaction appears to differ from the mechanism of the mammalian enzyme. Flux control coefficients for *D. melanogaster* ADH indicate that it is the rate-limiting enzyme for the conversion of ethanol into lipid. This conversion is important to ethanol utilization and tolerance of *D. melanogaster*.

Exposure to moderate levels of dietary ethanol induces parallel increases in ADH activity in third instar larvae. Most of the increase is due to an increase in ADH protein. ADH is encoded in a single structural gene, *Adh*, with both a proximal and distal promoter. Northern blot analysis indicates that the increase in ADH protein is accompanied by a relatively large increase in the amount of proximal transcript and a lesser increase in the distal transcript in wild-type larvae exposed to a moderate dose of ethanol. Deletion analysis of the 5' regulatory region of *Adh* using P-element mediated transformed lines suggests that a DNA sequence in the region -187 to -110 upstream from the proximal promoter is necessary for the ethanol-stimulated increase of the proximal transcript. The induction of ADH is accompanied by an increase in the flux through the ethanol to lipid pathway and represents a rapid means for larvae to adapt to environmental ethanol.

Studies of isochromosomal lines of *D. melanogaster* with chromosomes derived from females collected at the Tahbilk winery of Victoria, Australia suggest that individuals with particular combinations of the common alleles of *Adh* and *Gpdh* are more tolerant to ethanol. In these lines larvae with high relative amounts of long chain and unsaturated fatty acids in lipids survived a growth-limiting ethanol diet the best, suggesting that membrane composition is important to ethanol tolerance and utilization.

ACKNOWLEDGEMENTS. We would like to thank Joanne Ochoa, Dustin Joy, Karen Kandl, Gigi Barilla, Jessica Smiley, Rajeev Dadoo, Denise Johnson, Chris Baumgardner and Marilyn Langevin for their technical assistance. We are also grateful to Astrid Freriksen and Willem Scharloo for their comments and encouragement. This research was supported by National Institutes of Health Grant No. AA06702 to B.W.G. and a fellowship from the Netherlands Organization for Scientific Research (N.W.O.) to P.W.H.H.

References

Aquadro, C. F., Deese, S. F., Bland, M. M., Langley, C. H., and Laurie-Ahlberg, C. C., 1986, Molecular population genetics of the alcohol dehydrogenase gene region of *Drosophila melanogaster*, *Genetics* 114:1165-1190.

Benner, S. A., Nambiar, K. P., and Chambers, G. K., 1985, A stereochemical imperative in dehydrogenases: New data and criteria for evaluating function-based theories in bioorganic chemistry, *J. Am. chem. Soc.* 107:5513-5517.

Bewley, G. C., Nahmias, J. A., and Cooke, J. L., 1983, Developmental and tissue-specific control of catalase expression in *Drosophila melanogaster*: Correlations with rates of enzyme synthesis and degradation, *Dev. Genet.* 4:49-60.

Bloomfield, V., Peller, L., and Alberty, R., 1962, Multiple intermediates in steady-state enzyme kinetics. III. Analysis of the kinetics of some reactions catalyzed by dehydrogenases, *J. Am. chem. Soc.* 84:4375-4381.

Bodley, F., and Blair, A., 1971, Substrate characteristics of human aldehyde dehydrogenase, *Can. J. Biochem.* 49:1-5.

Bränden, C. H., Jörnvall, H., Eklund, H., and Fururgen, B., 1975, Alcohol dehydrogenase, in: *The Enzymes* Vol. XI (P. D. Boyer, ed.), Academic Press, New York, pp. 103-190.

Cavener, D. R., and Clegg, M. T., 1978, Dynamics of correlated genetic systems. IV. Multilocus effects of ethanol stress environments, *Genetics* 90:629-644.

Cavener, D. R., and Clegg, M. T., 1981, Multigenic response to ethanol in *Drosophila melanogaster*, *Evolution* 39:278-293.

Chambers, G. K., 1988, The *Drosophila* alcohol dehydrogenase gene-enzyme system, *Adv. Genet.* 25:40-107.

Cohan, F. M., and Graf, J., 1985, Latitudinal cline in *Drosophila melanogaster* for knockdown resistance to ethanol fumes and for rates of response to selection of further resistance, *Evolution* 39:278-293.

Cohan, F. M., and Hoffmann, A. A., 1986, Genetic divergence under uniform selection. II. Different responses to selection for knockdown resistance to ethanol among *Drosophila melanogaster* populations and their replicate lines, *Genetics* 114:145-164.

Corbin, V., and Maniatis, T., 1989, Role of transcriptional interference in the *Drosophila melanogaster Adh* promoter switch, *Nature, Lond.* 337:279-282.

Crabb, D., Bosron, W., and Li, T-K., 1983, Steady-state kinetic properties of purified rat liver alcohol dehydrogenase: Application to predicting alcohol elimination rates *in vivo, Archs Biochem. Biophys.* 224:299-309.

Dalziel, K., 1975, Kinetics and mechanism of nicotinamide-nucleotide-linked dehydrogenases, in: *The Enzymes* Vol. XI (P. D. Boyer, ed.), Academic Press, New York, pp. 2-60.

David, J. R., Bocquet, C., Arens, M., and Fouillet, P., 1976, Biological role of alcohol dehydrogenase in the tolerance of *Drosophila melanogaster* to aliphatic alcohols: Utilization of an ADH-null mutant, *Biochem. Genet.* 14:989-997.

David, J. R., Bocquet, C., van Herrewege, J., Fouillet, P., and Arens, M., 1978, Alcohol metabolism in *Drosophila melanogaster*: Uselessness of the most active aldehyde oxidase produced by the *aldox* locus, *Biochem. Genet.* 16:203-211.

David, J. R., Daly, K., and van Herrewege, J., 1984, Acetaldehyde utilization and toxicity in adults lacking alcohol dehydrogenase or aldehyde oxidase, *Biochem. Genet.* 22:1015-1029.

Deltombe-Lietaert, M. C., Delcour, J., Lenelle-Monfort, N., and Elens, A., 1979, Ethanol metabolism in *Drosophila melanogaster*, *Experientia* 35:579-581.

Dickinson, W., Rowan, R., and Brennan, M., 1984, Regulatory gene evolution: Adaptive differences in expression of alcohol dehydrogenase in *Drosophila melanogaster* and *D. simulans*, *Heredity* 52:215-225.

Eisses, K. Th., 1989, On the oxidation of aldehydes by alcohol dehydrogenase of *Drosophila melanogaster*: Evidence for the gem-diol as the reacting substrate, *Bioorg. Chem.* 17:268-274.

Eisses, K. Th., Schoonen, W. G., Aben, W., Scharloo, W., and Thörig, G. E. W., 1985, Dual function of the alcohol dehydrogenase of *Drosophila melanogaster*: Ethanol and acetaldehyde oxidation by two allozymes ADH-71k and ADH-F, *Mol. Gen. Genet.* 199:76-81.

Fast, P. G., 1966, A comparative study of the phospholipids and fatty acids of some insects, *Lipids* 1:201-215.

Garcin, F. J., Côté, J., Radouco-Thomas, S., Kasienczuk, D., Chawla, S., and Radouco-Thomas, C., 1983, Acetaldehyde oxidation in *Drosophila melanogaster* and *Drosophila simulans*: Evidence for the presence of an NAD$^+$-dependent dehydrogenase, *Comp. Biochem. Physiol.* 75B:205-210.

Garcin, F. J., Hin, G. L. Y., Côté, J., Radouco-Thomas, S., Chawla, S., and Radouco-Thomas, C., 1985, Aldehyde dehydrogenase in *Drosophila*: Developmental and functional aspects, *Alcohol* 2:85-89.

Geer, B. W., and Dolph, W. W., 1970, A dietary choline requirement for egg production in *Drosophila melanogaster*, *J. Reprod. & Fert.* 21:9-15.

Geer, B. W., Dolph, W. W., Maquire, J. A., and Dates, R. J., 1971, The metabolism of dietary carnitine in *Drosophila melanogaster*, *J. exp. Zool.* 176:445-460.

Geer, B. W., Dybas, L. K., and Shanner, L. J., 1989, Alcohol dehydrogenase and ethanol tolerance at the cellular level in *Drosophila melanogaster*, *J. exp. Zool.* 250:22-39.

Geer, B. W., Langevin, M. L., and McKechnie, S. W., 1985, Dietary ethanol and lipid synthesis in *Drosophila melanogaster*, *Biochem. Genet.* 23:607-622.

Geer, B. W., McKechnie, S. W., Bentley, M. M., Oakeshott, J. G., Quinn, E. M., and Langevin, M. L., 1988, Induction of alcohol dehydrogenase by ethanol in *Drosophila melanogaster*, *J. Nutr.* 118:398-407.

Geer, B. W., McKechnie, S. W., and Langevin, M. L., 1983, Regulation of sn-glycerol-3-phosphate dehydrogenase in *Drosophila melanogaster* by dietary ethanol and sucrose, *J. Nutr.* 113:1632-1642.

Geer, B. W., McKechnie, S. W., and Langevin, M. L., 1986, The effect of dietary ethanol on the composition of lipids of *Drosophila melanogaster* larvae, *Biochem. Genet.* 24:51-69.

Geer, B. W., McKechnie, S. W., Pyka, M. J., and Heinstra, P. W. H., 1990, Variation of biochemical traits that are important to the ethanol tolerance of *Drosophila melanogaster*. *Evolution*. in press.

Geer, B. W., and Perille, T. T., 1977, Effects of dietary sucrose and environmental temperature on fatty acid synthesis in *Drosophila melanogaster*, *Insect Biochem.* 7:371-379.

Geer, B. W., and Vovis, G. F., 1965, The effects of choline and related compounds on the growth and development of *Drosophila melanogaster*, *J. exp. Zool.* 158:223-236.

Geer, B. W., van der Zel, A., and Heinstra, P. W. H., 1989, The significance of catalase in alcohol metabolism in *Drosophila melanogaster*. In preparation.

Geer, B. W., Vovis, G. F., and Yund, M. A., 1968, Choline activity during the development of *Drosophila melanogaster*, *Physiol. Zool.* 41:280-292.

Gibson, J. B., 1970, Enzyme flexibility in *Drosophila*, *Nature, Lond.* 227:959-960.

Gibson, J. B., 1972, Differences in number of molecules produced by two allelic electrophoretic enzyme variants in *Drosophila melanogaster*, *Experientia* 28:975-976.

Green, P. R., and Geer, B. W., 1979, Changes in the fatty acid composition of *Drosophila melanogaster* during development and ageing, *Archs int. Physiol. Biochim.* 87:485-493.

Hall, J. G., and Koehn., R. K., 1983, The evolution of enzyme catalytic efficiency and adaptive inference from steady-state kinetic data, *Evol. Biol.* 16:53-96.

Heberlein, U., England, B., and Tjian, R., 1985, Characterization of *Drosophila* transcription factors that activate the tandem promoters of the alcohol dehydrogenase gene, *Cell* 41:965-977.

Heberlein, U., and Tjian, R., 1988, Temporal pattern of alcohol dehydrogenase gene transcription reproduced by *Drosophila* stage-specific extracts, *Nature, Lond.* 331:410-415.

Heinstra, P. W. H., Eisses, K. Th., Schoonen, W. G., Aben, W., de Winter, A. J., van der Horst, D. J., van Marrewijk, W. J. A., Beenakkers, A. M. Th., Scharloo, W., and Thörig., G. E. W., 1983, A dual function of alcohol dehydrogenase in *Drosophila*, *Genetica* 60:129-137.

Heinstra, P. W. H., Eisses, K. Th., Scharloo, W., and Thörig, G. E. W., 1986, Metabolism of secondary alcohols in *Drosophila melanogaster*: Effects on alcohol dehydrogenase, *Comp. Biochem. Physiol.* 83B:403-408.

Heinstra, P. W. H., Geer, B. W., Seykens, D., and Langevin, M. L., 1989, The metabolism of ethanol-derived acetaldehyde by alcohol dehydrogenase (EC 1.1.1.1.) and aldehyde dehydrogenase (EC 1.2.1.3.) in *Drosophila melanogaster* larvae, *Biochem. J.* 259:791-797.

Heinstra, P. W. H., Scharloo, W., and Thörig, G. E. W., 1987, Physiological significance of the alcohol dehydrogenase polymorphism in larvae of *Drosophila*, *Genetics* 117:75-84.

Heinstra, P. W. H., Scharloo, W., and Thörig, G. E. W., 1988a, Alcohol dehydrogenase polymorphism in *Drosophila*: Enzyme kinetics of product inhibition, *J. Mol. Evol.* **28**:145-150.

Heinstra, P. W. H., Thörig, G. E. W., Scharloo, W., Drenth, W., and Nolte, R. J. M., 1988b, Kinetics and thermodynamics of ethanol oxidation catalyzed by genetic variants of the alcohol dehydrogenase from *Drosophila melanogaster* and *D. simulans*, *Biochim. biophys. Acta* **967**:224-233.

Hovik, R., Winberg, J-O., and McKinley-McKee, J. S., 1984, *Drosophila melanogaster* alcohol dehydrogenase: Substrate specificity of the ADH-F alleleoenzyme, *Insect Biochem.* **14**:345-351.

Jörnvall, H., Bahr-Lindstrom, H., Jany, K. D., Ulmer, W., and Froschle, M., 1984, Extended superfamily of short alcohol-polyol-sugar dehydrogenases: Structural similarities between glucose and ribitol dehydrogenase, *FEBS Lett.* **165**:190-196.

Jörnvall, H., Persson, M., and Jeffrey, J., 1981, Alcohol and polyol dehydrogenases are both divided into protein types, and structural properties cross-relate the different enzyme activities within each type, *Proc. natn. Acad. Sci. USA* **78**:4226-4230.

Kacser, H., and Porteous, P., 1987, Control of metabolism: What do we have to measure? *Trends Biochem. Sci.* **12**:5-14.

Kapoun, A. M., Geer, B. W., Heinstra, P. W. H., Corbin, V., and McKechnie, S. W., 1990, Molecular control of the induction of alcohol dehydrogenase by ethanol in *Drosophila melanogaster* larvae. *Genetics* **124**: in press.

Kreitman, M., 1983, Nucleotide polymorphism at the alcohol dehydrogenase locus of *Drosophila melanogaster*, *Nature, Lond.* **304**:412-417.

Laurie-Ahlberg, C. C., Maroni, G., Bewley, G. C., Lucchesi, L. C., and Weir, B. S., 1980, Quantitative genetic variation of enzyme activities in natural populations of *Drosophila melanogaster*, *Proc. natn. Acad. Sci. USA* **77**:1073-1077.

Laurie, C. C., and Stam, L. F., 1988, Quantitative analysis of RNA produced by Slow and Fast alleles of *Adh* in *Drosophila melanogaster*, *Proc. natn. Acad. Sci. USA* **85**:5161-5165.

Lewis, N., and Gibson, J. B., 1978, Variation in amount of enzyme protein in natural populations, *Biochem. Genet.* **16**:159-170.

Lindsley, D. L., and Grell, E. H., 1968, *Genetic variations of* Drosophila melanogaster, Carnegie Institute of Washington, Publication No. 627, Washington D.C.

MacIntyre, R. J., and Davis, M. B., 1987, A genetic and molecular analysis of the alfa-glycerophosphate cycle in *Drosophila melanogaster*, *Isozymes. Current Topics in Biological & Medical Research* **14**:195-227.

Mahler, H., and Douglas, F., 1957, Mechanisms of enzyme-catalyzed oxidation-reduction reactions. I. An investigation of the yeast alcohol dehydrogenase reaction by means of the isotope rate effect, *J. Am. chem. Soc.* **79**:1159-1166.

Maroni, G., and Stamey, S. C., 1983, Developmental profile and tissue distribution of alcohol dehydrogenase, *Drosoph. Inf. Serv.* **59**:77-79.

McDonald, J. F., Chambers, G. K., David, J. R., and Ayala., F. J., 1977, Adaptive response due to changes in gene regulation: A study with *Drosophila*, *Proc. natn. Acad. Sci. USA* **74**:4562-4566.

McKechnie, S. W., and Geer, B. W., 1984, Regulation of alcohol dehydrogenase in *Drosophila melanogaster* by dietary alcohol and carbohydrate, *Insect Biochem.* **14**:231-242.

McKechnie, S. W., and Geer, B. W., 1986, sn-Glycerol-3-phosphate oxidase and alcohol tolerance in *Drosophila melanogaster* larvae, *Biochem. Genet.* **24**:859-872.

McKechnie, S. W., and Geer, B. W., 1988, The epistasis of *Adh* and *Gpdh* allozymes in ethanol tolerance of *Drosophila melanogaster* larvae, *Genet. Res.* **52**:179-184.

McKechnie, S. W., and Morgan, P., 1982, Alcohol dehydrogenase polymorphism

of *Drosophila melanogaster*: Aspects of alcohol and temperature variation in the larval environment, *Aust. J. biol. Sci.* **35**:85-93.

McKenzie, J. A., and Parsons, P. A., 1972, Alcohol tolerance: An ecological
parameter in the relative success of *Drosophila melanogaster* and
Drosophila simulans, *Oecologia* **10**:373-388.

Middleton, R. J., and Kacser, H., 1983, Enzyme variation, metabolic flux
and fitness: Alcohol dehydrogenase in *Drosophila melanogaster*, *Genetics*
105:633-650.

Moxom, L. N., Holmes, R. S., Parsons, P. A., Irving, M. G., and Doddrell,
D. M., 1985, Purification and molecular properties of alcohol dehydrogenase from *Drosophila melanogaster*: Evidence from NMR and kinetic
studies for function as an aldehyde dehydrogenase, *Comp. Biochem.
Physiol.* **80B**:525-535.

Oakeshott, J. G., Gibson, J. B., Anderson, P. R., Knibb, W. R., Anderson,
D. G., and Chambers, G. K., 1982, Alcohol dehydrogenase and glycerol-3-
phosphate dehydrogenase clines in *Drosophila melanogaster*, *Evolution*
36:86-96.

O'Brien, S. J., and MacIntyre, R. J., 1972, The alpha-glycerophosphate cycle in *Drosophila melanogaster*. I. Biochemical and developmental
aspects, *Biochem. Genet.* **7**:141-161.

Place, A. R., Powers, D. A., and Sofer, W., 1980, *Drosophila melanogaster*
alcohol dehydrogenase does not require metals for catalysis, *Fedn Proc.
Fedn Am. Socs exp. Biol.* **39**:1640.

Posakony, J. W., Fischer, J. A., and Maniatis, T., 1986, Identification of
DNA sequences required for the regulation of *Drosophila* alcohol dehydrogenase gene expression, *Cold Spring Harbor Symp. Quant. Biol.*
50:515-520.

Sacktor, B., 1970, Regulation of intermediary metabolism with special
reference to control mechanisms in insect flight muscle, *Adv. Insect
Physiol.* **7**:267-347.

Savakis, C., Ashburner, M., and Willis, J. H., 1986, The expression of the
gene coding for alcohol dehydrogenase during the development of
Drosophila melanogaster, *Devl Biol.* **114**:194-207.

Tilghman, J. A., and Geer, B. W., 1988, The effects of a choline deficiency
on the lipid composition and ethanol tolerance of *Drosophila
melanogaster*, *Comp. Biochem. Physiol.* **90C**:439-444.

Van Delden, W., Boerema, A. C., and Kamping, A., 1978, The alcohol dehydrogenase polymorphism in populations of *Drosophila melanogaster*. I.
Selection in different environments, *Genetics* **90**:161-191.

Winberg, J-O., Thatcher, D. R., and McKinley-McKee, J. S., 1982, Alcohol
dehydrogenase from the fruit fly *Drosophila melanogaster*. Substrate
specificity of alleloenzymes ADH-S and ADH-UF, *Biochim. biophys. Acta*
704:7-16.

Winberg, J-O., Thatcher, D. R., and McKinley-McKee, J. S., 1985, The alcohol dehydrogenase alleloenzymes ADH-S and ADH-F from the fruit fly
Drosophila melanogaster: An enzymatic rate assay to determine the
active-site concentration, *Biochem. Genet.* **23**:205-216.

Wratten, C., and Cleland, W. W., 1963, Product inhibition studies on yeast
and horse liver alcohol dehydrogenase, *Biochemistry* **2**:935-941.

Zera, A. J., Koehn, R. K., and Hall, J. G., 1985, Allozymes and biochemical
adaptation, in: *Comprehensive insect physiology, biochemistry and
pharmacology*, Vol. 10 (G. A. Kerkut, and L. I. Gilbert, eds), Pergamon
Press, New York, pp. 633-674.

Ziolo, L. K., and Parsons, P. A., 1982, Ethanol tolerance, alcohol dehydrogenase activity and *Adh* allozymes in *Drosophila melanogaster*, *Genetica*
57:231-237.

CHAPTER 14

Environmental Modulation of Alpha-Glycerol-3-Phosphate Oxidase (GPO) Activity in Larvae of Drosophila melanogaster

STEPHEN W. MCKECHNIE, JENNIFER L. ROSS, AND KERRIN L. TURNEY

1. INTRODUCTION

Evolution has led to a variety of adaptive processes which enable organisms to cope with an environment that changes. Internal homeostasis needs to be maintained for metabolic processes to continue and for DNA to replicate. Such adaptations occur at all levels of organisation. At one extreme, mobility and behavioural adaptations allow the organism to experience only optimal patches of a heterogeneous environment. Other adaptations enable the organism to thrive over a broad range of environmental extremes by making physiological/biochemical adjustments, which preserve "constant the conditions of life in the internal environment" (Claude Bernard, quoted from Haldane, 1932).

Some of the best examples of this latter type of adaptation can be seen in the enzymes. If the environmental change is relatively slow with respect to the lifetime of the organism, gene substitution leading to a more efficient allozyme is likely to occur over several or many generations (Watt et al., 1983; DiMichelle and Powers, 1984; Hilbish and Koehn, 1985). Alternatively, if temporal environmental change is relatively fast it can be tolerated and even taken advantage of by adaptations that adjust enzyme activity levels. This can be a genetic induction effect involving changes in mRNA level (Watson et al., 1987; Geer et al., 1988) or more direct, post-translational effects of metabolites on enzyme activities. The latter, mostly allosteric, effects on enzymes lead to changes in flux (Hochachka and Somero, 1984) and may involve isozyme conversion (McKechnie and Geer, 1984). Allosteric effects are implemented very quickly, perhaps in response to a particular dietary intake.

STEPHEN W. MCKECHNIE, JENNIFER L. ROSS, AND KERRIN L. TURNEY * Department of Genetics and Developmental Biology, Monash University, Clayton, Vic. 3168, Australia.

Ecological and Evolutionary Genetics of Drosophila, Edited by
J.S.F. Barker _et al._, Plenum Press, New York, 1990

One *Drosophila* gene enzyme system which has recently been the subject of scrutiny at the genetic and molecular levels (MacIntyre and Davis, 1987; Davis and MacIntyre, 1988) may potentially help in an in-depth understanding of adaptive enzyme modulation induced by the environment. Flavin-linked alpha-glycerol-3-phosphate oxidase (E.C.1.1.99.5; GPO) is a mitochondrial enzyme and it appears to have several functions in the interfacing of the larval insect with its environment. GPO is located on the surface of the inner mitochondrial membrane (Klingenberg, 1970). It is ubiquitous throughout the animal kingdom and has been described in insects (Estabrook and Sacktor, 1958), in many mammals (Lee and Lardy, 1965; Cottingham and Ragan, 1980; Shaw *et al.*, 1982; Garrib and McMurray, 1986), in yeast (von Jagow and Klingenberg, 1960) and in bacteria (Kistler *et al.*, 1969).

In dipteran flight muscle GPO is important as part of the glycerol-3-phosphate shuttle (Fig. 1) where it works in concert with the cytosolic enzyme glycerol-3-phosphate dehydrogenase (E.C.1.1.1.8; GPDH). The shuttle quickly regenerates cytosolic NAD^+ facilitating speedy aerobic glycolysis and ATP production (Sacktor, 1970). In larvae, GPO increases substantially in activity in response to two dietary factors. Addition of either sucrose or ethanol to a defined low-fat diet causes an elevation in GPO activity (Geer *et al.*, 1983). The simultaneous increase of cytoplasmic glycerol-3-phosphate dehydrogenase (E.C.1.1.1.8; GPDH), in both total tissue NAD^+ concentration, and the $NADH/NAD^+$ ratio strongly implicates a GPO role in larvae in the glycerol-3-phosphate shuttle. This shuttle operates in mammalian liver (Dawson, 1979), and Poso (1977) has suggested that GPO may be a rate limiting enzyme in shuttle function when ethanol is oxidised in rat liver. In dipteran larvae high sucrose and ethanol diets may place similar demands on cytoplasmic availability of NAD^+ for proficient glycolysis and ethanol oxidation. Indeed there is evidence that GPO activity influences survival of *D. melanogaster* larvae in an ethanol stress environment (McKechnie and Geer, 1986).

What other physiological/biochemical functions might GPO perform in *Drosophila* larvae? A number of roles for GPO have been established in vertebrates and these may provide a clue to its functions in dipteran larvae. Studies both *in situ* and *in vitro* suggest the activity of GPO affects the availability of glycerol-3-phosphate for phospholipid and triglyceride synthesis (Ohkawa *et al.*, 1969; Bukowiecki and Lindberg, 1974; Houstek and Drabota, 1975). GPO activity levels are elevated in fasting rats when triglycerides are presumedly being catabolised (Molaparast-Shahidsaless *et al.*, 1979). Thus GPO activity may influence the balance between lipid synthesis and breakdown. GPO has been detected at high levels in fast twitch muscle fibres in humans (Schantz and Henriksson, 1987) suggesting a role in aerobic energy production as occurs in insect flight muscle. Finally, GPO appears to be involved in cold tolerance, especially in non-shivering thermogenesis where the enzyme levels are regulated in response to cold shock, largely in brown adipose tissue, and are elevated in cold acclimatised animals (Yamada, 1973; Videla *et al.*, 1975;

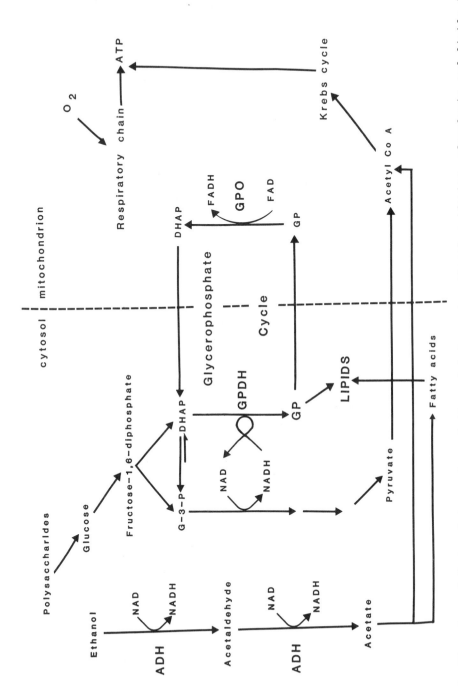

Figure 1. The glycerol-3-phosphate cycle and its relatedness to ethanol oxidation, glycolysis and lipids in *Drosophila* larvae.

Goglia *et al.*, 1983; Klein *et al.*, 1983). It is of interest
to determine whether or not GPO plays a role similar to any of
the above in *Drosophila* larvae.

A clue to environmental variable(s) that may influence
larval GPO expression comes from a recent field study. GPO
activity levels in larvae from strains of *D. melanogaster* isolated from a winery and its surrounds show a dependence on the
precise collection site. Strains originating from the winery
cellar had lower GPO levels than ones collected from a citrus
orchard approximately 100 meters outside (McKechnie and Geer,
1986). This observation has been repeated on independent samples taken 3 years later (Ross and McKechnie, unpubl.). Possible explanations for this microspatial differentiation in
enzyme activity are numerous. The possibility that natural
selection is involved makes this finding of interest and relevant to contempory issues in evolutionary genetics (see Hoffmann and O'Donnell, Chapter 10, this volume).

In the present study, we examine the effect of larval starvation, the addition of fat to the diet, and the effect of
cold exposure on GPO activity levels in an exploratory search
for new variables that influence larval GPO expression. We
also use a clone of the *Gpo* gene to ask whether or not *Gpo*
mRNA is involved in the mechanism by which dietary sugar and
ethanol elevate larval GPO activity; is this a genetic induction effect? Finally we ask if laboratory selection on a wine
"seepage" medium at cool temperatures, as occurs in winery
cellar habitats, produces a population with low levels of
larval GPO activity.

2. MATERIALS AND METHODS

2.1. Fly Maintenance

Basic medium contained 1000ml water, 19gm agar, 5.4gm (0.5%
w/v) sucrose, 32gm inactive Torula yeast and 5ml propionic
acid. The mixture was boiled for 3-5 mins, then cooled to
60°C before adding the propionic acid and dispensed into
bottles.

Three days prior to each experiment five to ten day old
flies were fed on sucrose medium (equal to basic medium with
sucrose increased to 5% w/v) and yeast then placed at 25°C
overnight in laying chambers with a thin disc of medium
(approximately 2 cm dia.) for egg collection. The larvae/eggs
were removed from the discs and transferred to the appropriate
experimental medium. The four media types used throughout
were basic, sucrose, ethanol (equal to basic medium supplemented with 3% v/v ethanol) and sucrose/ethanol (equal to
basic medium supplemented with both sucrose, at 5% w/v, and
ethanol, at 3% v/v).

2.2. GPO Enzyme Activity Determination

Batches of 20 larvae were ground to extract GPO enzyme as

described by McKechnie and Geer (1986). Enzyme activity was
measured in a Perkin-Elmer Lambda 3, UV/Vis spectrophotometer
with an Omniscribe chart recorder linked to a Haake FK circu-
lating water bath according to Geer *et al.* (1983). Protein was
determined by the method of Bradford (1976) using BSA as the
standard.   Specific activities (nmol/min/mg protein) were
analysed using SPSS^X Manova.

2.3. Starvation Experiment

Canton S flies were fed and eggs collected as described.
Eggs were transferred to sucrose medium in bottles at 20°C and
after four days batches of 40 larvae were collected and snap
frozen in liquid nitrogen. Duplicate samples were taken from
the three replicate bottles.

At the same time larvae from the same bottles were transferred into vials containing only 2% agar. Both the bottles
with remaining larvae and vials were left at 20°C for 24 hours
and larvae were again collected, 2 X 20 for each bottle and
vial. Following a further 24 hours at 20°C larvae were collected as above. Frozen larvae were stored at -70°C and
assayed for GPO activity.

2.4. Fatty Acid Feeding

Canton S flies were fed and eggs collected as described.
Eggs were transferred to bottles of modified Sang's medium C
(McKechnie and Geer, 1984) with 1% sucrose at 20°C for four
days when larvae were transferred using a small spatula to
vials with four different medium types. These were modified
Sang's medium C with: (i) 0.5% sucrose, (ii) 0.5%
sucrose/palmitate (1mg/ml), (iii) 5% sucrose, or (iv) 5%
sucrose/palmitate (1mg/ml). Larval density was kept approximately constant (approximately 50/vial) and the vials placed
at 20°C (5 vials per media type). After a further two days, 20
larvae were collected from each vial, snap frozen in liquid
nitrogen, stored at -70°C and assayed for GPO activity as
described.

2.5. Cold Exposure

Cantons S flies were fed and eggs collected as described.
Within 4 hours of hatching larvae were placed onto two types
of basic medium at (a) 25°C supplemented with 5% sucrose and
3% ethanol, and (b) supplemented with 0.5% sucrose. Four days
later, collections of 5 X 20 larvae from each of the two media
types were made and assayed for GPO activity, as a control.
Additional larvae were taken from the bottles and placed into
vials of the same media type and exposed to five different
temperature regimes (i) an additional control, 25°C for 24
hours; (ii) mild shock, 14°C for 24 hours; (iii) mild shock
14°C for 48 hours; (iv) harsh shock, 4°C for 5 hours, then
25°C for 24 hours, and (v) harsh shock 4°C for 10 hours, then
25°C for 24 hours. Five lots of 20 larvae from each of the

five conditions were collected, stored and assayed for GPO
activity as described.

2.6. Sucrose/Ethanol - mRNA Experiment

Eggs were collected from Canton S adults on discs of medium
as described. The following day each laying disc was cut into
four so each piece had approximately equal numbers of eggs.
One quarter was placed onto each of the four different media
in bottles (the contents of sucrose and ethanol were varied as
described above, the latter being added when the mixture
cooled to 60°C). This was repeated for four replicate sets of
bottles.

When larvae reached late third instar, batches of 20 and
200 for each bottle were harvested, washed twice with water
and snap frozen. Larvae were stored at -70°C until ground for
either GPO enzyme activity determination (20 larvae) or mRNA
extraction and dot blot analysis (200 larvae).

2.7. RNA Extractions

Larvae were homogenized at -70°C as they thawed in 7M urea;
135mM NaCl; 10mM Tris-HCl, pH 7; 2mM DTT; 7mM EDTA; 2% SDS
(w/v) (modified from Chandler et al., 1983). The homogenate
was extracted twice with phenol:chloroform and the nucleic
acid was precipitated with ethanol and ammonium acetate.
After being washed with 95% ethanol and 70% ethanol and redis-
solved in sterile water, the RNA was reprecipated with lithium
acetate before analysis.

PII65, a cDNA clone containing a 0.5 kb fragment of the GPO
gene (MacIntyre and Davis, 1987) inserted at the Hind III site
of pBR322 was used for a Gpo mRNA specific probe. As a standard, actin mRNA was probed with pHd-19, a derivative of
pBR322 with a 1.8 kb region inserted at the Hind III site that
includes the 5C actin gene (DmA2; Fyrberg et al., 1981). Both
probes were nick translated according to Maniatis et al.
(1982).

2.8. Dot Blots

Dot blot analysis of the mRNA was conducted with a Bio-Rad
vacuum manifold and Zetaprobe membrane according to Davis et
al. (1986). For detection of Gpo mRNA the filters were prehy-
bridized at 37°C for 4-6 hours in 50% formamide; 0.9M NaCl;
50mM sodium phosphate buffer, pH 7; 5mM EDTA; 0.5% blotto;
0.3% SDS; 500μg/ml sonicated herring sperm DNA; 5mg/ml yeast
mRNA and 20μg/ml polyA mRNA. Hybridization with pII65 was
performed at 37°C for 16-20 hours in the same mixture (fresh)
with addition of ^{32}P-labelled denatured probe. After
hybridization filters were washed with successive rinses of
2xSSC-0.1%SDS for 20 mins at 37°C, 2xSSC-0.1%SDS for 10 mins
at 37°C and 1xSSC-0.1%SDS for 10 mins at 65°C and autoradio-
graphed. GPO probe was then removed from the filters by wash-

ing according to the Amersham manual (1985) and hybridisation repeated with the actin probe pHd-19 as above.

Dot blots of 10µg and 6µg of total RNA were made from each of the 16 extracts on one filter which was duplicated. Autoradiographs were scanned with a Hoefer GS 300 Scanning Densitometer attached to the Apple II GS 340 Data System. Peak areas were calculated by integrating the area under the curve. Each filter was scanned twice and mean dot area was expressed as a percentage of the total area of the 16 dots at one RNA concentration. Analysis was on the mean of these four determinations of each extract. GPO standardised mRNA values were calculated by dividing *Gpo* mRNA values by the corresponding actin mRNA value.

2.9. Wine Selection

Three different laboratory populations, Canton S, and two derived from south-east Australian populations, the Tahbilk winery (collected outside of the cellar) and Silvan orchard were selected on wine medium. The latter two had been in the laboratory for eight months prior to the start of the experiment and were mass reared from large field collections (>100 females). To simulate larval growth conditions in the cellar habitat, the first 30 to 50 adults to eclose, were transferred each generation to bottles containing 20ml of simulated seepage on two Kleenex tissues and placed at 20°C for seven days to feed and lay. Three replicate selection lines were used for each population.

After removal of the adults the bottles were placed at 16°C until sufficient adults eclosed to establish the next generation. Simulated wine seepage was Tahbilk red (Shiraz) held at 37°C for 5 hours in a shallow tray to allow the alcohol to dissipate. This was an attempt to better simulate the Tahbilk larval habitat seepage which is low in ethanol (McKechnie and Morgan, 1982). For the control (three selection lines per population), more than 100 adults were placed on sucrose medium at 20°C (seeded with live yeast) and removed after 2 days. After 3 weeks at 20°C the next generation of control lines was similarly established. Specific activities of GPO were determined for larval samples of all 18 lines at the same time. After four generations on wine "seepage", larvae were reared from eggs to late third instar wandering larvae on instant medium (Carolina Biological Supply Co.) supplemented with raw sugar (5% w/v), before being washed and mitochondrial enzyme extracts prepared on duplicate groups of 20 larvae.

3. RESULTS

3.1. Induction as a Clue to the Role of GPO in Larvae

To explore the possibility of other functional roles of GPO in larvae, we have assayed levels of GPO activity in larvae subject to three different environmental conditions. This has been done in separate experiments each with appropriate con-

trols and each using Canton S larvae. The three environmental
changes, the absence of nutritional medium (starvation), the
presence of a fatty acid (at relatively high levels), and
exposure to cold stress are all known to modulate GPO activity
levels in mammals. Here we test for changes in GPO activity
levels in *Drosophila* larvae subjected to similar environmental
perturbations.

3.1.1. Starvation and Larval GPO

Mitochondrial enzyme extract was made from six replicate
batches (one per bottle) of 20 larvae for each of four conditions: (1) larvae removed from sucrose medium in bottles after
five days from egg laying at 20°C, (2) larvae removed at four
days and held 24 hours in vials containing only 2% agar before
grinding, (3) larvae removed from bottles after six days, and
(4) larvae removed at four days and held 48 hours on 2% agar
before grinding. To obtain larvae that had been starved at
20°C for 48 hours (without pupation occuring in the interim),
it was necessary to start the experiment with young larvae.
Extracts from 40 such zero-time larvae (four days from egg
laying; condition (2) above but without 24 hour on agar) prepared in the standard grind volume (used normally for 20
larvae) gave mean total protein estimates (Fig. 2) significantly less than from 20 larvae taken from sucrose medium 24
hours later (t=2.28, p<0.5) and 48 hours later (t=4.59,
p<0.001). Thus the 48 hours of this starvation experiment was
a period of significant larval growth if held on sucrose
medium. Specific activity data of GPO over this growth period
on sucrose medium suggested a decline in activity as larvae
approached pupation (Fig. 2), a result consistent with previous work on the developmental profile of the enzyme (O'Brien
and MacIntyre, 1972). The significantly lower protein content
in extracts from starved larvae after 48 hours, compared to 48
hours on sucrose medium (t=3.85, p<0.01) evidenced the negative effect of the fasting conditions on growth. Clearly significant was an increase in specific activity of GPO in
extracts from starved larvae (Fig. 2; Table I).

3.1.2. Dietary Palmitate and Larval GPO

To provide a diet high in fatty acid, palmitate was fed to
larvae on both a high sucrose (5% w/v) and low sucrose (0.5%
w/v) basic media, the appropriate media unsupplemented with
palmitate being included as controls. Late third instar larvae
were collected and GPO specific activity estimates made for
these four media. Protein contents of the larval extracts
indicated that on low sucrose, growth may have been inhibited
by palmitate since lower protein levels (t=3.11, p<0.01)
occurred in these extracts compared to larvae raised on low
sucrose in the absence of palmitate (Fig. 2). As expected,
higher dietary sucrose levels elevated GPO activity. However,
palmitate supplement had no detectable effect on enzyme specific activity, regardless of the sucrose level (Fig. 2; Table
I).

CHAPTER 13

Alcohol Dehydrogenase and Alcohol Tolerance in Drosophila melanogaster

BILLY W. GEER, PIETER W.H. HEINSTRA, ANN M. KAPOUN, AND ALEID VAN DER ZEL

1. THE PHYSIOLOGICAL EFFECTS OF ALCOHOLS

1.1. The Toxic Effects of Alcohols at the Cellular Level

Drosophila melanogaster has often been used as a model system for investigations of the genetic factors that underlie ethanol tolerance. The species not only is very tolerant to environmental ethanol (McKenzie and Parsons, 1972; David *et al.*, 1976), but it is able to use ethanol as a food source (McKechnie and Geer, 1984; Geer *et al.*, 1985). Judgement of the degree of ethanol tolerance in *D. melanogaster* is complicated by the complex nature of the trait, and by the diversity of the tests that have been used to measure tolerance. Ethanol tolerance is a composite of the abilities to grow and to survive in the presence of ethanol, and, as shown in this study, to utilize dietary ethanol as a foodstuff. The conditions for the tolerance test and the diagnostic traits in previous studies have varied. Many tolerance tests have been performed by adding ethanol to the medium (Gibson, 1970; McDonald *et al.*, 1977; Cavener and Clegg, 1978), but different life stages have been tested. Ethanol tolerance in adult *D. melanogaster* has been equated to the concentration of alcohol in sealed tubes that kills 50% of the individuals in a treatment of fixed duration (David *et al.*, 1978, 1984). Moreover, the knockdown time of adults by ethanol fumes has also been suggested to be an accurate indicator of ethanol tolerance (Cohan and Graf, 1985; Cohan and Hoffmann, 1986). Nonetheless, the sensitivities of *D. melanogaster* to ethanol may vary at different life stages,

BILLY W. GEER, ANN M. KAPOUN, AND ALEID VAN DER ZEL * Department of Biology, Knox College, Galesburg, IL 61401, USA; _PIETER W.H. HEINSTRA_ * Department of Population and Evolutionary Biology, University of Utrecht, The Netherlands.

Ecological and Evolutionary Genetics of Drosophila, Edited by
J.S.F. Barker *et al.*, Plenum Press, New York, 1990

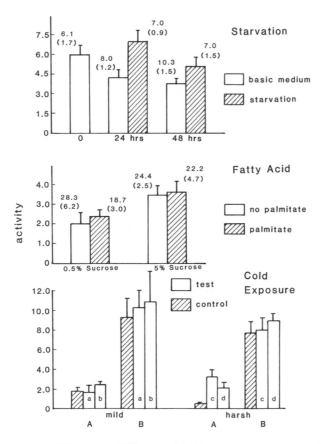

Figure 2. Larval GPO specific activity responses under various environmental changes: starvation, dietary fatty acid supplement, mild-cold and harsh-cold exposure. Activity is given in nmol/min/mg protein (+se). Numbers above the histogram bars give mean protein contents in mg/μl (se) of the respective larval mitochondrial enzyme extracts. Cold exposure at 14°C (mild) was given for either 24 hours (a) or 48 hours (b), and at 4°C (harsh) for either 5 hours (c) or 10 hours (d). Larvae were raised on either basic medium (A) or sucrose/ethanol medium (B).

3.1.3. Cold Exposure and Larval GPO

Our aim, to test for changes in specific activity levels of GPO in response to cold, required a number of preliminary cold exposure experiments to determine appropriate temperatures and times (as determined by mortality levels). Two conditions were finally employed, a mild and a harsh exposure, each with its own control. Both exposure conditions were carried out with larvae raised on basic medium and on an inducing medium (sucrose/ethanol). At four days from egg laying at 25°C, batches of larvae were transferred from bottles to vials of the same medium and placed at 14°C for either 24 hours or 48 hours at which time larval extracts were made on groups of 20

Table I

Factorial analyses of larval GPO enzyme activity under conditions of
starvation, high dietary fatty acid and cold exposure

Starvation				Mild cold exposure			
	DF	MS	F		DF	MS	F
Within	20	3.44		Within	18	12.17	
Starve	1	24.87	7.23**	Cold	1	2.34	0.19
Time[a]	1	8.99	2.61	Medium	2	395.30	32.50***
Starve X time	1	3.66	1.06	Cold X medium	2	0.75	0.06

Dietary palmitate				Harsh cold exposure			
	DF	MS	F		DF	MS	F
Within	16	1.12		Within	15	4.45	
Palmitate	1	9.11	0.26	Cold	1	4.40	0.99
Sucrose	1	0.31	7.53**	Medium	2	188.70	42.40***
Palmitate X sucrose	1	0.12	0.10	Cold X medium	2	2.99	0.67

[a]Zero time larvae not included.
**p = 0.01; ***p<0.001.

larvae. Samples at four days, directly from the bottles, provided controls for this mild-exposure experiment. Development
at 14°C in *D. melanogaster* is slow (McKenzie, 1978) and little
developmental change in GPO activity would be expected in the
48 hours of this experiment. For the harsh-exposure experiment, batches of larvae were transferred in vials to 4°C for
either 5 hours or 10 hours. These larvae were removed to 25°C
for 24 hours before enzyme preparations were made. Control
larvae were left the equivalent 24 hour period at 25°C without
cold exposure.

The basic and sucrose/ethanol medium resulted in very different levels of larval GPO activity as expected (Geer *et al.*,
1983), for both cold exposure tests (Fig. 2 and Table I).
However, no significant effect of cold on larval GPO levels
was observed under any of the test conditions.

3.2. Dietary Sucrose/Ethanol and *Gpo* mRNA Levels

To determine whether the levels of enzyme inducibility seen
for larvae are a result of regulation of mRNA levels, both GPO
enzyme activity and *Gpo* mRNA levels were determined and compared among diets. A clear elevation of enzyme activity level
occurred when the basic medium was supplemented with both
sucrose and ethanol together. Levels of enzyme activity on
the sucrose/ethanol medium were also higher than in larvae
raised on medium supplemented with ethanol only (Table II).
In the case of GPO mRNA the sucrose/ethanol diet produced
higher levels than basic, ethanol and sucrose media (Fig. 3

Table II

Mean larval GPO enzyme activities (±se) and mRNA dot blot intensities (±se)
on sucrose and ethanol diets

Diet[a]	n	Enzyme activity (nanomol/min/ mg protein)[b]	*Gpo* mRNA quantity[b]	Actin mRNA quantity[b]	Standardised *Gpo* mRNA quantity[c]
Basic	4	2.99±0.78*	3.42±0.73*	4.97±0.71*	0.72±0.10*
Sucrose	4	13.35±1.49	5.03±0.10*	5.97±0.37	0.87±0.01*
Ethanol	4	8.94±2.32*	6.95±0.16*	6.10±0.05	1.12±0.15[i]
Sucrose/ ethanol	3	24.06±6.27	11.95±1.04	7.56±0.56	1.60±0.06

[a]Diets are defined in Materials and Methods.
[b]Tukey's multiple mean comparison. Only sucrose/ethanol means showed
 significant (p<0.05) differences (to means indicated by an asterisk).
[c]Mann-Whitney U tests among ratio distributions. Asterisk indicates
 significantly different (p<0.05) to sucrose/ethanol distribution.
[i]Indicates different to basic medium distribution (p<0.05).

and Table II). Clearly maximal induction here is on the
medium containing both hydrocarbon supplements.

An increase in level of the actin mRNA was indicated in
response to the dietary supplements. However, these differences were not as marked as for the *Gpo* mRNA blots. The relative magnitude of this actin increase compared to *Gpo* mRNA on
supplemented media (Table II and Fig. 3) suggested that the
actin change may have been a general response to the increased
nutrition. Accordingly the *Gpo* mRNA quantities were standardised by dividing them by the actin mRNA levels determined for
the same blot. These standardised *Gpo* mRNA levels were
increased by ethanol and sucrose/ethanol supplement (Table II)
suggesting a specific response of *Gpo* mRNA levels over and
above any general nutritional effect. A scattergram indicated
an overall association between enzyme activities and *Gpo* mRNA
standardised levels (Fig. 4), supporting the idea that
ethanol-elevated larval GPO enzyme activity depended on elevated message and increased levels of translation.

3.3. Wine Selection

To test if the lower larval GPO activity level in cellar
derived strains is a consequence of natural selection operating on larvae during development in the cellar habitat a wine
selection experiment was initiated. Three discrete bottle
cultures, three lines, of each of three different *Drosophila*
strains, Canton S, Tahbilk and Silvan, were reared in the laboratory on simulated wine seepage. This was done at 16°C
since cellar temperatures are in general cooler than outside.
After four generations eggs were collected from adults, and
larvae raised on sucrose supplemented instant medium at 20°C
before estimation of larval GPO specific activity levels.
These levels were compared to those of larvae from control

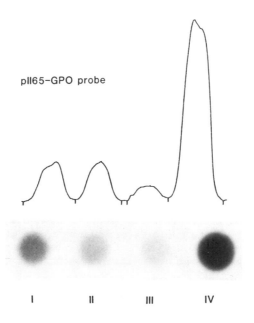

pII65-GPO probe

I   II   III   IV

pHd-19 Actin probe

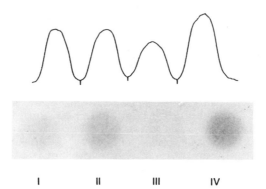

I   II   III   IV

Figure 3. Dot blots of RNA extracted from late third instar larvae raised on 4 different media (I - sucrose, II - ethanol, III - basic, IV - sucrose/ethanol), probed with a Gpo cDNA plasmid, with an actin plasmid and scanned.

lines not subject to the cooled wine seepage rearing (Fig. 5). A discordant response to the wine selection GPO activity levels among the populations was apparent, although an anova indicated this was of borderline significance (Table III). Wine selected Silvan populations appeared to have increased levels of larval GPO activity while the mean activity in Tahbilk and Canton S origin lines was less after wine rearing. The direction of the change in mean for the Tahbilk population is consistent with the field data.

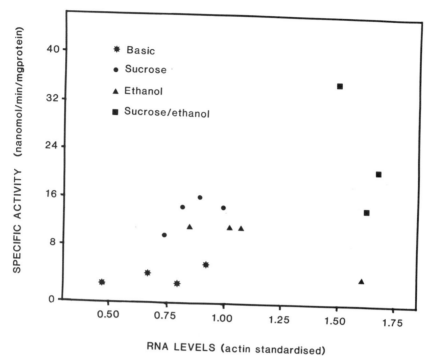

Figure 4. Scattergram of GPO specific activity versus *Gpo* RNA levels in extracts from larvae raised on 4 media types.

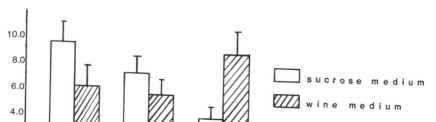

Figure 5. Mean larval GPO specific activity (+se) in wine selected and non-selected lines (on sucrose medium) from three populations.

Table III

Analysis (factorial-nested anova) of the effects of
wine-selection on GPO enzyme activity in 3 populations

Source	Df	MS	F
Within	18	11.44	
Population	2	8.39	0.75
Wine	1	0.03	0.00
Line (w Pop.)	6	11.24	0.98
Wine X Pop.	2	61.85	4.95*
Wine X line (w Pop.)	6	12.51	1.09

*$p = 0.05$.

4. DISCUSSION

A major interest in this study has been to identify new physiological/biochemical processes which depend on the activity of GPO in larvae. Our strategy has been to subject the larvae to specific environmental changes and to monitor for GPO elevation or depression. Such modulations of activity of specific enzymes by environmental shocks are usually adaptive responses indicating a functional role of that enzyme in a specific biochemical pathway, restoring the internal "status quo" of the organism (Blackstock, 1984). Any one enzyme in an organism is likely to be involved in a variety of different functions. Only by identification of all the different functions at different life stages and in different tissues can we hope to fully understand the nature of selective forces that impinge upon enzyme activity levels. Thus our approach here to elucidate GPO function, though an indirect one, is a useful starting point for a larval enzyme for which the only established insect function is to expedite glycolysis in adult flight muscle.

Under starvation conditions an increase in GPO specific activity of about 30% was observed in mitochondrial extracts, although no hint of enzyme modulation occurred with fatty acid feeding. We should interpret this cautiously. Since total protein content in the extracts declined with starvation, it will be important to assess specific activity changes in other mitochondrial and cytoplasmic enzymes under the same conditions. If little or no change occurs in levels of such control enzymes it would suggest a specific role for GPO, perhaps in starvation-promoted catabolism of lipids. Hence, in addition to a larval GPO function in the shuttle, a regulatory role in the balance between fatty acid oxidation and lipid synthesis is possible (Houstek et al., 1975; Ohkawa et al., 1969). Under starvation conditions a high GPO activity may result in low levels of glycerol-3-phosphate, levels insufficient for lipid synthesis. In larval tissue such as fat body, active synthesis of phospholipids and triglycerides draws on cytoplasmic glycerol-3-phosphate. Since this metabolite is the likely substrate for GPO, in what is reported to be an irreversible reaction (Dawson, 1979) cytoplasmic concentrations of glycerol-3-phosphate may be directly affected by GPO activity. High levels of enzyme might promote lipid catabolism and thus be adaptive during starvation (Molaparast-Shahidsaless *et al.*, 1979). Thus the possibility of a GPO role in *Drosophila* larval lipid metabolism should not be discounted. Interactions between glycerol-3-phosphate, fatty acids and lipids are likely to be complex.

In spite of the exothermic nature of the GPO reaction (Dawson, 1979) and the established functional role of GPO in vertebrate body heat regulation (Yamada, 1973; Klein *et al.*, 1984), no hint of induction occured in larvae exposed to a variety of cold conditions. If GPO plays a role in cold protection in *Drosophila* a different experimental approach would seem necessary.

Changes in GPO activity in larval mitochondrial extracts demonstrate the effectiveness of dietary sucrose and ethanol together to "induce" this enzyme. That this is a relatively specific effect, and not just the result of a general nutritional increase, such as enlargement of mitochondria, has been demonstrated previously (Geer *et al.*, 1983). Geer *et al.* (1983), also using Canton S, found that when larvae were fed on a sucrose/ethanol diet there was no increase in GPO enzyme activity over larvae fed on the sucrose diet or ethanol diet. However, our data suggest that these nutrients may be additive in their ability to induce GPO enzyme activity. Perhaps the difference between basic diets used in these two studies (defined low-fat versus dead yeast) affects the enzyme response to supplement. Despite this inconsistency, it is likely that the GPO increase allows an increase in glycerol-3-phosphate shuttle activity, thus regenerating cytoplasmic NAD^+ needed for elevated dehydrogenase activity on the sugar/ethanol supplement.

Quantification of *Gpo* mRNA levels in larvae raised on ethanol and/or sucrose, after standardisation for more general nutritional effects with actin mRNA probe, indicated that the enzyme activity response to sucrose/ethanol is mediated by increased levels of enzyme message and, hence by increased translation. Ethanol supplement alone appears to elevate *Gpo* message level but its effect on elevating enzyme activity levels was not convincing.

It is of interest to consider reasons for the lower levels of larval GPO activity in the cellar derived strains. Numerous explanations are possible including the association of selectively neutral GPO-modifier genes with genes which are better able to exploit or colonise the cellar habitat. The persistence of collection-site strain differences in larval GPO activity levels over 3 years in laboratory culture (McKechnie and Geer, 1986; Ross and McKechnie, unpubl.) argues for a genetic basis for this variation; diallel crosses are currently in progress to test this. If wine selection on larvae is at work, the effect is not general since the different populations differed in response. Our wine selection data suggests that the Tahbilk population may respond to continued culture on wine medium with a decrease in larval GPO activity levels. The laboratory selection is being continued to see if the trend continues. If so, it would need to be repeated with selection being on a defined medium, perhaps with low sucrose, no fat and with low ethanol levels (contenders for the crucial variables of a wine seepage habitat). If GPO activity limits the availability of glycerol-3-phosphate for larval lipid synthesis, low activity variants may be at an advantage on wine. In larvae with low GPO activity, cytoplasmic glycerol-3-phosphate levels may stabilise at high cell concentrations making it more readily available for lipid deposition. When the need for a shuttle is minimal, such as in the cellar seepage habitat with low free sugar and low ethanol levels, this GPO function may be the most important. We know that considerable lipid storage occurs in developing larvae and that much of the ethanol in a larval diet is laid down in body lipid (Geer, *et al.*, 1985).

Existing data support the involvement of larval GPO in a glycerol-3-phosphate shuttle. In this role high GPO activities would promote proficient oxidative degradation of substrates such as glucose and ethanol and lead to "fit" individuals. More speculatively, GPO in larvae may also influence cytoplasmic concentrations of its substrate, glycerol-3-phosphate, required for lipid synthesis. During development when rates of larval lipid deposition are high, an active GPO has the potential to limit availability of this substrate and perhaps slow development. In any given food resource a particular intermediate GPO level may be optimal. Such metabolic "tradeoffs" are likely to be important for our future understanding of the action of natural selection on systems of metabolic regulation. In any event the GPO enzyme is clearly an important component of metabolic plasticity in *Drosophila* larvae enabling them to adapt to short term variation in food resource.

5. SUMMARY

The dietary induction of glycerol-3-phosphate oxidase (GPO) in mitochondrial extracts from *D. melanogaster* larvae raised on 5% sucrose and 3% ethanol has been confirmed. Activity levels were raised at least 4-fold by the presence of both sucrose and ethanol in the larval diet. Three additional environmental variables were tested as signals for larval GPO modulation: (i) starvation conditions, (ii) high dietary fatty acid, and (iii) cold exposure. Only starvation lead to a change in GPO activity; it was elevated to approximately 1.3 times that in non-starved larvae.

Quantitative dot blot analysis of total RNA extracted from larvae raised on diets supplemented with ethanol and sucrose, after probing with a clone of the *Gpo* gene, demonstrated the involvement of elevated Gpo mRNA levels in the ethanol induction response.

Finally, three separate laboratory populations were selected for four generations on cooled simulated wine seepage, as may occur in the Tahbilk winery cellar, and larval GPO activity levels were compared to those from similar populations held on regular sucrose medium. Although a different response was indicated in different populations the GPO activity in larvae from the wine-selected Tahbilk population was less than in the control, a result consistent with previous estimates of low levels of this enzyme in Tahbilk cellar-captured strains.

ACKNOWLEDGEMENTS. Thanks to Ross MacIntyre and Mary Beth Davis for advice and providing the clone of the *Gpo* gene, and to Bill Geer and Ary Hoffmann for valuable discussion. We thank Sally Tobin for the actin gene clone. Financial support from the Australian Research Grants Scheme is gratefully acknowleged.

References

Amersham, 1985, Membrane Transfer and Detection Methods, Amersham International plc.

Blackstock, J., 1984, Biochemical metabolic regulation responses of marine invertebrates to natural environmental change and marine pollution, *Oceanogr. & Mar. Biol. Ann. Rev.* 22:263-313.

Bradford, M. M., 1976, A rapid and sensitive method for quantitation of microgram quantities of protein utilizing the principle of protein-dye binding, *Analyt. Biochem.* 72:248-254.

Bukowiecki, L. J., and Lindberg, O., 1974, Control of sn-glycerol-3-phosphate oxidation in brown adipose tissue mitochondria by calcium and acyl-CoA, *Biochim. Biophys. Acta* 348:115-125.

Chandler, P. M., Higgens, T. J. V., Randall P. J., and Spencer, D., 1983, Regulation of legumin levels in developing pea seeds under conditions of sulfur deficiency. Rates of legumin synthesis and levels of legumin mRNA, *Plant Physiol.* 71:46-54.

Cottingham, I. R., and Ragan, C. I., 1980, Purification and properties of L-3-glycerophosphate dehydrogenase from pig brain mitochondria, *J. Biochem.* 192:9-18.

Davis, L. G., Dibner, M. D., and Battey, J. F., 1986, *Molecular Cloning, A Laboratory Manual*, Cold Spring Harbor Laboratory, Cold Spring Harbor, New York.

Davis, M. B., and MacIntyre, R. J., 1988, A genetic analysis of the alpha-glycerol-3-phosphate oxidase in *Drosophila melanogaster*, *Genetics* 120:755-766.

Dawson, A. G., 1979, Oxidation of cytosolic NADH formed during aerobic metabolism in mammalian cells, *Trends Biochem. Sci.* 4:171-176.

DiMichelle, L., and Powers, D. A., 1984, Developmental and oxygen comsumption rate differences between lactate dehydrogenase-B genotypes of *Fundulus heteroclitus* and their effect on hatching time, *Physiol. Zool.* 57:52-56.

Estabrook, R. W., and Sacktor, B., 1958, Alpha-glycerophosphate oxidase in flight muscle mitochondria, *J. biol. Chem.* 233:1014-1019.

Fyrberg, E. A., Bond, B. J., Hershey, D., and Mixter, K. S., 1981, The actin genes of *Drosophila*: Protein coding regions are highly conserved but intron positions are not, *Cell* 24:107-116.

Garrib, A., and McMurray, W. C., 1986, Purification and characterization of glycerol-3-phosphate dehydrogenase (flavin-linked) from rat liver mitochondria, *J. biol. Chem.* 261:8042-8048.

Geer, B. W., Langevin, M. L., and McKechnie, S. W., 1985, Dietary ethanol and lipid synthesis in *Drosophila melanogaster*, *Biochem. Genet.* **23**:607-622.

Geer, B. W., McKechnie, S. W., Bentley, M. M., Oakeshott, J. G., Quinn, E. M., and Langevin, M. L., 1988, Induction of ADH by ethanol in *Drosophila melanogaster*, *J. Nutr.* **118**:398-407.

Geer, B. W., McKechnie, S. W., and Langevin, M. L., 1983, Regulation of sn-glycerol-3-phosphate dehydrogenase in *Drosophila melanogaster* larvae by dietary ethanol and sucrose, *J. Nutr.* **113**:1632-1642.

Golgia, F., Liverini, G., De Leo, T., and Barletta, A., 1983, Thyroid state and mitochondria population during cold exposure, *Pflugers Arch.* **396**:49-53.

Haldane, J. B. S., 1932, *The Causes of Evolution*, Harper, New York and London.

Hilbish, T. J., and Koehn, R. K., 1985, The physiological basis of natural selection at the Lap locus, *Evolution* **39**:1302-1317.

Hochachka, P. W., and Somero, G. N., 1984, *Biochemical Adaptation*, Princeton University Press.

Houstek, J. and Drahota, Z., 1975, The regulation of glycerol-3-phosphate oxidase of rat brown adipose tissue mitochondria by long-chain free fatty acids, *Mol. Cell. Biochem.* **7**:45-50.

Houstek, J., Cannon, B., and Lindberg, O., 1975, Glycerol-3-phosphate shuttle and its function in intermediary metabolism of hamster brown-adipose tissue, *Eur. J. Biochem.* **54**:11-18.

Kistler, N. S., Hirsch, C. A., Cozzarelli, N. R., and Lin, E. C., 1969, Second pyridine nucleotide-independent L-alpha-glycerophosphate dehydrogenase in *Escherichia coli* K-12, *J. Bact.* **100**:1133-1135.

Klein, A. H., Reviczky, A., Chou, P., Padbury J., and Fischer, D. A., 1983, Development of brown adipose tissue thermogenesis in the ovine fetus and newborn, *Endocrinology* **112**:1662-1666.

Klein, A. H., Reviczky, A., and Padbury, J. F., 1984, Thyroid hormones augment catecholamine-stimulated brown adipose tissue thermogenesis in the ovine fetus, *Endocrinology* **114**:1065-1069.

Klingenberg, M., 1970, Localization of GPDH in the outer phase of the mitochondrial inner membrane, *Eur. J. Biochem.* **13**:247-252.

Lee, Y. P., and Lardy, H. A., 1965, Influence of thyroid hormones on L-alpha-glycerophosphate dehydrogenase and other dehydrogenases in various organs of the rat, *J. biol. Chem.* **240**:1427-1436.

Maniatis, T., Fritsch, E. F., and Sambrook, J., 1982, *Molecular Cloning: A Laboratory Manual*, Cold Spring Harbor Laboratory, Cold Spring Harbor, New York.

MacIntyre, R. J., and Davis, M. B., 1987, A genetic and molecular analysis of the alpha-glycerophosphate cycle in *Drosophila melanogaster*, *Isozymes. Current Topics in Biological & Medical Research* **14**:195-227.

McKechnie, S. W., and Geer, B. W., 1984, Regulation of alcohol dehydrogenase in *Drosophila melanogaster* by dietary alcohol and carbohydrate, *Insect Biochem.* **14**:231-242.

McKechnie, S. W., and Geer, B. W., 1986, sn-Glycerol-3-phosphate oxidase and alcohol tolerance in *Drosophila melanogaster* larvae, *Biochem. Genet.* **24**:859-872.

McKechnie, S. W., and Morgan, P., 1982, Alcohol dehydrogenase polymorphism of *Drosophila melanogaster*: Aspects of alcohol and temperature variation in the larval environment, *Aust. J. biol. Sci.* **35**:85-93.

McKenzie, J. A., 1978, The effect of developmental temperature on population flexibility in *Drosophila melanogaster* and *D. simulans*, *Aust. J. Zool.* **26**:105-112.

Molaparast-Shahidsaless, F., Shrago, E., and Elson, C. E., 1979, Alpha-glycerophosphate and dihydroxyacetone phosphate metabolism in rats fed high-fat or high-sucrose diets, *J. Nutr.* **109**:1560-1569.

O'Brien, S. J., and MacIntyre, R. J., 1972, The alpha-glycerophosphate cycle in *Drosophila melanogaster*. I. Biochemical and developmental aspects, *Biochem. Genet.* **7**:141-161.

Ohkawa, K., Vogt, M. T., and Farber, E., 1969, Unusually high mitochondrial alpha-glycerophosphate dehydrogenase activity in rat brown adipose tissue, *J. Cell Biol.* **41**:441-449.

Poso, A. R., 1977, Influence of mitochondrial alpha-glycerophosphate oxidase on the alpha-glycerophosphate shuttle during ethanol oxidation, *FEBS Lett.* **83**:285-287.

Sacktor, B., 1970, Regulation of intermediary metabolism with special reference to control mechanisms in insect flight muscle, *Adv. Insect Physiol.* **7**:267-347.

Schantz, P. G., and Henriksson, J., 1987, Enzyme levels of the NADH shuttle systems: measurements in isolated muscle fibres from humans of differing physical activity, *Acta physiol. scand.* **129**:505-515.

Shaw, M. A., Edwards, Y. H., and Hopkinson, D. A., 1982, Human mitochondrial glycerol phosphate dehydrogenase (GPDm) isozymes, *Ann. Hum. Genet.* **46**:11-23.

Videla, L., Flattery, K. V., Sellers, E. A., and Israel, Y., 1975, Ethanol metabolism and liver oxidative capacity in cold acclimation, *J. Pharmac. exp. Ther.* **192**:575-582.

von Jagow, G., and Klingenberg, M., 1960, Pathways of hydrogen in mitochondria of *Saccharomyces carlsbergensis*, *Eur. J. Biochem.* **12**:583-592.

Watson, J. D., Hopkins, N. H., Roberts, J. W., Steitz, J. A., and Weiner, A.M., 1987, *Molecular Biology of the Gene*, Vol. 1, 4th ed., Benjamin/Cummings Publishing Company, Inc., Menlo Park, Calif.

Watt, W. B., Cassin, R. C., and Swan, M. S., 1983, Adaptation at specific loci. III. Field behavior and survivorship differences among *Colias* PGI genotypes are predictable from *in vitro* biochemistry, *Genetics* **103**:725-739.

Yamada, M., 1973, Effect of environmental temperature on activity of glycerophosphate dehydrogenases, *Am. J. Physiol.* **224**:1420-1424.

CHAPTER 15

Physiology, Biochemistry and Molecular Biology of the Est-6 Locus in Drosophila melanogaster

ROLLIN C. RICHMOND, KAREN M. NIELSEN, JAMES P. BRADY, AND ELIZABETH M. SNELLA

1. INTRODUCTION

An understanding of the effect of natural selection on a particular locus often depends upon a detailed knowledge of the function of the locus. Indeed, a mechanistic understanding of the biochemical and physiological role(s) of a gene product can reveal complexities which are vital to a full comprehension of the myriad effects of selection. This view is well expressed in the following quotation from Watt (1985, p. 124) which has its origins in a research strategy advocated by Clarke (1975) and Koehn (1978).

> "Variation at single loci sometimes has major effects on fitness components, but at other times does not, and the distinctions among those situations grow out of their mechanistic nature, rather than from a priori formal conceptions. The finding of such complexity is no end result, but rather a prod toward more comprehensive study."

In many senses, evolutionary biologists attempt to infer general principles from an assemblage of special cases. Such special cases are often a motivation for the development of new theoretical notions which are then extended to other applicable situations. We propose that the esterases of Drosophila and, in particular, the Est-6 locus (Est-6 = locus, EST 6 = protein) of Drosophila melanogaster provide a model system for investigating the role of various selective and non-selective processes in the evolution of the structure and regulation of protein coding loci. Here we review the background for our studies of Est-6 in D. melanogaster and our findings on the biochemical properties and physiological role of the EST 6 enzyme.

ROLLIN C. RICHMOND, KAREN M. NIELSEN, JAMES P. BRADY, AND ELIZABETH M. SNELLA * Department of Biology and Institute for Molecular and Cellular Biology, Indiana University, Bloomington, IN 47405, USA.

Ecological and Evolutionary Genetics of Drosophila, Edited by
J.S.F. Barker et al., Plenum Press, New York, 1990

2. EST 6 AS A MODEL SYSTEM FOR STUDYING MOLECULAR EVOLUTION

The *Est-6* locus, first described by Wright (1963), was the first of many biochemical polymorphisms which have been characterized in this species. Wright demonstrated that this locus was biochemically tractable, and MacIntyre and Wright (1966) provided the first indication that this locus would be highly polymorphic when they identified thermostability in addition to electrophoretic variation at the locus. We now know that the *Est-6* locus is one of the most polymorphic enzyme systems described in any organism (Oakeshott *et al.*, Chapter 19, this volume), and is a representative of the most variable class of allozyme loci as described by Lewontin (1985). The majority of variants at such loci are presumed to be little affected by natural selection and such loci represent an opportunity to evaluate the relative importance of stochastic processes and natural selection on the frequencies of alleles at a single locus. Such loci have not been widely studied in *Drosophila* with the exception of the *Xdh* locus of *D. pseudoobscura* (Riley, 1989).

When we began our studies of the *Est-6* locus about 15 years ago, there was little reason to believe that polymorphic variants at the *Est-6* locus would be subject to the action of natural selection. A null allele of the locus had been described (Johnson *et al.*, 1966) and flies homozygous for this allele were perfectly viable and fertile. Moreover, MacIntyre and Wright (1966) had completed several population cage studies following the frequencies of the two major allozymic forms of *Est-6* and were unable to demonstrate that selection was acting on these allelic variants. Yamazaki (1971) studied the *Est-5* locus of *D. pseudoobscura,* which we now know is homologous to the *Est-6* locus (Brady *et al.*, 1990), and concluded that allelic variation was neutral under population cage conditions. However, Kojima and Huang (1972) and later Birley and Beardmore (1977) presented evidence that the *Est-6* locus was subject to selection under some conditions, in particular high density. Kojima and Huang suggested that frequency-dependent selection for larval viability at high densities explained their observations. Birley and Beardmore reported that the production of isoamyl alcohol and p-hydroxybenzoic acid by homozygous fast and slow *Est-6* genotypes respectively could provide a molecular mechanism for the effects of *Est-6* genotypes on larval viability. As far as we know, this intriguing hypothesis has never been verified. Unlike the *Adh* locus which has been extensively studied in *Drosophila* and for which there is ample reason to suppose that selection operates on the allozyme variants (Simmons *et al.*, 1989), the *Est-6* locus is not a vital enzyme and is unlikely to be involved in any central metabolic pathway. Thus the finding of selective differences among variants at this locus provides particularly significant evidence that natural selection affects the frequencies of allozyme alleles, even at loci which are highly polymorphic.

There are several other aspects of the biology of EST 6 and its relatives which make this system a particularly good paradigm for evolutionary studies. Among the eight species of the *melanogaster* subgroup, there is a wide range of interspecific variation in the quaternary structure, tissue, sex, and temporal patterns of *Est-6* expression. These findings are summarized in Figure 1. The primary site of expression of the enzyme in *D. pseudoobscura* is in the adult eye of both sexes while in *D. melanogaster* the enzyme is expressed primarily in the adult male reproductive tract. This strongly suggests that the function of this enzyme has been altered substantially among relatively recently diverged species. Figure 1 shows similar though less dramatic shifts in the regulation, and perhaps function, of EST 6 among the sibling species of the *melanogaster* group. Thus the *Est-6* gene provides an opportunity to investigate molecular variation in both the coding and regulatory regions of a locus which appears to be evolving rapidly both in structure and function.

EST 6 is a member of a large class of enzymes categorized as serine hydrolases. Each member of this class utilizes the hydroxyl group of a serine residue in the active site of the enzyme. This class includes the serine proteases which are well characterized with respect to their evolutionary rela-

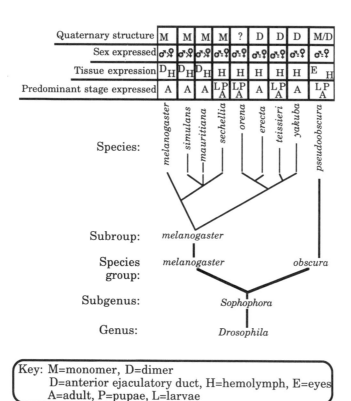

Figure 1. Summary of variability in the quaternary structure, sex, stage, and tissue specific expression of *Est-6* and its homologues in several related species. This table suggests that the structure, regulation and perhaps function of this locus are evolving rapidly.

tionships. In comparison, there are only fragmentary data on the serine esterases (Myers *et al.*, 1988). Our molecular studies of the *Drosophila* esterases have already begun to elucidate their evolutionary history and future work should delineate the broad family of serine esterases and their relationships to other hydrolytic enzymes.

A further reason to study the molecular evolution of esterases stems from their involvement in the evolution of resistance to organophosphate insecticides in insects (Mouches *et al.*, 1986). As the papers by McKenzie *et al.* (Chapter 3, this volume) and Russell *et al.*, (Chapter 16, this volume) indicate, insecticide resistance is a fertile field which has been relatively untouched by investigators concerned with the mechanisms controlling the level of molecular variation in natural populations. The *Est-6* locus and its relatives in *Drosophila* are likely to provide a particularly tractable system for studying the evolution and population dynamics of insecticide resistance genes.

3. VARIATION AT THE EST-6 LOCUS IN NATURAL AND LABORATORY POPULATIONS

The *Est-6* locus exhibits a world-wide polymorphism for two major electrophoretic allozymes, EST 6^F and EST 6^S (Oakeshott *et al.*, 1981). Null alleles for the *Est-6* locus have been recovered from laboratory stocks (Johnson *et al.*, 1966; Richmond *et al.*,1980) and induced by mutagenesis (Bell *et al.*, 1972; Richmond and Sheehan, unpublished). However, a careful search has revealed no naturally occurring *Est-6* null alleles (Voelker *et al.*, 1980; Langley *et al.*, 1981). Oakeshott *et al.* (1981) have shown that world-wide latitudinal clines occur in the frequencies of the EST 6^F and EST 6^S electromorphs, such that EST 6^F becomes rarer and EST 6^S more frequent with increasing distance from the equator. A similar set of clines has been found for the *Est-6* locus in *D. simulans*, a sibling species of *melanogaster* (Anderson and Oakeshott, 1984). The extent of the clines (over $40°$ latitude in each hemisphere), and more importantly, the repeatability across continents and species provide compelling evidence for the action of natural selection on the *Est-6* polymorphism.

Recent molecular analyses (Oakeshott *et al.*, 1987; Cooke and Oakeshott, 1989; Collet *et al.*, 1989) of variation at the *Est-6* locus have revealed an even greater wealth of variants at the DNA sequence level. Cooke and Oakeshott (1989) estimate from their study of the nucleotide sequences of 13 alleles of *Est-6* that a single population of *D. melanogaster* may be segregating for as many as 60 allelic forms of the enzyme. These investigators detected even more silent nucleotide polymorphisms which may affect the expression of this locus. Among the 16 amino acid polymorphisms present among the 13 alleles of *Est-6* sequenced, the substitutions which do occur tend to be conservative with respect to the physicochemical properties of the amino acids. These findings add credence to our conclusions, based on less precise methods of analysis, that allozyme variation at the *Est-6* locus may be adaptive.

Evidence of selection operating on the *Est-6* locus from studies of natural populations is bolstered by our studies of cage populations of *D. melanogaster* segregating for the EST 6^F and EST 6^S allozymes (Gilbert and Richmond, 1982b). In three independently derived sets of populations maintained at either 18 or 25°C, the EST 6^F allozyme is predominant in cages at 25°C. While the effect is significant, it is not strong, and it is complicated by large differences between cages, sexes and deviations from Hardy-Weinberg expectations. Nevertheless, one would not expect a pronounced effect since selection coefficients between these variants are expected to be very small.

The finding that the frequency of the EST 6^F allozyme is consistently higher at 25°C than at 18°C is as expected from the clinal analyses of Oakeshott and colleagues, and leads to the prediction that temporal studies of a single population should reveal seasonal shifts in *Est-6* frequencies if temperature is a major factor maintaining allozyme variation in natural populations.

Temporal studies of natural populations in our laboratory (Gilbert and Richmond, 1982b), by Cavener and Clegg (1981), Franklin (1981) and by Oakeshott *et al.* (1988) generally fail to demonstrate significant seasonal shifts in EST 6 allozyme frequencies. Of 10 experimental populations studied by Oakeshott *et al.* under natural conditions, all but one showed an increase in the frequency of *Est-6*^{S} over the single winter studied. This result, although not statistically significant, is in accord with the clinal data on *Est-6*. Franklin's study of three natural populations did yield evidence that *Est-6* allozyme frequencies vary on a seasonal basis, but the direction of the change is contrary to that predicted by the clinal data. These results suggest that some environmental variable other than temperature may play an important role in regulating allozyme frequencies at this locus in natural populations. Factors influencing allozyme frequencies may vary with temperature on a larger scale. Thus temperature is involved, although indirectly. Indeed Gilbert and Richmond (1982a) noted that in populations of *D. simulans* collected at the same time and place as *D. melanogaster*, the predominant EST 6 allozyme is opposite in the two species. These two species differ in a number of characteristics including sexual behavior and responses to various environmental parameters (Manning, 1959; Parsons, 1973). Moreover, Tepper *et al.* (1982) showed that effects of the X chromosome on the regulation of *Est-6* expression also differ between the two species. The dominance of opposite EST 6 allozymes in these two species suggests that a character important to the species difference may be affected by EST 6.

In summary, studies of natural and laboratory populations of *D. melanogaster* have provided strong evidence that natural selection affects the frequencies of at least the major allozymes of EST 6. However, the nature of the selective forces and their interactions is as yet far from clear.

4. BIOCHEMICAL CHARACTERIZATION OF THE ALLOZYMES OF EST 6

Biochemical differences between allozymes, especially if they can be related to environmental variables, provide a potential basis for the action of natural selection at an allozyme locus. Mane *et al.* (1983) purified EST 6 and characterized the major allozymic forms. They found the enzyme to be a monomeric glycoprotein with a molecular weight of about 62–65,000. Kinetic characterization of the fast and slow forms of the enzyme at 30°C using several naphthyl and nitrophenol esters as substrates showed that the allozymes did not differ significantly from one another for K_m, V_{max}, K_{cat} and the specificity constant, K_{cat}/K_m, which measures enzyme efficiency at substrate levels which are presumed to approximate those *in vivo*. However, at 30°C, the fast form of EST 6 was consistently the more active enzyme and had a consistently higher specificity constant, but these differences were not statistically significant.

The observation that the fast allozyme of EST 6 might have somewhat greater efficiency than the slow at a temperature which was at the upper end of the range experienced by *D. melanogaster* in nature, led us to investigate the possibility that the fast and slow allozymic forms of EST 6 might differ in their kinetic characteristics over the range of temperatures experienced by the species under natural conditions. Accordingly, we undertook a series of kinetic studies of the two major EST 6 allozymes using purified preparations (White *et al.*, 1988) and the artificial substrate, ß-naphthyl acetate. Fast and slow allozymes of EST 6 were purified to homogeneity from two highly inbred lines of *D. melanogaster*. It is possible that differences among the biochemical properties of the two allozymes might merely reflect the relative abundance of active enzyme within the preparations. To exc-

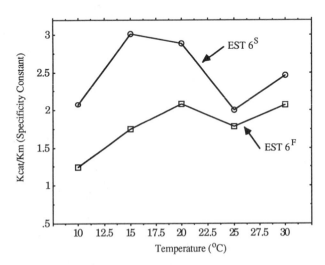

Figure 2. Catalytic efficiencies (K_{cat}/K_m) of purified preparations of the slow (EST 6[S]) and fast (EST 6[F]) allozymes of EST 6 over a range of temperatures.

lude this possibility, the purified samples were analyzed for
the concentration of active sites and kinetic parameters were
adjusted accordingly. The results of the kinetic analyses are
summarized in Figure 2. The slow form of EST 6 is
significantly more efficient than the fast form at temperatures ranging from 10-20°C . However at 25 and 30°C, the catalytic efficiencies of the two allozymes do not differ statistically. If selective effects at the *Est-6* locus are related
to the catalytic efficiencies of the allozymes, then cooler
temperatures should favor the slow allozyme and selective differences should narrow at temperatures above 20°C. The clinal
data of Anderson and Oakeshott (1984) are consistent with this
hypothesis.

In addition to comparisons of catalytic efficiencies among
the allozymes of EST 6, White *et al.* (1988) noted that the
allozymes differed in their thermostabilities, responses to
pH, activation by alcohols, and inhibition by one of five
organophosphate inhibitors which were tested with purified
EST 6. Unfortunately these data were obtained before the
molecular characterization of EST 6 was complete (Oakeshott *et
al.*, 1987; Collet *et al.*, 1989; Cooke and Oakeshott, 1989),
thus we do not know the exact nature of the amino acid substitutions which characterize the slow and fast variants of EST 6
analyzed by White *et al.* However, it is clear from the study
of Cooke and Oakeshott (1989) that the major allozymes of
EST 6 are differentiated by two amino acid substitutions. One
substitution accounts for the mobility difference between the
allozymes and a second nearby substitution is found within a
loop created by a disulfide bond. Cooke and Oakeshott (1989)
have speculated that this latter substitution may disrupt the
structure of the protein around the disulfide bond in some
way, thereby altering the folding of the molecule as a whole
or the geometry of the active site. This possibility is supported by the kinetic and other biochemical differences found
by White *et al.* (1988) and by the observation that the variant
residue within the disulfide loop is conserved between EST 6,
Drosophila acetylcholinesterase and human pseudocholinesterase
(Myers *et al.*, 1988).

One of the major problems plaguing attempts to characterize
the biochemical differences between esterase allozymes is our
ignorance of their *in vitro* substrates. EST 6 is categorized
as a carboxylesterase on the basis of its substrate specificity and its pattern of inhibition by a series of
organophosphate inhibitors. Esterases are capable of hydrolyzing carboxylester, carboxyamide (peptide) and carboxythioester
bonds (Heyman, 1980). Carboxylesterases have been studied
routinely using artificial carboxylester substrates, such as
ß-naphthyl acetate, for which the enzymes have high affinity.
White *et al.* (1988) noted that EST 6 is capable of hydrolyzing
acetanilide, a substrate which is regularly hydrolysed by amidases (Heyman, 1980). Other workers have noted that proteases
often have esterolytic activity (Peters, 1982). In order to
explore the possibility that EST 6 might act as a protease, we
have made use of an agar diffusion assay in which the agar
contains bovine casein, a general substrate for proteases. The
presence of the casein in a thin agar gel renders the normally

transparent gel turbid. The sample to be tested is placed in a well in the agar gel and allowed to diffuse into the agar. If casein is a substrate for a protease in the sample, it is hydrolyzed and the agar becomes transparent in a ring around the sample well. The size of the ring is linearly related to the concentration of protease in the sample (Biorad Bulletin 1111, 1982). Figure 3 shows such a test using purified EST 6 prepared as described by White *et al.* (1988). Increasing concentrations of commercially prepared trypsin give rings of correspondingly larger sizes. Heat-denatured trypsin or a sample well containing only buffer produce no transparent ring. Both crude homogenates of *D. melanogaster* and three concentrations of purified EST 6 provide evidence for proteolytic activity in flies and in the EST 6 preparation.

A possible explanation for the result observed in Figure 3 is that the EST 6 preparation is contaminated with a low level of a protease. However, this does not seem likely since the purified EST 6 used in this experiment was incubated with ^{3}H-labelled DFP, a compound which binds covalently to the hydroxyl group of the active site serine in serine hydrolases (Aldridge and Reiner, 1972). DFP should bind to both EST 6 and any contaminating serine proteases in such a sample.

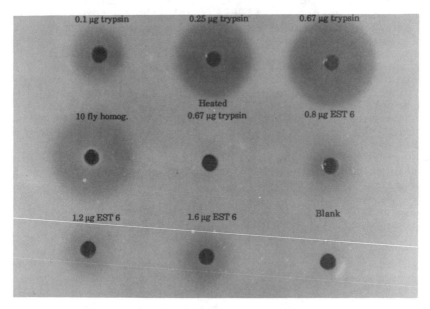

Figure 3. Hydrolysis of bovine casein by purified preparations of EST 6. 20 μl volumes of each the substances indicated were added to a well in 1% agar containing bovine casein in a Tris buffered physiological saline solution, pH 7.2, and incubated overnight at 25°C. The trypsin sample was commercial bovine trypsin (Sigma). The 10 fly homogenate is a crude homogenate (adult male Oregon R homogenized in 0.1M sodium phosphate buffer, pH 6.8, and centrifuged for 5 min at 16,500g to remove debris). The EST 6 preparations were obtained as described by White *et al.* (1988). Controls are heat-treated (boiled for 5 min.) commercial trypsin, and a buffer blank.

Autoradiographic analyses of PAGE gels of this batch of purified EST 6 reveal the presence of only a single band of DFP-
labelled material with a molecular weight identical to EST 6.
Thus we conclude that EST 6 has some proteolytic activity,
although in comparison with the commercial trypsin we tested,
the specific activity of EST 6 against casein would seem to be
substantially less.

5. PHYSIOLOGICAL ROLE OF EST 6

5.1. Developmental Profile

EST 6 is expressed predominantly in the adult male
although the enzyme is detected at a low level in all other
stages as well as in females. Figure 4 shows the developmental profile of EST 6. Enzyme activity remains at a low level
until adults eclose, and this pattern is duplicated when *Est-6*
mRNA is assayed (Oakeshott *et al.*, 1987). There is a small
peak of activity in mid-larval development whose significance
we have not studied. After eclosion, activity levels rise
slightly in females, but male activity begins to increase during the first day after eclosion and by three days after eclosion may reach a level which is two to ten times that found in
females (Tepper *et al.*, 1982).

5.2. Tissue Distribution

The predominance of EST 6 in adult males led us to examine

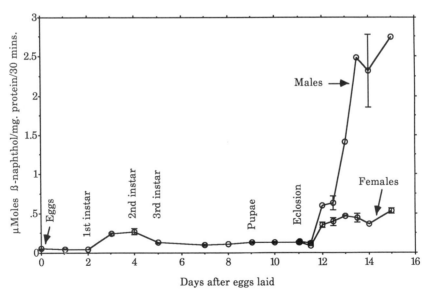

Figure 4. EST 6 activity (μMoles ß-naphthol/mg protein/30 min) of
homogenates prepared from various developmental stages of an Oregon R
strain of *Drosophila melanogaster*. Data are taken from Sheehan *et al.*
(1979).

the enzyme's tissue distribution in this sex.   The enzyme is largely localized to the anterior ejaculatory duct (AED) of the male reproductive system and to hemolymph (Stein *et al.*, 1984).   We have used two approaches to demonstrate that EST 6 is localized to the AED and is also synthesized there.   Stein *et al.* (1984) cultured AEDs in the presence of ^{3}H-leucine and were able to immunoprecipitate labelled EST 6 from homogenates of cultured AEDs.   More recently we have used *in situ* hybridization to whole mount preparations of male reproductive systems to show that EST 6 mRNA is restricted to the AED in the adult male reproductive system.   Figure 5 shows whole mount preparations hybridized with an RNA probe prepared from a cDNA clone of EST 6.   In panel A, the preparation was

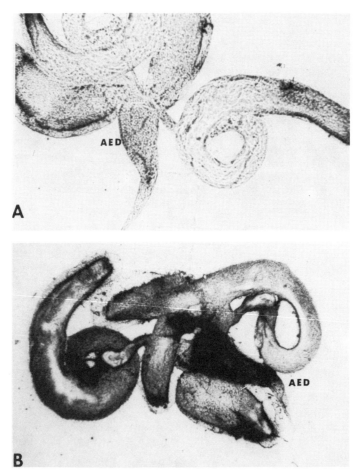

Figure 5. Hybridization of whole mount preparations of male reproductive tracts from the Dm 145 (EST 6[S]) strain of *Drosophila melanogaster* with RNA probes specific for *Est-6* mRNA.   Panel A: Preparation hybridized with a sense strand control probe which is not complementary to the *Est-6* mRNA. Panel B: Preparation hybridized with a nonsense strand probe which is complementary to the *Est-6* mRNA.   See Hafen and Levine (1986) for details of methodology.

hybridized with RNA from the cDNA clone which is not complementary to the *Est-6* message. There are only a few areas of background hybridization primarily around the periphery of tissues. In panel B, the preparation was hybridized with RNA complementary to the *Est-6* message. Strong hybridization to the anterior two-thirds of the AED is clearly apparent. While the AED is probably not the only site of *Est-6* expression in adult males because of its presence in hemolymph, it is clearly an important one.

5.3. Transfer to Females

The synthesis and localization of EST 6 to a male reproductive tract organ whose morphology suggests modification for the synthesis of seminal fluid components (Bairaiti, 1968) immediately suggests that the enzyme might be a component of the seminal fluid in *D. melanogaster*. We tested this possibility by mating females homozygous for a null allele of *Est-6* to males which produce active enzyme. Immediately after mating was completed, females were homogenized and the presence of active enzyme assayed by standard electrophoretic technologies (Richmond *et al.*, 1980). After mating, null females clearly have active EST 6. Thus we conclude that the enzyme is transferred from males to females at mating. Richmond and Senior (1981) showed that EST 6 is transferred within the first three minutes after the initiation of copulation, and can be detected in homogenates of whole EST 6° (=EST 6 null) females for 1-2 hrs after mating using a spectrophotometric assay.

The disposition of male-derived EST 6 in *Est-6°* females might give an indication of its function, thus we have attempted to determine where seminal-fluid-derived EST 6 is localized in females. Our expectation was that male-derived EST 6 would be found in the female's reproductive system. However, electrophoretic analysis of dissected reproductive systems from *Est-6°* females which had just completed mating with *Est-6^S* males revealed that no enzyme was present. Examination of the remaining carcass showed that EST 6 activity was indeed present. After examination of a number of tissues, we (Meikle *et al.*, 1990) determined that the only tissue having significant EST 6 activity in mated females was the hemolymph. This result is shown in Figure 6. Removal of female reproductive systems within seconds of the initiation of mating showed that EST 6 is transferred into the female reproductive tract, but within 30 seconds of the beginning of copulation, the bulk of the male-derived enzyme has been translocated to the female hemolymph. The movement of a large, presumably globular protein from the female reproductive tract into the hemolymph suggests the effects of an active transfer process. Similar phenomena have been previously noted in houseflies (Terranova *et al.*, 1972), but the mechanism is unknown. Figure 6 also shows EST 6 activity in the hemolymph of *Est-6^S* males thereby demonstrating the point made above that EST 6 is found in male hemolymph as well as the AED.

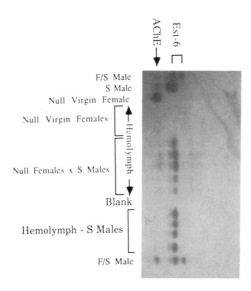

Figure 6. Starch gel zymogram stained for esterase activity using the methods of Richmond (1972). Hemolymph samples were obtained using the methods
of Kambysellis (1984).

Although Richmond and Senior (1981) had demonstrated that
activity from male-derived EST 6 is present in females for 1-2
hrs after mating, their assay procedure was not particularly
sensitive and is complicated by the presence of other
esterases in females. Meikle *et al.* (1990) undertook a study
of the length of time EST 6 protein is present in the female
hemolymph using Western blot technology. *Est-6°* females were
mated to *Est-6°* males and proteins from the hemolymph of the
females separated on denaturing SDS-PAGE gels and blotted to
nitrocelluose membranes for detection with rabbit polyclonal
antibodies. The results are shown in Figure 7. Male-derived
EST 6 can be detected in *Est-6°* females for as long as 4 days
after mating. The loss of EST 6 protein from female hemolymph
is likely due to complete degradation of the enzyme since
fragments of smaller size which bind the EST 6 antibody are
not apparent from females even several days after mating. We
turn now to the effects which EST 6 has on male and female
behavior and female productivity.

5.4. Behavioral and Physiological Effects of EST 6

We have investigated the effects of male *Est-6* genotype on
six fitness-related measures summarized in Table I. Lines
from which males were derived for the experiments summarized
in Table I had been made largely isogenic by repeated backcrossing to our standard *Est-6°* line (Scott, 1986). Males producing active EST 6 mate sooner and copulate for a shorter
time both at 18 and 25°C. The *Est-6* genotype of a female's
mate has a pronounced effect on her remating time which is
dependent upon temperature and conditions under which remating

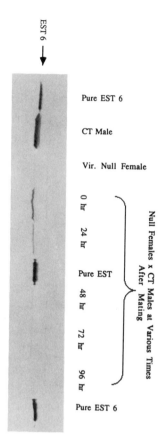

EST 6 →

Pure EST 6

CT Male

Vir. Null Female

0 hr

24 hr

Pure EST

48 hr

72 hr

96 hr

Null Females x CT Males at Various Times After Mating

Pure EST 6

Figure 7. Western blot of homogenates of wild-type, male *Drosophila melanogaster* (CT = Chateau Tahbilk, Victoria, Australia) and virgin females or females of the Dm 100 (EST 6°) strain mated to CT males and homogenized after mating at the times indicated. Purified EST 6 was obtained using the methods of White *et al.* (1988). See Meikle *et al.* (1990) for details of methodology.

is measured. Scott (1986) determined the proportion of females which remate within 18 hrs after mating to *Est-6*^{S} or *Est-6*^{O} males at 25°C. For a period of about 4 hrs after mating, females mated to males of either EST 6 type show a much reduced probability of remating. However, between 6 and 18 hrs after mating, females mated to *Est-6*^{S} males remate with a significantly lower probability than females initially mated to *Est-6*^{O} males. This effect is at least partly dependent on the presence of sperm in the females' seminal receptacle, since the probabilities of remating within 6 to 18 hrs after mating are altered in females mated to sterile males of the two EST 6 types.

When the probability of female remating is measured daily beginning 24 hrs after an initial mating to *Est-6*^{O} or *Est-6*^{S}

Table I

Summary of the behavioral and physiological effects of homozygous male *Est-6* genotype on characters measured for either males or *Est-6*-null females mated to males of the given genotype. $6^S = Est\text{-}6^S/Est\text{-}6^S$, $6^\circ = Est\text{-}6^\circ/Est\text{-}6^\circ$

Measure	18°C	25°C	Ref.
Male mating speed	$6^S > 6^\circ$	$6^S > 6^\circ$	a
Copula duration	$6^S < 6^\circ$	$6^S < 6^\circ$	a
Female remating time (0-18hr)	?	$6^S > 6^\circ$	b
Female remating time (1-7d)	$6^S = 6^\circ$	$6^S < 6^\circ$	a,d
Female sperm load (1-4d)	?	$6^S > 6^\circ$	d
Fertile eggs laid	$6^S > 6^\circ$	$6^S = 6^\circ$	a,c

[a]Gilbert and Richmond, 1982a
[b]Scott, 1986
[c]Gilbert *et al.*, 1981a
[d]Gilbert *et al.*, 1981b

males, a very different pattern results. At 25°C, females mated to *Est-6^S* males remate about 0.5 to 1 day sooner than females remated to Est-6° males. This is the converse of that observed by Scott (1986) for remating tests made during the first 18 hrs after mating. The differences in EST 6 effects on remating between these two experiments may be due to an EST 6 effect on the numbers of sperm remaining in the female's seminal receptacle. Gilbert *et al.* (1981b) showed that females inseminated by *Est-6^S* males lose sperm at a significantly greater rate than females mated to *Est-6°* males. However, this effect is not apparent until 2 to 3 days after mating. Thus the short-term (<18hr) effects of male-derived EST 6 on a female's probability of remating may be governed by an interaction between EST 6 and other factors (probably sperm) in the female. The long-term effects (1-7 days) of male-derived EST 6 on a female's probability of remating are most likely a result of a direct or indirect effect of EST 6 on the pattern of sperm use by females. The mechanism of these effects is unknown, although our observation that the long-term effects of EST 6 on female remating are not apparent if the experiment is conducted at 18°C suggests that EST 6 may exert its effects by influencing sperm motility. Evidence that temperature affects sperm motility has been found both *in vitro* and *in vivo* (Richards, 1963; Hartl, 1973).

In addition to its effects on male and female reproductive behavior, male-derived EST 6 has a temperature dependent influence on the number of progeny produced by females. At 25°C, Gilbert *et al.* (1981a) demonstrated that the productivity of females mated to *Est-6°* or *Est-6^S* males was equivalent. At 18°C, there is a significant difference in productivity which favors females inseminated by *Est-6^S* rather than *Est-6°* males. These data are summarized in Table II. The physiological mechanism(s) for this effect is not known. However, our observations that male-derived EST 6 appears to interact with stored sperm suggest a possibility. EST 6 may act in some

Table II

Total numbers of progeny (± SEM) produced by *forked* females mated
to *Est-6⁰* or *Est-6ˢ* males and brooded at 18 or 25°C. t values
test the significance of differences in the means of progeny
numbers between the two male types

Male type	Total progeny produced at	
	18°C	25°C
EST 6^S	153±14.9	321±46
EST 6^0	110±14.8	332±46
t value	2.93*	0.17

*P<0.01

direct or indirect way to influence sperm motility (Gilbert,
1981). There is evidence that both cold and hot temperature
shocks to *Drosophila* males influence the productivity of subsequent matings (Iyengar and Baker, 1962). Moreover there is
evidence that temperature affects sperm function both *in vitro*
and *in vivo* (Gilbert and Richmond, 1982). Thus it is reasonable to hypothesize that EST 6 may act in the female to adjust
for temperature mediated changes in sperm motility which might
otherwise result in a decrement in a male's fitness at temperatures of 18°C and below.

The behavioral and physiological effects of EST 6 summarized in Table I all suggest that temperature should interact
with EST 6 allozyme function to select for different frequencies of alleles in natural populations. Indeed the effect of
the EST 6⁰ male genotype is consistently to reduce fitness at
18°C and for some measures at 25°C. Based on the productivity
and long-term remating data summarized in Table I, we would
expect the frequency of *Est-6⁰* alleles to decline in 18°C populations and to increase in 25°C populations. Gilbert (1981)
has demonstrated that the *Est 6⁰* allele is maintained at
frequencies near 50% for several generations in population
cages also segregating for the *Est-6ˢ* allele. Moreover, null
alleles of *Est-6* are apparently absent in natural populations
(Voelker *et al.*, 1980; Langley *et al.*, 1981). These findings
suggest that the behavioral and physiological effects of EST 6
may be counter balanced by other selective pressures resulting
in a complex set of interacting forces which act to maintain
the polymorphism present at this locus in most natural populations.

6. CONCLUSIONS

The analysis of adaptation begins with efforts to understand the functional significance of the character under
study. The *Est-6* locus in *Drosophila melanogaster* clearly has
a role in reproductive processes and thus influences fitness
at several points. Our analyses suggest that the selective
forces impinging on this locus are complex and result in conflicting pressures which may be responsible for maintaining

some of the allelic variants in natural populations. However, we are far from a full understanding of the function of EST 6 and know little of the proximate mechanisms by which its effects on reproductive processes are achieved. Despite some initially promising results (Richmond *et al.*, 1980; Vander Meer *et al.*, 1986), we do not know the *in vivo* substrate(s) of EST 6. This information will be critical for a complete understanding of the physiological role(s) of the enzyme and to infer the selective forces in operation.

Esterase genes and the *Est-6* locus in particular are among the most polymorphic loci in *Drosophila* and indeed in many other organisms. Lewontin's (1985) classification of protein coding loci postulates that genes such as *Est-6* are likely to be subject to a mixture of selective and stochastic forces. These agents will generate an allele frequency distribution which consists of several variants maintained at significant frequencies by some form of balancing selection, and a selection of rare alleles whose frequencies are governed primarily by the balance between mutation and genetic drift. The selective forces involved are likely to vary between species, and we would expect the identities and frequencies of major alleles to differ between species. Our analysis of the *Est-6* homologue in *D. pseudoobscura* (Brady *et al.*, 1990) provides strong support for this contention and adds yet more evidence that natural selection is responsible for maintaining some of the allelic variants at this locus. Synonymous substitutions between *melanogaster* and *pseudoobscura* are evenly distributed over the two exons suggesting that mutational pressures are equivalent throughout the locus. However, the distribution of amino acid differences over the two exons in the two species differs from intraspecific comparisons among *Est-6* alleles suggesting that selection pressures are different in these species. Given the very different sex, tissue and temporal pattern of expression of these loci (Fig. 1), this conclusion is not surprising. Our current efforts to obtain a complete set of sequence data for all the members of the *melanogaster* subgroup will provide a series of other tests of the possibility that interspecific differences in the expression and perhaps functions of *Est-6* homologues result in variable patterns of selection at this locus (Hudson *et al.*, 1987). Thus we are reminded of our original premise that a full understanding of adaptation at the molecular level depends upon mechanistic analyses of functional differences which arise from molecular polymorphisms. The *Est-6* system has proven to be a valuable model for such studies.

7. SUMMARY

The *esterase-6* locus in *Drosophila melanogaster* encodes the structure of a beta-carboxylesterase. This locus is highly polymorphic in natural populations and comparisons of its pattern of expression reveal that its regulation has also changed, sometimes dramatically, even between closely related *Drosophila* species. The *esterase-6* locus has been cloned and analyses of nucleotide variation reveal a level of polymorphism unmatched by previously studied loci. Studies of natural and laboratory populations have provided strong evidence that natural selection operates on the major allozymic variants of esterase 6. Biochemical characterization of the major allozymes of esterase 6 reveals significant kinetic differences between them which may account, at least in part, for clinal variation observed in latitudinal samples from three continents. We have also shown that esterase 6 may have proteolytic activity by demonstrating that it hydrolyzes bovine casein. Studies of the physiological role of esterase 6 in *D. melanogaster* show that the enzyme is synthesized in the anterior ejaculatory duct of the male and transferred to females at mating. Within seconds of its transfer to females, the enzyme is translocated to the female hemolymph where it can be detected for as long as 4 days after mating. Male-derived esterase 6 has effects on the remating time of females, female sperm load and a temperature-dependent effect on the number of progeny produced by females. Our analyses suggest that the selective forces impinging on this locus are complex and result in conflicting pressures which may be responsible for maintaining some of the allelic variants in natural populations.

ACKNOWLEDGEMENTS. Over the course of almost twenty years many individuals have contributed their intellectual and practical skills to an understanding of the meaning of genetic variation at the *Est-6* locus. Without these many efforts, we would know little today. Much of the molecular analysis of variation at the *Est-6* locus has been made possible by an enjoyable, fruitful and ongoing collaboration with the molecular biology laboratory in the Division of Entomology at the CSIRO in Canberra, Australia headed by Dr. John Oakeshott. We are grateful for the contributions of all these individuals. Financial support for research in our laboratory has been provided by grants from the US National Institutes of Health and the National Science Foundation.

References

Aldridge, W. N., and Reiner, E., 1972, *Enzyme Inhibitors as Substrates*, North-Holland Publishing Co., Amsterdam.

Anderson, P. R., and Oakeshott, J. G., 1984, Parallel geographic patterns of allozymic variation in two sibling *Drosophila* species, *Nature, Lond.* **308**:729-731.

Bairati, A., 1968, Structure, and ultrastructure of the male reproductive system in *Drosophila melanogaster* Meig. 2. The genital duct and accessory glands, *Monitore zool. ital.* **2**:105-182.

Bell, J. B., MacIntyre, R. J., and Olivieri, A. P., 1972, Induction of null-activity mutants for the acid phosphatase-1 gene of *Drosophila melanogaster*, *Biochem. Genet.* **6**:205-216.

Birley, A. J., and Beardmore, J. A., 1977, Genetical composition, temperature, density and selection in an enzyme polymorphism, *Heredity* **39**:133-144.

Brady, J. P., Richmond, R. C., and Oakeshott, J. G., 1990, Cloning of the *esterase-5* locus from *Drosophila pseudoobscura* and molecular analysis of interspecific evolutionary changes at this locus, *Evolution* (submitted).

Cavener, D. R., and Clegg, M. T., 1981, Temporal stability of allozyme frequencies in a natural population of *Drosophila melanogaster*, *Genetics*
98:613-623.

Clarke, B., 1975, The contribution of ecological genetics to evolutionary
theory: Detecting the direct effects of natural selection on particular
loci, *Genetics* (Suppl.) 79:101-113.

Collet, C., Nielsen, K. M., Russell, R. J., Karl, M. J., Oakeshott, J. G.,
and Richmond, R. C., 1990, Molecular analysis of duplicated esterase
genes in *Drosophila melanogaster*, *Mol. Biol. Evol.* (in press).

Cooke, P. H., and Oakeshott, J. G., 1989, Amino acid polymorphisms for
esterase-6 in *Drosophila melanogaster*, *Proc. natn. Acad. Sci. USA*
86:1426-1430.

Franklin, I. R., 1981, An analysis of temporal variation at isozyme loci in
Drosophila melanogaster, in: *Genetic Studies of Drosophila Populations*
(J. B. Gibson, and J. G. Oakeshott, eds), Australian National Univer-
sity Press, Canberra, pp. 217-236.

Gilbert, D. G., 1981, Function and adaptive significance of esterase 6
allozymes in *Drosophila melanogaster* reproduction, Ph.D. Dissertation,
Indiana University, Bloomington, Indiana.

Gilbert, D. G., and Richmond, R. C., 1982a, Esterase 6 in *Drosophila
melanogaster*: Reproductive function of active and null alleles at low
temperature, *Proc. natn. Acad. Sci. USA* 79:2962-2966.

Gilbert, D. G., and Richmond, R. C., 1982b, Studies of esterase 6 in
Drosophila melanogaster. XII. Evidence for temperature selection of Est
6 and Adh alleles, *Genetica* 58:109-119.

Gilbert, D. G., Richmond, R. C., and Sheehan, K. B., 1981a, Studies of
esterase 6 in *Drosophila melanogaster*. V. Progeny production and sperm
use in females inseminated by males carrying active or null alleles,
Evolution 35:21-37.

Gilbert, D. G., Richmond, R. C., and Sheehan, K. B., 1981b, Studies of
esterase 6 in *Drosophila melanogaster*. VII. The timing of remating in
females inseminated by males having active or null alleles, *Behav.
Genet.* 11:195-208.

Hafen, E., and Levine, M., 1986, The localization of RNAs in *Drosophila*
tissue sections by *in situ* hybridization, in: *Drosophila A Practical
Approach* (D. B. Roberts, ed.), IRL Press, Oxford, pp. 139-174.

Hartl, D., 1973, The mechanism of a brooding effect associated with segregation distortion in *Drosophila melanogaster*, *Genetics* 74:619-631.

Heyman, E., 1980, Carboxylesterases and amidases, in: *Enzymatic Basis of
Detoxification*, Vol.2 (W. Jakoby, ed.), Academic Press, New York, pp.
291-323.

Hudson, R. R., Kreitman, M., and Aguadé, M., 1987, A test of neutral molecular evolution based on nucleotide data, *Genetics* 116:153-159.

Iyengar, S. V., and Baker, R. M., 1962, The influence of temperature on the
pattern of insemination by *Drosophila* males, *Genetics* 47:963-964.

Johnson, F. M., Wallis, B. B., and Dennison, C., 1966, Recessive esterase
deficiencies controlled by alleles of Est C and Est 6 in *Drosophila
melanogaster*, *Drosoph. Inf. Serv.* 41:159.

Kambysellis, M. P., 1984, A highly efficient method for collection of
hemolymph, hemocytes, or blood-borne organisms from *Drosophila* and
other small insects, *Drosoph. Inf. Serv.* 60: 219-220.

Koehn, R. K., 1978, Physiology and biochemistry of enzyme variation: The
interface of ecology and population genetics, in: *Ecological Genetics:
The Interface* (P.F. Brussard, ed.), Springer-Verlag, New York, pp. 51-
72.

Kojima, K-I., and Huang, S. L., 1972, Effects of population density on the
frequency-dependent selection in the esterase-6 locus of *Drosophila
melanogaster, Evolution* 26:313-321.

Langley, C. H., Voelker, A. J., Leigh-Brown, A., and Ohnishi, S., 1981, Null allele frequencies at allozyme loci in natural populations of *Drosophila melanogaster*, *Genetics* 99:151-156.

Lewontin, R. C., 1985, Population genetics, *A. Rev. Genet.* 19:81-102.

MacIntyre, R. J., and Wright, T. R. F., 1966, Response of esterase-6 alleles of *Drosophila melanogaster* and *D. simulans* to selection in experimental populations, *Genetics* 53:371-387.

Mane, S. D., Tepper, C. S., and Richmond, R. C., 1983, Purification and characterization of esterase 6, a polymorphic carboxylesterase of *Drosophila melanogaster*, *Biochem. Genet.* 21:1019-1040.

Manning, A., 1959, The sexual behavior of two sibling *Drosophila* species, *Behaviour* 15:123-145.

Meikle, D., Sheehan, K., Phillis, D., and Richmond, R. C., 1990, Studies of esterase 6 in *Drosophila melanogaster*. XIX. Localization and longevity of male derived enzyme in female hemolymph, *J. Insect Physiol.* (submitted).

Mouches, C., Pasteur, N., Bergé, J., Hyrien, O., Raymond, M., de Saint Vincent, B. R., de Silvestri, M., and Georghiou. G. P., 1986, Amplification of an esterase gene is responsible for insecticide resistance in a California *Culex* mosquito, *Science* 233:778-780.

Myers, M., Richmond, R. C., and Oakeshott, J. G., 1988, On the origins of esterases, *Mol. Biol. Evol.* 5:113-119.

Oakeshott, J. G., Collet, C., Phillis, R. W., Nielson, K. M., Russell, R. J., Chambers, G. K., Ross, V., and Richmond. R. C., 1987, Molecular cloning and characterization of esterase 6, a serine hydrolase of Drosophila, *Proc. natn. Acad. Sci. USA* 84:3359-3363.

Oakeshott, J. G., Chambers, G. K., Gibson, J. B., and Willcocks, D. A., 1981, Latitudinal relationships of esterase 6 and phosphoglucomutase gene frequencies in *Drosophila melanogaster*, *Heredity* 47:385-396.

Oakeshott, J. G., Wilson, S. R., and Knibb, W. R., 1988, Selection affecting enzyme polymorphisms in enclosed *Drosophila* populations maintained in a natural environment, *Proc. natn. Acad. Sci. USA* 85:293-297.

Parsons, P., 1973, *Behavioral and Ecological Genetics*, Oxford University Press, London.

Peters, J., 1982, Nonspecific esterases of *Mus musculus*, *Biochem. Genet.* 20:585-606.

Richards, A. G., 1963, The rate of sperm locomotion in the cockroach as a function of temperature, *J. Insect Physiol.* 9:545-549.

Richmond, R. C., 1972, Enzyme variability in the *Drosophila willistoni* group. III. Amounts of variability in the superspecies, *D. paulistorum*, *Genetics* 70:87-112.

Richmond, R. C., and Senior, A., 1981, Esterase 6 of *Drosophila melanogaster*: Kinetics of transfer to females, decay in females and male recovery, *J. Insect Physiol.* 27:849-853.

Richmond, R. C., Gilbert, D. G., Sheehan, K. B., Gromko, M. H., and Butterworth, F. M., 1980, Esterase 6 and reproduction in *Drosophila melanogaster*, *Science* 207:1483-1485.

Riley, M. A., 1989, Nucleotide sequence of the Xdh region in *Drosophila pseudoobscura* and an analysis of the evolution of synonymous codons, *Mol. Biol. Evol.* 6:33-52.

Scott, D., 1986, Inhibition of female *Drosophila melanogaster* remating by a seminal fluid protein (esterase 6), *Evolution* 40:1084-1091.

Sheehan, K., Richmond, R. C., and Cochrane, B. J., 1979, Studies of esterase 6 in *Drosophila melanogaster*. III. The developmental pattern and tissue distribution, *Insect Biochem.* 9:443-450.

Simmons, G. M., Kreitman, M. E., Quattlebaum, W. F., and Miyashita, M., 1989, Molecular analysis of alleles of alcohol dehydrogenase along a cline in *Drosophila melanogaster*. I. Main, North Carolina, and Florida, *Evolution* **43**:393–409.

Stein, S. P., Tepper, C. S., Able, N. D., and Richmond, R. C., 1984, Studies of esterase 6 in *Drosophila melanogaster*. XVI. Synthesis occurs in the male reproductive tract (anterior ejaculatory duct) and is modulated by juvenile hormone, *Insect Biochem.* **14**:527–532.

Tepper, C. S., Richmond, R. C., Terry, A. L., and Senior, A., 1982, Esterase 6 in *Drosophila melanogaster*: Modification of Esterase 6 activity by unlinked genes, *Genet. Res.* **40**:109–125.

Terranova, A. C., Leopold, R. A., Degrugillier, M. E., and Johnson, J. R., 1972, Electrophoresis of the male accessory secretion and its fate in the mated female, *J. Insect. Physiol.* **18**:1573–1591.

Vander Meer, R., Obin, M., Sheehan, K. S., Zawistowski, S., and Richmond, R., 1986, A reevaluation of the role of cis-vaccenyl acetate, cis vaccenyl alcohol and esterase 6 in the regulation of mated female sexual attractiveness in *Drosophila melanogaster*, *J. Insect Physiol.* **32**:681–686.

Voelker, R. A., Langley, C. H., Leigh-Brown, A. J., Ohnishi, S., Dickson, B., Montgomery, E., and Smith, S. C., 1980, Enzyme null alleles in natural populations of *Drosophila melanogaster*: Frequencies in a North Carolina population, *Proc. natn. Acad. Sci. USA* **77**:1091–1095.

Watt, W., 1985, Allelic isozymes and the mechanistic study of evolution, *Isozymes. Current Topics in Biological & Medical Research* **12**:89–132.

White, M. W., Mane, S. D., and Richmond, R. C., 1988, Studies of esterase-6 in *Drosophila melanogaster*. XVIII. Characterization of the slow and fast allozymes, *Mol. Biol. Evol.* **5**:41–62.

Wright, T. R. F., 1963, The genetics of an esterase in *Drosophila melanogaster*, *Genetics* **48**:787–801.

Yamazaki, T., 1971, Measurement of fitness at the Est-5 locus in *Drosophila pseudoobscura*, *Genetics* **67**:579–603.

CHAPTER 16

Insecticide Resistance as a Model System for Studying Molecular Evolution

ROBYN J. RUSSELL, MIRA M. DUMANCIC, GEOFFREY G. FOSTER, GAYE
L. WELLER, MARION J. HEALY, AND JOHN G. OAKESHOTT

1. INTRODUCTION

Apart from its applied significance, insecticide resistance
is an excellent model system for studying the molecular basis
of evolutionary change, in particular, the acquisition of a
qualitatively different phenotype. It also has the unusual
advantage in evolutionary biology that the change has been
widespread and rapid enough to be amenable to analysis;
approximately 450 species of insects and mites have developed
resistance to chemical insecticides over the past forty years
(Georghiou, 1986).

There are five major classes of chemical insecticides:
organophosphates (e.g., diazinon and parathion), carbamates
(e.g., carbaryl and aldicarb), DDT and its analogues (e.g.,
methoxychlor), cyclodienes (e.g., heptachlor and dieldrin) and
pyrethroids (e.g., permethrin and fenvalerate) (Hutson and
Roberts, 1985). All five classes affect the nervous system in
some way, the organophosphates and carbamates affecting the
metabolism of the neurotransmitter, acetylcholine, the cyclodienes apparently affecting processes in synapses, and DDT and
pyrethroids affecting the sodium channels in nerve membranes
(Hama, 1983; Narahashi, 1983; Oppenoorth, 1985). Many different mechanisms of resistance to insecticides have been identified, and these can vary with the insecticide, the species of
insect and, sometimes, among populations within particular
species. Nevertheless, multiple cases of most mechanisms have
been found; most mechanisms confer resistance to all chemicals
within a given class, and many mechanisms confer cross-resis-
tance to members of other classes (for review see Oppenoorth,
1985).

Clearly, then, insecticide resistance involves a finite
number of biochemical and genetic options, each with some generality as a model of evolutionary change. It is equally

ROBYN J. RUSSELL, MIRA M. DUMANCIC, GEOFFREY G. FOSTER, GAYE L. WELLER,
MARION J. HEALY, AND JOHN G. OAKESHOTT * CSIRO Division of Entomology,
G.P.O. Box 1700, Canberra, A.C.T. 2601, Australia.

Ecological and Evolutionary Genetics of Drosophila, Edited by
J.S.F. Barker *et al.*, Plenum Press, New York, 1990

clear, however, that the elucidation of any one model system
at the molecular genetic level first requires detailed knowledge of the genetics and biochemistry of resistance for the
particular chemical(s) and insect population(s). The criteria
have been met only for a limited number of systems and major
advances are now being made in the biochemical and molecular
genetics of these systems. This article will review these systems giving particular emphasis to mechanisms of resistance to
the organophosphate class of insecticides. Evidence from several Dipteran species will be drawn together to show that
organophosphate resistance in this order, at least, often
involves a tightly linked cluster of carboxyl-esterase genes
in the same esterase multigene family. For a comprehensive
overview of the molecular mechanisms of resistance to all
classes of insecticide, the reader is referred to Soderlund
and Bloomquist (1989).

2. MAJOR MECHANISMS OF INSECTICIDE RESISTANCE

Three broad classes of biochemical adaptation have been
found to confer insecticide resistance: decreased absorption
of the insecticide, modification of the target site or
increased detoxication.

2.1. Decreased Penetration

Decreased penetration or absorption of the insecticide by
the insect is usually only of secondary importance to other
mechanisms of resistance. Genes controlling reduced penetration, such as the *pen* gene on chromosome 3 of the house fly
(Hoyer and Plapp, 1968; Sawicki and Farnham, 1968a,b), confer
a low level of resistance to most insecticides by delaying
their penetration.

2.2. Target Site Modification

Alterations in the sensitivity of the target molecule(s)
for a particular insecticide can confer high levels of resistance. The primary target of organophosphate and carbamate
insecticides is acetylcholinesterase (ACE; Hama, 1983), so
mutations in the ACE gene which render the enzyme insensitive
to the action of these insecticides will confer resistance.
The nature of these ACE mutations will be discussed in more
detail below.

The primary targets for pyrethroids and DDT are the sodium
channels in nerve membranes (Narahashi, 1983). Electrophysio-
logical studies have shown that pyrethroids modify a small
fraction of the total number of sodium channels to prolong
their open time, without altering channel conductance
(Yamamoto *et al.*, 1983; Lund, 1984). House flies (*Musca domestica*) bearing the *kdr* (*knock-down resistance*) mutation are
resistant to both pyrethroids and DDT (for review see Sawicki,
1985) and one biochemical study suggested that *kdr* mutant
house flies have a reduced number of sodium channels

(Rossignol, 1988). However, since similar studies using well-defined susceptible, *kdr* and *super-kdr* house fly strains have failed to repeat this observation (Grubs *et al.*, 1988; Pauron *et al.*, 1989; Sattelle *et al.*, 1988), it is unlikely that a reduction in sodium channel density is the mechanism of resistance in *kdr* mutants.

It is interesting to note that the *nap*[ts] (*no action potential*) mutant of *Drosophila melanogaster* has a reduced number of sodium channels and has low levels of resistance to pyrethroids (Kasbekar and Hall, 1988), raising the possibility that a reduction in the number of modified open channels may confer slight resistance. It is unlikely that the *nap*[ts] and *kdr* mutants define the same genetic loci since they have been mapped to non-homologous chromosomes: the *kdr* mutation maps to chromosome 3 in the house fly (Tsukamoto, 1983), which is homologous to the X chromosome in *D. melanogaster* (Foster *et al.* 1981), whereas the *nap*[ts] mutation maps to chromosome 2R in *D. melanogaster* (Wu *et al.*, 1978).

2.3. Increased Detoxication of Insecticides

Three of the five major groups of insecticides, namely the organophosphates, carbamates and pyrethroids, are esters. Carboxylesterases are often implicated in metabolic resistance to organophosphate insecticides and are sometimes involved in resistance to pyrethroids, but play practically no role at all in resistance to carbamates (Oppenoorth, 1985). Carboxylesterase mediated resistance will be discussed in more detail below in relation to organophosphate insecticides.

The cytochrome P-450-mediated monooxygenase (mixed-function oxidase) systems play an important role in the metabolism of all classes of insecticides except the cyclodienes, and the activity of these systems depends on one or more cytochrome P-450s, cytochrome *c* reductase and the concentration of NADPH (Nakatsugawa and Morelli, 1976). An increase in mixed-function oxidase activity is one of the most common mechanisms of resistance to a wide variety of different insecticides (for a review see Oppenoorth, 1985). Recent studies of the cytochrome P-450 system in *D. melanogaster* (Waters and Nix, 1988) mapped the gene(s) required for expression of cytochrome P-450-B, which is unique to certain strains, to chromosome 2R, at, or in the vicinity of, a major locus (*Rst*) conferring resistance to a broad range of insecticides (genetic map position II-65; Ogita, 1958). Waters and Nix (1988) showed that cytochrome P-450-B expression was related to insecticide resistance and therefore raised the possibility that cytochrome P-450-B might be the product of the major resistance locus on chromosome 2R. Maximum expression of cytochrome P-450-B was regulated by two loci, one on each arm of chromosome 3. (One of these regulatory loci is in the vicinity of and may be identical to a locus conferring resistance to nicotine sulphate and possibly phenylthiourea and phenylurea) (Ogita, 1958). The molecular mechanism underlying increased cytochrome P-450-B activity is unknown. Structural gene amplification or modification, or alterations in *cis*-acting regulatory sequences or genes encoding *trans*-acting regulatory proteins are all possibilities.

In the house fly, at least two genes controlling increased mixed-function oxidase activity are involved in insecticide resistance, and these have been mapped to chromosomes 2 and 5 (for a review see Oppenoorth, 1985). It is interesting to note that chromosome 5 in the house fly is postulated to be homologous to chromosome 2R in *D. melanogaster* (Foster *et al.*, 1981), the location of the gene required for cytochrome P-450-B expression (Waters and Nix, 1988). Recently, Feyereisen (1988, 1989) reported the isolation of cDNA clones of the genes encoding house fly cytochrome P-450 and P-450 reductase. The availability of these cloned probes will now allow a molecular analysis of mixed-function oxidase mediated insecticide resistance.

Glutathione S-transferases are involved in detoxication of organophosphates and perhaps DDT (see below) (Motoyama and Dauterman, 1980; Clark and Shamaan, 1984). Increased glutathione S-transferase activity has been implicated in conferring organophosphate resistance in several insect species, including the house fly and *Heliothis virescens*, although one or more other mechanisms are usually operating as well (for a review see Oppenoorth, 1985). At least three different glutathione S-transferases have been identified in the house fly, and each may be produced preferentially in different strains to produce different types of resistance (Clark *et al.*, 1984). It is not known whether these enzymes are encoded by different genes or by the same genetic locus. However, one gene (*g*) controlling increased glutathione S-transferase activity in resistant house flies has been mapped to chromosome 2 (Motoyama *et al.*, 1977). Whether increased glutathione S-transferase activity in resistant insects involves mutations in structural genes, regulatory loci or gene amplification is not known at present. However, molecular genetic analyses are now possible with the availability of glutathione S-transferase cloned probes from species such as rat and humans (Pickett *et al.*, 1984; Telakowski-Hopkins *et al.*, 1985; Board and Webb, 1987).

Also on chromosome 2 of the house fly is a gene controlling the activity of the enzyme, DDT-dehydrochlorinase. This soluble enzyme degrades DDT and its analogues into non-toxic products and plays an important role in DDT resistance in at least this species of insect (for review see Oppenoorth, 1985). There is strong evidence to suggest that DDT-dehydrochlorinase is actually a glutathione S-transferase (Clark and Shamaan, 1984).

The esterases apart (see below), it is thus clear that many genes that confer resistance to various insecticides by altering the activity levels of mixed-function oxidases, glutathione S-transferases, or DDT-dehydrochlorinase, map to chromosome 2 in the house fly. Are these separate loci which just happen to map to the same region of the genome? Or are they, as Plapp and Wang (1983) have suggested, simply different mutations of the one gene encoding a major regulatory protein? Specifically, Plapp and Wang (1983) have postulated that such a gene might encode a cytoplasmic receptor which binds inducers of a variety of different detoxifying enzymes. Plapp (1988) suggested that this cytoplasmic receptor may be a juvenile hormone receptor which recognizes and binds insecticides. These authors also cited as evidence for their hypothesis that mixed-function oxidases, glutathione S-transferases and DDT-dehydrochlorinase are all coordinately induced (Ottea and Plapp, 1981).

While it is entirely possible that mutations in genes encoding *trans*-acting regulatory proteins or receptors will lead to resistance, there is no evidence at present that mutations in a single gene can affect different biochemical mechanisms of resistance. The critical genetic experiments necessary to test the Plapp and Wang (1983) hypothesis have so far been frustrated by the segregation of both several different chromosome 2 inversions and of several different *combinations* of resistance mechanisms (for example, increased levels of DDT-dehydrochlorinase, mixed-function oxidases and glutathione S-transferases and reduced sensitivity of ACE) among the house fly strains available for study. However, the hypothesis may be amenable to critical testing by a molecular biological approach, given the availability of clones for ACE, glutathione S-transferase and cytochrome P-450 genes from various species.

3. RESISTANCE TO ORGANOPHOSPHATE INSECTICIDES

3.1. Mechanism of Action of Organophosphates

Figure 1 shows a schematic diagram of the mechanism of action of organophosphate insecticides. As esters, organophosphate insecticides are substrates for esterase enzymes. However, while organophosphates readily bind the catalytic site of esterases, their rates of hydrolysis are generally very slow. Indeed, it is by binding the catalytic site of acetyl

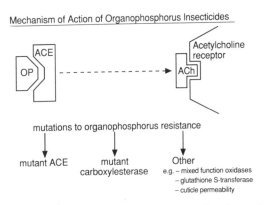

Figure 1. A schematic diagram of the mechanism of action of organophosphate insecticides. See text for details.

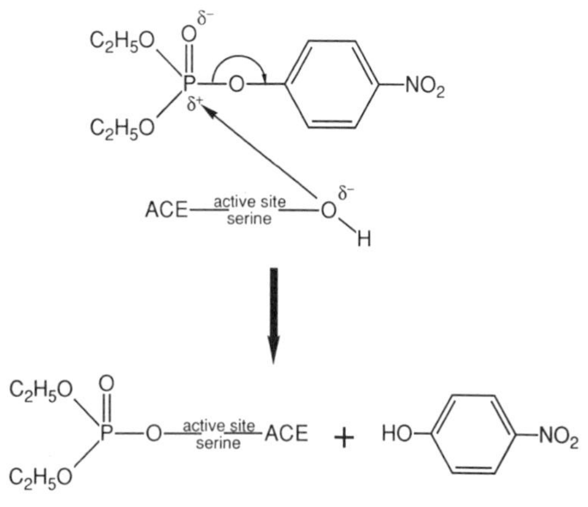

Figure 2. The reaction between acetylcholinesterase (ACE) and the organophosphate, paraoxon (Hutson and Roberts, 1985). See text for details.

cholinesterase that organophosphates exert their insecticidal effect; ACE is then prevented from carrying out its normal function, which is the rapid hydrolysis of acetylcholine, the essential neurotransmitter (for review see Hutson and Roberts, 1985).

Details of the reaction between ACE and a representative organophosphate, paraoxon, are shown in Figure 2. Inhibition of ACE occurs via phosphorylation of the serine hydroxyl group at the active site of the enzyme. It should be noted that phosphorothionates (P=S compounds) such as parathion and diazinon must be first bioactivated by oxidative desulphurization to produce the oxygen analogues, paraoxon and diazoxon, respectively, before the toxic interaction with ACE can occur (Hutson and Roberts, 1985).

About one third of the cases of organophosphate resistance that have been studied in insects involve mutant ACE, the target for these insecticides. A further third involve mutant carboxylesterases, while the remainder involve other mechanisms such as mixed-function oxidases, glutathione S-transferases and altered cuticle permeability (Georghiou, 1986) (Fig. 1).

3.2. Target Site Resistance: Mutations in Acetylcholinesterase

Organophosphate resistance due to alterations in the sensitivity of ACE to the insecticide has been well-documented for different insect and mite species, including the Australian cattle tick, house flies and *Drosophila* (for review see Oppenoorth, 1985). The rate of the reaction between the mutant ACE and the organophosphate inhibitor is reduced, while the ability of the enzyme to hydrolyze acetylcholine is either unimpaired or reduced but still sufficient to allow survival of the insect. Different ACE mutations have been noted across different resistant strains of several insect species (Oppenoorth, 1985).

The molecular basis of the ACE mutation has been investigated recently in *D. melanogaster* (Berge and Fournier, 1988). While the sensitive and resistant ACE enzymes could not be distinguished by physico-chemical parameters, DNA sequence analysis of the genes encoding both enzymes revealed, besides silent site substitutions, the replacement of a highly conserved phenylalanine by a tyrosine in the resistant enzyme. The availability of the *Drosophila* ACE probe will enable the cloning of ACE genes from resistant and susceptible strains of other insect species. DNA sequence comparisons will then provide valuable information concerning those alterations in the ACE enzyme which result in the inhibition of binding to organophosphate inhibitors without appreciable impairment of catalytic function.

3.3. Metabolic Resistance: Carboxylesterase Mutations

Mutations in genes encoding carboxylesterases can lead to increased detoxication of organophosphates, either by enhanced synthesis of an otherwise unaltered carboxylesterase or by alterations in the structure of the enzyme.

3.3.1. Enhanced Synthesis of a Structurally Unaltered Enzyme

Enhanced synthesis of an otherwise unaltered carboxylesterase can lead to detoxication of organophosphates in one of two ways (Devonshire, 1987). The large amounts of carboxylesterase produced in resistant insects may sequester the organophosphate and reduce the level of the insecticide to below the toxic dose. Alternatively, certain organophosphates, such as dimethoate, are actually hydrolyzed (albeit slowly) by the large amounts of carboxylesterase produced in resistant insects.

As yet no cases have been characterized in which enhanced synthesis of carboxylesterases results from mutations in *cis*-acting regulatory sequences or *trans*-acting regulatory proteins. However, organophosphate resistance resulting from carboxylesterase gene amplification has been well-characterized in *Culex* mosquitoes (Mouches *et al.*, 1986,1987) and the peach-potato aphid, *Myzus persicae* (Field *et al.*, 1988). In mosquitoes, esterases A1 (an α-carboxylesterase) and B1 (a ß-carboxylesterase) are over-produced by factors of at least 70 and 500-fold in resistant strains of *C. pipiens* and *C. quinquefasciatus*, respectively (Mouches *et al.*, 1987) [α- and ß-carboxylesterases have preferences for substrates such as α- and ß-naphthyl acetate (NA), respectively]. cDNA clones of esterase B1 have been isolated and used to investigate the molecular basis of its overproduction in resistant *C. quinquefasciatus* (Mouches *et al.*, 1986, 1988). Resistant strains possess approximately 250 copies of a constant 22 kilobase (kb) "core amplification unit" containing the esterase B1 gene of less than 3 kb. Interestingly, these authors noted the presence of two transposable elements in the amplification unit.

Organophosphate resistance in *Myzus persicae* is associated with an increase of esterase E4 production of up to 60-fold (Devonshire, 1977). Esterase E4 cDNA clones have been isolated and used to show that increased esterase synthesis by resistant aphids results from amplification of the esterase E4 structural gene (Field *et al.*, 1988). The degree of amplification correlates with the increase in esterase activity and the level of organophosphate resistance. The structure of the amplification unit has yet to be reported. It is interesting to note that esterases A1 and B1 in *Culex* mosquitoes and esterase E4 in the aphid are not immunologically cross-reactive (Mouches *et al.*, 1987).

3.3.2. Alterations in the Structure of the Enzyme

Mutations affecting the structure of a carboxylesterase can lead to enhanced hydrolysis of organophosphate insecticides, either by increased phosphotriester hydrolase activity or, in the case of malathion, hydrolysis of the α and β carboxylester groups (Kao *et al.*, 1984). Malathion has an unusual carboethoxy structure (Fig. 3) and its α and β carboxylester groups can be attacked rapidly by carboxylesterases (Oppenoorth, 1985).

Several house fly strains resistant to parathion, diazinon and related compounds, as well as at least two strains resistant to malathion, have very low carboxylesterase activities to esters such as methyl- and phenylbutyrate and α-NA (Van Asperen and Oppenoorth, 1959; Van Asperen, 1962). This low esterase activity mapped to a gene, *a*, on chromosome 2 (Franco and Oppenoorth, 1962). Backcrossing experiments involving a strain resistant to parathion-diazinon indicated that the low esterase activity was associated with increased phosphotriester hydrolase activity and resistance (Oppenoorth and Van Asperen, 1960). These observations led to the formulation of the "mutant ali-esterase hypothesis", which proposes that the mutant *a* gene encodes an esterase with increased organophosphate hydrolytic activity and concomitant decreased activity for α-NA and related esters. It was supposed that the malathion carboxylesterase activity in malathion resistant strains derived from esterase *a*, by a mutation different from that resulting in hydrolysis of parathion related compounds. Furthermore, the mutant enzymes present in strains resistant

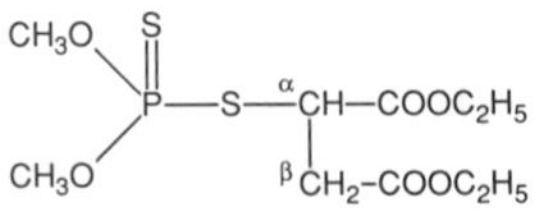

Figure 3. The chemical structure of malathion (from Oppenoorth, 1985).

to parathion-diazinon were shown to possess a very high affinity for the oxygen analogues of these compounds (Oppenoorth and Van Asperen, 1960).

Direct proof of the mutant ali-esterase hypothesis is lacking at present. Studies of phosphotriester hydrolase and carboxylesterase activities in resistant insects have been hampered by the presence of multiple biochemical mechanisms of resistance in the strains under study. For example, malathion resistance in the Hirokawa strain of the house fly is associated with altered ACE and high glutathione S-transferase activity, as well as high carboxylesterase activity (Motoyama et al., 1980). Unlike the mutant ali-esterase described above, this malathion carboxylesterase was associated with normal activity for α-NA. Indeed, malathion hydrolyzing carboxylesterases have been identified in malathion resistant strains of many different insects, such as *Culex tarsalis* and *Triatoma infestans* (Wood et al., 1985; Ziegler et al., 1987), and the biochemical properties of these enzymes vary in the different insects. For example, the malathion carboxylesterases from both *C. tarsalis* and *T. infestans* do not hydrolyze α-NA and the *C. tarsalis* enzyme is extremely unstable and is even inactivated by agents used routinely to polymerize acrylamide gels.

Clearly, detailed studies of the purified enzymes are required. Furthermore, cloning the genes encoding these enzymes from both resistant and susceptible insects would enable comparisons of DNA sequences and elucidation of possible active site modifications.

4. ORGANOPHOSPHATE RESISTANCE IN LUCILIA CUPRINA

We have seen that insect species have evolved resistance to particular insecticides by many different mechanisms. Many genes can mutate to confer resistance, which might account for its rapid evolution. However, in a given field population, only one or two of these mechanisms are generally operating, that is, only one or two loci are involved (Roush and McKenzie, 1987).

Two major-gene organophosphate resistance loci have been characterized in some detail in *Lucilia cuprina*: R_{OP-1}, which confers resistance to a broad range of organophosphates, and R_{MAL}, which confers resistance only to malathion and a small group of structurally related compounds such as phenthoate (Arnold and Whitten, 1976; Hughes, 1982; Hughes et al., 1984; Hughes and Raftos, 1985; Raftos and Hughes, 1986). Notwithstanding the difficulties of mapping genes of incomplete penetrance, allelism tests suggested that R_{OP-1} and R_{MAL} reside at separate, albeit closely-linked, loci on chromosome 4. This conclusion was corroborated by linkage analysis of each resistance locus to the flanking markers, *singed vibrissae* (*sv*) and *tangerine eye* (*tg*). The total evidence available, while insufficient for precise mapping, nevertheless suggested that R_{OP-1} and R_{MAL} could be separated by as much as two map units (Raftos and Hughes 1986) (Fig. 4). Biochemical and genetic evidence

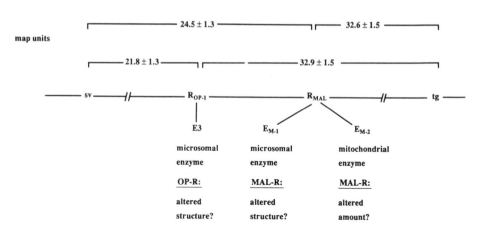

Figure 4. Proposed linkage map of the R_{OP-1}-R_{MAL} region of chromosome 4 of *L. cuprina*. The map distances (in map units) of the R_{OP-1} and R_{MAL} loci from the visible markers, *sv* (*singed vibrissae*) and *tg* (*tangerine eyes*), are indicated. Data are taken from Raftos and Hughes (1986). See text for details.

reviewed in detail below suggest that both of these resistance loci encode carboxylesterases (Hughes, 1982; Hughes and Devonshire, 1982; Hughes *et al.*, 1984; Hughes and Raftos, 1985; Raftos, 1986; Raftos and Hughes, 1986).

As with the house fly, increased mixed-function oxidase and glutathione S-transferase activities are also associated with organophosphate resistance in *L. cuprina* (Hughes and Devonshire, 1982; Terras *et al.*, 1983; Hughes and Raftos, 1985). However, these enzyme systems are only minor components of resistance. Mixed-function oxidase activity is associated with the rare resistance locus, R_{OP-2}, on chromosome 6 (Arnold and Whitten, 1976 ; Hughes and Devonshire, 1982; Terras *et al.*, 1983).

4.1. The R_{OP-1} Locus

There is strong evidence to suggest that the R_{OP-1} locus encodes the microsomal esterase, E3. Hughes and Raftos (1985) performed polyacrylamide gel electrophoresis on extracts from individual flies and stained the gels for esterase activity using a mixture of α- and ß–NA. Esterase E3, an α-carboxylesterase, was found consistently in both organophosphate susceptible and heterozygously resistant flies, but was absent from homozygously resistant flies. The difference between the staining and non-staining forms of E3 (E3+ and "E3 null") mapped to the same region on chromosome 4 as R_{OP-1}, and no recombinants were detected between the characters in 650 back-cross progeny. Furthermore, there was a close association between the frequency of E3 null and the proportion of flies resistant to organophosphate insecticides in field populations of *L. cuprina.*

These experiments were interpreted in terms of the mutant ali-esterase hypothesis formulated for organophosphate resistance in the house fly (Oppenoorth and Van Asperen, 1960). It was suggested that a mutant esterase E3 present in resistant flies is able to hydrolyze organophosphates more effectively than the E3 enzyme from susceptible flies, but has lost the ability to hydrolyze α- and ß-NA (Hughes and Raftos, 1985).

The small number of resistant *L. cuprina* strains studied by Hughes and Raftos (1985) presents some difficulty to the hypothesis of congruence between the R_{OP-1} and E3 null genes, given the finding that multiple resistance factors commonly distinguish different house fly strains (Motoyama *et al.,* 1980). In particular, the problems with the genetic analyses caused by incomplete penetrance of R_{OP-1} do not allow rejection of the alternate hypothesis that R_{OP-1} and E3 null are non-

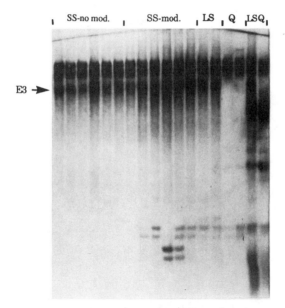

Figure 5. Polyacrylamide gel electrophoresis of *L. cuprina* individual adults. Sugar-fed adults were homogenized in extraction buffer (0.7 ml for females or 1 ml for males) containing 0.5% v/v Triton X-100 to solublize membranes (Hughes and Raftos, 1985), and 5µl aliquots were applied to 7% native polyacrylamide gels (see Hughes and Raftos, 1985, for details of the gel system used). After electro0phoresis for 3 hours at 4°C, gels were stained for esterase activity using α- and ß-NA. Lanes 1-6 and 7-12 contain extracts of two diazinon susceptible strains, SS-no modifier and SS-modifier (obtained from J. McKenzie, Melbourne University), respectively; lanes 13-14 and 15-16 contain extracts from the LS (susceptible) and Q (resistant) strains (obtained from P. Hughes, Biological and Chemical Research Institute, Department of Agriculture, N.S.W.), respectively; lanes 17 and 18 contain extracts of LS and Q individuals, respectively, which had been fed liver rather than sugar prior to electrophoresis.

allelic but closely linked loci in strong disequilibrium in the few strains studied. For this reason we have extended the study of Hughes and Raftos (1985) to include several independently isolated resistant and susceptible strains.

We have surveyed 10-30 individuals from each of three susceptible and three resistant strains of *L. cuprina* on native polyacrylamide gels stained for esterase activity using α- and ß-NA. The samples of the three susceptible strains, LS (Hughes and Raftos, 1985), SS-no modifier and SS-modifier (McKenzie and Game, 1987) were all fixed for the E3+ phenotype, as would be predicted from the Hughes and Raftos hypothesis (see Fig. 5 for an illustrative gel). Furthermore, samples of the resistant strain, Q (Hughes and Raftos, 1985), were fixed for the E3 null phenotype, again in agreement with the Hughes and Raftos hypothesis. However, individuals from resistant strains RX and RY (obtained from J. McKenzie, Melbourne University), expected to be homozygous for the R_{OP-1} locus, were found to segregate for E3+ and E3 null (Fig. 6A and B). Approximately 70% of strain RX and 12% of strain RY individuals had an E3+ phenotype which, according to Hughes and Raftos (1985), would either suggest that R_{OP-1} and E3 are not congruent, or that diazinon susceptible or heterozygously resistant flies segregate in these strains.

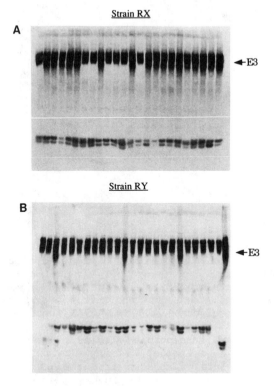

Figure 6. Polyacrylamide gel electrophoresis of *L. cuprina* individual adults of strains RX (panel A) and RY (panel B). See the legend to Figure 5 for details.

RX or RY ♂   X   (i) LS (susceptible) ♀
                            (ii) Q (resistant) ♀

PAGE on all ♂'s and score for the presence of E3

Test progeny of matings to LS for *resistance to diazinon*

PAGE on progeny of matings to Q to distinguish between E3 +/- and +/+ parental ♂'s

Figure 7. The experiment designed to test the concordance between the E3 and R_{OP-1} loci in *L. cuprina* strains, RX and RY. See text for details. PAGE refers to polyacrylamide gel electrophoresis.

To resolve these possibilities, the experiment outlined in Figure 7 was performed. Individual males of the RX or RY strains were mated to females of both the LS and Q strains, and then scored for the presence of E3 by polyacrylamide gel electrophoresis. Polyacrylamide gel electrophoresis was performed on progeny of matings to Q to distinguish between E3+ homozygotes and heterozygotes among the original parental males (all progeny from the homozygotes would have the E3+ phenotype, whereas half from the E3+ heterozygotes would have the E3 null phenotype). Progeny of each of the matings to LS were tested for resistance to diazinon at a dose known to kill homozygous susceptible individuals (McKenzie *et al.*, 1980). Survival of all the progeny from a given cross would indicate that the parental male had the RR genotype; survival of approximately half would indicate that he had the RS genotype; the survival of no progeny would indicate that he had the SS genotype.

The results of this experiment are given in Table I. In summary, the two males assigned the SS genotype were independently scored as E3+ homozygotes; the six males assigned the RS genotype were E3+/E3 null heterozygotes; the 24 males assigned the RR genotype were all E3 null homozygotes. Thus, these results indicate not only that strains RX and RY were indeed segregating for both the E3 and diazinon resistance phenotypes, but also that the R_{OP-1} and E3 genotypes are completely concordant. However, it remains a formal possibility that the concordance in these population samples simply reflects strong gametic disequilibrium between two distinct but tightly linked loci. Direct proof of the hypothesis that this non-staining E3 esterase enzyme is causally involved in resistance will await data to show that it can hydrolyze organophosphates more effectively than the enzyme from susceptible flies.

4.2. The R_{MAL} Locus

There is also strong evidence that R_{MAL} encodes a carboxylesterase (E_M) (Hughes *et al.*, 1984; Raftos, 1986; Raftos and Hughes, 1986). Crude extracts prepared from a malathion

Table I

Results of the progeny-testing experiment to test the concordance between
E3 and diazinon resistance in strains RX and RY (Fig. 7)

Parental male	Diazinon treatment[a] of progeny from mating to LS		R_{OP-1} genotype of parental male[b]	E3 genotype of parental male[c]
	No. tested	No. survivors		
Strain RX:				
1	20	20	RR	null/null
3	21	11	RS	null/+
7	3	3	?RR	null/null
8	20	20	RR	null/null
12	20	20	RR	null/null
13	20	10	RS	null/+
14	12	12	RR	null/null
20	20	5	RS	null/+
21	20	13	RS	null/+
22	20	–	SS	+/+
24	19	2	SS	+/+
Strain RY:				
1	20	20	RR	null/null
2	20	20	RR	null/null
3	20	10	RS	null/+
4	20	20	RR	null/null
5	20	20	RR	null/null
7	17	17	RR	null/null
9	20	20	RR	null/null
10	17	17	RR	null/null
11	20	20	RR	null/null
12	20	10	RS	null/+
13	20	20	RR	null/null
14	8	8	?RR	null/null
15	20	19	RR	null/null
16	22	22	RR	null/null
17	20	20	RR	null/null
18	20	20	RR	null/null
20	20	20	RR	null/null
21	18	18	RR	null/null
22	20	19	RR	null/null
23	20	20	RR	null/null
24	5	5	?RR	null/null

[a]Approximately 0.5µl diazinon (0.01% v/v in deodorized kerosine) was applied topically to the thoraces of males and survival was scored after 24 hours at 27°C (McKenzie *et al.*, 1980).

[b]R = resistant, S = susceptible, ? = uncertain assigment due to the low number of progeny available for testing.

[c]No further progeny testing was carried out if the parental male was scored as E3 null/E3 null. Otherwise, at least 8 progeny from matings to E3 null/E3 null ♀ females were typed in order to distinguish between an E3+/E3+ homozygous and E null/E3+ heterozygous parent.

resistant strain of *L. cuprina* (strain RM) showed both a three-fold increase in the rate of hydrolysis of malathion and a three-fold higher carboxylesterase activity for the substrate, 4-nitrophenylacetate. Interestingly, little difference was found between the LS and RM extracts in their activities for α-NA. No recombinants were detected between R_{MAL} and a single locus controlling this high carboxylesterase activity for 4-nitrophenylacetate. Moreover, the esterase inhibitor, triphenyl phosphate, decreased the degradation of malathion *in vitro*, and eliminated malathion resistance *in vivo*.

Raftos (1986) found enhanced malathion hydrolyzing activity in both the mitochondrial and microsomal fractions of resistant flies. The microsomal preparation from the RM strain had a five-fold higher K_m than that from the LS strain, whereas the mitochondrial preparations from both strains had the same K_m. Both fractions had increased carboxylesterase activities. The interpretation of these results was that the mitochondrial fraction of resistant flies contains increased amounts of an otherwise unaltered esterase, whereas the microsomal fraction contains an altered esterase activity. The relative importance of each of these enzymes to the overall resistance phenotype is unknown. If both are involved then the results might imply the presence of two tightly-linked esterase genes involved in malathion resistance. If they are not then it remains to be determined which gene is involved in malathion resistance and where the other gene maps.

4.3. Molecular Analysis of the R_{OP-1} and R_{MAL} Loci

It can be seen from the above discussion that chromosome 4 of *L. cuprina* possibly contains a cluster of three (or more) esterase genes, all of which are involved in organophosphate resistance, albeit in different ways (Fig. 4). It is thought that the R_{OP-1} locus encodes a microsomal esterase, E3, mutations in the coding sequences of which would lead to increased hydrolysis of a broad range of organophosphates. The R_{MAL} locus might in fact comprise two genes, one (E_{M-1}) encoding a microsomal and the other (E_{M-2}) a mitochondrial esterase. Mutations in the structural region of the E_{M-1} gene might lead to increased hydrolysis of malathion and, consequently, malathion resistance. Alternatively, a regulatory mutation or gene amplifaction might lead to the over-production of an otherwise unaltered E_{M-2} esterase and thus, malathion resistance. P. Smith and J. Lai Fook (personal communication) have recently mapped a locus encoding electrophoretic variants of an hemolymph α-carboxylesterase to chromosome 4. Its linkage relationships with the R_{OP-1}-R_{MAL} cluster are not yet known but are presently under investigation.

Clustering of esterase genes involved in organophosphate resistance has also been observed on chromosome 2 in the house fly, the homologue of chromosome 4 in *L. cuprina* (Foster *et al.*, 1981). In fact, many of the esterase bands identified by Ogita and Kasai (1965) among 10 strains of house fly were controlled by loci on the second chromosome. As noted earlier, ali-esterase *a* , which in certain strains is associated with low esterase activity for α-NA and increased organophosphate hydrolysis, has been mapped to chromosome 2 (Van Asperen and Oppenoorth, 1959; Franco and Oppenoorth, 1962; Van Asperen, 1962). Moreover, Sawicki *et al.* (1984) found that organophosphate resistance in field strains of the house fly was associated with the presence of an α-NA hydrolyzing esterase, $E_{0.39}$,

which also mapped to chromosome 2. This esterase, which was
not found in susceptible strains, is apparently not allelic to
ali-esterase *a*. Malathion carboxylesterase activity associated
with normal α-NA activity and malathion resistance was found
in the nuclear, mitochondrial and microsomal fractions of the
Hirokawa strain of the housefly (Motoyama *et al.*, 1980;
Picollo de Villar *et al.*, 1983). However, it is not known
whether the same carboxylesterase is present in each of these
subcellular fractions, or where these activities map.

Our laboratory is undertaking a molecular analysis of
organophosphate resistance in *L. cuprina*. Our approach is
three-fold. First, we plan to isolate an organophosphate
hydrolyzing E3 enzyme from parathion-diazinon resistant flies,
since direct proof of the congruence between the R_{OP-1} and E3
null loci is lacking at present. Second, we will attempt to
purify the E3 and E_{M-1}-E_{M-2} proteins from organophosphate susceptible and malathion resistant flies, respectively, in order to
obtain protein sequence data to aid our cloning studies.
Third, we aim to clone the E3 and E_{M-1}/E_{M-2} genes from both susceptible and resistant strains of *L. cuprina*.

One aim of these molecular analyses is to describe the
organization of the cluster of structural genes encoding
esterases implicated in organophosphate resistance. In particular, we wish to know how many loci are involved, how closely
linked they are, whether they are members of the serine
esterase multi-gene family and whether other esterases not
implicated in organophosphate resistance reside in the region.
A second general aim is to determine the nature of the mutations which confer resistance in the relevant esterase genes.
Against the background of the Plapp and Wang (1983) and mutant
ali-esterase (Oppenoorth and Van Asperen, 1960) hypotheses, we
wish to know whether the mutations affect the structure of the
enzymes or the regulation of esterase gene expression. If, as
seems likely for E3 and E_{M-1}, the structure of the enzyme is
altered, we wish to know how the mutations relate to the body
of knowledge from other serine esterases (Myers *et al.*, 1988)
concerning the sequence-structure-function relationships of
the enzyme domains involved in substrate binding and the catalytic mechanism. Answers to questions such as these will provide a unique picture of the molecular basis of a major adaptive evolutionary change.

One of our strategies for cloning the chromosome 4 esterase
genes of *L. cuprina* is to clone first the *Est C* gene of *D.
melanogaster*. *Est C* has been mapped cytogenetically to polytene chromosome position 84D3-5 on chromosome 3R (Cavener *et
al.*, 1986), the homologous chromosome to chromosome 4 of *L.
cuprina* (Foster *et al.*, 1981). Like E3, the EST C protein is
an α-carboxylesterase (Triantaphyllidis and Christodoulou,
1973). It is likely, therefore, that the *Est C* clone would
provide a probe for isolating genes from the E3-E_M cluster of
L. cuprina.

Since *Est C* is the major soluble esterase present in
embryos of *D. melanogaster* (Healy, Dumancic and Oakeshott, in
preparation), a 3-12 hour embryo cDNA library (obtained from

T. Kornberg, University of California, San Francisco) has been screened at low stringency with the peach aphid E4 α-carboxylesterase cDNA clone (obtained from A. Devonshire, Rothamsted Experimental Station, U.K.). Several clones ranging in size from 1.0 to 1.8 kb with varying degrees of homology have been obtained. Preliminary *in situ* hybridization analyses of two of the more strongly hybridizing clones have revealed that these cDNAs, while not *Est C*, are possibly esterase clones. One clone hybridized *in situ* to polytene chromosome position 87E-88A, the cytological site of ACE (Hall and Spierer, 1986); the second hybridized to position 64A-B, a site of hybridization to oligonucleotide probes corresponding to regions of a ß-carboxylesterase gene of *D. melanogaster*, *Est 6* (J. Oakeshott, R. Richmond and R. Phyllis, unpublished). Analysis of the remaining cDNA clones is in progress.

A second approach being undertaken to clone the *Est C* gene of *D. melanogaster* is a chromosome walk from the cloned *rotund (rn)* gene (obtained from R. Griffin-Shea and S. Kerridge, C.N.R.S., France), which occupies the complementation group adjacent to *Est C* at 84D (Cavener *et al.*, 1986). This region of the chromosome has been saturated for visible and lethal complementation groups. To date overlapping clones spanning approximately 45 kb of genomic DNA have been isolated in the direction of *Est C*. Low stringency Southern blot hybridization of this DNA using the peach aphid E4 and *D. melanogaster Est 6* cDNA clones as probes has failed to detect any esterase genes. Continuation of the chromosome walk and *in situ* hybridization analysis of the extreme ends of the cloned region are in progress.

Once identified, the *Est C* clone will be used to isolate clones from the E3/E_M cluster of *L. cuprina,* using genomic and cDNA libraries of both resistant and susceptible flies. Alternative cloning strategies such as direct screening of *L. cuprina* cDNA libraries with the available peach aphid, mosquito and *D. melanogaster* esterase clones are also being investigated. Identification of the E3 and E_M clones will be made by comparison of the DNA sequences with tryptic peptide sequence data obtained from the purified proteins. It is hoped that a comparison of the DNA sequences of both the mutant and susceptible enzymes will shed some light on the organophosphate and malathion hydrolyzing activities of the mutant enzymes.

5. SUMMARY

About 450 species of insects and mites have evolved major gene resistance to chemical insecticides over the last 40 years. The mechanism of resistance generally involves a modification to the target site for the chemical or increased detoxication of the insecticide by metabolic enzymes like carboxylesterases, mixed-function oxidases or glutathione-S-transferases.

Organophosphate resistance in *Lucilia cuprina* involves at least two and possibly three tightly linked genes on chromosome 4. One gene, R_{OP-1}, confers resistance to a broad range of organophosphates; another, R_{MAL}, confers resistance only to malathion. Biochemical and genetic evidence suggests R_{OP-1} encodes a mutant form of a microsomal enzyme, the α-carboxylesterase, E3. It is hypothesised that the mutation alters the structure of E3 such that it acquires the ability to degrade organophosphates while losing its ability to hydrolyse other esters like α-naphthyl acetate. Limited genetic data suggest that R_{MAL} resides at a separate locus from R_{OP-1} but is no more, and probably substantially less than, 2 map units away. The R_{MAL} mutation is correlated with changes in two carboxylesterases, one a microsomal enzyme which shows increased affinity for malathion as a substrate, and the other a mitochondrial enzyme which has unaltered affinity for malathion but is present in greater amounts in susceptible flies. Either or both of these changes could be causally connected with the R_{MAL} lesion and the malathion resistance phenotype.

Experiments are in progress to clone and characterize the cluster of carboxylesterases on chromosome 4 from various R_{OP-1} and R_{MAL} genotypes. These experiments will elucidate the number and nature of the molecular changes involved in the evolution of organophosphate resistance in *L. cuprina*.

ACKNOWLEDGEMENTS. The authors would like to thank P. Hughes and J. McKenzie for supplying *L. cuprina* strains, A. Devonshire, R. Griffin-Shea, S. Kerridge and C. Mouches for making various cloned DNAs available to us, and T. Kornberg for the *D. melanogaster* embryo cDNA library. We also thank A. Game for her assistance with the progeny testing experiment. We are grateful to A. Game, P. Hughes, J. McKenzie, A. Parker, R. Roush and M. Whitten for their critical reading of the manuscript.

References

Arnold, J. T. A., and Whitten, M. J., 1976, The genetic basis for organophosphorus resistance in the Australian sheep blowfly, *Lucilia cuprina* (Wiedemann) (Diptera, Calliphoridae), *Bull. ent. Res.* 66:561-568.

Berge, J. B., and Fournier, D., 1988, Advances in molecular genetics of acetylcholinesterase insensitivity in insecticide - resistant insects, Abstract, *XVIII International Congress of Entomology*, Vancouver, July 3-9, p. 461.

Board, P. G., and Webb, G. C., 1987, Isolation of a cDNA clone and localization of human glutathione S-transferase 2 genes to chromosome band 6p12, *Proc. natn. Acad. Sci. USA* 84:2377-2381.

Cavener, D. R., Otteson, D. C., and Kaufman, T. C., 1986, A rehabilitation of the genetic map of the 84B-D region in *Drosophila melanogaster*, *Genetics* 114:111-123.

Clark, A. G., and Shamaan, N. A., 1984, Evidence that DDT-dehydrochlorinase from the house fly is a glutathione S-transferase, *Pestic. Biochem. & Physiol.* 22:249-261.

Clark, A. G., Shamaan, N. A., Dauterman, W. C., and Hayoaka, T., 1984, Characterization of multiple glutathione transferases from the housefly, *Musca domestica* (L), *Pestic. Biochem. & Physiol.* 22:51-59.

Devonshire, A. L., 1977, The properties of a carboxylesterase from the peach-potato aphid, *Myzus persicae* (Sulz.), and its role in conferring insecticide resistance, *Biochem. J.* **167**:675-683.

Devonshire, A. L., 1987, Biochemical studies of organophosphorus and carbamate resistance in house flies and aphids, in: *Combating Resistance to Xenobiotics : Biological and Chemical Approaches* (M. G. Ford, D. W. Holloman, B. P. S. Khambay, and R. M. Sawicki, eds), Weinheim, VCH, Chichester, Ellis, Horwood, pp. 239-255.

Feyereisen, R., 1988, Isolation and sequence of cDNA clones for cytochrome P-450 from an insecticide-resistant strain of *Musca domestica*, Abstract, *XVIII International Congress of Entomology*, Vancouver, July 3-9, p. 465.

Feyereisen, R., Koener, J. F., Farnsworth, D. E., and Nebert, D. W., 1989, Isolation and sequence of cDNA encoding a cytochrome P-450 from an insecticide-resistant strain of the house fly, *Musca domestica, Proc. natn. Acad. Sci. USA* **86**: 1465-1469.

Field, L. M., Devonshire, A. L., and Forde, B. G., 1988, Molecular evidence that insecticide resistance in peach-potato aphids (*Myzus persicae* Sulz.) results from amplification of an esterase gene, *Biochem. J.* **251**:309-312.

Foster, G. G., Whitten, M. J., Konovalov, C., Arnold, J. T. A., and Maffi, G., 1981, Autosomal genetic maps of the Australian sheep blowfly, *Lucilia cuprina dorsalis* R.-D. (Diptera : Calliphoridae), and possible correlations with the linkage maps of *Musca domestica* L., and *Drosophila melanogaster* (Mg.), *Genet. Res.* **37**:55-69.

Franco, M. G., and Oppenoorth, F. J., 1962, Genetical experiments on the gene for low ali-esterase activity and organophosphate resistance in *Musca domestica* L., *Entomol. exp. Appl.* **5**:119-123.

Georghiou, G. P., 1986, The magnitude of the resistance problem, in: *Pesticide Resistance : Strategies and Tactics for Management*, National Academy of Sciences, Washington, D.C., pp. 14-43.

Grubs, R. E., Adams, P. M., and Soderlund, D. M., 1988, Binding of [3H] saxitoxin to head membrane preparations from susceptible and *knockdown-resistant* house flies, *Pestic. Biochem. & Physiol.* **32**: 217-223.

Hall, L. M. C., and Spierer, P., 1986, The Ace locus of *Drosophila melanogaster*: structural gene for acetylcholinesterase with an unusual 5' leader, *EMBO J.* **5**:2949-2954.

Hama, H., 1983, Resistance to insecticides due to reduced sensitivity of acetylcholinesterase, in: *Pest Resistance to Pesticides* (G. P. Georghiou, and T. Saito, eds), Plenum, New York, pp. 299-331.

Hoyer, R. F., and Plapp, F. W., 1968, Insecticide resistance in the house fly: identification of a gene that confers resistance to organotin insecticides and acts as an intensifier of parathion resistance, *J. econ. Ent.* **61**:1269 -1276.

Hughes, P. B., 1982, Organophosphorus resistance in the sheep blowfly, *Lucilia cuprina* (Wiedemann) (Diptera : Calliphoridae) : a genetic study incorporating synergists, *Bull. ent. Res.* **72**:573-582.

Hughes, P. B., and Devonshire, A. L., 1982, The biochemical basis of resistance to organophosphorus insecticides in the sheep blowfly, *Lucilia cuprina, Pestic. Biochem. & Physiol.* **18**:289-297.

Hughes, P. B., Green, P. E., and Reichmann, K. G., 1984, Specific resistance to malathion in laboratory and field populations of the Australian sheep blowfly, *Lucilia cuprina* (Diptera : Calliphoridae), *J. econ. Ent.* **77**:1400-1404.

Hughes, P. B., and Raftos, D. A. 1985, Genetics of an esterase associated with resistance to organophosphorus insecticides in the sheep blowfly, *Lucilia cuprina* (Wiedemann) (Diptera : Calliphoridae), *Bull. ent. Res.* **75**:535-544.

Hutson, D. H., and Roberts, T. R., 1985, Insecticides, in: *Insecticides* (D. H. Hutson, and T. R. Roberts, eds), John Wiley and Sons Ltd, New York, pp. 1-34.

Kao, L. R., Motoyama, N., and Dauterman, W. C., 1984, Studies on hydrolases in various house fly strains and their role in malathion resistance, *Pestic. Biochem. & Physiol.* 22:86-92.

Kasbekar, D. P., and Hall, L. M., 1988, A *Drosophila* mutation that reduces sodium channel number confers resistance to pyrethroid insecticides, *Pestic. Biochem. & Physiol.* 32:135-145.

Lund, A. E., 1984, Pyrethroid modification of sodium channel: current concepts, *Pestic. Biochem. & Physiol.* 22:161-168.

McKenzie, J.A., and Game, A.Y., 1987, Diazinon resistance in *Lucilia cuprina*; mapping of a fitness modifier, *Heredity* 59:371-381.

McKenzie, J. A., Dearn, J. M., and Whitten, M. J., 1980, Genetic basis of resistance to diazinon in Victorian populations of the Australian sheep blow fly, *Lucilia cuprina*, *Aust. J. biol. Sci.* 33:85-95.

Motoyama, N., and Dauterman, W. C., 1980, Glutathione S - transferases : their role in the metabolism of organophosphorus insecticides, *Rev. Biochem. Toxicol.* 2: 49-69.

Motoyama, N., Dauterman, W. C., and Plapp, F. W., 1977, Genetic studies on glutathione-dependent reactions in resistant strains of the house fly, *Musca domestica* L., *Pestic. Biochem. & Physiol.* 7:443-450.

Motoyama, N., Hayaoka, T., Nomura, K., and Dauterman, W. C., 1980, Multiple factors for organophosphorus resistance in the housefly, *Musca domestica* L., *J. Pesticide Sci. (Noyaku Kagaku Kenkyukai)* 5:393-402.

Mouches, C., Pasteur, N., Berge, J. B., Hyrien, O., Raymond, M., De Saint Vincent, B. R., De Silvestri, M., and Georghiou, G. P., 1986, Amplification of an esterase gene is responsible for insecticide resistance in a California *Culex* mosquito, *Science* 233:778-780.

Mouches, C., Magnin, M., Berge, J. B., De Silvestri, M. Beyssat, V., Pasteur, N., and Georghiou G. P., 1987, Overproduction of detoxifying esterases in organophosphate resistant *Culex* mosquitoes and their presence in other insects, *Proc. natn. Acad. Sci. USA* 84:2113-2116.

Mouches, C., Pasteur, N., Lemieux, L., Poplin, Y., Abadon, M., and Georghiou, G. P., 1988, Advances in molecular genetics of organophosphate - detoxifying esterases in insects, Abstract, *XVIII International Congress of Entomology*, Vancouver, July 3-9, p. 460.

Myers, M., Richmond, R. C., and Oakeshott, J. G., 1988, On the origin of esterases, *Mol. Biol. Evol.* 5:113-119.

Nakatsugawa, T., and Morelli, M. A., 1976, Microsomal oxidation and insecticide metabolism, in: *Insecticide Biochemistry and Physiology* (C. F. Wilkinson, ed.), Plenum, New York, pp. 61-114.

Narahashi, T., 1983, Resistance to insecticides due to reduced sensitivity of the nervous system, in: *Pest Resistance to Pesticides* (G. P. Georghiou, and T. Saito, eds), Plenum, New York, pp. 333-352.

Ogita, Z., 1958, The genetical relation between resistance to insecticides in general and that to phenylthiourea (PTU) and phenylurea (PU) in *Drosophila melanogaster*, *Botyu-kagaku* 23:188-205.

Ogita, Z., and Kasai, T., 1965, Genetic control of multiple *esterases* in *Musca domestica*, *Jap. J. Genet.* 40:1-14.

Oppenoorth, F. J., 1985, Biochemistry and genetics of insecticide resistance, in: *Comprehensive Insect Physiology, Biochemistry and Pharmacology, vol. 12, Insect Control* (G. A. Kerkut, and L. I. Gilbert, eds), Pergamon, London, pp. 731-770.

Oppenoorth, F. J., and Van Asperen, K., 1960, Allelic genes in the house fly producing modified enzymes that cause organophosphate resistance, *Science* 132:298-299.

Ottea, J. A., and Plapp, F. W., 1981, Induction of glutathione S. aryl transferase by phenobarbital in the house fly, *Pestic. Biochem. & Physiol.* **15**:10-13.

Pauron, D., Barhanin, J., Amichot, M., Pralavorio, M., Berge, J. B., and Lazdunski, M., 1989, Pyrethroid receptor in the insect Na$^+$ channel: alteration of its properties in pyrethroid-resistant flies, *Biochemistry* **28**: 1673-1677.

Pickett, C. B., Telakowski-Hopkins, C. A., Ding, G. J.-F., Argenbright, L., and Lu, A. Y. H., 1984, Rat liver glutathione S-transferases. Complete nucleotide sequence of a glutathione S-transferase mRNA and the regulation of the Ya, Yb, and Yc mRNAs by 3-methylcholanthrene and phenobarbital, *J. biol. Chem.* **259**:5182-5188.

Picollo de Villar, M. I., Van Der Pas, L. J. T., Swissaert, H. R., and Oppenoorth, F. J., 1983, An unusual type of malathion-carboxylesterase in a Japanese strain of house fly, *Pestic. Biochem. & Physiol.* **19**:60-65.

Plapp, F. W., 1988, Major role for a regulatory gene in metabolic resistance to insecticides in the house fly *Musca domestica* L. (Diptera: Muscidae), Abstract, *XVIII International Congress of Entomology*, Vancouver, July 3-9, p. 460.

Plapp, F. W., and Wang, T. C., 1983, Genetic origins of insecticide resistance, in: *Pest Resistance to Pesticides* (G. P. Georghiou, and T. Saito, eds), Plenum, New York, pp. 47-70.

Raftos, D. A., 1986, The biochemical basis of malathion resistance in the sheep blowfly, *Lucilia cuprina, Pestic. Biochem. & Physiol.* **26**:302-309.

Raftos, D. A., and Hughes, P. B., 1986, Genetic basis of a specific resistance to malathion in the Australian sheep blow fly, *Lucilia cuprina* (Diptera : Calliphoridae), *J. econ. Ent.* **79**:553-557.

Rossignol, D. P., 1988, Reduction in number of nerve membrane sodium channels in pyrethroid resistant house flies, *Pestic. Biochem. & Physiol.* **32**:146-152.

Roush, R. T., and McKenzie, J. A., 1987, Ecological genetics of insecticide and acaricide resistance, *A. Rev. Ent.* **32**:361-380.

Sattelle, D. B., Leech, C. A., Lummis, S. C. R., Harrison, B. J., Robinson, H. P. C., Moores, G. D., and Devonshire, A. L., 1988, Ion channel properties of insects susceptible and resistant to insecticides, in: *Neurotox '88: Molecular Basis of Drug and Pesticide Action* (G. G. Lunt, ed.), Elsevier, Amsterdam, pp. 563-582.

Sawicki, R. W., 1985, Resistance to pyrethroid insecticides in arthropods, in: *Insecticides* (D. H. Hutson, and T. R. Roberts, eds), Wiley, New York, pp. 143-192.

Sawicki, R. M., and Farnham, A. W., 1968a, Genetics of resistance to insecticides of the SKA strain of *Musca domestica* III. Location and isolation of the factors of resistance to dieldrin, *Entomol. exp. Appl.* **11**:133-142.

Sawicki, R. M., and Farnham, A. W., 1968b, Examination of the isolated autosomes of the SKA strain of house flies (*Musca domestica* L) for resistance to several insecticides with and without pretreatment with sesamex and TBTP, *Bull. ent. Res.* **59**:409-421.

Sawicki, R. M., Devonshire, A. L., Farnham, A. W., O'Dell, K. E., Moores, G. D., and Denholm, I., 1984, Factors affecting resistance to insecticides in house-flies, *Musca domestica* L. (Diptera : Muscidae). II. Close linkage on autosome 2 between an esterase and resistance to trichlorphon and pyrethroids, *Bull. ent. Res.* **74**:197-206.

Soderlund, D. M., and Bloomquist, J. R., Molecular mechanisms of insecticide resistance, in: *Pesticide Resistance in Arthropods* (R. T. Roush, and B. E. Tabashnik, eds), Chapman and Hall, New York, (in press).

Telakowski-Hopkins, C. A., Rodkey, J. A., Bennet, C. D., Lu, A. Y. H., and
    Pickett, C. B., 1985, Rat liver glutathione S-transferases.
    Construction of a cDNA clone complementary to a Yc mRNA and prediction
    of the complete amino acid sequence of a Yc subunit, *J. biol. Chem.*
    260:5820-5825.
Terras, M. A., Rose, H. A., and Hughes, P. B., 1983, Aldrin epoxidase
    activity in larvae of a susceptible and a resistant strain of the sheep
    blowfly, *Lucila cuprina* (Wiedemann), *J. Aust. Entomol. Soc.* **22**:256.
Triantanphyllidis, C. D., and Christodoulou, C., 1973, Studies of a
    homologous gene-enzyme system, *Esterase C*, in *Drosophila melanogaster*
    and *Drosophila simulans*, *Biochem. Genet.* **8**:383-390.
Tsukamoto, M., 1983, Methods of genetic analysis of insecticide resistance,
    in: *Pest Resistance to Pesticides* (G. P. Georghiou, and T. Saito, eds),
    Plenum, New York, pp. 71-98.
Van Asperen, K., 1962, A study of house fly esterases by means of a
    sensitive colorimetric method, *J. Insect Physiol.* **8**:401-416.
Van Asperen, K., and Oppenoorth, F. J., 1959, Organophosphate resistance
    and esterase activity in house flies, *Entomol. exp. Appl.*, **2**:48-57.
Waters, L. C., and Nix, C. E., 1988, Regulation of insecticide resistance -
    related cytochrome P-450 expression in *Drosophila melanogaster*, *Pestic.
    Biochem. & Physiol.* **30**:214-227.
Wood, E. J., De Villar, M. I. P., and Zerba, E. N., 1985, Role of a
    microsomal carboxylesterase in reducing the action of malathion in eggs
    of *Triatoma infestans*, *Pestic. Biochem. & Physiol.* **23**:24-32.
Wu, C.-F., Ganetzky, B., Jan, L. Y., Jan, Y.-N., and Benzer, S., 1978, A
    Drosophila mutant with a temperature-sensitive block in nerve
    conduction, *Proc. natn. Acad. Sci. USA* **75**:4047-4051.
Yamamoto, D., Quandt, F. N., and Narahashi, T., 1983, Modification of
    single sodium channels by the insecticide tetramethrin, *Brain Res.*
    274:344-349.
Ziegler, R., Whyard, S., Downe, A. E. R., Wyatt, G. R., and Walker, V. K.,
    1987, General esterase, malathion carboxylesterase, and malathion
    resistance in *Culex tarsalis*, *Pestic. Biochem. & Physiol.* **28**:279-285.

CHAPTER 17

Temperature Effects on Epicuticular Hydrocarbons and Sexual Isolation in Drosophila mojavensis

THERESE ANN MARKOW, AND ERIC C. TOOLSON

1. INTRODUCTION

Drosophila mojavensis is found in several geographically separate regions of the Sonoran Desert of North America. Within each area, it utilizes different host plants as its primary breeding site. When flies from two of these regions, Baja California and Sonora (including southern Arizona) are raised in the laboratory and then placed together to mate, a significant degree of sexual isolation is observed (Zouros and D'Entremont, 1980; Koepfer, 1987a, b). This isolation is due largely to a failure of Baja males to mate with Sonora females. The factors responsible for the observed isolation have not yet been identified, but they have been demonstrated to have a genetic component by the selection experiments of Koepfer (1987a, b), which makes *D. mojavensis* a particularly attractive subject for studies of speciation. We have been interested in both the proximate and ultimate causes underlying the sexual isolation between these populations. Our approach to this problem has been to first identify where the behavioral breakdown occurs and then to seek the characters involved. Our long term goal is to understand the evolutionary forces which caused the responsible characters to diverge in the first place.

The basic courtship pattern of *D. mojavensis* is presented in Figure 1. Krebs and Markow (1989) have compared the qualitative and quantitative characteristics of the courtships of the unsuccessful combinations of males and females (Baja and Sonora, respectively), in order to identify the point or points in this pattern at which the pairings fail. Their results reveal that the largest breakdown in courtship occurs because Sonora females fail to become receptive following courtship by Baja males. We are interested in the qualitative and/or quantitative nature of the information exchanged during courtship which is responsible for this breakdown and

THERESE ANN MARKOW * Department of Zoology, Arizona State University, Tempe, Arizona 85287, USA; **ERIC C. TOOLSON** * Department of Biology, University of New Mexico, Albuquerque, New Mexico 87131, USA.

Ecological and Evolutionary Genetics of Drosophila, Edited by
J.S.F. Barker *et al.*, Plenum Press, New York, 1990

in the evolutionary factors underlying its divergence in nature.

Drosophila use visual, auditory, and chemical cues during courtship, but the relative importance of each varies with species (Ewing, 1983). Courtship in *D. mojavensis* is similar to the other *repleta* group species in that during 90% of the time spent courting, the male is positioned behind the female with his proboscis and foretarsi touching the distal portion of the female's abdomen. No visually detectable differences in courtship exist among members of the *repleta* group looked at so far in the laboratory. The courtship songs of male *D. mojavensis* from Baja and Sonora do not show significant differences (Ewing and Miyan, 1986), leaving chemical cues as the most likely mode of information exchange between flies during courtship. In a number of Dipteran species, pheromone activity has been attributed to chemical constituents of the epicuticle. Especially in *Drosophila*, hydrocarbons have been shown to be behaviorally active, mediating aggregation (Bartelt and Jackson, 1984) and sexual behavior (Carlson *et al.*, 1978; Uebel *et al.*, 1975; Antony and Jallon, 1982). Our efforts to identify the factors controlling sexual isolation between Baja and Sonora *D. mojavensis* have thus focused on this class of chemical compounds.

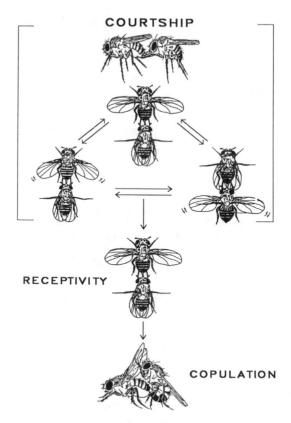

Figure 1. Basic courtship pattern of *D. mojavensis*.

2. CUTICULAR HYDROCARBONS OF INSECTS

Hydrocarbons (HC's) are the predominant epicuticular component of most terrestrial arthropods (Hadley, 1984). As their
name implies, HC's are composed solely of carbon and hydrogen.
Two general types of HC have been identified in the cuticular
lipids of insects: alkanes (HC's lacking double bonds), which
predominate in most insect species, and olefins (HC's containing one or more double bonds). In most insects, alkanes are
usually straight-chain molecules (n-alkanes) containing 19-33
carbon atoms, although methyl branched alkanes are ubiquitous
and comprise the majority of the alkane fraction in *Drosophila*
species so far examined. With respect to the olefins, alkenes
(one double bond), alkadienes (two double bonds), and even
alkatrienes (three double bonds) have been reported from a
number of species. The epicuticular HC's of most Diptera contain relatively low proportions of alkenes and alkadienes. In
addition to the differences in types of epicuticular HC
(alkanes *vs* olefins), interspecific comparisons reveal considerable diversity in patterns of methyl branching and posi-
tional isomers of the various olefinic components.

The most extensively studied function of epicuticular HC's
is their role in restricting transcuticular water flux.
Although they account for only about 0.1% of the mass in a
typical insect, the epicuticular HC's reduce permeability to
water by as much as 1300% (Edney, 1977; Toolson, 1982). The
epicuticular HC's have thus played an essential role in the
invasion of terrestrial environments by arthropods. Indeed,
much of the variability in epicuticular HC profile appears to
reflect evolutionary response to the problem of regulating
cuticular permeability in the face of spatiotemporal variation
in ambient thermal and/or desiccation regimes that characterizes terrestrial environments.

The rate of water flux through a lipid layer depends on the
diffusion gradient and on the layer's permeability. In the
case of a terrestrial arthropod, the diffusion gradient for
water molecules is determined primarily by temperature and the
water vapor content of the air surrounding the arthropod
(Toolson, 1978). The permeability of a lipid layer is dependent on a number of parameters. The type of lipid is important, with HC's typically being the most resistant to diffusion of water molecules (Wilson, 1975). Structural characteristics of the component molecules are also important. *Ceteris
paribus*, larger molecules tend to produce layers of low permeability. This is because the stabilizing (fluidity-reducing)
effect of weak intermolecular forces such as van der Waals
interactions is more pronounced in larger molecules. Methyl
branches or double bonds, by virtue of the fact that they disrupt weak intermolecular interactions, tend to increase permeability.

In general, arthropod species inhabiting environments in
which temperatures are high or in which desiccation stress is
relatively intense are characterized by lower cuticular permeabilities than related species collected from less extreme
habitats (Hadley, 1984). The low permeabilities are generally

associated with higher proportions of larger HC molecules. Comparable changes in permeability, reflecting corresponding shifts in epicuticular HC profile, have been observed on a seasonal basis in some long-lived species such as scorpions (Toolson and Hadley, 1979) and tenebrionid beetles (Hadley, 1977). The inference that spatiotemporal patterns of epicuticular HC profile and cuticular permeability are the outcome of natural selection has been corroborated by results of a laboratory study of *D. pseudoobscura* (Toolson and Kuper-Simbron, 1989).

3. CUTICULAR HYDROCARBONS AND CACTOPHILIC *DROSOPHILA*

We have recently completed a preliminary analysis of the epicuticular HC's of the four Sonoran Desert endemic cactophilic *Drosophila*: *D. mojavensis*, *D. mettleri*, *D. pachea*, and *D. nigrospiracula* (Table I). In each species, dienes unlike those heretofore reported for any insect comprise the major class of HC's. As is often, but not always (Toolson, 1988), observed in hot desert insect species, the epicuticular HC profiles are characterized by relatively high proportions of long chain molecules. More detailed analysis (Toolson *et al.*, 1989) of *D. mojavensis* HC's revealed that not only are the diene molecules considerably larger than those seen in other insects, including *Drosophila* (Bartelt *et al.*, 1986; Jallon and David, 1987; Toolson, 1982), but mass spectroscopic analysis revealed that the locations of the double bonds are unique, as well. This uniqueness of the epicuticular dienes led us initially to focus on them in our search for compounds with sex pheromone activity.

Table I

Distribution of HC in µg/fly for the four species of *Drosophila* endemic to the Sonoran Desert[a]

Species	Sex	Hydrocarbon chain length						
		27	29	31	33	35	37	39
D. mojavensis	F	–	4	14	8	51	15	15
	M	–	3	13	7	56	10	–
D. pachea	F&M	–	13	19	10	30	26	–
D. nigrospiracula	F	2	12	12	22	34	5	–
	M	2	13	12	28	30	2	–
D. mettleri	F	–	17	21	30	19	12	2
	M	–	18	69	13	2	–	–

[a]Flies were collected at the following localities: *D. mojavensis*, San Carlos, Sonora; *D. pachea*, San Carlos, Sonora; *D. nigrospiracula*, Superstition Mountains, Arizona; *D. mettleri,* Superstition Mountains, Arizona.

Two additional features of the dienes drew our attention to them for future investigation as possible sex pheromones in *D. mojavensis*. First of all, there is sexual dimorphism in the relative amounts of the C_{35} and C_{37} dienes. The ratio of the amounts of the two dienes, hereafter referred to as R (= $\mu g\ C_{35:2}/\mu g\ C_{37:2}$) is greater in males. Secondly, in *D. mojavensis*, age at sexual maturity is sexually dimorphic (Markow, 1982), with females maturing in three days and males not until about eight days, and an ontogenetic study revealed that the R-values of newly eclosed flies are low and do not increase to values characteristic of sexually mature flies until they are old enough to exhibit sexual behavior (Toolson *et al.*, 1989). Both of these characteristics are consistent with what we would expect of epicuticular HC's if they are, in fact, involved in mating behavior of *D. mojavensis*.

We have, therefore, conducted a series of experiments to assess whether the epicuticular dienes of *D. mojavensis* are involved in mating success of male or female flies. In addition, since epicuticular HC composition in many species is extremely labile in the face of variations in ambient temperature and humidity regimes, we are investigating the potential effect of varying abiotic conditions in mating behavior. In this paper, we report the results of studies of the effects of temperature on the composition of the epicuticular dienes and on sexual behavior in *D. mojavensis*. We document temperature-induced shifts in epicuticular HC profile that are associated with differential mating success of male flies, and present data that suggest that males utilize epicuticular dienes as a mate choice criterion as well. Our results suggest proximate and ultimate explanations for the asymmetrical sexual isolation of the Baja and Sonora populations.

3.1. Temperature Shifts and Hydrocarbon Profiles

Since many studies have demonstrated significant effects of temperature changes on epicuticular HC profiles (see Hadley, 1984), we first asked if exposure to different temperatures could shift HC's in any consistent way. A strain of flies was established from a multifemale (n=68) collection of flies made in San Carlos, Sonora in the Spring of 1988 and maintained in mass culture until experiments commenced in Fall 1988. Experimental cultures were started with 20 pairs per half pint bottle, with standard cornmeal medium. The parents were removed after 5 days to avoid larval crowding. Cultures were maintained at 24°C. Emerging males and females were separated with an aspirator and stored 10 flies/vial at either 17°C, 24°C, or 34°C until they reached eight days of age. Twenty 8-day old individuals of a given sex were pooled for extraction. Four replications of each treatment were extracted. Epicuticular HC's were obtained by placing the flies on a mini-column of BioSil A and washing with 7 ml of hexane. The alkadienes were then recovered separately and analyzed by capillary gas liquid chromatography as discussed in Toolson *et al.* (1989).

The epicuticular dienes of males and females at the three experimental temperatures are presented in Table II. Of particular interest is the shift in the amounts of C_{35} and C_{37} dienes and, therefore, in values of R. A decrease in R as temperature increases is the result of increases in the amount of $C_{37:2}$, apparently at the expense of $C_{35:2}$. This shift toward longer HC molecules at higher temperatures has been reported in many other species (Toolson, 1982; Toolson and Hadley, 1977, 1979; Hadley, 1984; Toolson and Kuper-Simbron, 1989). Similar shifts were found in flies undergoing preimaginal development and maturation at 17°C and 31°C (Riedy et al., 1989). It is now clear that adults can acclimate epicuticular HC composition in response to temperature changes despite having been through their preimaginal development at 24°C.

In spite of the strong temperature-dependence of R (especially in males), males were characterized at each acclimation temperature by greater R-values than were females reared at the same temperature (Table II). Given our earlier finding that the dimorphism in R is also observed in populations of D. mojavensis in nature (Toolson et al., 1989), we felt this was further evidence that the epicuticular dienes play a role in courtship of D. mojavensis. We formulated the working hypothesis that the value of R conveyed information about the sex of a particular fly, with relatively high R-values characterizing males whereas low R-values are more typical of females.

To test this hypothesis, we asked if these same temperature treatments could lead to some degree of sexual isolation between flies reared at 17°C and 34°C. We made two predictions. First, in mating experiments involving 17°C females, we predicted that 34°C males, because their R-values were so similar to those of 17°C females, would suffer reduced mating success as compared with 17°C males. Second, we predicted that 34°C females would mate equally readily with 17°C or 34°C males, since the R-values for both groups of males were greater than 34°C females.

Table II

Amounts of the predominant alkadienes in adult D. mojavensis reared at 24°C and acclimated to 17°C, 24°C, or 34°C for eight days

Acclm. Temp.	Sex	Alkadienes (µg/fly)				R[a]		
		C_{35}		C_{37}				
		Mean	se	Mean	se	Mean	se	n
17°C	F	0.39	0.021	0.16	0.004	2.64	0.997	3
	M	0.41	0.097	0.05	0.015	8.03	1.20	4
24°C	F	0.38	0.202	0.19	0.036	1.91	0.729	3
	M	0.35	0.006	0.09	0.006	3.82	0.160	3
34°C	F	0.23	0.039	0.25	0.015	0.91	0.121	4
	M	0.31	0.034	0.16	0.029	1.98	0.242	4

[a]R is the ratio C_{35}/C_{37}.

3.2. Temperature Shifts and Sexual Isolation

The influence of maturation temperature on sexual isolation was examined using flies which had been reared at 24°C and matured for eight days at the two extreme temperatures, 17°C and 34°C. Eggs, larvae, and pupae were kept at 24°C to avoid any developmental effects of temperature on the flies. Ten pairs from each of the two temperatures were moved from the incubators to 24°C and placed in chambers 30 minutes later. Flies from the two temperature treatments had been marked with differently colored fluorescent dusts (Markow, 1985) to identify the types of flies mating. Ten virgin males and ten virgin females from each of the two temperatures were placed in the same observation chamber and scored as to the types of flies mating. Replicates were homogeneous and the pooled numbers of flies mating are shown in Table III. Flies, especially males, which had matured at 17°C participated in significantly more matings than the 34°C flies.

To examine the possibility that 17°C males were simply courting more vigorously, we placed a male from each temperature with a single female, either from 17°C or 34°C. Under these conditions, courtship vigor is reflected in the percentage of time males spend courting a female and these results are presented in Table IV. In twenty observations, the per cent time spent courting a 17°C female was 37.1% ± 4.1% (for 17°C males) and 36.8% ± 12.4% (for 34°C males). While 34°C males showed the same average percentage of time courting, the data were more variable. When a 34°C female was used, the courtship vigor indices were still equal: 23.6% ± 5.1% (17°C males) and 24.5% ± 9.7% (34°C males), but the total percentage of time these females were courted was less than for 17°C females. Thus, it seems unlikely that differences in courtship vigor on the part of 17°C and 34°C males can account for their differential mating success.

3.3. Hydrocarbons and Sexual Isolation

Acclimation to different thermal regimes produces significant changes in the epicuticular dienes of *D. mojavensis*, and

Table III
Numbers of Matings Between Flies Matured at 17°C and 34°C[a]

		Males		
		17°C	34°C	
Females	17°C	54	4	χ^2 = 58.23
	34°C	22	5	

[a]Ten virgin males and ten virgin females of each type were placed in a chamber. Six replications were homogeneous, and were pooled.

Table IV
Proportion of time males spent courting after courtship
initiation

Female	% time spent courting (se)	
	17°C Male	34°C Male
17°C	37.1 (4.1)	36.8 (12.4)
34°C	23.6 (5.1)	24.5 (9.7)

results in a significant degree of sexual isolation between
flies matured at the two extreme temperatures. In tests for
behavioral isolation, males with lower R-values are less successful in mating. However, this association between HC's and
mating success in no way constitutes proof that this epicuticular HC parameter (i.e., relative R-value) controls sexual
isolation. This conclusion is warranted only after appropriate behavioral bioassays of the relevant HC components.

A number of studies have developed successful behavioral
bioassays for dipteran sex pheromones (Antony *et al.*, 1985;
Tompkins and Hall, 1981; Carlson *et al.*, 1978) by observing
behavioral responses to (i) a pheromone-impregnated piece of
filter paper placed in an observation chamber or (ii) an immo-
bilized target fly or dummy to which a known amount of putative pheromone has been applied. We took a slightly different
approach, because, in the present case, we wished to test
whether modifying epicuticular HC profile of male flies would
affect their mating success. We addressed this by determining
whether 34°C (low R-value) males treated with 17°C dienes
(high R-value) exhibited an increase in mating success. However, our attempts to modify epicuticular diene profiles of
live male *D. mojavensis* by directly applying diene extracts of
known composition were unsuccessful. Because of their large
molecular size, these dienes must be dissolved in a nonpolar
solvent (hexane) prior to application. The hexane proved to
be so toxic that even 1 μl drastically impaired the behavior
of treated flies.

We attempted to get around this problem by employing
another method of treating flies with putative pheromones.
Vials of a high ratio (R = 10.36) and low ratio (R = 1.44)
extracts were placed at 34°C for thirty minutes to "soften".
Mature virgin males, acclimated at 34°C, were then aspirated
into those vials, ten males per vial, and gently tapped to the
bottoms for 15 minutes. We anticipated that the flies would
pick up the warmed dienes on their feet and spread them on
their bodies by preening. High ratio (HR) and low ratio (LR)
treated males were then placed individually with single 17°C
females in observation vials. The number of each type of male
mating in one hour was scored. The particular combination of
17°C females and 34°C males was selected because the low probability of matings among untreated flies (see Table III) would

enhance the likelihood of detecting increased mating by HR treated males.

We tested flies on four different mornings and obtained the matings reported in Table V. As expected, the frequency of matings remained low. However, totaling the matings from the four replications shows that HR males participated in three times as many matings as the LR males. While small numbers of matings and the discrete nature of the data play against statistical significance, the results are in the direction predicted by Table III. Although it was impossible to control the amount of extract acquired by each fly, the enhanced mating success of HR-treated males supports our hypothesis that this diene ratio plays a role in *D. mojavensis* courtship.

During the exposure of males to the extracts, a remarkable behavior was noticed when flies were in the vials. Males in the HR vials had to be continually tapped down to the bottoms of the vials while the males in the LR vials remained on the bottoms which is where the dienes were concentrated during sample preparation. On closer inspection, the male flies were seen to have their probosci extended and touching the vial floor. This observation suggested another experiment. Because the HR extract is more characteristic of male flies, and the LR more so of female flies, we predicted that, if the LR profile is indeed behaviorally active in a sexual way for males, the proportion of males on the bottom of an LR vial should be higher than for an HR vial. Females, on the other hand, would not be expected to spend more time on the bottom of an LR vial.

To test this prediction we introduced ten 24°C males to an LR vial and ten to an HR vial and allowed them to distribute themselves for five minutes. At that time, the number on the bottom of each vial was counted. This experiment was repeated

Table V
Males reared at 24°C, acclimated for
8 days at 34°C, and treated with
dienes from males matured at 17°C
(HR) or 34°C (LR) prior to pairing
with 17°C females[a]

	Number of males mating	
Rep	HR	LR
1	3	0
2	2	1
3	1	0
4	3	2
Total	9	3

[a]In a sign test, assuming that HR and
LR males were equally likely to mate
in 4 of 4 trials, p = 0.0625.

Table VI

Number of flies on the bottoms of vials containing dienes of different R values[a]

Sex of flies	Rep	R values of vials			
		10.36	4.94	1.63	1.44
Males	1	2	3	6	10
	2	1	1	9	9
	3	3	3	9	10
	4	1	1	8	9
	5	2	2	8	8
	6	2	1	10	9
	Total	11	11	50	55
Females	1	2	2	3	2
	2	3	1	1	1
	3	3	4	2	2
	4	2	2	2	3
	Total	10	9	8	8

[a]Ten flies, either males or females, were gently aspirated into vials and allowed to settle for five minutes. Heterogeneity chi squares between columns: males $\chi^2 = 54.64$, $p < 0.0001$, females $\chi^2 = 0.314$, $p > 0.95$.

six times for males and four times for females (Table VI). In vials with lower ratios, males congregated on the extracts evporated on the vial bottoms; females did not. Thus, we conclude that the behavioral activity of the LR diene extract is clearly sex-specific.

Males must sense these dienes with their foretarsi and probosci, responding by maintaining contact with a surface coated with them. This surface may be the glass onto which the dienes have been evaporated, or the tip of a female's abdomen. While the observations strongly argue for the importance of "female"-like HC's, namely a low ratio of C_{35} to C_{37} dienes, in maintaining contact during courtship, it is apparent from our data that the dienes by no means form the entire basis of mate recognition and differential courtship success in this species.

Females, on the other hand, do not appear to employ their legs or probosci in assessing courting males. The anatomical portion of a female which is in contact with a courting male is the tip of the abdomen, including the ovipositor, which, in dipterans, is covered with chemoreceptors (Dethier, 1962). In the test vials, the ovipositors do not contact the extract on the bottom. Only the legs do. We are, therefore, not surprised that females do not show any differential responses to the LR and HR extracts. However, an association between HR and male success is supported by the mating experiments (Tables III and V). Clearly, these substances require contact, and with the appropriate male or female anatomical part, in order to provide behavioral stimulation. While epicuticular HC's are clearly involved in the sexual behavior and mating success of *D. mojavensis*, these chemicals only act effectively in the context of additional and complex sexual behaviors.

4. TEMPERATURE EFFECTS IN THE LABORATORY AS A MODEL SYSTEM FOR SEXUAL ISOLATION IN DROSOPHILA

Our laboratory experiments have demonstrated that acclimation to different abiotic environments can lead to strong sexual isolation. Could these results provide insight into the incipient speciation observed between Sonora and Baja populations of *D. mojavensis*?

Under models of allopatric speciation, organisms become separated into two or more isolated populations. Some degree of environmental differences exist between the habitats of the separate populations, and these differences act as selective forces which eventually cause genetic divergence between the populations. In the event that contact is reestablished, the genetic differentiation may have resulted in some degree of reproductive isolation at the pre- or postzygotic level. Studies of reproductive isolation between races of organisms have been utilized to test this model as well as the genetic basis of speciation and, in many cases, have revealed significant isolation between populations. Natural selection on different phenotypes has been invoked to explain the general genetic divergence of allopatric populations, but the actual characters under selection usually go unidentified. In our study with laboratory populations of *D. mojavensis*, we have succeeded in identifying a character which appears to be exposed to different natural selection regimes in different areas as well as having a role in sexual selection.

The enhanced mating success of males with high R-values suggests that females are more receptive to such males. This led us to make two predictions. First, that Baja males, because they are less acceptable to Sonora females, should exhibit lower R-values than Sonora males. Second, that R-values of Baja males should be greater than those of Baja females. In other words, we predicted that Baja populations would also be sexually dimorphic for R-values, but that the R-values would be lower than the corresponding R-values of Sonora males and females. These predictions were based on the fact that the interpopulational sexual isolation is asymmetrical, with Sonora females discriminating against Baja males, while Baja females show little, if any, preference (Zouros and d'Entremont, 1980; Koepfer, 1987a, b; Krebs and Markow, 1989).

Comparison of the epicuticular diene profiles of Sonora flies with those of Baja flies (Table VII) shows that both predictions were confirmed. Compared with Sonora flies, Baja flies have less epicuticular dienes, and the R-values of both male and female Baja flies are significantly less than the respective values for Sonora flies. Also note that the range in R-values of Baja males shows significant overlap with that

Table VII

The two major dienes and their ratios in male and female *D. mojavensis* from Baja (Punta Prieta and San Lucas) and Sonora (San Carlos and Santa Rosa Mts)[a]

Locality	Sex	μg dienes per fly				R		
		C_{33} mean ± se	C_{35} mean ± se	C_{37^+} mean ± se	Total mean ± se	Mean	se	Range
Baja	F	0.01 ±0.004	0.27 ±0.088	0.15 ±0.057	0.50 ±0.140	1.83	0.458	1.38 to 2.28
	M	0.01 ±0.005	0.26 ±0.090	0.09 ±0.025	0.40 ±0.082	2.83	0.509	2.36 to 3.28
Sonora	F	0.03 ±0.010	0.58 ±0.103	0.24 ±0.058	0.90 ±0.160	2.56	0.633	2.02 to 3.47
	M	0.03 ±0.012	0.43 ±0.152	0.10 ±0.033	0.60 ±0.234	4.31	0.801	3.26 to 5.67

[a]The Student-Newman-Keuls multiple range test for unequal sample sizes, at a=0.05, gives two subsets (A-Sonora males, B-all others) for R values.

of Sonora females. Since high relative R-values are important for male mating success (Tables III and V), our data offer an explanation for the reduced success of Baja males with Sonora females. We suggest that this overlap would result in many Baja males being rejected by Sonora females because their epicuticular diene profiles were too close to those of Sonora females. In other words, courting Baja males would be so female-like to Sonora females that they would not elicit a receptive response.

What factors might have driven the intersexual and interpopulational differentiation in epicuticular diene profiles? Our demonstration that high R-value males enjoy enhanced mating success leaves little doubt that sexual selection is a major force in shaping the sexual dimorphism in epicuticular diene profiles. However, other selective pressures are likely to be involved in driving both the sexual and interpopulational differences in diene profile. Regulation of the alkadiene biosynthetic system of *D. mojavensis* is obviously responsive to temperature and would, therefore, theoretically be capable of evolutionary response to ambient thermal regimes. We believe that other biotic and abiotic variables may be involved as well.

Male and female *D. mojavensis* show distinct differences in how they utilize their host cacti. *Drosophila mojavensis* were observed and collected on three different organ pipe cacti in San Carlos, Sonora. Males and females were found to be distributed differently on parts of the cacti, namely moist necrotic regions, hollow pockets, and non-necrotic surfaces of arms of the same plant (Table VIII). A preponderance of females was found both inside rot pockets and feeding on juice oozing from necrotic tissue. On the dry skin of arms (either directly adjacent to drips or several feet away), groups of males were standing or walking slowly, maintaining a distance

Table VIII

Distribution of *D. mojavensis* on three organ pipe cacti in San Carlos, Sonora, Mexico[a]

Location	Numbers of flies		
	Single females	Single males	Copulating pairs
Drips	108	52	7
Holes and pockets	77	39	4
Arms (1) near drip	23	59	16
Arms (2) away	13	59	49
Total flies	221	209	77

[a]Flies were collected from dripping necroses, hollow necroses, and healthy arms of cacti. Their gender and whether or not they were mating were scored for each collection place. Homogeneity in distribution between cacti allowed the data to be pooled.

of about 1 to 2 inches from each other. If a female landed near a male in this area, she was courted immediately. Most of the matings in *D. mojavensis* occurred in these areas away from necrotic, moist tissue. Males, by virtue of the greater amount of time spent at these exposed sites, face greater risk of desiccation than females. Given the differences in the way male and female flies are partitioning their environment, it would not be surprising if they have evolved differences in the genetic system underlying regulation of epicuticular HC composition.

There is reason to believe that similar considerations can account for the differences in epicuticular HC profiles that characterize Baja and Sonora populations of *D. mojavensis*. Among the most obvious are general climatic differences and host plant differences. Baja and Sonora differ with respect to temperature, rainfall and humidity (Markham, 1972; Hastings and Turner, 1965; Burgess, 1988), although south of Guaymas, Sonora, these climatic differences tend to disappear. These climatic differences are associated with differential distri- butions of plants (Burgess, 1988), and the terrestrial arthro- pods most likely experience different selection pressures as well. The primary host plants within Baja and Sonora display different characteristics. Agria (*Stenocereus gummosus*), is the primary host plant in Baja while organ pipe (*Stenocereus thurberi*), is used throughout most of southern Arizona and northern Mexico. Both are columnar cacti, but the arms of agria tend to be smaller in diameter. Agria rots have been reported to occur with a higher density than organ pipe (Ruiz and Heed, 1988; Mangan, 1982), but may give up their moisture content and, therefore, dry out more quickly than organ pipe rots (Etges, 1989). These differences in host plant charac- teristics have been invoked to explain differences between the Baja and Sonora *D. mojavensis* populations with respect to development time, body size, and ovariole number. It would be surprising if the host plant characteristics did not also expose the resident flies to different selective pressures on epicuticular diene composition. Finally, the possibility that sympatry with *D. arizonensis*, the sibling species of *D. mojavensis*, in Mexico and southern Arizona, exerts a selective force on the mate recognition system, i.e., pheromones, of *D. mojavensis*, cannot be overlooked.

In this study, we focused on the ability of one abiotic variable, temperature, to alter the epicuticular diene pro- files and mating behavior of *D. mojavensis*. Our future research plans include evaluating the relative roles of the other biotic and abiotic factors mentioned above in shaping epicuticular diene profiles both in the laboratory and in the field. We feel that epicuticular diene composition in *D. mojavensis* offers an ideal system for examining the interplay between sexual and natural selection in that this character represents a physiological system of known adaptive signifi- cance as well as being a mediator of successful reproduction.

5. MODEL FOR SEXUAL SELECTION/SEXUAL ISOLATION IN DROSOPHILA MOJAVENSIS

1. Successful mating requires sexual dimorphism in epicuticular diene profile between members of the courting pair.
   a. females must have low relative R-values in order to be courted by males.
   b. males must have relatively high R-values in order to stimulate receptivity by females.
2. Shifts in relative R-values can result in behavioral isolation: when a low R-value male courts a high R-value female, there will be a reduced probability of a mating.
   a. Baja males and Sonora females exemplify this.
   b. 34°C males and 17°C females exemplify this.
3. At least one abiotic variable, temperature, can shift R values in the laboratory. In nature, it is likely that epicuticular dienes are selected by a suite of abiotic and biotic variables.

6. SUMMARY

Hydrocarbons are the predominant class of lipids in the epicuticle of most insects. They are responsible for regulating cuticular permeability, enabling survival in terrestrial habitats. There is a general relationship between epicuticular hydrocarbon composition and the potential degree of desiccation associated with different habitats: a higher proportion of longer chain hydrocarbons occurs in organisms occupying hot, dry environments.

Drosophila mojavensis and the three other desert-adapted *Drosophila* species, *D. nigrospiracula*, *D. mettleri*, and *D. pachea* are shown to have higher relative proportions of long chain epicuticular hydrocarbons than do non-desert-adapted *Drosophila* species, suggesting that this relationship between the potential for desiccation and increased hydrocarbon chain length holds for this genus as well. The epicuticular hydrocarbon composition of *D. mojavensis* is seen to shift following exposure to high and low temperatures. At 34°C the relative proportion of the $C_{37:2}$ alkadiene increases, while at 17°C it decreases.

Sexual dimorphism exists in *D. mojavensis* for the ratio (R) of the amounts of $C_{35:2}$ to $C_{37:2}$ alkadienes, such that R is greater in males. Baja and Sonora populations of *D. mojavensis* both show sexual dimorphism in R-values, although these two geographic host races show significant differences in the ranges of R-values. We have shown, using temperature induced shifts in R-values, that this feature of epicuticular hydrocarbon composition mediates mating behavior in this species, and we offer a model for the influence of population differences in epicuticular hydrocarbon composition in the behavioral isolation observed between Baja and Sonora flies.

ACKNOWLEDGEMENTS.   Mr.   Michael   Riedy   and   Ms.   Diane   Gallagher
provided   excellent   technical   assistance   throughout   the
project.   This research was funded by NSF grants BSR 8600105
and BSR 8708531 to T.A.M.

References

Antony, C., and Jallon, J. M., 1982, The chemical basis for sex recognition
    in *Drosophila melanogaster*, *J. Insect Physiol.* **28**:873-880.
Antony, C., Davis, T. L., Carson, D. A., Pechine, J.-M., and Jallon, J. M.,
    1985, Compared behavioral responses of male *Drosophila melanogaster*
    (Canton-S) to natural and synthetic aphrodisiacs, *J. Chem. Ecol.*
    11:1617-1629.
Bartelt, R. J., Arnold, M. T., Schaner, A. M. and Jackson, L. L., 1986,
    Comparative analysis of the cuticular hydrocarbons in the *Drosophila
    virilis* species group, *Comp. Biochem. Physiol.*, **83B**:731-742.
Bartelt, R. J., and Jackson, L. L., 1984, Hydrocarbon component of the
    Drosophila virilis (Diptera: Drosophilidae) aggregatin pheromone: (Z)-
    10-heneicosene, *Ann. ent. Soc. Am.* **77**:364-371.
Burgess, T., 1988, The relationship between climate and leaf shape in the
    Agave cerulata complex, Ph.D Thesis, University of Arizona.
Carlson, D. A., Langely, P. A., and Huyton, P., 1978, Sex pheromone of the
    tsetse   fly: isolation,   identification   and   synthesis   of   contact
    aphrodisiacs, *Science* **201**:750-753.
Dethier, V. G., 1962, *To Know a Fly*, Holder-Day, San Francisco.
Edney, E. B., 1977, *Water Balance in Land Arthropods*, Springer-Verlag,
    Berlin.
Etges, W. E., 1989, Evolution of developmental homeostasis in *Drosophila*,
    Evol. Ecol., in press.
Ewing, A. W., 1983, Functional aspects of *Drosophila* courtship, *Biol. Rev.*
    58:275-292.
Ewing, A. W., and Miyan, J. A., 1986, Sexual selection, sexual isolation,
    and the evolution of song in the *Drosophila repleta* group of species,
    Anim. Behav. **34**:421-429.
Hadley, N. F., 1977, Epicuticular lipids of the desert tenebrionid beetle,
    Eleodes armata: seasonal and acclimatory effects on composition, *Insect
    Biochem.* **7**:277-283.
Hadley, N. F., 1984, Cuticle: Ecological significance, in: *Biology of the
    Integument*, Vol. 1-*Invertebrates* (J. Berieter-Hahn, A. G. Matolsky, and
    K. S. Richards, eds), Springer-Verlag, Heidelberg, pp. 685-693.
Hastings, J. R., and Turner, R. M., 1965, Seasonal precipitation regimes in
    Baja California, Mexico, *Geografiska Annales Ser. A.* **47**:204-223.
Jallon, J. M., and David, J., 1987, Variations in cuticular hydrocarbons
    among the eight species of the *Drosophila melanogaster* group, *Evolution*
    41:294-302.
Koepfer, H. R., 1987a,   Selection for sexual isolation between geographic
    forms of *Drosophila mojavensis*. I. Interactions between the selected
    forms, *Evolution* **41**:37-48.
Koepfer, H. R., 1987b, Selection for sexual isolation between geographic
    forms of *Drosophila mojavensis*. II. Effects of selection of mating
    preference and propensity, *Evolution* **41**:1409-1413.
Krebs, R., and Markow, T. A., 1989, Courtship behavior and control of
    reproductive isolation in *Drosophila*. *Evolution*, in press.

Mangan, R. L., 1982, Adaptations to competition in cactus breeding *Drosophila*, in: *Ecological Genetics and Evolution. The Cactus-Yeast-Drosophila Model System* (J. S. F. Barker, and W. T. Starmer, eds), Academic Press Australia, Sydney, pp. 257-272.

Markham, C. G., 1972, Baja California Climate, *Weatherwest* 25:64-76.

Markow, T. A., 1982, Mating systems of Cactophilic *Drosophila*. in: *Ecological Genetics and Evolution. The Cactus-Yeast-Drosophila Model System* (J. S. F. Barker, and W. T. Starmer, eds), Academic Press Australia, Sydney, pp. 273-287.

Markow, T. A., 1985, A comparative investigation of the mating system of *Drosophila hydei*, *Anim. Behav.* 33:775-781.

Riedy, M., Toolson, E., and Markow, T., 1989, Rearing temperature and epicuticular lipid composition in *Drosophila*, *Drosoph. Inf. Serv.*, in press.

Ruiz, A., and Heed, W. B., 1988, Host plant specificity in the cactophilic *Drosophila mulleri* species complex, *J. Anim. Ecol.* 57:237-249.

Tompkins, L., and Hall, J. C., 1981, The different effects on courtship of volatile compounds from mated and virgin *Drosophila* females, *J. Insect Physiol.* 27:17-21.

Toolson, E. C., 1978, Diffusion of water through the arthropod cuticle: thermodynamic consideration of the transition phenomenon, *J. Thermal Biol.* 3:69-73.

Toolson, E. C., 1982, Effects of rearing temperature on cuticle permeability and epicuticular lipid composition in *Drosophila pseudoobscura*, *J. exp. Zool.* 222:249-253.

Toolson, E. C., 1988, Cuticle permeability and epicuticular hydrocarbon composition of Sonoran Desert *Drosophila pseudoobscura*, in: *Endocrinological Frontiers in Insect Physiological Ecology* (F. Schnal, A. Zabza, and D. L. Denlinger, eds), Wroclaw Technical Univ. Press, Wroclaw, Poland, pp. 505-510.

Toolson, E. C., and Hadley, N. F., 1977, Cuticular permeability and epicuticular lipid composition in two Arizona *Vijovid* scorpions, *Physiol. Zool.* 50:323-330.

Toolson, E. C., and Hadley, N. F., 1979, Seasonal effects on cuticular permeabiity and epicuticular lipid composition in *Centuroides sculpteratus* Ewing 1928 (Scorpiones: Buthidae), *J. Comp. Physiol.* 129:319-325.

Toolson, E. C., and Kuper-Simbron, R., 1989, Laboratory evolution of epicuticular hydrocarbon composition and cuticular permeability in *Drosophila pseudoobscura*: effects on sexual dimorphism and thermal-acclimation ability, *Evolution* 43:468-472.

Toolson, E. C., Howard, R., Jackson, L., and Markow, T. A., 1989, Epicuticular hydrocarbon composition of wild and laboratory-reared *Drosophila* (Diptera: Drosophilidae), *Ann. ent. Soc. Am.* (submitted).

Uebel, E. C., Sonnet, P. E., Bierl, B. A., and Miller, R. W., 1975, Sex pheromone of the tsetse fly isolation and preliminary identification of compounds that reduce mating strike behavior, *J. Chem. Ecol.* 1:377-385.

Wilson, L., 1975, Wax components as a barrier to aqueous solutions, Ph.D. Dissertation, University of California, Davis, 123 pp.

Zouros, E., and D'Entremont, J., 1980, Sexual isolation among *Drosophila* populations: response to pressure from a related species, *Evolution* 34:421-430.

Molecular Evolution: Introduction

Variation in proteins and nucleic acid sequences can be used in a variety of ways to study evolutionary patterns and processes. In recent years, DNA sequences have largely superseded allozyme frequencies as the raw data for *Drosophila* molecular evolutionary biologists. DNA sequencing has been applied to three areas of investigation: the estimation of population genetic parameters, phylogenetic reconstructions, including estimates of rates of change in specific gene products, and the molecular basis of structural and regulatory gene mutations.

Restriction enzyme analysis, direct sequencing of cloned DNA and more recently, polymerase chain reaction technology (PCR) have considerably sharpened our measurements of intraspecific genetic variation. This variation includes nucleotide substitutions and insertion/deletion polymorphisms, and can be used to estimate many parameters of natural populations. For example, the array of alleles detected at several loci, such as *Esterase-6* and *Adh*, in natural populations indicates an effective population size for the species *D. melanogaster* of 3-7 X 10^6. With appropriate primers and PCR sequencing, it should be possible to measure accurately the frequencies and population distributions of Y chromosome specific variants and/or maternally inherited sequences in the mitochondrial genome.

In the past, the raw data for phylogenetic reconstructions of *Drosophila* species has come from DNA-DNA hybridization (ΔTm's) or allozyme surveys (genetic distances). This has effectively limited the reconstructions to species groups. Only a few molecules have been identified which allow the examination of the phylogenetic relationships between species groups, even those groups from the same subgenus. The ease and speed with which sequences can be obtained with PCR should soon rectify this situation.

The measurement of rates of change in different *Drosophila* genes has suffered from a paucity of reliable dates for speciation events in the evolution of the genus. In its 60 million year history, the times of only a few major phylogenetic divisions can be estimated, e.g. the *repleta-tripunctata* split at ca. 40 million years (MY), the *virilis-montana* split at 15 MY, several Hawaiian *Drosophila* phylads within the last 6 MY and the *melanogaster-simulans* divergence at 1-2 MY. Immunological studies on several enzymes, which allow relative rate tests to be applied, indicate those *Drosophila* proteins at least, are evolving at fairly constant rates, but more data are clearly needed. Once again, with appropriate primers from enough genes, PCR should come to the rescue.

Ecological and Evolutionary Genetics of Drosophila, Edited by
J.S.F. Barker *et al.*, Plenum Press, New York, 1990

The final area in which DNA sequence data have been very informative concerns the molecular basis of structural and regulatory gene arrangement and mutations. *Drosophila* genes, like those of most eukaryotes, are usually split into several exons, and codon bias within homologous structural genes can vary within the genus. In some species transposable elements may be the primary agent for spontaneous mutation.

It is significant that several cases of duplicate genes, both tandem and non-tandem, have been uncovered in different *Drosophila* species. Studies on the expression of these duplicate copies should provide important information on the evolution of gene regulation and its adaptive significance. Recent studies on "enhancer traps" in *Drosophila* indicate that the genome is littered with cognate sequences with enhancer activity. Thus, the acquisition of new temporal and tissue specific patterns of expression by a duplicate gene may be a very frequent evolutionary event.

The chapters in this final section exemplify the power and promise of molecular data with regard to evolutionary questions. John Oakeshott and his collaborators discuss DNA sequence variation in the *Esterase-6* gene in relation to its structure and its regulation. With regard to phylogenetic reconstruction, Bill Heed, using allozyme data, examines the relationships between West Indian species of the *mulleri* complex. Glen Collier and colleagues reinforce the value of allozyme data when constructing within species group dendrograms by quantitatively comparing genetic distance based phylogenies with those constructed from more traditional data. They also identify several proteins which are evolving slowly and steadily enough to be potentially useful in building molecular phylogenies of Dipteran families. Tony Howells and colleagues, in comparing the organization of the homologous scarlet genes, find some fascinating differences between the genome structures of *Lucilia* and *Drosophila*. *Lucilia* has a very low G+C content, due mainly to extensive interspersed A-T rich repeats, located within introns and between structural genes. Whether there is an effect of these repeats on gene expression is an open and fascinating question.

Peter East and his collaborators and David Sullivan and Peter Atkinson present cases of gene duplication in *Drosophila*, in which the duplicate copies exhibit different tissue and temporal patterns of gene expression. In Sullivan and Atkinson's study of *Adh*, the use of a "distal" promoter by one gene and a "proximal" promoter by its duplicated partner in the *repleta* group species is functionally analogous to the situation in *D. melanogaster*, where a single gene uses two alternative promoters. This study also exemplifies the use of P element mediated transformation to test the functional role of specific nucleotide sequences. The analysis by East and his collaborators of the duplicated esterase genes in *D. buzzatii*, related species of the *mulleri* subgroup, and in *D. melanogaster* suggests that genes within a species are more similar to each other than are the presumably orthologous homologues between species.

These studies clearly point the way toward the future.   The adaptive significance of interspecific differences in gene regulation and the differences between duplicate gene expression, needs to be examined, thus dovetailing the research projects of molecular biologists and ecological geneticists. There is also a need to examine the molecular evolution not just of gene-enzyme systems, but of genes with more obvious and fundamental effects on fitness.   Among the latter are the genes for early embryonic pattern formation, genes controlling behavior, as well as genes whose products affect signal transduction, cell-cell interactions, transcription factors, etc. *Drosophila* molecular evolutionists should also expand the lexicon of DNA sequences to include those which can be used to reconstruct phylogenies of major *Drosophila* phylads, as well as the families of the order Diptera.   In any event, it seems clear that molecular evolutionary biologists will continue to exploit the genus *Drosophila*, with its rich and well understood array of species.   In the relatively new arena of molecular evolution, *Drosophila* has again proven to be the "queen of biology".

CHAPTER 18

Gene and Genome Structure in Diptera: Comparative Molecular Analysis of an Eye Colour Gene in Three Species

ABIGAIL ELIZUR, YGAL HAUPT, RICHARD G. TEARLE, AND ANTONY J. HOWELLS

1. INTRODUCTION

Drosophila melanogaster and *Drosophila buzzatii* belong to the same family (Drosophilidae) within Order Diptera but to different species groups, which are thought to have diverged about 50-60 MY ago; the sheep blowfly (*Lucilia cuprina*), how- ever, belongs to a different family (Calliphoridae), which is thought to have diverged from Drosophilidae at least 100 MY ago (Beverley and Wilson, 1984). We have been analyzing the similarities and differences in gene and genome organization at the molecular level between these species using eye colour genes as our model system.

The wild-type eye colour of adults of *D. melanogaster* and *D. buzzatii* is red-brown due to the presence within the pig- ment cells of each ommatidial unit of the compound eye of both red (drosopterins) and brown (xanthommatin) pigments. In *L. cuprina*, however, the drosopterins are not present, being replaced by a yellow pteridine (sepiapterin) (Summers and How- ells, 1980); consequently the wild type eye colour is orange- brown. Mutations which block xanthommatin synthesis therefore give a red eye colour phenotype in the *Drosophila* species but a yellow eye colour in *L. cuprina*.

Xanthommatin is synthesised from tryptophan by a pathway which includes formyl-kynurenine, kynurenine and 3-hydrox- ykynureuine as intermediates (for a review see Summers *et al.*, 1982). In *D. melanogaster* four genes have been identified as being essential for xanthommatin production - *vermilion*, *cinnabar*, *scarlet* and *white*. Two of these genes (*vermilion* and *cinnabar*) code for enzymes of the pathway while the other two (*scarlet* and *white*) provide proteins which appear to be

---

ABIGAIL ELIZUR, and RICHARD G. TEARLE * Department of Biochemistry, University of Adelaide, S.A. 5000, Australia. *YGAL HAUPT* * Walter and Eliza Hall Institute of Medical Research, PO Royal Melbourne Hospital, Victoria 3050, Australia. *ANTONY J. HOWELLS* * Department of Biochemistry, Faculty of Science, Australian National University, Canberra, ACT, Australia.

Ecological and Evolutionary Genetics of Drosophila, Edited by J.S.F. Barker *et al.*, Plenum Press, New York, 1990

involved in transporting the intermediates of the pathway into
the cells in which the pathway operates. That the *scarlet* and
white proteins might interact to provide a membrane embedded
permease was proposed on biochemical grounds by Sullivan and
Sullivan (1975) and Sullivan *et al.*(1980), a proposal which is
strongly supported by the DNA sequence data obtained for these
genes (O'Hare *et al.*, 1984; Mount, 1987; Tearle *et al.*, 1989).

The *scarlet* (*st*) gene of *D. melanogaster* and its homologues in *D. buzzatii* and *L. cuprina* have been the major focus
of our study; (the gene is also called *scarlet* in *D. buzzatii*
but *topaz* (*to*) in *L. cuprina*, in view of the yellow mutant
phenotype). The *scarlet* gene of *D. melanogaster* was isolated
by cloning the breakpoints of a chromosome inversion which
broke in 87D13-14 (within the region covered by the *rosy-Ace*
chromosome walk - Bender *et al.*, 1983) and in 73A3-4 (the
cytological location of *st*) (Tearle *et al.*, 1989). Clones
carrying the homologues of *st* from *D. buzzatii* and *L. cuprina*
were obtained by screening genomic DNA libraries from these
species using a sub-cloned restriction fragment from the *D.
melanogaster* gene as the probe.

2. MATERIALS AND METHODS

2.1. Insects

2.1.1. *D. melanogaster*

The st^1 and st^{sp} stocks were obtained from the Bowling Green
stock centre; st^{dct} was provided by M.M. Green (it was found in
flies caught in the field in California).

2.1.2. *D. buzzatii*

The mutant *st* stock was provided by P.D. East (it was found
in flies caught in the field near Armidale, NSW).

2.1.3. *L. cuprina*

Wild type and mutant lines were obtained from the CSIRO
Division of Entomology, Canberra. The to^1 and to^2 mutant
stocks were derived from flies caught in the field.

2.2. Preparation of Insect DNA and Southern Analysis

Genomic DNA was usually prepared from adults although occasionally embryos were used; both the larger scale method of
Miklos *et al.*(1984) and a smaller scale method (essentially
that of Coen *et al.*, 1982) were used. Digestions with
restriction enzymes were carried out according to the manufac-
turer's specifications. Restriction digests of genomic DNA
(1-4 μg DNA per lane) were separated by electrophoresis on
agarose gels (0.7-1%) and then transferred either to nitrocellulose (Southern, 1975) or to zetaprobe membrane (Biorad)
using 0.4M NaOH (Reed and Mann, 1985). Stringency conditions
of the hybridizations were as follows: for nitrocellulose,

3xSSC, 60°, 16h followed by two washes in 0.1 x SSC + 0.1% SDS, 50°, 30 min; for zetaprobe, 1.5 x SSPE, 10% dextran sulphate, 68°, 16h followed by two washes in 0.1 x SSC + 1% SDS, 50°, 30 min (Reed and Mann, 1985). Filters were probed with recombinant phage or plasmid DNA nick translated using ^{32}P-dCTP (Amersham, 3000 Ci/mmol) to 1-2 x 10^8 cpm/μg.

2.3. Construction and Screening of Genomic DNA Libraries

Genomic DNA libraries were prepared from the Canton S strain of D. melanogaster and the wild type and to^1 and to^2 mutant strains of L. cuprina. The genomic DNA was partially digested with Sau3A, size-fractionated on NaCl gradients and ligated into the BamH1 sites of the λ-derived cloning vector EMBL3A (Frischauf et al., 1983). The preparation of the vector DNA, restriction enzyme digestions, ligations and the in vitro packaging of the recombinant molecules were carried out by standard procedures (Maniatis et al., 1982). Recombinant phage were plated on E. coli strain LE392. The wild type genomic DNA library of D. buzzatii was provided by J. G. Oakeshott; it was also prepared in EMBL3A by procedures identical to those outlined above.

Unamplified preparations of libraries were always used for screening; aliquots were plated on large petri dishes (approximately 10,000 plaques per plate), lifted onto nitrocellulose filters and hybridized in 3 x SSC. In the initial heterologous screenings (D. buzzatii and L. cuprina libraries probed with D. melanogaster st subclones), hybridizations were at 55° for 24h; in homologous screenings (when subcloned fragments of the original clones were used as probes), hybridizations were at 65° for 24h. Probes were labelled by nick translation with ^{32}P - dCTP.

2.4. Subcloning into Plasmid Vectors

A variety of restriction fragments from the λ-clones were subcloned. The following vectors were used: pAT153 (grown in E. coli strain RR1); pUC13 and pUC19 (in JM 83); pGEM-1 and pGEM-2 (in MC1061.1). Plasmid DNA was prepared by the small-scale SDS/NaOH procedure described in Maniatis et al.(1982).

2.5. DNA Sequencing and Computer Analysis

Restriction fragments from the λ-clones or plasmid subclones were ligated into the M13 sequencing vectors mp8, mp18 or mp19 (grown in E. coli strain JPA101). Initially restriction enzymes with 6bp recognition sites were used to generate the fragments but later, as sequence data accumulated, suitable enzymes with 4bp recognition sites were used (Sau3A, HpaII and TaqI). Single-stranded M13 DNA for sequencing was prepared by the method of Messing (1983). All sequencing was carried out by the dideoxy-nucleotide chain termination method developed by Sanger et al.(1977). Sequences were analysed

with the DNA Inspector II program using a Macintosh Plus
microcomputer.

3. RESULTS

3.1. Structures of the *scarlet* and *topaz* Genes

Restriction maps and positions of exons in the *scarlet* (*st*)
genes of *D. melanogaster* and *D. buzzatii* and of the *topaz* (*to*)
gene of *L. cuprina* are shown in Fig. 1.   In no case has the
complete gene structure been determined; for *D. melanogaster*
4.4 kb of DNA has been sequenced (from the *Pst*1 site at -3.0
to the *EcoR*1 site at 1.4). The exons, which contain a continuous open reading-frame and which are bounded in all cases by
consensus splice donor and acceptor sequences, have been
highly conserved between the three species.   By contrast the
introns, which usually contain stop codons in all three read-
ing-frames, show little or no sequence conservation.

For *D. melanogaster* seven exons have been defined; in
length these total 1904bp (see Table I).   Since the size of

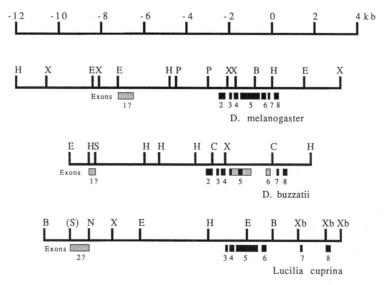

Figure 1.   Structures of the *scarlet* gene homologues in the
three species.   Restriction enzymes are abbreviated as
follows: B = *Bam*HI;   C = *Cla*I;   E = *Eco*RI;   H = *Hind*III;   N
= *Nhe*I;   P= *Pst*I;   S = *Sal*I;   X = *Xho*I;   Xb = *Xba*I.   The
scale is in kb;   positions of those exons that have been
fully sequenced are shown as black boxes beneath the maps
(the dotted boxes show tentative positions of exons based on
cross-species homologies).   Map coordinate 0 for the *D.
melanogaster* gene was arbitrarily assigned to the *Hind*III
site in intron 7 (Tearle *et al.*, 1989) and the *D. buzzatii*
and *L. cuprina* genes have been approximately aligned to it
using suitably placed restriction sites to set coordinate 0
(*Cla*I and *Bam*HI sites respectively).

Table I
Exons of the *scarlet* gene homologues in three Dipteran species

Exon No.	Length (bp)		
	D. melanogaster	*D. buzzatii*	*L. cuprina*
2	235	235	–
3	50	50	50
4	137	137	137
5	966	–	966
6	242	–	242
7	101	101	101
8	173	164	173
Overall			
Length (bp)	1904	687	1669
% Homology to *D. melanogaster*			
Nucleotides	–	77	70
Amino Acids	–	81	81
% G+C	54	54	39
% G+C (3rd position of codons)	62	64	32

the *st* mRNA, as detected on Northern blots (Tearle *et al.*, 1989), is about 2.4 kb, it appears possible that a single additional exon of 4–500 bp remains to be characterized. This must be located between the *Eco*R1 site at -7.3 and the *Pst*I site at -3.0, since the 8.7 kb *Eco*R1/*Eco*R1 fragment (-7.3 to 1.4) has been shown by P element-mediated transformation to restore *st*⁺ function to *st*⁻ individuals.

For *D. buzzatii*, only two regions have been sequenced – from the *Hind*III site at -3.6 to the *Xho*I at -2.2 and from the *Cla*I site at 0 downstream for about 1 kb. This has given the complete sequence of five exons (2,3,4,7, and 8); in addition two small regions of sequence from exon 5 have been obtained. More sequence data has been obtained from the *to* region of *L. cuprina* – almost continuous sequence from the *Hind*III site at -3.0 to the *Xba*I site at 3.2; this has provided the complete sequences of six exons (3–8). Exon 2 of *to* has not yet been located although, on the basis of weak hybridization between the 0.9 kb *Pst*I/*Xho*I fragment of *D. melanogaster* which contains exon 2 and a *Sal*I/*Nhe*I fragment (-9.6 to -8.8) of *to*, it appears to lie at least 6 kb upstream from exon 3.

3.2. Comparison of Exon Sequences

The data for the comparison of the sequences of the exons for the three species are given in Table I. Those that have been characterized so far are identical in length between the species (with the sole exception of exon 8 of *D. buzzatii* which appears to be three codons shorter than in *D. melanogaster* or *L. cuprina*) and are also interrupted in exactly the same places by introns. The sequences are highly homologous between *D. melanogaster* and *D. buzzatii* (77%) and only slightly less homologous between *D. melanogaster* and *L. cuprina* (70%). When the derived amino acid sequences are compared, the homology is even higher – 81% for both the *D. melanogaster*/*D. buzzatii* and the *D. melanogaster*/*L. cuprina* comparisons.

Despite the highly conserved nature of the nucleotide sequences, a striking difference in the G + C composition between the *Drosophila* species on the one hand (54% for both) and *L. cuprina* on the other (39%) is apparent. This comes about because of the markedly different patterns of codon usage in the *st* genes as compared with *to* (Table II). The *Drosophila* species show a much greater tendency to use codons that have G or C in the third position (62% for *D. melanogaster* and 64% for *D. buzzatii*), whereas codons ending in A or T are used much more frequently by *L. cuprina* (only 32% of codons end with G or C).

3.3. Comparison of Intron Sequences

In contrast to the exons, the introns almost invariably differ in length (Table III). Those so far characterized in *D. melanogaster* are short (the longest being intron 4, 185bp) and this also appears to be the case in *D. buzzatii*. (Note, however, that intron 1 of the *st* genes may be quite long – 3 to 6 kb; see Fig. 1). In contrast, although three introns of the *L. cuprina* gene are also short (introns 3, 4 and 5) the other two (6 and 7) are substantially longer (1–2 kb). It also seems very likely that intron 2 of *to* is quite long (approx. 6 kb) as compared with only 179 and 225 bp for the *st* genes. The introns also vary substantially in nucleotide sequence; the homology between *D. melanogaster* and *D. buzzatii* is only 30% and is even lower (28%) between *D. melanogaster* and *L. cuprina*. Other than at the splice junctions (which conform to the normal consensus sequences and which therefore show higher levels of homology), it seems that the sequences of the introns have been subject to little or no selective pressure. It should also be noted that the nucleotide compositions of the introns are quite different; those of the *to* gene are extremely AT-rich (only 27% G + C) whereas those of the *st* genes are much more GC-rich (39 and 50% for *D. melanogaster* and *D. buzzatii* respectively).

3.4. Repeated Sequence DNA in the *scarlet* and *topaz* Regions

In Fig. 2(a) and 2(b) a comparison is shown between the results of homologous genomic Southern blots using the labelled DNA of lambda clones of *D. melanogaster st* or *L. cuprina to* as the probe. In the case of the *st* blot (Fig. 2a), the *D. melanogaster* DNA was double-digested with *Hind*III and *Eco*R1 prior to electrophoretic fractionation and the blot shows a series of discrete bands which correspond in length (generally) to the *Hind*III/*Eco*R1 fragments in the clone. (The

Table II
Patterns of codon usage in the *scarlet* and *topaz* genes

Codons with A or T in the 3rd position	% Usage			Codons with G or C in the 3rd position	% Usage		
	Dm	Db	Lc		Dm	Db	Lc
AAA(lys)	36	20	69	AAG(lys)	64	80	31
AAT(asn)	58	62	100	AAC(asn)	42	38	0
ACA,ACT(thr)	22	62	69	ACG,ACC(asn)	78	38	31
AGA,CGA, CGT(arg)	38	21	78	AGG,CGG, CGC(arg)	62	79	22
AGT,TCA TCT(ser)	38	23	70	AGC,TCG, TCC(ser)	62	77	30
ATA,ATT(ile)	70	50	76	ATC(ile)	30	50	24
CAA(gln)	33	36	69	CAG(gln)	67	64	31
CAT(his)	50	67	78	CAC(his)	50	33	22
CCA,CCT(pro)	23	50	67	CCG,CCC(pro)	77	50	33
CCA,CCT, TTA(leu)	26	25	52	CTG,CTC, TTG(leu)	74	75	48
GAA(glu)	13	33	91	GAG(glu)	87	67	9
GAT(asp)	68	88	88	GAC(asp)	32	12	12
GCA,GCT(ala)	30	43	67	GCG,GCC(ala)	70	57	33
GGA,GGT(gly)	57	24	79	GGG,GGC(gly)	43	76	21
GTA,GTT(val)	21	31	58	GTG,GTC(val)	79	69	42
TAT(tyr)	44	55	79	TAC(tyr)	56	45	21
TGT(cys)	31	0	75	TGC(cys)	69	100	25
TTT(phe)	46	25	75	TTC(phe)	54	75	25

Table III
Introns of the *scarlet* gene homologues in three Dipteran species

Intron No.	Length (bp)		
	D. melanogaster	*D. buzzatii*	*L. cuprina*
2	179	225	>6000
3	109	70	109
4	185	179	55
5	77	–	183
6	59	–	~1700
7	57	82	~1050

Overall

Length (bp)	666	556	>9000
% Homology to *D. melanogaster*	–	30	28
% G+C	38	50	27

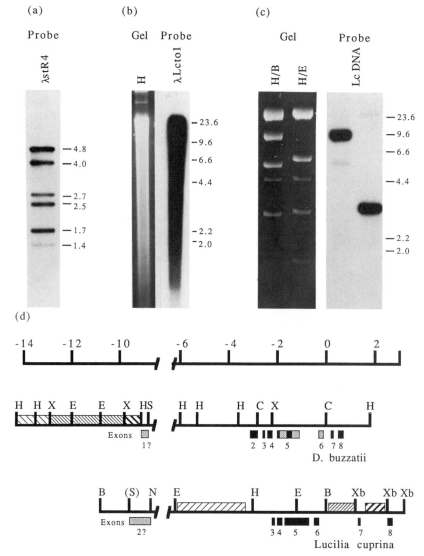

Figure 2. Repeated sequence DNA in the *scarlet* gene homologues in the three species.
(a)   Southern blot analysis of genomic DNA from the Canton S strain of *D. melanogaster*.   The DNA was double-digested with *Eco*RI and *Hin*dIII, electrophoretically separated on an agarose gel, blotted and probed with radiolabelled DNA from the recombinant phage λstR4 which carries the *st* gene (coordinates -11.4 to 2.8 on the map - Fig. 1).   Lengths (kb) of the major bands (shown beside the panel) were determined from the mobilities of marker DNA fragments run on the gel.
(b)   Southern blot analysis of genomic DNA from a wild type strain of *L. cuprina*.   The DNA was digested with *Hin*dIII, separated on an agarose gel (stained gel shown on the left), blotted and probed with radiolabelled DNA from a recombinant phage λLcto1 which carries the *to* gene (coordinates -11.7 to 7.7 on the map).   The

pattern is complicated slightly by restriction site polymorphism in the Canton S stock, as has been described in Tearle
et al., 1989). For *to* however, a completely different result
is obtained; there are no discrete bands but instead an
intense smear of hybridization involving fragments varying in
size from greater than 20 to less than 1 kb (Fig. 2b). Such a
result is indicative of the presence, within *to*, of sequences
which are repeated in thousands of other locations in the
genome. The locations of these dispersed repeated sequences
within the cloned *to* DNA were determined by carrying out
"reverse genomic blots" (Fig. 2c). In this type of experiment, the cloned DNA is digested with restriction enzymes,
electrophoretically fractionated, blotted and the blot then
probed with labelled *L. cuprina* genomic DNA. Only restriction
fragments which contain repeated DNA sequences bind sufficient
labelled probe to give hybridization signals. Three restriction fragments have been identified as containing repeated DNA
and their positions are indicated on the restriction map shown
in Fig. 2d; the *Eco*RI/*Hind*III fragment (-6.2 to -3.0) produces
a strong hybridization signal and hence contains sequences
which are repeated at a high frequency in the genome, whereas
the *Bam*H1/*Xba*I (0 to 1.2) and the *Xba*I/*Xba*I (1.2 to 2.5) fragments give weaker signals and therefore contain sequences
which are repeated at lower frequencies. In all three cases,
the restriction fragments correspond to intron regions and, in
fact, coincide with those introns which are markedly longer in
the *to* compared to the *st* genes (introns 2, 6 and 7 - see
Table III).

Sequence data has been obtained from each of the repeated
DNA regions and a summary of this data is given in Table IV.
A number of different types of repeated sequences have been
identified which appear to fall into three categories. The
simplest of these is exemplified by the sequences in intron 6;
the region contains about 200 tandemly repeated copies of a
tetramer which is predominantly GACA at the left-hand end
(with the occasional GATA variant) and TACA at the right. We
estimate that this tandem repeat is present at about 100 other
sites in the genome. The second type also contains an internal repeating unit but the units are usually not tandemly
repeated; the core units are somewhat variable in both length
and sequence and are separated from each other by 10 - 50 bp

positions on the gel of marker fragments (*Hind*III-digested λ DNA)
are indicated.
(c) Reverse genomic Southern blot of DNA from the recombinant
phage λLcto1. The DNA was digested with the restriction enzymes
indicated, electrophoretically separated on an agarose gel
(stained gel on the left), blotted and probed with radiolabelled
genomic DNA from *L. cuprina*. Marker fragments as in (b).
(d) Restriction maps of the *to* region of *L. cuprina* and the *st*
region of *D. buzzatii* showing the locations of repeated sequence
DNA. Restriction enzyme abbreviations, positions of exons and
coordinates are as given in Fig. 1. Hatched regions above the
map show the locations of the repeated sequences; different types
of cross-hatching indicate different repetitive frequences in the
genome for the sequences.

Table IV
Repeated DNA sequences in the *topaz* introns

Map coordinate (Fig.1)	Intron	Repeat unit (bp)	Most common sequence	Comments
-6.2 to -6.0	2	13	TATNGN$^{C}_{T}$TAGTC$^{T}_{C}$	Some tandem arrays; some dispersed
0 to 0.8	6	4	GA$^{C}_{T}$A, TACA	Tandem arrays
0.9 to 1.3	6,7	13	ACTTTT$^{T}_{C}$GTTCTT	Dispersed units
1.8 to 2.2	7	14	AGTTGAATNGACAT	Some tandem arrays; some dispersed

Tandem Arrays (Intron 6)
<u>GAC</u>AGACAGACAGATAGACAGACAGACAGATA
Dispersed Units (Intron 6)
GGT<u>ACTTTTTCGTTCTT</u>TTCTTAAAATTAAAGCTAAAAACATGTAACGTATA
AAAAGA<u>ACTTTTTGTTCTT</u>TAAAATGT<u>ACTTTTTGAGTTTC</u>CATAATATTGA

of non-core sequence. Several different examples of this type of repeat organization were found in the *to* region in which the core sequence was different (introns 2, 6 and 7); an example (from intron 6) is given at the bottom of Table IV. The third type of repeats shows no obvious internal repetitive structure but they are, nevertheless, sequences which are which are present at hundreds to thousands of other sites in the genome; repeats of this type are present in intron 2.

As indicated by the discrete banding pattern of the Southern blot (Fig. 2a), no repeated DNA sequences have been detected within the 40 kb region which includes and surrounds the *st* gene of *D. melanogaster*. In the case of *D. buzzatii* some restriction fragments from the *st* region do give hybridization signals of different strengths when "reverse genomic blots" are carried out; however, in all cases the signals are relatively weak indicating that the repetition frequency of these sequences in the genome is low (30-100 copies). The positions of these fragments are shown in Fig. 2d; as yet we have no sequence data from these regions.

3.5. Natural Mutation at the *scarlet* and *topaz* Loci

Several natural mutants of the *st* and *to* genes have been examined with a view to identifying the molecular lesions which have disrupted gene function. In the case of *D. melanogaster*, we have examined three natural alleles - st^{1}, st^{sp} and st^{dct}. As has been described in Tearle *et al.*(1989), the st^{1} and st^{sp} genes contain insertions of non-*st* DNA (Fig. 3). For st^{1} a 7.8. kb insert is located within the 0.9 kb

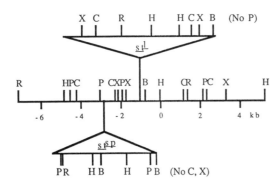

Figure 3. Molecular changes in the st^1 and st^{sp} mutant alleles of *D. melanogaster*. Partial restriction maps of the insertions in st^1 and st^{sp} are given. Restriction enzyme abbreviations and map coordinates are as given in Fig. 1. Restriction sites within the inserts were determined from the lengths of altered restriction fragments (relative to wild-type) on Southern blots; note that a maximum of two sites can be determined for each enzyme since additional sites lying between them will not be detected. Restriction enzymes that do not cut are indicated in parentheses.

*Xho*1/*Bam*H1 fragment (-1.7 to -0.8), possibly within exon 5 (see Fig. 1). In the case of st^{sp} the insertion is 5 kb in length and is located within the *Pst*I/*Xho*I fragment (-3.0 to -2.1), probably within intron 2. The third allele st^{dct} contains an insertion of about 1.6 kb also within the 0.9 kb *Pst*I/*Xho*I fragment. In no case have we obtained sequence data for the insertions and neither have we been able to identify them as known transposable elements on the basis of the partial restriction maps we have assembled. The *st* allele of *D. buzzatii* was also examined by genomic Southern blotting, but the pattern of bands obtained was so different from wild type that we were unable to make deductions about the possible lesion. Probably the extensive differences reflect the extent of sequence polymorphism in wild populations.

Two mutant alleles of *topaz*, to^1 and to^2, have been examined; the former allele appears to be a null for *to* activity since adults contain less than 5% of the wild type level of xanthommatin, while the latter is leaky (about 20% wild type xanthommatin) (Summers and Howells, 1978). Because of the difficulties which the presence of repeated sequences within the *to* clones cause, little of the analysis involved genomic Southern blots; instead the mutant genes were cloned and extensive sequence data obtained.

For to^1 5.7 kb of almost continuous sequence was determined from near the *Hind*III site (-3.0) upstream of exon 3 to the *Xba*I site (3.2) downstream of exon 8. Numerous sequence differences were detected relative to wild type but since the data does not cover exons 1 or 2 (for either wild

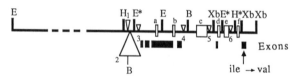

Figure 4. Molecular changes in the *to*[1] allele of *L. cuprina*.

Restriction enzyme abbreviations, positions of exons and map coordinates are as given for *to*[+] in Fig. 1. The sequence data for *to*[1] extends from the *Hin*dIII site at -3.0 to the *Xba*I site at 3.2. The positions of 4 single bp changes are shown - three that create restriction sites relative to *to*[+] (indicated by asterisks) and one that causes a conservative amino acid substitution in exon 8. (There are an additional 109 single bp changes in introns and exons). Positions of deletions are indicated by the open boxes and their lengths (bp) are as follows: a - 18; b - 2; c - approx. 500; d - 8; e - approx. 120; f - 2. Insertions are indicated by triangles - the two larger ones (1 and 2) are approximately to scale (0.24 and 1.05 kb respectively); for the small ones the lengths are as follows: 3 - 7 bp; 4, 5, 6 - 1bp. The E/H fragment (3.2 kb in *to*[+], coordinates -6.2 to -3.0) is considerably longer in *to*[1] - as indicated by the dashed line; this region contains repeated sequence DNA but has not yet been sequenced in *to* .

type or mutant) we are unable to state which (if any) of the changes results in defective *to* function. The positions within the gene of the more prominent changes are shown in Fig. 4 and can be summarized as follows:

1. Single base pair changes. 107 single bp changes have been found in intron regions, three of which create *Eco*R1 or *Hin*dIII sites (the *Eco*R1 sites at -2.4 and 1.7, the *Hin*dIII site at 2.2) giving the *to*[1] restriction map a somewhat altered appearance relative to wild type. Six further single bp changes occur within the exons, five of which are silent; the sixth causes a conservative ile -> val change in exon 8 but it seems unlikely that such a substitution would affect protein function.

2. Deletions. Two restriction fragments are shorter in *to*[1] compared with *to*[+], the *Bam*H1/*Xba*I fragment which is 1.2 kb in *to*[+] (0-1.2) but only 0.7 kb in *to*[1] and the adjacent *Xba*I/*Xba*I fragment (1.2-2.5) which is 0.1 kb shorter in *to*[1]. Both fragments contain repeated sequence DNA and sequencing has shown that the deletions involve the repeats. However, in both cases the

reduction in size is more complicated than a simple deletion of repeat units. For the BamH1/XbaI fragment the number of copies of the tandemly repeating tetramer is reduced from about 200 in to^+ to less than 100 in to^l but the order of tetramer units is also changed - and the TACA variant seen in to^+ is not present in to^l. In the case of the XbaI/XbaI fragment both a deletion and an insertion appear to have occurred, the net result of which is a region of 240 bp of altered sequence (relative to to^+) and a length reduction of about 0.1 kb. A third more straight-forward deletion of 18 bp has been identified within exon 6; such a deletion would remove six amino acids but would not alter the reading frame of translation. It is possible that the removal of these six amino acids would disrupt the structure of the protein and cause loss of function but this is not necessarily the case. In addition to these major deletions several other small (2-8 bp) deletions are found in introns.

3.   Insertions.   Two insertions have been found in intron 2;   one is about 0.2 kb and the other about 1.1kb and they are separated from each other by only 22 bp (at about coordinate -2.4).   Both insertions have been completely sequenced - neither shows any internal repetitive structure, nor do they have repeated ter-mini; in neither case have they caused duplications of genomic sequence at their sites of insertion.   Never-theless they are repeated elsewhere in the genome,   the smaller insertion appearing to have a repetition fre-quency of about 2000.   The situation for the larger insertion is more complicated;   sequences at the left-hand end also appear to be repeated about 2000 times but those at the right-hand end only about 100 times. Obviously more than one repeat element is involved, one apparently being embedded in the other. In addition to these larger insertions several smaller insertions of 1-7 bp have been found within the introns.

   The analysis of to^2 is less complete, partly as a result of cloning difficulties.   Despite the fact that three separate libraries have been made from to^2 DNA, and eight different to^2 clones have been isolated, clones carrying the 3'-end of the gene (i.e., carrying sequences from beyond the BamH1 site at 0) have not been obtained.   The region beyond the BamH1 site contains repeated sequences and it is interesting to note that genomic Southern blots indicate that the BamH1/XbaI fragment (1.2kb in to^+), which contains the tandemly repeating tetramer, is very much longer in to^2 (about 3.2 kb).   This suggests that the repeat number of the tetramer may be very much higher in to^2 compared with to^+, which may cause instability during replication in *E. coli*.   As a consequence of cloning difficul-ties only about 2 kb of to^2 has been sequenced,   concentrated in those regions which show major differences between to^l and to^+.   The results can be summarized as follows:

   1.   Insertion region in intron 2.   Insertions are pre-sent in to^2;   the sequence data is incomplete but show

that they are not identical to those in to^1. They do, however, share some sequences in common, suggesting that a different order of multiple insertion events may have occurred.

2.   Deletion in exon 5. The 18 bp deletion detected in to^1 is not present in to^2.

3.   Single base pair changes. Base pair changes are common in the intron regions, some of which are the same as those in to^1 and some of which are different. Of the four single bp changes in exon regions, three are silent and one causes a non-conservative gln -> his substitution in exon 5; since to^2 is a leaky mutant it is possible that the reduction in to function could be due to such a single amino acid change.

4. DISCUSSION

The structure of the st/to gene has been strongly conserved during dipteran evolution.   Not only are the exon regions homologous in sequence, but they are also interrupted by introns at precisely the same sites in all three species. When the homology of the derived amino acid sequences of the proteins are compared exon by exon for *D. melanogaster* and *L. cuprina*, it is clear that the conservation of sequence is quite uniform - from a low of 80% (exon 4) to a high of 87% (exon 3).   Despite this overall uniformity, it is clear that certain regions within exons are more conserved than others. The product of the st/to gene appears to belong to a family of ATP-dependant membrane proteins with permease function (Mount, 1987;   Dreesen *et al.*, 1988;   Tearle *et al.*, 1989).   Not surprisingly the ATP-binding sites, which are likely to be crucial for function, have been highly conserved.   Overall, the rate of change in the amino acid sequence for the *D. melanogaster/L. cuprina* comparison has been about 1% per 5 MY, which is slower than that for some proteins (e.g., *Adh* and larval haemolymph protein within genus *Drosophila* - 3 to 4 MY) but much faster than the rate for highly conserved proteins like glyceraldehyde-3-phosphate dehydrogenase and arginine kinase (36 and 59 MY respectively)(R.J.McIntyre, personal communication).   It is interesting that the overall extent of amino acid change between the *st* genes of *D. melanogaster* and *D. buzzatii* is much the same as between *st* and *to*, suggesting that the rate of change has been faster within Family Drosophilidae;   however, since the sequence data is incomplete it is premature to draw such conclusions.

In marked contrast to the exons, there is little evidence for conservation of intron sequences (in any of the three interspecies comparisons).   Other than at the splice junctions, which fit the normal consensus sequence patterns (Mount, 1982), no other obvious regions of sequence conservation have been detected.   This suggests that the introns which have been examined so far do not contain sequences involved in transcriptional regulation.   In addition to different sequences, the introns also differ in length between species;

those characterized for the *st* genes are all short (less than 250 bp) - although intron 1 may turn out to be much longer - while in *to*, two introns (6 and 7) are much longer than their *st* counterparts (1-2 kb) while a third (intron 2) appears to be quite long (about 6 kb). As a consequence of the introns, the *to* transcription unit is clearly much longer than those of the *st* genes - possibly 2-3 times longer - which correlates with the larger genome size of *L. cuprina* compared with *D. melanogaster*. Preliminary Cot analysis which we have carried out using *L. cuprina* DNA indicates that its genome is about 4 times greater than that of *D. melanogaster*, probably similar in size to that of the blowfly *Calliphora erythrocephala* (Bier and Muller, 1969).

Despite the overall structural similarity between the three genes, two major differences in sequence organization have been detected. The first involves the nucleotide composition. Both of the *st* genes are relatively GC-rich, with the *D. buzzatii* gene (overall 52% G+C) being slightly richer than that of *D. melanogaster* (51%), whereas the *to* gene is much less so (only 29% G+C). The differences are particularly obvious in the introns (27% G+C in *L. cuprina* compared with 38% in *D. melanogaster* and 50% in *D. buzzatii*) but are also apparent in the exons (38% compared with 54% for both *Drosophila* species). This difference in the exons reflects the markedly different patterns of codon usage in *to* compared with the *st* genes, in which codons ending in A or T are commonly used in *to* but more rarely used in the *st* genes. A bias towards codons ending in G or C has been noted for other *D. melanogaster* genes (O'Connell and Rosbash, 1984; Bodmer and Ashburner, 1984). Differences in patterns of codon usage have been noted previously between genes within the one organism; in the cases of *E. coli* (Holm, 1986) and yeast (Sharp *et al.*, 1986) strong correlations exist between the codon preference pattern of genes, their level of expression and the abundance of the different tRNA species - in highly expressed genes the preferred codons correspond with abundant tRNA species. Such correlations may help to fine-tune the rates of translations and maximize the energetic efficiency. In mammals differences in codon usage have been found for genes expressed in different tissues (Newgard *et al.*, 1988) which may be important in controlling the tissue-specificity of expression. However, it is difficult to understand why the same gene, which apparently provides an identical function in two species, would show such different patterns of codon usage or to imagine the series of evolutionary events which might have led to such differences. Of particular interest in the present case is that the two *Drosophila* species show similar patterns of codon usage while it is the *L. cuprina* pattern that is different. It will be interesting to discover whether other genes in *L. cuprina* are also AT-rich and show similar patterns of codon usage to that found in *to* and, if so, whether other species belonging to Family Calliphoridae also prefer codons ending in A or T. An indication that this might be the case can be found in the published partial sequence of the xanthine dehydrogenase gene of *C. vicina* (Rocher-Chambonnet *et al.*, 1987), in which G + C in the third position of codons is only 36%.

The second major difference in sequence organization between the genes involves the frequency of interspersed repeated sequence DNA. In the case of *D. melanogaster* no repeated sequence DNA has been found either within the *st* gene or in its vicinity (for at least 30 kb upstream and downstream). This is typical of the long period interspersion pattern which characterizes the genome of this species (Manning *et al.*, 1975). For *D. buzzatii*, the majority of the cloned DNA including and surrounding *st* (approximately 30 kb) is unique sequence, so the general organization of the *D. buzzatii* genome is probably of the long period interspersion pattern. However, a 5.5 kb region that does contain repeated sequences has been found; it lies almost 6 kb upstream from exon 2 and is almost certainly upstream of the *st* transcription unit. The frequency with which sequences from this region are repeated elsewhere in the genome is low (30-100) but, interestingly, there appear to be at least three different sequence elements in the region which have different repetition frequencies. It may be a region of scrambled repeats of the type which have been found in the *D. melanogaster* genome (Wensink *et al.*, 1979).

The organization of the genome in *L. cuprina* appears to be considerably different from that of the *Drosophila* species. Within the *to* clones three different regions containing interspersed repeated sequence DNA have been located and all appear to lie within the *to* transcriptional unit. Two of the regions have been shown definitely to lie within introns by sequencing (introns 6 and 7) while the third is located in a region that is almost certainly part of intron 2. This latter region contains a number of different types of repeats, some of which are repeated at as many as 10^3 to 10^4 other sites in the genome. The *L. cuprina* genome therefore appears to have a short period interspersion pattern which is common for the genomes of vertebrates, but which has also been found for a number of other insects (Crain *et al.*, 1976; Efstradiatis *et al.*, 1976). Of the three different types of repeat structures we have detected by sequencing, the most striking are the tandem repeats (intron 6) which are similar in many ways to highly repeated satellite DNA. Like satellite DNA the repeat units show some sequence heterogeneity and also appear to expand and contract in copy number over relatively short genetic distances. Thus the number of copies of the tetrameric repeat unit in intron 6 varies in the three different strains of *L. cuprina* that we have examined (*to*$^+$, *to*^{1} and *to*^{2}) and in the case of the *to*$^+$/*to*^{1} comparison, the distribution of the tetrameric sequence variants is different. Like the tandem repeats of satellite DNA, these interspersed tandem repeats also appear to be replicated poorly in *E. coli* so that clones carrying them often appear to be under-represented in genomic DNA laboratories (e.g., the 3'-region of *to*^{2}) and subcloning such regions can be difficult. The frequency with which the particular tandem repeat of intron 6 is present at other sites in the genome is not great (about 10^2) but we have found another tandemly repeating 11-mer in clones carrying the *white* gene of *L. cuprina* which shows a much higher repetition frequency in the genome (about 10^4).

An interesting feature of the repeating tetramer in intron 6 is that in its most common forms (GACA and GATA) it is the same as the so-called "Garden of Eden" repeat, discovered first to be associated with the sex-determining chromosome in the snake (*Bungarus fasciatus*) and subsequently with the sex-determining chromosome in a variety of other organisms (Singh *et al.*, 1980; Epplen *et al.*, 1982). This association lead Singh and Jones (1982) and Jones (1983) to propose that the "Garden of Eden" repeat region might be involved in the sex-determining mechanism, but subsequent work in which these repeats have been found on the X-chromosome and also to be autosomally located in some organisms (Traut, 1987; Nanda *et al.*, 1988) has not supported this proposal. Our finding that GACA/GATA repeats form part of the interspersed moderately repeated DNA of *L. cuprina* is certainly not compatible with a function in sex determination. GACA repeats have also been detected in the genomes of other blowflies (*Calliphora erythrocephala* and *Chrysomya rufifacies*) by Southern blotting (Kirchhoff, 1988). These data indicates that in these blowflies, as in *L. cuprina*, the repeats occur in tandem blocks interspersed amongst the unique sequence DNA, are present at a relatively low repetition frequency in the genome and also that the copy number of the repeat unit in each cluster can vary even in DNA isolated from different tissues.

In addition to the tandem repeats, we have found two other types of interspersed repeated sequences. One type consists of relatively short (10-14 bp) repeating core units which are somewhat variable in both length and sequence; occasionally core units occur in short tandem groups but much more commonly are separated from each other by variable lengths of non-core sequence. The organization of these repeating DNA structures is similar in some respects to that of the interspersed mini-satellite repeats found in the human genome by Jeffreys *et al.*, (1985). Repeated sequence DNA of this type may be relatively common in the *L. cuprina* genome since we have detected several different core sequences within the *to* introns.

The third type of repeating element seems to have a more complex structure and perhaps comprises the majority of the interspersed repeated sequence DNA in the *L. cuprina* genome. These types of elements are best exemplified by the insertions found in intron 2 of the *to[1]* gene. They have no obvious internal repeating structure, they are unlike the DNA of transposable elements in that no repeated termini are present and neither do they have poly dA-dT tracts at one end, as is characteristic of reverse transcribed mRNA. The small (240 bp) insertion could be a *L. cuprina* SINE but it is not flanked by duplications of the genomic sequence which is a feature of SINES in most organisms (Singer, 1982). Our understanding of the structure of the larger insertion is still quite incomplete. We know that the repetition frequency of the sequences at the two ends are very different, indicating clearly that at least two different repeating elements are involved; however, we have not investigated the repetition frequency of internal regions, so it is quite conceivable that other elements are represented there. Its structure may be somewhat similar to that of the "scrambled repeats" of *D. melanogaster* (Wensink *et al.*, 1979) which consist of multiple repetitive elements inserted within each other; similar types of structures have been found in the genomes of other organisms (Musti *et al.*, 1981; Peoples *et al.*, 1985; Daniels and Deininger, 1985).

Interspersed repeated sequence DNA consisting of tandem repeats or of repeating core units need not necessarily be transposable in the genome, but could arise from reiterative copying errors of particular sequences by a polymerase during DNA synthesis, followed, perhaps, by unequal crossing-over events. However, it does seem likely that the third type of repeat structure must be mobile within the *L. cuprina* genome since they can be present at a particular genomic site in some individuals but absent from that site in others. However, their structure provides no clues about the mechanism by which such transposition might occur. Another aspect for specula- tion, about interspersed repeated sequence DNA generally, con- cerns the potential evolutionary effect that such sequences might have by generating diversity in the genome. The fact that, for *to*, the repeated sequence DNA is located only in non-coding regions tends to suggest that generally it is func- tionally inert. However, it is quite conceivable that chance reiterative replication of sequences in a promoter region or the transposition of sequences with enhancer properties could change the regulation of genes. Even more powerfully, inter- spersed repeated sequences might affect chromosome stability, acting as foci for recombination between non-homologous regions of chromosomes and potentially giving rise to chromo- some deletions, duplications, inversions and transpositions. In the human genome gene duplications (Lehrman *et al.*, 1987a), gene deletions (Lehrman *et al.*, 1987b; Nicholls *et al.*, 1987) and chromosome transpositions (Rouyer *et al.*, 1987) involving recombinational events between Alu repeats have been discov- ered. As yet, however, there is no evidence that the chromo- somes of *L. cuprina*, which contain much interspersed repeated sequence DNA, are less stable than those of *D. melanogaster* which contain little.

Related to the topic of interspersed repeated sequence DNA is the one of natural mutation. In the case of *D. melanogaster* it is now well established that many spontaneous mutations arise by the insertion into genes of transposable elements. For example, in the case of the mutations of *white* studied by Zachar and Bingham (1982), inserts of non-*white* DNA were detected in seven of the 13 alleles. Our data for *st*, in which all three spontaneous mutant genes contain insertions, provides a further illustration of the powerful mutagenic effect of mobile DNA in this species. In the case of *L. cup- rina*, despite extensive sequencing, we are still unable to come to definite conclusions about the nature of the lesions in the mutant *to* genes. It is possible that the insertions present in intron 2 of *to*[1] and *to*[2] could affect transcription; however this seems unlikely to be the case since preliminary Northern blot analysis indicates that *to* transcripts are pre- sent in polyA$^+$-RNA isolated from *to*[1] and *to*[2] pupae. It is also possible that the mutagenic events involve fairly simple sequence changes – perhaps the 18 bp deletion of exon sequence for *to*[1] and the single bp change for *to*[2] which causes a non-

conservative amino acid substitution. However, until the complete sequences of the wild type and mutant alleles are available such proposals are purely speculative. Nevertheless it
is interesting that so far our investigations of the two
mutant alleles of *to* (and also several of *white*) have produced
no evidence that transposable elements (of the types found in
D. melanogaster) play a major role in natural mutation in *L.
cuprina*.

Although our studies on *to^1* and *to^2* have failed so far to
give definite information about the molecular nature of the
lesions, they have been extremely valuable since they have
given us further insights into the nature of interspersed moderately repeated DNA in the *L. cuprina* genome and, also, have
highlighted the extent to which sequence polymorphism exists
in natural populations of this species. In the case of the
to$^+$/to^1 comparison 107 single bp changes were present in about
4 kb of intron sequence; in addition a further seven small
insertions or deletions were detected. Assuming that the
latter changes were the result of single events, the level of
polymorphic variation between the alleles within intron
regions is therefore about 3%, which is similar to the level
of intronic sequence variation observed between alleles of the
Adh gene of *D. melanogaster* (Kreitman, 1983). However, in
addition to the sequence polymorphism caused by small changes,
there are the larger scale polymorphisms in *to* involving the
expansion and contraction of the tandemly repeated and repeating core regions, as well as that brought about by the insertion (or deletion) of complex structures like those found in
intron 2. The net result of the sequence polymorphism are
restriction maps for *to^1* and *to^2* which are markedly different
from that of the *to$^+$* allele. It seems likely that a similar
level of sequence polymorphism also exists in populations of
D. buzzatii since our preliminary analysis of the mutant *st*
gene, by Southern blotting, gave patterns of bands which were
very different from those obtained with the *st$^+$* strain.

5. SUMMARY

Extensive sequencing of the *scarlet* (*st*) gene of *D.
melanogaster* and of its homologue the *topaz* (*to*) gene of *L.
cuprina*, and some preliminary sequencing of the *st* gene of *D.
buzzatii*, have been undertaken. On the basis of cross-species
comparisons of sequences, it has been possible to identify
seven of the exons of the *D. melanogaster* gene, six exons for
the *L. cuprina* and five for the *D. buzzatii* gene, which have
strongly conserved sequences (70-80% homology). By contrast,
virtually no conservation of sequence can be found in intron
regions. Two major differences in gene and genome organization have been found between the *Drosophila* species and *L.
cuprina*. Firstly, the nucleotide compositions are markedly
different, being relatively GC-rich (about 50% G+C) in the
former species but much less so in the latter (only 29% G+C)
In exons this is accompanied by dramatic changes in the pattern of codon usage, with codons ending in G or C being used
frequently in the *st* genes (62%) but much less so in *to* (32%).
The second difference involves the prevalence of interspersed

repeated sequence DNA. No such DNA has been found within or adjacent to *st* of *D. melanogaster* but in *to*, repeated sequences (of several different types, some of which are repeated at least 10^4 times in the genome) have been found in three of the introns. The *st* gene of *D. buzzatii* seems to be more like its *D. melanogaster* counterpart in that no repeated sequences are located in introns (although some are located in a region which is probably upstream of the gene). An analysis of two spontaneous mutants of *to* by sequencing has highlighted the extent to which sequence polymorphism exists in natural populations of *L. cuprina*; in the comparison between *to*[+] and *to*[7] alleles, 114 small differences (single bp changes or insertions or deletions of less than 20 bp) were found in 4 kb of intron sequence and a further 7 (including an 18 bp deletion) in exons. In addition to the small-scale changes, large-scale differences involving the presence or absence or repetition frequency of the repeated sequence regions were found.

ACKNOWLEDGEMENTS. This work was supported by a Wool Industry Postgraduate Award to A.E., a Commonwealth Postgraduate Award to R.G.T. and research grants from the Rural Credits Development Fund and the Australian Research Grants Committee. We are indebted, in particular, to Donna Boyle who did most of the sequencing of the *st* gene of *D. melanogaster* and also to a number of other former Research Assistants and Honours students who have contributed to the work: Vanessa Corrigan, Catherine Landsberg, Juliet Miller, Fiona Morris and Barton Wicksteed.

References

Bender, W., Spierer, P., and Hogness, D. S., 1983, Chromosomal walking and jumping to isolate DNA from the *Ace* and *rosy* loci and the *bithorax* complex in *D. melanogaster*, *J. molec. Biol.* **168**:17-33.

Beverley, S. M., and Wilson, A. C., 1984, Molecular evolution in *Drosophila* and Higher Diptera II. A time scale for fly evolution, *J. Mol. Evol.* **21**:1-13.

Bier, K., and Muller, W., 1969, DNS-Messungen bei Insekten und eine Hypothese uber retartiderte Evolution und besonderen DNS-Reichtum im Tierreich, *Biol. Zbl.* **88**:425-449.

Bodmer, M., and Ashburner, M., 1984, Conservation and change in the DNA sequences coding for alcohol dehydrogenase in sibling species of *Drosophila*, *Nature, Lond.* **309**:425-430.

Coen, E. S., Thoday, J. M., and Dover, G., 1982, Rate of turnover of structural variants in the rDNA family of *Drosophila melanogaster*, *Nature, Lond.* **295**:564-568.

Crain, W. R., Davidson, E. H., and Britten, R. J., 1976, Contrasting patterns of DNA sequence arrangement in *Apis mellifera* (Honeybee) and *Musca domestica* (Housefly), *Chromosoma* **59**:1-12.

Daniels, G. R., and Deininger, P. L., 1985, Integration site preferences of the Alu family and similar repetitive DNA sequences, *Nucl. Acids Res.* **13**:8939-8954.

Dreesen, T. D., Johnson, D. H., and Henikoff, S., 1988, The *brown* protein of *Drosophila melanogaster* is similar to the *white* protein and to components of active transport complexes, *Mol. Cell. Biol.* **8**:5206-5215.

Efstradiatis, A., Crain, W. R., Britten, R. J., and Davidson, E. H., 1976,

DNA sequence organization in the lepidopteran *Antheraea pernyi*, *Proc. natn. Acad. Sci. USA* **73**:2289-2293.

Epplen, J. T., McCarrey, J. R., Sutou, S., and Ohno, S., 1982, Base sequence of a cloned snake W-chromosome fragment and identification of a male-specific putative mRNA in the mouse, *Proc. natn. Acad. Sci. USA* **79**:3798-3802.

Frischauf, A. M., Lehrach, H., Poustaka, A., and Murray, N., 1983, Lambda replacement vectors carrying polylinker sequences, *J. molec. Biol.* **170**:827-842.

Holm, L., 1986, Codon usage and gene expression, *Nucl. Acids Res.* **14**:3075-3087.

Jeffreys, A. J., Wilson, V., and Thein, S. L., 1985, Hypervariable "minisatellite" regions in human DNA, *Nature, Lond.* **317**:67-73.

Jones, K. W., 1983, Evolutionary conservation of sex specific DNA sequences, *Differentiation* **23(S)**:S56-S59.

Kirchhoff, C., 1988, GATA tandem repeats detect minisatellite regions in blowfly DNA (Diptera: Calliphoridae), *Chromosoma* **96**:107-111.

Kreitman, M., 1983, Nucleotide polymorphism at the alcohol dehydrogenase locus of *Drosophila melanogaster*, *Nature, Lond.* **304**:412-417.

Lehrman, M. A., Goldstein, J. L., Russell, D. W., and Brown, M. S., 1987a, Duplication of seven exons in LDL receptor gene caused by Alu-Alu recombination in a subject with familial hypercholesterolemia, *Cell* **48**:827-835.

Lehrman, M. A., Russell, D. W., Goldstein, J. L., and Brown, M. S., 1987b, Alu-Alu recombination deletes splice acceptor sites and produces secreted low density lipoprotein receptor in a subject with familial hypercholesterolemia, *J. biol. Chem.* **262**:3354-3361.

Maniatis, T., Fritsch, E. F., and Sambrook, J., 1982, *Molecular cloning: a laboratory manual*, Cold Spring Harbor Laboratory, New York.

Manning, J. E., Schmid, C. W., and Davidson, N., 1975, Interspersion of repetitive and non-repetitive DNA sequences in the *Drosophila melanogaster* genome, *Cell* **4**:141-155.

Messing, J., 1983, New M13 vectors for cloning, *Meth. Enzym.* **101**:20-78.

Miklos, G. L. G., Healy, M. J., Pain, P., Howells, A. J., and Russell, R. J., 1984, Molecular and genetic studies on the euchromatin-heterochromatin transition region of the X chromosome of *Drosophila melanogaster*: I. A cloned entry point near to the *uncoordinated* locus, *Chromosoma* **89**:218-227.

Mount, S. M., 1982, A catalogue of splice junction sequences, *Nucl. Acids Res.* **10**:459-472.

Mount, S. M., 1987, Sequence similarity, *Nature, Lond.* **325**:487.

Musti, A. M., Sobieski, D. A., Chen, B. B., and Eden, F. C., 1981, Repeated deoxyribonucleic acid clusters in the chicken genome contain homologous sequence elements in scrambled order, *Biochemistry* **20**:2989-2999.

Nanda, I., Neitzel, H., Sperling, K., Studer, R., and Epplen, J. T., 1988, Simple GAT/CA repeats characterize the X chromosomal heterochromatin of *Microtus agrestis*, European field vole (Rodentia, Cricetidae), *Chromosoma* **96**:213-219.

Newgard, C. B., Nakano, K., Hwang, P. K., and Fletterick, R. J., 1986, Sequence analysis of the cDNA encoding human liver glycogen phosphorylase reveals tissue-specific codon usage, *Proc. natn. Acad. Sci. USA* **83**:8132-8136.

Nicholls, R. D., Fischel-Ghodsian, N., and Higgs, D. R., 1987, Recombination at the human α-globin gene cluster: sequence features and topological constraints, *Cell* **49**:369-378.

O'Connell, P., and Rosbash, M., 1984, Sequence, structure, and codon preference of the *Drosophila* ribosomal protein 49 gene, *Nucl. Acids Res.* **12**:5495-5513.

O'Hare, K., Murphy, C., Levis, R., and Rubin, G. M., 1984, DNA sequence of the *white* locus of *Drosophila melanogaster*, *J. molec. Biol.* **180**:437-455.

Peoples, O. P., Whittaker, P. A., Pearston, D., and Hardman, N., 1985, Structural organization of a hypermethylated nuclear DNA component in *Physarum polycephalum*, *J. gen. Microbiol.* **131**:1157-1165.

Reed, K. C., and Mann, D., 1985, Rapid transfer of DNA from agarose gels to nylon membranes, *Nucl. Acids Res.* **13**:7207-7221.

Rocher-Chambonnet, C., Berreur, P., Houde, M., Tiveron, M. C., Lepesant, J. A., and Bregegere, F., 1987, Cloning and partial characterization of the xanthine dehydrogenase gene of *Calliphora vicina*, a distant relative of *Drosophila melanogaster*, *Gene* **59**:201-212.

Rouyer, F., Simmler, M-C., Page, D. C., and Weissenbach, J., 1987, A sex chromosome rearrangement in a human XX male caused by Alu-Alu recombination, *Cell* **51**:417-425.

Sanger, F., Nicklen, S., and Coulsen, A. R, 1977, DNA sequencing with chain-terminating inhibitors, *Proc. natn. Acad. Sci. USA* **74**:5463-5467.

Sharp, P. M., Tuohy, T. M. F., and Mosurski, K. R., 1986, Codon usage in yeast: Cluster analysis clearly differentiates highly and lowly expressed genes, *Nucl. Acids Res.* **14**:5125-5143.

Singer, M. F., 1982, SINEs and LINEs: Highly repeated short and long interspersed sequences in mammalian genomes, *Cell* **28**:433-434.

Singh, L., Purdom, I. F., and Jones, K. W., 1980, Conserved sex-chromosome-associated nucleotide sequences in eukaryotes, *Cold Spring Harbor Symp. Quant. Biol.* **45**:805-813.

Singh, L., and Jones, K. W., 1982, Sex reversal in the mouse (*Mus musculus*) is caused by a recurrent nonreciprocal crossover involving the X and an aberrant Y chromosome, *Cell* **28**:205-216.

Southern, E., 1975, Detection of specific sequences among DNA fragments separated by gel electrophoresis, *J. molec. Biol.* **98**:503-517.

Sullivan, D. T., and Sullivan, M. C., 1975, Transport defects as the physiological basis for eye color mutants of *Drosophila melanogaster*, *Biochem. Genet.* **13**:603-613.

Sullivan, D. T., Bell, L. A., Paton, D. R., and Sullivan, M. C., 1980, Genetic and functional analysis of tryptophan transport in Malpighian tubules of *Drosophila*, *Biochem. Genet.* **18**:1109-1130.

Summers, K. M., and Howells, A. J., 1978, Xanthommatin biosynthesis in wild type and mutant strains of the Australian sheep blowfly, *Lucilia cuprina*, *Biochem. Genet.* **16**:1153-1163.

Summers, K. M., and Howells, A. J., 1980, Pteridines in wild type and eye colour mutants of the Australian sheep blowfly *Lucilia cuprina*, *Insect Biochem.* **10**:151-154.

Summers, K. M., Howells, A. J., and Pyliotis, N. A., 1982, Biology of eye pigmentation in insects, *Adv. Insect Physiol.* **16**:119-166.

Tearle, R. G., Belote, J. M., McKeown, M., Baker, B. S., and Howells, A. J., 1989, Cloning and characterization of the *scarlet* gene of *Drosophila melanogaster*, *Genetics* **122**:595-606.

Traut, W., 1987, Hypervariable Bkm DNA loci in the moth, *Ephestia kuhniella*: Does transposition cause restriction fragment length polymorphism (RFLP)? *Genetics* **115**:493-498.

Wensink, P. C., Tabata, S., and Pachl, C., 1979, The clustered and scrambled arrangement of moderately repetitive elements in *Drosophila* DNA, *Cell* **18**:1231-1246.

Zachar, Z., and Bingham, P. M., 1982, Regulation of *white* locus expression: The structure of mutant alleles at the *white* locus of *Drosophila melanogaster*, *Cell* **30**:429-441.

CHAPTER 19

Regulatory Evolution of β-Carboxyl Esterases in Drosophila

JOHN G. OAKESHOTT, MARION J. HEALY, AND ANNE Y. GAME

1. INTRODUCTION

The last decade has seen great advances in our understanding of the *cis* acting elements controlling expression of several *Drosophila* genes. Promoters for genes like alcohol dehydrogenase (*Adh*) and dopa decarboxylase (*Ddc*) have been dissected into several separable elements regulating different aspects of the level and cell-type specificity of transcription. As yet, however, relatively few studies have examined the population and evolutionary genetics of these *cis* control elements.

Moreover most of those few population genetic studies which have addressed this issue have not been able to find associations between variation in expression and nucleotide differences in 5' control elements. For example, analyses of restriction endonuclease site polymorphisms in the promoter regions of the *Adh,* α-amylase (*Amy*) and white genes among field-collected lines of *D. melanogaster* have yielded little evidence for associations between promoter polymorphisms and the whole organism activities of the respective gene products (Birley, 1984; Aquadro *et al.*, 1986; Templeton *et al.*, 1987; Langley *et al.*, 1988; Miyashita and Langley, 1988).

One explanation for these negative results may be that little promoter polymorphism affecting the transcription of these genes exists. Large differences in the whole organism activities of many enzymes do segregate in natural populations (Laurie-Ahlberg *et al.*, 1980, 1982). However, the contribution of *cis* controlled transcriptional variation to these differences is generally unknown, and, in a few cases like ADH where it is known, the contribution is small. Much of the whole organism ADH activity variation is due to stability or

JOHN G. OAKESHOTT, MARION J. HEALY, AND ANNE Y. GAME * CSIRO Division of Entomology, GPO Box 1700, Canberra, ACT 2601, Australia.

Ecological and Evolutionary Genetics of Drosophila, Edited by
J.S.F. Barker *et al.*, Plenum Press, New York, 1990

specific activity differences among polymorphic ADH allozymes
(Templeton *et al.*, 1987; Laurie and Stam, 1988, but see also
Schott *et al.*, 1988).

An alternative explanation for the failure to detect associations between 5' restriction site polymorphisms and transcriptional variation may be that the whole organism activity
of the gene product is an inappropriate index of transcriptional activity. Significantly, in two studies where intraspecific activity variation could be correlated with 5'
restriction fragment length polymorphism, the activity was
measured for specific tissues, rather than the whole organism.
One of these studies examined the causes of variation among
field-collected lines of *D. melanogaster* in the ratio of DDC
activities in neural *versus* epidermal tissues. The variation
was found to be due to *cis* inherited differences in the level
of *Ddc* transcription in the two tissue types (Estelle and Hodgetts, 1984a) and was associated with insertional polymorphism
in the *Ddc* promoter (Estelle and Hodgetts, 1984b). The second
study of this type concerned a rare *Adh* variant in *D.
melanogaster* containing a *copia* insert in the *Adh* promoter;
the ADH activity of this variant was depressed in those tissues and life stages at which *copia* was maximally transcribed,
presumably as a result of transcriptional interference (Strand
and McDonald, 1989).

Interspecific comparisons have provided some more examples
of tissue-specific activity variation caused by promoter
differences. ADH is a notable example, with qualitative
variation in the tissues expressing ADH now documented among
Hawaiian species, among melanogaster group species, and
between *D. mojavensis* and other cactophilic repleta group
species (Dickinson, 1983; Batterham *et al.*, 1984; Chambers,
1988). The results of interspecific gene transfer experiments
suggest that most of these tissue-specific differences in ADH
expression are due to differences in *cis* acting promoter
elements regulating the expression of the *Adh* genes (Brennan
et al., 1988; Sullivan *et al.*, Chapter 21, this volume). As a
further example, *D. melanogaster* and *D. virilis* differ
slightly in the timing of DDC expression and analysis of
interspecific transformants again shows that the variation is
due to promoter differences controlling transcription of the
Ddc gene (Bray and Hirsh, 1986).

Thus no clear picture emerges from the work to date on the
role of 5' regulatory polymorphisms in generating enzyme
activity variation. On one hand, little role at all has been
found from intraspecific analysis of whole organism ADH and
AMY activity variation. On the other hand, however, promoter
differences are largely responsible for the differences in
tissue and temporal specificity of ADH and DDC activities
across species, and, to some extent at least, within species
as well. Clearly further analyses of other model systems will
be needed before generalities emerge. Here we present the
credentials and some early results for another such model, the
ß-carboxyl esterases.

2. INTERSPECIFIC VARIATION IN ß-CARBOXYL ESTERASES

2.1. Structural Properties

Carboxyl esterase is the generic term for hydrolytic enzymes that cleave carboxyl acid esters (Heymann, 1980). All insects investigated to date express a number of carboxyl esterases and these have been implicated in diverse metabolic processes, including pheromone reception (Vogt and Riddiford, 1981), juvenile hormone catabolism (Hammock, 1985), digestion (Kapin and Ahmad, 1980), reproduction (Richmond *et al.* , Chapter 15, this volume) and insecticide resistance (Hughes and Devonshire, 1982; Russell *et al.*, Chapter 16, this volume). Seldom however are the *in vivo* substrates of individual carboxyl esterases precisely identified and most esterases exhibit activity against a broad range of substrates *in vitro*. Those which show such a broad range of substrates *in vitro* but nevertheless exhibit a preference for ß- over α-naphthyl acetates are termed ß-carboxyl esterases (Richmond *et al.*, Chapter 15, this volume).

All *Drosophila* characterised so far have at least one and sometimes several electrophoretically distinct ß-carboxyl esterases. Evidence of three types indicates that those characterised to date are structurally related. The three types of evidence concern their quaternary structure, immunological cross reactivity and, for a few, primary sequence data.

Two types of quaternary structure have been reported among the ß-carboxyl esterases characterised to date. The enzymes isolated from most species studied from either the *Drosophila* or *Sophophora* subgenera are homodimers of molecular weight 100 - 120 KDa (Fig. 1, and see also Sasaki and Narise, 1978). Within the *Sophophora*, one ß-carboxyl esterase characterised from *D. melanogaster* and some sibling species is a 50 - 60 KDa monomer (Morton and Singh, 1985). Evidence from *D. pseudoobscura* and some of its siblings indicates that enzymes with the two quaternary structures are closely related; the major adult ß-carboxyl esterase of these species is generally a 100 - 120 KDa homodimer but some strains are polymorphic for a 50 - 60 KDa monomeric form (Townsend and Singh, 1984; Arnason and Chambers, 1987).

Some widespread if not completely consistent structural similarities are also indicated by the patterns of immunological cross-reactivity found among the enzymes isolated from different *Drosophila* species (Fig. 1). On one hand, antisera raised against the major adult ß-carboxyl esterase of *D. melanogaster* (EST6) react not only with whole fly extracts from other species in the melanogaster and obscura groups of *Sophophora*, but also with representatives of the virilis and repleta groups of subgenus *Drosophila* (D. Morris and R.C. Richmond, pers. comm.). On the other hand, antisera against a major adult *D. virilis* ß-carboxyl esterase (EST2) cross-react with homogenates of other members of the virilis species group and with representatives of the melanica, robusta and repleta groups, also in the subgenus *Drosophila*, but not with extracts from species in the quinaria section of this subgenus, nor

Taxon				Quaternary Structure (a),(b),(c),(d),(e)	virilis EST 2 (f)	melanogaster EST 6 (g)
repleta-virilis section	repleta group	mulleri subgroup	mojavensis	D		+
			arizonensis	D		
			mulleri			
			aldrichi			
			buzzatii	D		
		mercatorum subgroup	mercatorum		+	
		melanopalpa subgroup	repleta		++	
		hydei subgroup	hydei		+	
	robusta group		moriwakii		+	
			sordidula		+	
			pseudosordidula		+	
			lacertosa		+	
			robusta		+	
	melanica group		pengi		+	
	virilis group		americana	D		
			texana	D		
			virilis	D	+++	+
			ezoana		++	
			montana		++	
			imeretensis		++	
quinaria section	bizonata historio group		bizonata		−	
			historio		−	
	testacae group		testacae		−	
	quinaria group		nigro-maculata		−	
	funebris group		funebris	D	−	
			multispina		−	
			macrospina		−	
			subfunebris		−	
	immigrans group		immigrans		−	
			curviceps		−	
			albomicans		−	
Sophophora Subgenus	melanogaster group		melanogaster	M	−	++
			simulans	M		++
			mauritiana	M		++
			erecta	D		+
			teisseri	D		+
			yakuba	D		++
			ananassae	D	−	
			lutea		−	
			auraria		−	
	obscura group		pseudoobscura	D	−	+
			bifasciata		−	

Drosophila Subgenus

Immunological Studies: virilis EST 2 (f), melanogaster EST 6 (g)

with any species tested from the subgenus *Sophophora* (Sasaki, 1975). In addition, antisera against the EST4 larval ß-carboxyl esterase of *D. mojavensis* only react with extracts from other species in the repleta group of *Drosophila*, and not with those from the virilis group of that subgenus or the melanogaster group of *Sophophora* (Pen *et al.*, 1986, and see Fig. 3). While a lack of immunological cross-reactivity does not necessarily imply a lack of relationship, the results of these immunological comparisons at least suggest some significant sequence divergence among ß-carboxyl esterases in the genus.

The available nucleotide and amino acid sequence data indeed confirm a clear structural relationship but nevertheless some considerable sequence divergence among the enzymes. Nucleotide sequence data are now available for the *Est6* and *EstP* genes from several species in the melanogaster group (Oakeshott *et al.*, 1987; Myers *et al.*, 1988; Collet *et al.*, 1990; J. Karotam, E.A. van Papenrecht, and J.G. Oakeshott unpublished); for the *Est5* gene from *D. pseudoobscura* in the obscura group (J. P. Brady, and R. C. Richmond pers. comm.); for the putative *Est1* and *EstJ* genes of *D. buzzatii* in the repleta group (East *et al.*, Chapter 20, this volume); and for *EstS* of *D. virilis* in the virilis group (L. I. Korochkin pers. comm.). Partial amino acid sequence data have also been reported for the EST4 and EST5 enzymes of *D. mojavensis* in the repleta group (Pen *et al.*, 1989). A common origin for all these systems is indicated by their significant overall sequence similarities. The similarities are strongest in some specific structural regions like the active site and the locations of cysteine residues. However, while overall sequence similarities exceed 90% among some of the systems, they fall below 60% among others. We see below that much of this structural divergence is paralleled by qualitative changes in the expression and, presumably therefore, in the functions of the enzymes.

2.2. Regulatory Variation

Temporal patterns of ß-carboxyl esterase activities have been determined for several *Drosophila* species. However, their tissue specific patterns of expression are generally

---

Figure 1. Comparison of the structural properties of major adult ß-carboxyl esterases from different *Drosophila* species. Quaternary structure indicates that the enzyme exists as a dimer (D) or a monomer (M). Immunological studies indicate a positive (+) or negative (-) reaction between antibodies against EST2 from *D. virilis* or EST6 from *D. melanogaster* and esterases from other species. Data are compiled from: (a) Morton and Singh (1985); (b) Zouros *et al.* (1982); (c) McReynolds (1967); (d) East (1984); (e) Narise and Hubby (1966); (f) Sasaki (1975); (g) D. Morris and R. C. Richmond (pers. comm.). Phylogenetic relationships are as determined by Lemeunier and Ashburner (1984) for the melanogaster group; Throckmorton (1975) for the quinaria section; Throckmorton (1982) for the virilis group; Wasserman (1982) for the repleta group; Levitan (1982) for the robusta and melanica groups.

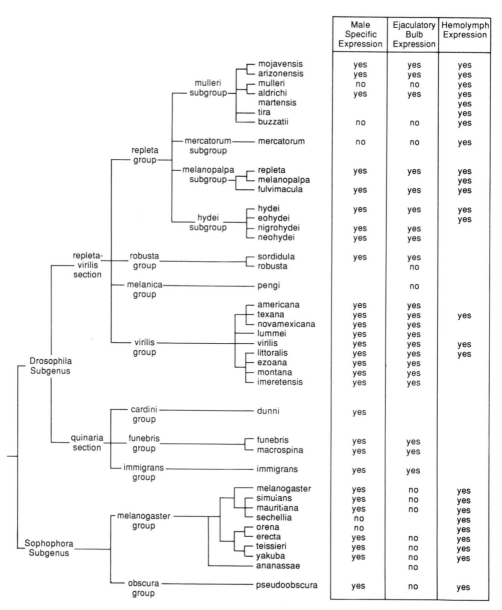

Figure 2. Comparison of the patterns of tissue specific expression of the major adult ß-carboxyl esterases from different *Drosophila* species. The presence or absence of male-specific ejaculatory bulb and hemolymph activity is indicated. Data are compiled from the sources listed in Tables II, III, IV and V and Carrasco *et al.* (1984). Phylogenetic relationships are derived from the sources listed in Figure 1.

known only for adults. The major sites of activity in adults are hemolymph and the male reproductive tract (Fig. 2), although work on a few species also reveals minor activities in the gut, eyes, mouthparts and proboscis (see below).

The hemolymph activity has been detected in virtually all the *Drosophila* and *Sophophora* species tested. While not as extensively surveyed, the minor activities have also been reported from disparate species covering both subgenera. However, male reproductive tract activity, at least as assessed by ß-naphthyl acetate hydrolysis, proves to be highly variable. It is confined to the ejaculatory bulb in most species tested from the subgenus *Drosophila*, but expression in species surveyed from the *Sophophora* ranges from the ejaculatory duct, the testes and accessory glands, to none of the male reproductive tissues at all. Little is known of the function of the hemolymph and minor activities but the male reproductive tract esterase of some species is transferred to the female and affects her subsequent egg laying and remating behaviors (Johnson and Bealle, 1968; Kambysellis *et al.*, 1968; Korochkin *et al.*, 1976; Richmond *et al.*, Chapter 15, this volume).

Most of the species surveyed for tissue specific ß-carboxyl esterase activity derive from four species groups, namely the melanogaster and obscura groups in the subgenus *Sophophora* and the repleta and virilis groups in the subgenus *Drosophila*. Below we give detailed comparisons of the expression phenotypes and genetics of the ß-carboxyl esterases among species from these groups.

2.3. The Melanogaster Group

ß-carboxyl esterases have been characterised to some degree from six species in this group, three each from the melanogaster and yakuba complexes. Adults of each of these species express one major ß-carboxyl esterase, termed EST6 (Morton and Singh, 1985).

The *Est6* gene has been localised to 69A1 on chromosome arm 3L in *D. melanogaster* (Oakeshott *et al.*, 1987) and a similar location is indicated in *D. simulans*, the only other species in the group for which *Est6* has been mapped (Table I and Wright and MacIntyre, 1963). Purified EST6 enzyme is a 63 KDa monomeric glycoprotein in *D. melanogaster* (Mane *et al.*, 1983) and it seems to have essentially the same structure in *D. simulans* and *D. mauritiana*, also in the melanogaster complex (Morton and Singh, 1985). In the three species in the yakuba complex, *D. yakuba*, *D. erecta* and *D. teissieri*, EST6 is a homodimeric protein of about twice the molecular weight (Morton and Singh, 1985).

The three members of the melanogaster complex all show substantially less EST6 activity in pre-adults than in adults, particularly males (Table II). Little is known about the tissue distribution of pre-adult activity, but adults of both

Table I
Chromosomal locations of genes encoding ß-carboxyl esterases

Species	Chromosome	Homologous *D. melanogaster* chromosome[a,b,c,d]
melanogaster group		
D. melanogaster EST6[e]	3L	3L
D. simulans EST6[f]	3L	3L
obscura group		
D. pseudoobscura EST5[g]	X/R	3L
repleta group		
D. mojavensis EST4/EST5[h]	2	3R
D. arizonensis EST4/EST5[h]	2	3R
D. buzzatii EST1/ESTJ[i]	2	3R
D. mulleri ESTD[j]	2	3R
virilis group		
D. virilis EST2[k]	2	3R
D. virilis ESTS[l]	2	3R
D. americana EST2[k]	2-3	3R
D. texana EST2[m]	2	3R
D. lummei EST2[m]	2	3R
D. littoralis EST2[m]	2	3R
D. montana EST2[m]	2	3R

[a]Patterson and Stone, 1952.
[b]Wasserman, 1982.
[c]M. M. Green, P. D. East, and J. S. F. Barker (pers. comm.).
[d]Whiting *et al.*, 1989.
[e]Wright, 1963.
[f]Wright and MacIntyre, 1963.

[g]Hubby and Lewontin, 1966.
[h]Zouros *et al.*, 1982.
[i]Knibb *et al.*, 1987.
[j]Zouros, 1976.
[k]McReynolds, 1967.
[l]Korochkin *et al.*, 1978.
[m]Korochkin *et al.*, 1987.

sexes from all three species show significant EST6 activity in the hemolymph (Table II) and, at least in *D. melanogaster*, low levels in the digestive system, mouthparts, antennae and virgin female reproductive tract as well (M.J. Healy, M.M. Dumancic, and J.G. Oakeshott, unpublished). Significant levels of EST6 occur in the reproductive tract of males from all three species, but in *D. melanogaster* and *D. simulans* it is confined to the ejaculatory duct, whereas in *D. mauritiana* some EST6 is found in the testes and some in either the ejaculatory duct or accessory glands (Table III).

Tables II and III reveal several differences in EST6 expression in the yakuba complex. As with the melanogaster complex, *D. yakuba* and *D. erecta* show most EST6 activity in adults; however relatively high levels are found in larvae and pupae of *D. teissieri*. Sex differences are less pronounced among adults of all three species, with only 6-12% of male activity localised to the reproductive tract. In *D. erecta* the reproductive tract enzyme is confined to the testes but in *D. yakuba* and *D. teissieri* some activity is also found in the accessory glands or ejaculatory duct. All three species again express EST6 in adult hemolymph, however the hemolymph and reproductive tract activities together only account for 25-35%

Table II

Characterisation of adult ß-carboxyl esterases in the melanogaster and obscura groups

	D. melanogaster EST6[a,b]	D. simulans EST6[c,d,e]	D. mauritiana EST6[d,e]	D. erecta EST6[d,e]	D. teissieri EST6[d,e]	D. yakuba EST6[d,e]	D. pseudo-obscura EST5[d,e,f,g]
Temporal expression	Mainly Adults	Mainly Adults	Mainly Adults	Mainly Adults	Larvae Pupae Adults	Mainly Adults	Mainly Adults
Sex specific expression	Male $\gg$ Female	Male $\gg$ Female	Male $\gg$ Female	Male $\geq$ Female	Male $\geq$ Female	Male $\geq$ Female	Male $\geq$ Female
Tissue specific expression: Larvae	Hemolymph	nr[h]	nr	nr	nr	nr	nr
Adults	Hemolymph Gut Mouthparts Antennae Ejaculatory duct	Hemolymph Ejaculatory duct	Hemolymph Testes Accessory gland or Ejaculatory duct	Hemolymph Testes	Hemolymph Testes Accessory gland or Ejaculatory duct	Hemolymph Testes Accessory gland or Ejaculatory duct	Hemolymph Eye Testes Accessory gland or Ejaculatory duct

[a]Sheehan et al., 1979.
[b]M. J. Healy, M. M. Dumancic, and J. G. Oakeshott (unpublished).
[c]Hubby and Lewontin, 1966.
[d]Morton and Singh, 1985.
[e]R. C. Richmond, and D. Morris (pers. comm.).
[f]Berger and Canter, 1973.
[g]Lunday and Farmer, 1983.
[h]Not reported.
Note also that the lists of tissues with activity which are given in this and Tables IV and V are probably not complete, because many tissues have not been tested in each species.

Table III
Male specific expression of β-carboxyl esterases in the melanogaster and obscura groups

	D. melanogaster EST6	D. simulans EST6	D. mauritiana EST6	D. erecta EST6	D. teissieri EST6	D. yakuba EST6	D. pseudo-obscura EST5
Male activity	0.087	0.342	nr[c]	0.022	nr	0.106	0.094
Male tissue distribution (%)[a]							
reproductive tract	28	41	23	6	11	12	17
hemolymph	43	42	42	18	24	25	19
rest of body	29	17	35	76	65	63	64
Reproductive organ distribution (%)[a,b]							
testes	0	0	39	100	27	58	47
accessory glands	0	0	61	0	73	42	53
ejaculatory duct	100	100	0	0	0	0	0
ejaculatory bulb	0	0	0	0	0	0	0

[a] Morton and Singh, 1985.
[b] Johnson and Bealle, 1968.
[c] Not reported.

of total activity and the location of the remaining activity has yet to be identified.

The early electrophoretic analyses of *D. melanogaster* revealed several fainter bands of ß-carboxyl esterase activity in addition to EST6 (Wright, 1963). These bands now assume greater significance, with the recent finding of a tandem duplication of the *Est6* gene in the six melanogaster group species above (Collet *et al.*, 1990; J. Karotam, E. A. van Papenrecht, and J. G. Oakeshott, unpublished). The second esterase gene, called *EstP,* was discovered through the molecular analysis of the *Est6* region. The coding region of *EstP* begins only about 200 bp 3' to the *Est6* termination codon and shows about 60% similarity with the *Est6* coding region. However, the 5' non-coding region of *EstP* shows negligible similarity with that for *Est6* and Northern analysis of *D. melanogaster* shows *EstP* and *Est6* to have very different patterns of expression. *EstP* mRNA is found most abundantly in late larvae, with a small amount in the adult eye (Collet *et al.*, 1990; K. M. Nielsen and R. C. Richmond, pers. comm.). We have not yet determined which of the fainter ß-carboxyl esterase bands resolved by electrophoresis corresponds to the ESTP protein.

2.4. The Obscura Group

D. pseudoobscura is the only species in the obscura group in which ß-carboxyl esterases have been investigated in any detail. As with the melanogaster group, electrophoresis of *D. pseudoobscura* reveals one strong and several faintly staining ß-carboxyl esterase bands. The strong band, termed EST5, is polymorphic for about 60 KDa monomeric and 100 - 120 KDa homodimeric forms (Townsend and Singh, 1984; Arnason and Chambers, 1987). The chromosomal location of the *Est5* gene in *D. pseudoobscura* is at least approximately homologous to that for *Est6* in *D. melanogaster* (Table I).

Tables II and III show that there are also several similarities between the pattern of expression of EST5 in *D. pseudoobscura* and that for EST6 in the melanogaster group. EST5 is expressed most abundantly in adults and, within the adult, one of the locations of EST5 activity is the hemolymph. About 17% of the total EST5 activity in adults is male specific and this 17% is distributed across the testes and the ejaculatory duct or accessory glands. However, the other major site of EST5 activity is the adult eye and in this respect EST5 more closely resembles ESTP than EST6 in *D. melanogaster.*

A further complication is revealed by molecular analysis of the *Est5* region in *D. pseudoobscura* (J. P. Brady, and R. C. Richmond, pers. comm.), which reveals three adjacent esterase genes, not two as in *D. melanogaster.* The coding regions of all three genes show strong sequence similarity with each other and with *Est6* and *EstP* in *D. melanogaster.* Electrophoretic analysis of germ-line transformants shows *Est5* to be the central gene in the triplet but the expression profiles of the other two genes have yet to be determined.

Table IV

Characterisation of adult β-carboxyl esterases in the repleta group

	D. aldrichi[a]			D. mulleri[a]	D. buzzatii[b]	D. mojavensis[b]	D. arizonensis[b]
	ESTD[c,d]	ESTF[c,d]	ESTG[c,d]	ESTD[c,d]	EST1[d,e]	EST5[d,f]	EST5[d,f]
Temporal expression	Adults	Adults	Adults	Adults	Pupae Adults	Late larvae Adults	Late larvae Adults
Sex specific expression	Male = Female	Male only	Male only	Male = Female	Male = Female	nr	nr
Tissue specific expression: Larvae	nr[g]	nr	nr	nr	Hemolymph	Hemolymph Fat body	Hemolymph Fat body
Adults	Hemolymph Eye Proboscis Antennae	Ejaculatory bulb	Ejaculatory bulb	Hemolymph	nr	nr	nr

[a] Adults only tested.
[b] Significant levels throughout development with peak levels at the stages indicated.
[c] Kambysellis et al., 1968.
[d] Johnson and Bealle, 1968.
[e] East, 1982.
[f] Zouros et al., 1982.
[g] Not reported.

2.5. The Repleta Group

ß-carboxyl esterases have been studied to varying extents in five species of the mulleri subgroup of the repleta group (Table IV).

Three electrophoretically distinct enzymes termed ESTD, ESTF and ESTG have been identified in *D. aldrichi* (Kambysellis *et al.*, 1968; Johnson and Bealle, 1968). The chromosomal locations of the presumptive *EstF* and *EstG* genes are unknown, but it seems likely that the *EstD* gene maps to a chromosome not homologous to those bearing *Est6* in *D. melanogaster* and *Est5* in *D. pseudoobscura* (Table I). Despite the non-homologous locations of the encoding genes, the expression of ESTD shows strong similarities with that for EST5 in *D. pseudoobscura*, with greatest ESTD activity in the adult hemolymph and lesser amounts in the eye, proboscis and antennae. The activities of both ESTF and ESTG are largely confined to the male ejaculatory bulb, and ESTG is transferred to the female during mating and may have functions in common with EST6 of *D. melanogaster*.

Although *D. mulleri* is in the same species complex as *D. aldrichi* it only expresses one major adult ß-carboxyl esterase (Table IV). This enzyme, called ESTD (or EST5 in Zouros *et al.*, 1982), has very similar biochemical and genetic properties to ESTD in *D. aldrichi* and is presumably homologous. Again expression of the ESTD enzyme is similar in the two sexes, being mainly confined to the hemolymph. No minor sites of ESTD activity have yet been reported in *D. mulleri* and, in particular, no ß-carboxyl esterase activity is detectable in the male reproductive tract.

Only one major ß-carboxyl esterase is found in adults or pre-adults of *D. buzzatii* (EST1), *D. mojavensis* (EST5) and *D. arizonensis* (EST5) (Table IV). The chromosomal location of the encoding gene in each species is homologous with that for *EstD* in *D. aldrichi* and *D. mulleri* but not with that for *Est6* in *D. melanogaster* or *Est5* in *D. pseudoobscura* (Table I). The enzyme in *D. buzzatii*, *D. mojavensis* and *D. arizonensis* is found in larval hemolymph and is likely to be expressed in the hemolymph of their adults as well. Although *D. mojavensis* and *D. arizonensis* express ejaculatory bulb esterase activity, it is not known whether it is the same as the EST5 found in their hemolymph, and there is no ß-carboxyl esterase activity in the ejaculatory bulb of *D. buzzatii*.

Electrophoretic analyses reveal a second, less abundant ß-carboxyl esterase in larvae of *D. buzzatii* (ESTJ), *D. mojavensis* (EST4) and *D. arizonensis* (EST4). Expression of the second esterase is confined to the larval carcass (Fig. 3). Genetic studies suggest that the genes encoding the two ß-carboxyl esterases in these species (*i.e.*, EST4 and EST5 in *D. mojavensis* and *D. arizonensis* and EST1 and ESTJ in *D. buzzatii*) are part of a small tandem duplication (Zouros *et al.*, 1982; East *et al.*, Chapter 20, this volume). Preliminary molecular data corroborate this interpretation for *Est1* and *EstJ* in *D. buzzatii* (East *et al.*, Chapter 20, this volume).

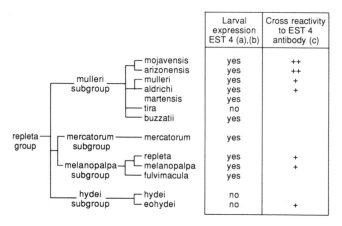

	Larval expression EST 4 (a),(b)	Cross reactivity to EST 4 antibody (c)
mojavensis	yes	++
arizonensis	yes	++
mulleri	yes	+
aldrichi	yes	+
martensis	yes	
tira	no	
buzzatii	yes	
mercatorum	yes	
repleta	yes	+
melanopalpa	yes	+
fulvimacula	yes	
hydei	no	
eohydei	no	+

Figure 3. Comparison of the structure and expression of EST4 in *D. mojavensis* with its putative homologue in other repleta group species. The presence of EST4 activity in the larval carcass and the extent of cross-reactivity between antibodies against EST4 from *D. mojavensis* and esterases from the other species are shown. Data are compiled from: (a) Zouros *et al.* (1982); (b) East (1982); (c) Pen *et al.* (1984). Phylogenetic relationships are as determined by Wasserman (1982).

Although comparable genetic or molecular analyses have not been carried out in other repleta group species, electrophoretic analyses suggest that the duplication is widely conserved within the repleta group. Apparently homologous enzymes are present in most species of the mulleri, melanopalpa and mercatorum subgroups, although not in the hydei subgroup. The patterns of expression of the duplicated esterases are conserved within the repleta group and immunological studies demonstrate that the EST4 (equivalent to ESTJ in *D. buzzatii*) proteins from different species are structurally related (Fig. 3).

Both their close linkage to the gene encoding the major adult ß-carboxyl esterase and their predominant expression in larvae suggest that the *Est4* and *EstJ* genes above closely resemble the *EstP* gene of *D. melanogaster*. Partial amino acid sequence data for the purified products of the *Est4* and *Est5* genes of *D. mojavensis* (Pen *et al.*, 1986, 1989) and partial DNA sequence data for putative clones of the *Est1* and *EstJ* genes of *D. buzzatii* (East *et al.*, Chapter 20, this volume; P.D. East, pers. comm.) have recently confirmed their similarities with the tandem *Est6* and *EstP* genes of *D. melanogaster*. Given all these indications of homology, we suggest that the non-homologous chromosomal locations of the duplicated genes in the different species groups reflect a chromosomal rearrangement rather than independent evolutionary origins.

One structural feature of the EST4 enzyme of *D. mojavensis* is also informative more generally about the evolution of ß-carboxyl esterases. This enzyme is polymorphic for variants which differ qualitatively in their *in vitro* substrate speci-

ficity; some variants show typical ß-carboxyl esterase speci-
ficities but others show a preference for α- over ß-naphthyl
esters (Zouros and van Delden, 1982). This suggests a close
genetic relationship between the ß-carboxyl esterases and at
least some α-carboxyl esterases. We concur with Myers *et al.*
(1988), who suggested that any multi-gene family containing ß-
carboxyl esterase genes may also contain some other esterase
genes as well.

2.6. The Virilis Group

Twelve species in the virilis group have been characterised
for ß-carboxyl esterases (Korochkin *et al.*, 1987 for a review)
and two major adult ß-carboxyl esterases termed EST2 and ESTS
have been identified in each.

Genetic analysis of *D. virilis* indicates that the *Est2* and
EstS genes are about 8cM apart on chromosome 2 (Yenikolopov *et
al.*, 1983, and see Table I). Therefore, the major adult ß-
carboxyl esterases of both the virilis and repleta groups are
located on homologous chromosomes, but these chromosomes are
not homologous to the 3L arm bearing *Est6* of *D. melanogaster.*
However other evidence suggests that both *Est2* and *EstS* are
nevertheless closely related, if not directly homologous to
Est6. In the case of *EstS*, molecular evidence (L. I.
Korochkin, pers. comm.) indicates that it is part of a tandem
duplication of esterase genes whose sequences and organisation
bear striking similarities to those of *Est6* and *EstP* in *D.
melanogaster*. Molecular data are not yet available for *Est2*
but, on the basis of close similarities in its expression pattern outlined below, together with its immunological cross-
reactivity discussed in an earlier section, it would also seem
to be closely related to *Est6* and its homologues.

Table V
Characterisation of adult ß-carboxyl esterases in *D. virilis*

	EST2[a,b,c]	ESTS[d,e]
Temporal expression	Mainly adults	Mainly adults
Sex specific expression	Male ≥ female	Male » female
Tissue specific expression	Hemolymph Eye Proboscis Thorax Ejaculatory bulb Spermathecae	Ejaculatory bulb

[a]Korochkin and Matveeva, 1974. [d]Korochkin *et al.*, 1978.
[b]Sasaki, 1974. [e]Korochkin *et al.*, 1976.
[c]Korochkin *et al.*, 1973a.

The patterns of expression of EST2 and ESTS have been
determined for all twelve species in the virilis group, and
essentially the same patterns are obtained for each (Table V
and Korochkin et al., 1987). EST2 activity is first detected
in second instar larvae, and is then present throughout development, peaking in adults. Activity levels in males and females are similar, with most activity being located in the
hemolymph, thorax and proboscis and smaller amounts in the
eye. In addition, small amounts of activity are found in the
ejaculatory bulb of the male reproductive tract and the sper-
mathacae of the female reproductive tract (Korochkin et al.,
1973a; Korochkin and Matveeva, 1974). There are thus extensive similarities between the patterns of expression of this
enzyme and those of ESTD in D. aldrichi (Table IV) and EST6 in
D. melanogaster (M. J. Healy, M. M. Dumancic, and J. G.
Oakeshott, unpublished).

ESTS is essentially a male specific enzyme. It is mainly
expressed in adult males several days after eclosion and has
been localised to the waxy plug of their ejaculatory bulb
(Korochkin et al., 1976). The waxy plug is transferred to the
female during mating and it has been proposed that ESTS plays
a role in the insemination reaction (Patterson and Stone,
1952), by inducing swelling of the vaginal cavity (Korochkin
et al., 1976). Traces of the male-donated ESTS have also been
found in the seminal receptacle of the female, suggesting a
second function for the enzyme.

Members of the virilis group can be categorised by the subcellular fraction in which ESTS is expressed. The enzyme from
D. virilis and D. texana has been localised to the microsomal
fraction (Korochkin et al., 1973b), while in D. montana and D.
imeretensis it is localised in the aqueous fraction (Korochkin
et al., 1976). Although ESTS from each species appears to be
immunologically related (Korochkin et al., 1987) and is
expressed in the same tissue, it is not known how the
differences in subcellular location might affect the function
of the enzyme.

3. EST6 VARIATION WITHIN D. MELANOGASTER

3.1. Structural Polymorphisms

The esterase 6 enzyme of D. melanogaster has long been used
as a model system for analysing intraspecific variation in
protein structure. The EST6-F/EST6-S polymorphism was among
the first to be identified by electrophoresis (Wright, 1963)
and again was one of the first to be subdivided into additional variants by thermostability (Cochrane, 1976), higher
resolution electrophoresis (Cooke et al., 1987) and DNA
sequence analysis (Cooke and Oakeshott, 1989).

Twenty one polymorphic EST6 variants have been identified
so far by combinations of high resolution electrophoretic and
thermostability analyses and 17 of these lie within the major
EST6-F and EST6-S mobility classes (Labate et al., 1989).
Genetic analyses map this variation to the EST6 structural

gene and DNA sequencing of this gene reveals polymorphisms for
16 amino acid replacements among just 13 isolates covering 10
of the protein variants (Cooke and Oakeshott, 1989).   The
amino acid polymorphisms segregate within as well as among the
known protein variants, suggesting that the actual number of
protein variants could be considerably higher than 21.   These
results make EST6 in *D. melanogaster* one of the most polymorphic enzymes characterised.

Biochemical data are not available for most of the EST6
variants.   However both Danford and Beardmore (1979) and White
et al. (1988) have recorded biochemical differences between
purified EST6 preparations which they describe as EST6-F and
EST6-S but do not distinguish further on high resolution electrophoretic or thermostability criteria.   The magnitudes of
the differences are small but they occur in several properties, including K_m, catalytic efficiency and susceptibility to
some organophosphate inhibitors.   Oakeshott *et al.* (1989) and
Richmond *et al.* (Chapter 15, this volume) review several lines
of evidence indicating that at least some of these structural
differences in EST6 are not neutral to natural selection.

3.2. Activity Variation

No true *Est6* null alleles have been isolated from field
collected flies (Cambissa *et al.*, 1982) but several fold differences in EST6 amount are found and these differences are
inherited.   This first became evident in a chromosome substitution study by Tepper *et al.* (1982, 1984), who found that the
V_{max} of homogenates of whole virgin males was affected substantially by the X and third chromosomes, but not the second.
The third chromosome effects may reside directly at the *Est6*
locus, which also maps to this chromosome (Wright, 1963).   The
X chromosome effects may relate to the inducibility of the
enzyme by juvenile hormone and ecdysone (Richmond and Tepper,
1983), since the X is the only chromosome to which factors
affecting the titres of these hormones have been mapped (Kiss
et al., 1978; Klose *et al.*, 1980).   Interestingly also, a similar chromosome substitution study of *D. virilis* showed that
effects on the V_{max} of its ejaculatory bulb ß-carboxyl esterase
(EST-S) were again restricted to the homologues of the X and
third chromosomes of *D. melanogaster* (Korochkin *et al.*, 1978,
1987).

Subsequent work by Game and Oakeshott (1989) has further
characterised the genetic variation in EST6 activity in *D.
melanogaster.*   Forty two third chromosome isoallelic lines
were extracted from a single field population.   For each line
the V_{max} was determined for whole fly homogenates of 4-5 day old
virgins of each sex from several replicate cultures.   Among
the lines V_{max} varied several fold in each sex and the variation
in the two sexes was significantly but not strongly correlated
(r = 0.32, P<0.05).   Most EST6 in males is located in the
anterior ejaculatory duct and most in females in their
hemolymph; therefore the weakness of the correlation between
the sexes implies that a substantial proportion of the V_{max}
variation is tissue-specific.

The V_{max} differences in males were only weakly associated with electrophoretic differences ($F_{5,36}$ = 3.28, P<0.05) and the association in females was negligible ($F_{5,36}$ = 0.84, P>0.05). However, V_{max} in males was strongly associated with the amount of EST6 enzyme (r = 0.80, P<0.001), as assessed from radial immunodiffusion assays with polyclonal anti-EST6 antisera (amounts in females were too low to assay reliably). While not discounting possible differences in translational efficiency or in the stability of the mRNA or protein, Game and Oakeshott (1989) proposed that the tissue specificity of much of the V_{max} variation, the strong association with EST6 amount in males, and the weak association with electromorphs in either sex, were all most easily explained by differences in levels of *Est6* transcription.

3.3. Molecular Bases of Activity Variation

Game and Oakeshott (1990) have begun testing the above proposition by screening for associations between the V_{max} variation and restriction site polymorphisms around the *Est6* gene among the 42 isoallelic lines. Six restriction site polymorphisms have been identified in the 1.15 kbp 5' of *Est6* and three within its 1.7 kbp coding region (Table VI). The six 5' polymorphisms are located from 200 to 900 bp upstream of the initiation codon and are also well upstream of the TATA box and transcription start site (about 70 and 35 bp 5' of the initiation codon respectively; Collet *et al.*, 1990). The three coding region polymorphisms are located from 200 to 800 bp downstream of the initiation codon in exon I. One is a silent substitution but the other two involve amino acid replacements; the Thr->Ala at amino acid residue 247 is a charge conservative polymorphism, while the Asn->Asp at amino acid residue 237 produces the charge change responsible for the major EST6-F/EST6-S electrophoretic mobility difference.

We have already seen from the various electrophoretic and sequencing analyses of the *Est6* coding region that it is an unusually variable locus (Cooke and Oakeshott, 1989; Oakeshott *et al.*, 1989). The restriction site analysis now shows its regulatory region to be even more variable. The proportion of polymorphic nucleotides in the 1.15 kbp 5' region (0.065 ± 0.004) is more than three times higher than that for the coding region (0.019 ± 0.002). The 5' level corresponds to an average of one polymorphic nucleotide site for every 15 bp, or about 75 polymorphic sites across the 1.15 kbp. The six restriction site polymorphisms identified are thus only a minority of those likely to lie in the 1.15 kpb 5' region.

High levels of gametic disequilibrium occur among several of the polymorphisms detected, with the six 5' polymorphisms only yielding ten haplotypes and the three coding region polymorphisms only three (Table VI). The effects of the restriction site polymorphisms on EST6 activity must therefore be examined not only in terms of the associations of individual polymorphisms but also in terms of the joint associations of groups of polymorphisms represented in different haplotypes.

Table VI
Mean male and female EST6 activity and male EST6 amount
(± standard error) for lines (n) possessing given haplotypes for the
regions 1.15 kbp 5' of *Est*6 and the *Est*6 coding region. Also shown are the
F values and their significance for the association of haplotypes with
activity and amount values

Haplotypes						n	Male activity	Female activity	Male amount

For 5' sites

*Dde*I	*Hinf*I	*Hinf*I	*Taq*I	*Rsa*I	*Sau*3A	n	Male activity	Female activity	Male amount
-0.90	-0.80	-0.68	-0.63	-0.53	-0.20				
−	+	+	+	+	−	1	3194	515	2722
−	+	+	−	+	−	2	3099±346	351±40	2328±735
+	+	+	+	−	−	1	2295	348	2033
+	+	+	+	+	+	4	2255±146	364±22	1804±40
−	−	+	−	−	−	1	2231	468	1642
+	+	+	+	+	−	1	2086	459	1983
+	+	−	−	+	−	1	2070	488	1800
+	+	+	+	−	+	1	1930	399	1633
−	+	+	−	−	−	28	1852±64	358±13	1567±64
−	+	+	+	+	+	2	1687±65	336±29	1073±107
						$F_{8,33}$ =	5.07***	1.53	2.78*

For coding region sites

*Rsa*I	EST6-F/EST6-S (Asp/Asn)	*Taq*I (Thr/Ala)				n	Male activity	Female activity	Male amount
+0.20	+0.77	+0.80							
+	−	−				10	2261±173	385±19	1793±150
−	−	−				1	2086	459	1983
+	+	+				31	1921±74	362±3	1607±74
						$F_{2,39}$ =	2.21	1.22	1.00

*P<0.05, ***P<0.001.

Table VI shows that the three haplotypes found for the coding region do not differ significantly in male EST6 activity or amount, or female EST6 activity. Likewise, of the three individual coding region polymorphisms, the silent site change (*Rsa* I at +0.20 kbp) does not significantly affect any of the measures of EST6 activity or amount (data not shown). The Thr->Ala replacement (*Taq* I at +0.80 kbp) and the Asn->Asp replacement change (EST6-F/EST6-S, at +0.77 kbp) do show weak associations with male EST6 activity ($F_{1,40}$ = 4.37, P<0.05 for each) but none with female EST6 activity or male EST6 amount. Game and Oakeshott (1990) concluded that the coding region variation represented by these three polymorphisms at least makes little contribution to the large differences in EST6 activity and amount segregating among the 42 lines.

The ten haplotypes in the 5' region do not vary significantly in female EST6 activity either. This may indicate that the variation in female activity is due to polymorphisms outside the region Game and Oakeshott (1990) scored, or it could simply reflect their low level of ascertainment of polymorphisms in the 5' region. Full sequence analysis of this region in various lines differing in female activity would be needed to resolve this issue.

In contrast to the results for female activity, the ten 5' haplotypes detected by Game and Oakeshott (1990) do show large and significant differences in both male EST6 activity and amount (Table VI). There is about a two-fold range of values among the haplotypes for both these measures, which accounts for 59% and 44%, respectively, of the total variation among the 42 lines. As might be expected from the strong correlation between male activity and amount across all 42 of the lines (r_{40} = 0.80, P<0.001), there is also a strong correlation between the two variables across the ten haplotypes (r_8 = 0.91, P < 0.001).

These data suggest that the associations of the 5' haplotypes with male EST6 activity reflect their primary associations with EST6 amounts. Moreover, they support our earlier prediction (Game and Oakeshott, 1989) that substantial proportions of EST6 activity variation are mediated by polymorphisms regulating transcription from the *Est6* gene. Significantly, they are also consistent with preliminary analyses of germ-line transformants bearing various deletions 5' to *Est6*; these analyses indicate that the *cis* acting sequences controlling major aspects of the level and tissue specificity of *Est6* transcription lie in the first 1.15 kbp 5' of the coding region (J. P. Brady and R. C. Richmond, pers. comm.; M. J. Healy and J. G. Oakeshott unpublished).

In order to identify the regulatory polymorphisms more specifically, Game and Oakeshott (1990) then dissected the 5' haplotype effects into the contributions of individual restriction site differences. Only one of the six individual 5' restriction site differences shows an association with EST6 activity. Presence of the *Rsa* I site at -0.53 kbp is associated with 25% higher male activity than is its absence ($F_{1,40}$ = 10.9, P<0.01). Importantly, the same pattern of differences is evident in the values for male EST6 amount ($F_{1,40} = 3.7$, $P<0.10$), but not for female EST6 activity ($F_{1,40} = 1.3$, $P>0.25$).

One interpretation of these results is that the *Rsa* I site is part of a male-specific (perhaps ejaculatory duct-specific) positive regulatory element. It may be significant in this context that the *Rsa* I site is part of a 16 bp palindrome (AAATATGTACATATTT), since palindromes have been found in several regulatory elements characterised to date (Wingender, 1988). However, there are two reasons why this interpretation must be regarded with caution. Firstly, we showed earlier that the six 5' polymorphisms detected are probably only a minority of those present in the region. Secondly, strong gametic disequilibrium occurs among several of the polymorphisms detected, and presumably involves some of the undetected differences as well. The *Rsa* I polymorphism may therefore simply act as a marker for other as yet unknown differences which are the primary causes for the activity variation among the haplotypes described above.

Clearly this issue, like those concerned with the female activity variation cited earlier, will only be resolved by full sequence analysis of the 1.15 kbp 5' region in several representative lines. Despite its limited power, however, the restriction site analysis by Game and Oakeshott (1990) has still provided some of the best evidence yet that *cis* acting promoter polymorphisms contribute significantly to the generation of intraspecific enzyme activity variation.

4. CONCLUSIONS

The phenotypic data available clearly show several qualitative regulatory shifts during the evolution of the major adult ß-carboxyl esterases of *Drosophila*, although the evidence as yet is generally insufficient to determine the molecular bases of the changes. We can already identify the following six major features of the evolutionary process but further data will undoubtedly reveal several more.

The first point is that in all cases investigated to date the coding regions of the relevant genes have proved to be related, if not directly homologous to one another. Direct molecular data to demonstrate this have now been obtained for the *Est6* genes from six melanogaster group species, *Est5* from *D. pseudoobscura* in the obscura group, *Est1* from *D. buzzatii* in the repleta group, and *EstS* from *D. virilis* in the virilis group. Other, less definitive measures suggest that the relationship or homology extends to most of the other relevant genes and species as well. These measures include the quaternary structures and immunological cross-reactivity of the proteins, some aspects of their tissue specific expression and, as discussed below, the involvement of these genes in tandem duplications with larval ß-carboxyl esterase genes.

One notable exception to the biochemical and genetic evidence for homology concerns the chromosomal locations of the encoding genes. At least approximately similar locations are found for all species investigated in the melanogaster and obscura groups within the subgenus *Sophophora*. The same holds true for all species investigated in the repleta and virilis groups within the subgenus *Drosophila*. However the chromosomes implicated in the two subgenera are not homologous. As stated earlier, the weight of other evidence for the homologies of the major adult ß-carboxyl esterase genes across the two subgenera suggests that their non-homologous chromosomal locations are best explained as the result of a chromosomal rearrangement.

The second major point about the adult ß-carboxyl esterases is that they are generally part of a tandem duplication or triplication. Evidence for such an array has been obtained for all species where appropriate molecular or genetic data are available and this includes members of all four species groups. The molecular data for *D. melanogaster* clearly show only two closely related esterase coding regions in the array, but comparable data for *D. pseudoobscura* and *D. buzzatii* indicate an array of three. Thus the number of copies of tandemly repeated ß-carboxyl esterase genes has changed at least once during the evolution of *Drosophila*. Interestingly, genetic data for α-carboxyl esterases in the virilis group also suggest changes in the number of esterase genes in a tightly linked array (Baker, 1980), although in this case confirmatory molecular data are not yet available.

One particular aspect of the ß-carboxyl esterase array is that the component genes are expressed at different life stages, one predominantly in adults and another mainly in larvae. Although not yet studied in most species, this has been clearly established for *D. melanogaster* and species within the repleta group. These species cover both subgenera, suggesting that the divergent expression of the tandem genes may be a general feature with a common origin among *Drosophila* species.

A third major feature of the evolution of the adult ß-carboxyl esterases is that the two subgenera *Drosophila* and *Sophophora* have diverged in the number and tissue specificity of these enzymes. All sophophorans studied express a single major ß-carboxyl esterase in both the hemolymph and the male reproductive tract. An enzyme with a similar expression profile is also found in most species from the subgenus *Drosophila* studied. However, several of the latter species also express a second ß-carboxyl esterase at high levels in the male reproductive tract. Moreover, this second enzyme is essentially confined to the ejaculatory bulb, whereas the reproductive tract activity of the first is localised to the ejaculatory duct, the testes, or the accessory glands.

The limited molecular data so far available for the two enzymes in the subgenus *Drosophila* suggest that they are closely related to each other and to the single enzyme found in the *Sophophora*. The relationship is evident both in terms of overall sequence similarities and in terms of their locations in tandem arrays of esterase genes. There thus seem to be at least two tandem arrays in the subgenus *Drosophila*, whereas only one has been recovered in the *Sophophora*. Unfortunately the molecular evidence for the subgenus *Drosophila* is presently limited to the hemolymph/reproductive tract EST1 enzyme of *D. buzzatii* and the ejaculatory bulb ESTS enzyme of *D. virilis*. Therefore, a high priority for future research will be to analyse the molecular biology of the two enzymes in the one species. *D. virilis* would be a good model for such a study because it is the only species in which both enzymes have been mapped genetically.

The fourth point we wish to highlight relates specifically to the ejaculatory bulb ß-carboxyl esterases in the subgenus *Drosophila*. Within the repleta group of this subgenus the number of detectable ejaculatory bulb ß-carboxyl esterases varies from none in *D. buzzatii*, through one in *D. mojavensis* and *D. arizonensis*, to two in *D. aldrichi*. This suggests the possibility of still further changes in the number of adult ß-carboxyl esterase genes in a relatively recent lineage within the subgenus. The changes could reflect changes in gene number within an array, or changes in the number of arrays. Alternatively they may simply reflect the production of various izozymic forms from a single gene; for example, we do not yet know the genetic relationship between the two ejaculatory bulb enzymes ESTF and ESTG of *D. aldrichi*. We also stress that the lack of detectable ejaculatory bulb enzyme in *D. buzzatii* using an artificial *in vitro* substrate need not reflect the absence of active enzyme against *in vivo* substrates.

A fifth feature of adult ß-carboxyl esterase evolution concerns the reproductive tract expression of the hemolymph/reproductive tract enzyme in the other subgenus, *Sophophora*. Species of the melanogaster group show at least three regulatory changes for this enzyme, EST6. The direction of these changes can be traced by reference to an outgroup, the homologous enzyme EST5 in *D. pseudoobscura* in the obscura group. One change involves the loss of expression in the testes of *D. melanogaster* and *D. simulans*. The second involves the loss of expression in the ejaculatory duct or accessory glands of *D. erecta*. The third involves the marked increase in the total reproductive tract EST6 activity in *D. melanogaster, D. simulans* and *D. mauritiana*. Intensive molecular analysis of *Est6* in these species suggests that these changes are not due to changes in gene number. We propose that the changes lie in the regulation of *Est6* expression, although interspecific transformation experiments will be necessary to determine whether the differences lie in *cis* or *trans* acting elements.

The final feature of adult ß-carboxyl esterase evolution on which we wish to comment is the variation in the regulation of *Est6* within *D. melanogaster*. This variation produces three-fold differences in EST6 activity in both hemolymph and ejaculatory duct and the activity differences in these two tissues are largely uncorrelated. Although full sequence analysis of variant lines has yet to be carried out, a limited restriction site survey has already shown that some of the variation is associated with nucleotide polymorphism in sequences 5' of the *Est6* gene. Moreover, this 5' polymorphism is sex, and, by inference, tissue specific in its effects on EST6 activity. We suggest that natural selection has exploited such polymorphisms to establish the diverse patterns of expression now evident among different ß-carboxyl esterase genes and different *Drosophila* species.

5. SUMMARY

The ß-carboxyl esterases of adult *Drosophila* could provide an informative model system for investigating the molecular basis of evolutionary change in gene expression. Extensive physiology, biochemistry and genetics on many species reveal several qualitative shifts in ß-carboxyl esterase expression. Molecular data for a few species are now providing our first insights into the molecular bases of these shifts. From the data available we can identify six particular features of the system.

First, all the available biochemical and molecular genetic data point to a common origin for the coding regions of the adult ß-carboxyl esterase genes. Second, the genes are generally part of a tandem array of two and sometimes three related esterase genes, one of the other genes being mainly expressed in larvae. Third, species in the subgenus *Sophophora* only express a single major ß-carboxyl esterase gene in both the adult hemolymph and the adult male reproductive tract, whereas several members of the subgenus *Drosophila* carry an additional ß-carboxyl esterase gene whose expression is essentially restricted to the ejaculatory bulb, within the male reproductive tract. The limited data available suggest that the two genes are parts of different tandem arrays. Fourth, the number of esterases with detectable ejaculatory bulb specific expression can vary from none to two among closely related species in the subgenus *Drosophila*, although the genetic and molecular bases of this difference are not yet known. Fifth, the male reproductive tract expression of the hemo-lymph/reproductive tract enzyme in the subgenus *Sophophora* can vary substantially, both in the level and in the site of expression within the reproductive tract. Sixth, intensive analysis of one species in the latter subgenus reveals specific nucleotide polymorphism in the 5' region of the gene encoding the hemolymph/reproductive tract enzyme and this polymorphism has sex and, by inference, tissue specific effects on the expression of the gene.

ACKNOWLEDGEMENTS. We thank Peter Atkinson, Peter East, Rollin Richmond and Robyn Russell for valuable discussions and for comments on early drafts of this paper. We are also indebted to Leonid Korochkin for providing us with some unpublished molecular data on *EstS* in *D. virilis*; to Mel Green, Peter East and Stuart Barker for unpublished data on linkage group homologies in the repleta group; and to Karen Nielsen, Rollin Richmond and Deb Morris for data on tissue specific expression of *Est6* and *EstP* in species in the melanogaster group.

References

Aquadro, C. F., Desse, S. F., Bland, M. M., Langley, C. H., and Laurie-Ahlberg, C. C., 1986, Molecular population genetics of the alcohol dehydrogenase gene region of *Drosophila melanogaster, Genetics* **114**:1165-1190.

Arnason, E., and Chambers, G. K., 1987, Macromolecular interaction and the electrophoretic mobility of esterase 5 from *Drosophila pseudoobscura, Biochem. Genet.* **25**:287-307.

Baker, W. K., 1980, Evolution of the alpha-esterase duplication within the montana subphylad of the virilis species group of *Drosophila, Genetics* **94**:733-748.

Batterham, P., Chambers, G. K., Starmer, W. T., and Sullivan, D. T., 1984, Origin and expression of an alcohol dehydrogenase gene duplication in the genus *Drosophila, Evolution* **38**:644-657.

Berger, E., and Canter, R., 1973, The esterases of *Drosophila.* I. The anodal esterases and their possible role in eclosion, *Devl. Biol.* **33**:48-55.

Birley, A. J., 1984, Restriction endonuclease map variation and gene activity in the *Adh* region in a population of *Drosophila melanogaster, Heredity* **52**:103-112.

Bray, S. J., and Hirsh, J., 1986, The *Drosophila virilis* dopa decarboxylase gene is developmentally regulated when integrated into *Drosophila melanogaster, EMBO J.* **5**:2305-2311.

Brennan, M. D., Wu, C-Y., and Berry, A.J., 1988, Tissue-specific regulatory differences for the alcohol dehydrogenase genes of Hawaiian *Drosophila* are conserved in *Drosophila melanogaster* transformants, *Proc. natn. Acad. Sci. USA* **85**:6866-6869.

Cambissa, V., Nigro, L., Danieli, G. A., and Costa, R., 1982, Occurrence of a rare allele at the *Est6* locus in natural populations of *Drosophila melanogaster, Boll. Zool.* **49**:229-239.

Carrasco, C. E., Perez-Chiesa, Y., and Bruck, D., 1984, The esterases of *Drosophila dunni, Comp. Biochem. Physiol.* **79B**:375-378.

Chambers, G. K., 1988, The *Drosophila* alcohol dehydrogenase gene-enzyme system, *Adv. Genet.* **25**:39-107.

Cochrane, B. J., 1976, Heat stability variation of esterase 6 in *Drosophila melanogaster, Nature, Lond.* **263**:131-132.

Collet, C., Nielsen, K. M., Russell, R. J., Karl, M., Oakeshott, J. G., and Richmond, R. C., 1990, Molecular analysis of duplicated esterase genes in *Drosophila melanogaster, Mol. Biol. Evol.* **7**:9-28.

Cooke, P. H., and Oakeshott, J. G., 1989, Amino acid polymorphisms for esterase 6 in *Drosophila melanogaster, Proc. natn. Acad. Sci. USA* **86**:1426-1430.

Cooke, P. H., Richmond, R. C., and Oakeshott, J. G., 1987, High resolution electrophoretic variation at the esterase 6 locus in a natural population of *Drosophila melanogaster, Heredity* **59**:259-264.

Danford, N. D., and Beardmore, J. A., 1979, Biochemical properties of esterase 6 in *Drosophila melanogaster, Biochem. Genet.* **17**:1-22.

Dickinson, W. J., 1983, Tissue specific allelic isozyme patterns and *cis* acting developmental regulators, *Isozymes . Current Topics in Biological & Medical Research* **9**:107-122.

East, P. D., 1982, Non-specific esterases of *Drosophila buzzatii*, in: *Ecological Genetics and Evolution; the Cactus-Yeast-Drosophila Model System* (J. S. F. Barker, and W. T. Starmer, eds), Academic Press Australia, Sydney, pp. 323-338.

East, P. D., 1984, Biochemical genetics of two highly polymorphic esterases in *Drosophila buzzatii*, Ph.D. Thesis, University of New England.

Estelle, M. A., and Hodgetts, R. B., 1984a, Genetic elements near the

structural gene modulate the level of dopa decarboxylase during
    Drosophila development, *Mol. Gen. Genet.* 195:434-441.
Estelle, M. A., and Hodgetts, R. B., 1984b, Insertion polymorphisms may
    cause stage specific variation in mRNA levels for dopa decarboxylase in
    Drosophila, Mol. Gen. Genet. 195:442-451.
Game, A. Y., and Oakeshott, J. G., 1989, Variation in the amount and activ-
    ity of esterase 6 in a natural population of *Drosophila melanogaster,
    Heredity* 62:27-34.
Game, A. Y., and Oakeshott, J. G., 1990, Associations between restriction
    site polymorphism and enzyme activity variation for esterase 6 in
    Drosophila melanogaster (submitted).
Hammock, B. D., 1985, Regulation of juvenile hormone titer: degradation,
    in: *Comprehensive Insect Physiology, Biochemistry and Pharmacology*,
    Vol. 7 (G. A., Kerkut, and L. I. Gilbert, eds), Pergamon Press, Oxford,
    pp. 431-472.
Heymann, E., 1980, Carboxylesterases and amidases. in: *Enzymatic Basis of
    Detoxication.* Vol. 2 (W. Jakoby, ed.), Academic Press, New York, pp
    291-323.
Hubby, J. L., and Lewontin, R. C., 1966, A molecular approach to the study
    of genic heterozygosity in natural populations. I. The number of alle-
    les at different loci in *Drosophila pseudoobscura, Genetics* 54:577-594.
Hughes, P. B., and Devonshire, A. L., 1982, The biochemical basis of resis-
    tance to organophosphorus insecticides in the sheep blowfly *Lucilia
    cuprina, Pestic. Biochem. & Physiol.* 18:289-297.
Johnson, E. M., and Bealle, S., 1968, Isozyme variability in species of the
    genus *Drosophila.* V. Ejaculatory bulb esterases in *Drosophila* phy-
    logeny, *Biochem. Genet.* 2:1-18.
Kambysellis, M. P., Johnson, F. M., and Richardson. R. H., 1968, Isozyme
    variability in species of the genus *Drosophila.* IV. Distribution of the
    esterases in the body tissues of *D. aldrichi* and *D. mulleri* adults,
    Biochem. Genet. 1:249-265.
Kapin, M. A., and Ahmad, S., 1980, Esterases in larval tissues of gypsy
    moth, *Lymantria dispar* (L): optimum assay conditions, quantification
    and characterisation, *Insect Biochem.* 10:331-337.
Kiss, I., Szabad, J., and Major, J., 1978, Genetic and developmental analy-
    sis of puparium formation in *Drosophila, Mol. Gen. Genet.* 164:77-83.
Klose, W., Gateff, E., Emmerich, H., and Beikirch, H., 1980, Developmental
    studies on two ecdysone deficient mutants of *Drosophila melanogaster,
    Wilh. Roux. Arch. Dev. Biol.* 189:57-68.
Knibb, W. R., East, P. D., and Barker, J.S.F., 1987, Polymorphic inversion
    and esterase loci complex on chromosome 2 of *Drosophila buzzatii.* I.
    Linkage disequilibria, *Aust. J. biol. Sci.* 40:257-269.
Korochkin, L. I., Belyaeva, E. S., Matveeva, N. M., Kuzin, B. A., and
    Serov, O. L., 1976, Genetics of esterases in *Drosophila.* IV. Slow-
    migrating S- esterase in *Drosophila* of the virilis group, *Biochem.
    Genet.* 14:161-182.
Korochkin, L. I., Ludwig, M. Z., Poliakova, E. V., and Philinova, M. R.,
    1987, Some molecular-genetic aspects of cellular differentiation in
    Drosophila, Sov. Sci. F. Physiol. Gen. Biol. 1:411-466.
Korochkin, L. I., and Matveeva, N. M., 1974, Genetics of esterases in
    Drosophila of the *virilis* group. II. Sequential expression of paternal
    and maternal esterases in ontogenesis, *Biochem. Genet.* 12:1-7.
Korochkin, L. I., Matveeva, N. M., Golubovsky, M. D., and Evgeniev, M. B.,
    1973a, Genetics of esterases in *Drosophila* of the *virilis* group. I.
    Characteristics of esterase patterns in *D. virilis, D. texana, D. lit-
    toralis* and their hybrids, *Biochem. Genet.* 10:363-393.
Korochkin, L. I., Matveeva, N. M., and Kerkis, A. Y., 1973b, Subcellular
    localisation of esterases in *D. virilis* and *D. texana, Drosoph. Inf.*

Serv. **50**:130-131.

Korochkin, L. I., Matveeva, N. M., Kuzin, B. A., Karasik, G. I., and Maximovsky, L. F., 1978, Genetics of esterases in *Drosophila*. VI. Gene system regulating the phenotypic expression of the organ-specific esterase in *Drosophila virilis, Biochem. Genet.* **16**:709-726.

Labate, J., Bortoli, A., Game, A. Y., Cooke, P. H., and Oakeshott, J. G., 1989, The number and distribution of esterase 6 alleles in populations of *Drosophila melanogaster, Heredity* **63**:203-208.

Langley, C. H., Shrimpton, A. E., Yamazaki, T., Miyashita, N., Matsuo, Y., and Aquadro, C. F., 1988, Naturally occurring variation in the restriction map of the *Amy* region of *Drosophila melanogaster, Genetics* **119**:619-629.

Laurie, C. C., and Stam, L. F., 1988, Quantitative analysis of RNA produced by Slow and Fast alleles of *Adh* in *Drosophila melanogaster, Proc. natn. Acad. Sci. USA* **85**:5161-5165.

Laurie-Ahlberg, C. C., Maroni, C., Bewley, G. C., Lucchesi, J. C., and Weir, B. S., 1980, Quantitative genetic variation of enzyme activities in natural populations of *Drosophila melanogaster, Proc. natn. Acad. Sci. USA.* **77**:1073-1077.

Laurie-Ahlberg, C. C., Williamson, J. H., Cochrane, B. J., Wilton, A. N., and Chasalow, F. I., 1982, Autosomal factors with correlated effects on the activities of glucose-6-phosphate and 6-phosphogluconate dehydrogenases in *Drosophila melanogaster, Genetics* **99**:127-150.

Lemeunier, F., and Ashburner, M., 1984, Relationships within the *melanogaster* species subgroup of the genus *Drosophila (Sophophora).* IV. The chromosomes of two new species, *Chromosoma* **89**:343-351.

Levitan, M., 1982, The robusta and melanica groups, in: *The Genetics and Biology of Drosophila*, Vol. 3B (M. Ashburner, H. L. Carson, and J. N. Thompson, eds), Academic Press, London, pp. 141-192.

Lunday, A. J., and Farmer, J. L., 1983, Tissue localisation of esterase-5 in *Drosophila pseudoobscura, Biochem. Genet.* **21**:453-463.

Mane, S. D., Tepper, C. S., and Richmond, R. C., 1983, Studies of esterase 6 in *Drosophila melanogaster*. XIII. Purification and characterisation of the two major isozymes, *Biochem. Genet.* **21**:1019-1040.

McReynolds, M. S., 1967, Homologous esterases in three species of the *virilis* group of *Drosophila, Genetics* **56**:527-540.

Miyashita, N., and Langley, C. H., 1988, Molecular and phenotypic variation of the white locus region in *Drosophila melanogaster, Genetics* **120**:199-212.

Morton, R. A., and Singh, R. S., 1985, Biochemical properties, homology and genetic variation of *Drosophila* 'nonspecific' esterases, *Biochem. Genet.* **23**:959-973.

Myers, M., Oakeshott, J. G., and Richmond, R. C., 1988, On the origins of esterases, *Mol. Biol. Evol.* **5**:113-119.

Narise, S., and Hubby, J. L., 1966, Purification of esterase-5 from *Drosophila pseudoobscura, Biochim. biophys. Acta* **122**:281-288.

Oakeshott, J. G., Collet, C., Phillis, R. W., Nielsen, K. M., Russell, R. J., Chambers, G. K., Ross, V., and Richmond, R. C., 1987, Molecular cloning and characterisation of esterase 6, a serine hydrolase of *Drosophila, Proc. natn. Acad. Sci. USA* **84**:3359-3363.

Oakeshott, J. G., Cooke, P. H., Richmond, R. C., Bortoli, A., Game, A. Y., and Labate, J., 1989, Molecular population genetics of structural variants of esterase 6 in *Drosophila melanogaster, Genome* (in press).

Patterson, J. T., and Stone, W. S., 1952, *Evolution in the Genus Drosophila*, Macmillan, New York.

Pen, J., van Beeumen, J., and Beintema, J. J., 1986, Structural comparison of two esterases from *Drosophila mojavensis* isolated by immunoaffinity chromotography, *Biochem. J.* **278**:691-699.

Pen, J., Bolks, G. J., Hoeksema-Du Pui, M. L. L., and Beintema, J. J., 1989, Variability in enzymic properties and structural conservation of serine esterases (submitted).

Pen, J., Rongen, H. A. H., and Beintema, J. J., 1984, Purification and properties of esterase-4 from *Drosophila mojavensis*, *Biochim. biophys. Acta* **789**:203-209.

Richmond, R. C., and Tepper, C. S., 1983, Genetic and hormonal regulation of esterase 6 activity in male *Drosophila melanogaster*, *Isozymes. Current Topics in Biological & Medical Research* **9**:91-106.

Sasaki, F., 1974, Esterase isozymes of *Drosophila virilis*: developmental aspects of the α- and ß-esterases and their distribution in the body parts, *Jap. J. Genet.* **49**:223-232.

Sasaki, F., 1975, Immunological comparison of the α- and ß-esterases in *Drosophila* species, *Jap. J. Genet.* **50**:217-233.

Sasaki, M., and Narise, S., 1978, Molecular weight of esterase isozymes of *Drosophila* species, *Drosoph. Inf. Serv.* **53**:123-124.

Schott, D. R., East, P. D., and Paigen, K., 1988, Characterization of the *Adh*^{SL} regulatory mutation in *Drosophila melanogaster*, *Genetics* **119**:631-637.

Sheehan, K., Richmond, R. C., and Cochrane, B. J., 1979, Studies of esterase-6 in *Drosophila melanogaster*. II. The developmental pattern and tissue distribution, *Insect Biochem.* **9**:443-450.

Strand, D. J., and McDonald, J. F., 1989, Insertion of a *copia* element 5' to the *Drosophila melanogaster* alcohol dehydrogenase gene (*Adh*) is associated with altered developmental and tissue-specific patterns of expression, *Genetics* **121**:787-794.

Templeton, A. R., Boerwinkle, E., and Sing, C. F., 1987, A cladistic analysis of phenotypic associations with haplotypes inferred from restriction endonuclease mapping. I. Basic theory and an analysis of alcohol dehydrogenase activity in Drosophila, *Genetics* 117:343-351.

Tepper, C. S., Richmond, R. C., Terry, A. L., and Senior, A., 1982, Studies of esterase 6 in *Drosophila melanogaster*. XI. Modification of esterase 6 activity by unlinked genes, *Genet. Res.* **40**:109-125.

Tepper, C. S., Terry, A. L., Holmes, J. E., and Richmond, R. C., 1984, Studies of esterase 6 in *Drosophila melanogaster*. XIV. Variation of esterase 6 levels controlled by unlinked genes in natural populations, *Genet. Res.* **43**:181-190.

Throckmorton, L., 1975, The phylogeny, ecology and geography of *Drosophila*, in: *Handbook of Genetics*, Vol. 3 (R. C. King, ed.), Plenum, New York, pp. 421-469.

Throckmorton, L., 1982, The virilis species group, in: *The Genetics and Biology of Drosophila*, Vol. 3B (M. Ashburner, H. L. Carson, and J. N. Thompson, eds), Academic Press, London, pp. 227-297.

Townsend, D. R., and Singh, R. S., 1984, Genetic variation for monomer-dimer equilibria of esterase-5 in *Drosophila pseudoobscura*, *Drosophila persimilis* and *Drosophila miranda*, *Can. J. Genet. Cytol.* **26**:374-381.

Vogt, R. G., and Riddiford, L. M., 1981, Pheromone binding and inactivation by moth antennae, *Nature, Lond.* **293**:161-163.

Wasserman, M., 1982, Evolution of the repleta group, in: *The Genetics and Biology of Drosophila*, Vol. 3B (M. Ashburner, H. L. Carson, and J. N. Thompson, eds), Academic Press, London, pp. 61-139.

White, M. M., Mane, S. D., and Richmond, R. C., 1988, Studies of esterase 6 in *Drosophila melanogaster*. XVIII. Biochemical differences between the slow and fast allozymes, *Mol. Biol. Evol.* **5**:41-62.

Whiting, J. H., Pliley, M. D., Farmer, J. L., and Jeffery, D. E., 1989, *In situ* hybridisation analysis of chromosomal homologies in *Drosophila melanogaster* and *Drosophila virilis*, *Genetics* **122**:99-109.

Wingender, E., 1988, Compilation of transcription regulating proteins,

Nucl. Acids Res. **16**:1879-1902.

Wright, T. R. F., 1963, The genetics of an esterase in *Drosophila melanogaster, Genetics* **48**:787-801.

Wright, T. R. F., and MacIntyre, R. J., 1963, A homologous gene-enzyme system, esterase 6, in *Drosophila melanogaster* and *D. simulans, Genetics* **48**:1717-1726.

Yenikolopov, G. N., Kuzin, B. A., Evgen'ev, M. B., Ludwig, M. Z., Korochkin, L. I., and Georgiev, G. P., 1983, The cloning and expression of the gene encoding organ specific esterases from the genome of *Drosophila virilis, EMBO J.* **2**:1-7.

Zouros, E., 1976, The distribution of enzyme and inversion polymorphism over the genome of *Drosophila:* evidence against balancing selection, *Genetics* **83**:169-179.

Zouros, E., and van Delden, W., 1982, Substrate preference polymorphism at an esterase locus of *Drosophila mojavensis, Genetics* **100**:307-314.

Zouros, E., van Delden, W., Odense, R., and van Dijk, H., 1982, An esterase duplication in *Drosophila*: differences in expression of duplicate loci within and among related species, *Biochem. Genet.* **20**:929-942.

CHAPTER 20

Molecular Isolation and Preliminary Characterisation of a Duplicated Esterase Locus in Drosophila buzzatii

PETER EAST, ANNE GRAHAM, AND GILLIAN WHITINGTON

1. INTRODUCTION

Most plants and animals possess a great variety of esterase isozymes of broad substrate specificity and diverse patterns of developmental expression. In the house mouse *Mus musculus* several loci encoding esterases have been mapped in two clusters on chromosome 8 (Peters, 1982). Some of these loci are very tightly linked and their protein products have similar subunit molecular weights. Similar clusters of apparently homologous loci are found in related species (Peters, 1982), notably the laboratory rat, *Rattus norvegicus* (Hedrich and von Deimling, 1987), where two clusters are found 17 map units apart on linkage group V. The occurrence of very tightly linked clusters suggests an origin by repeated tandem duplication for some of these esterases. Interestingly, individual members of a cluster are characterised by essentially unique patterns of developmental expression (Peters, 1982; Hedrich and von Deimling, 1987). This implies a relatively rapid alteration of the elements controlling the spatial and temporal expression of these proteins.

There also is abundant genetic evidence for the existence of esterase duplications in insects. Baker (1980) and Baker and Kaeding (1981) have reported a duplicated esterase locus in *Drosophila montana* and Zouros and co-workers have identified esterase duplications in other species of *Drosophila* (Zouros et al., 1982) and in the Mediterranean fruit fly *Dacus oleae* (Zouros and Krimbas, 1970).

There are two documented instances in which insect pest species, the peach potato aphid, *Myzus persicae* (Devonshire, 1977) and the mosquito *Culex quinquefasciatus* (Mouches et al., 1986) have evolved resistance to organophosphate insecticides through amplification of an esterase gene. These amplification events are the result of repeated tandem duplications of a single ancestral locus encoding a non-specific esterase.

PETER EAST, ANNE GRAHAM, AND GILLIAN WHITINGTON * Department of Animal Science, University of New England, Armidale NSW 2351, Australia.

Ecological and Evolutionary Genetics of Drosophila, Edited by
J.S.F. Barker *et al.*, Plenum Press, New York, 1990

The apparent propensity of esterase genes to rapidly evolve novel patterns of developmental expression and presumably altered *in vivo* functions make them attractive loci for the study of molecular events associated with these phenomena. In this paper we present preliminary results of our attempts to isolate and characterise a duplicated carboxylesterase locus from *Drosophila buzzatii*, a member of the *mulleri* subgroup of the *repleta* species cluster.

2. EVIDENCE FOR A ß-ESTERASE DUPLICATION IN THE *MULLERI* SUBGROUP

The ß-esterases of one member of the *mulleri* subgroup, *D. mojavensis* have been studied in considerable detail, both genetically (Zouros, 1973; Zouros and van Delden, 1982; Zouros *et al.*, 1982) and biochemically (Pen *et al.*, 1984, 1986a, b). Initial studies by Zouros *et al.* (1982) suggested that the loci designated *Est-4* and *Est-5* of *D. mojavensis* represented a single tandem duplication. The evidence for a duplication was based on both genetic and biochemical criteria. A number of biochemical results suggested that the esterase-4 and esterase-5 proteins were products of genetically distinct loci, notably the occurrence of substrate preference polymorphism at the *Est-5* locus which was not accompanied by a comparable alteration at the *Est-4* locus in the same individual. At the genetic level, a study of the linkage relationships between several esterase loci in *D. mojavensis* failed to produce a recombinant between *Est-4* and *Est-5* in a sample of 1849 chromosomes tested (Zouros *et al.*, 1982). These studies also reported electrophoretic evidence for a weak zone of esterase activity, representing an interlocus heterodimer, from individuals genetically homozygous at the two esterase loci. There is, however, a high degree of homology between the two loci as shown by micro-sequence analysis of the amino terminal tryptic peptides of purified esterase-4 and esterase-5 protein (Pen *et al.*, 1986b). In a sequence of 34 amino acid residues, only six were different between the two proteins, and of these, four represented chemically conservative substitutions. Given the extremely high degree of homology between these amino terminal sequences, it is of interest to note that Pen *et al.* (1984, 1986b) observed little immunological cross-reactivity between esterase-4 and esterase-5, although one, or occasionally two, proteins immunologically related to esterase-4 were readily detected in a variety of species of the *mulleri* subgroup using the Ochterlony double-diffusion technique (Pen *et al.*, 1984).

The phylogenetic distribution of the duplication was examined in 15 species of the *repleta* group of *Drosophila*, comprising members of most of the major species complexes of the *mulleri* subgroup and two near relatives in the *hydei* subgroup (Zouros *et al.*, 1982). In that study, no *Est-4* homologue was electrophoretically detectable in *D. tira*, a member of the *mulleri* complex, and in neither of the two *hydei* subgroup species tested. In contrast, using immunological methods, Pen *et al.* (1984) reported a protein in *D. eohydei* which strongly cross-reacted with antibody to purified esterase-4 of *D.*

mojavensis. These results provide a cautionary note, that electrophoretic detectability may not be a reliable criterion for the presence of these enzymes in the absence of any knowledge of their *in vivo* substrates.

In the context of protein evolution, these duplicated loci are particularly noteworthy for their patterns of temporal and spatial expression. Zouros *et al.* (1982) observed that esterase-4 activity could not be detected prior to 96 hours post-oviposition. Activity peaked around the time of pupation and declined steadily thereafter, being undetectable in newly eclosed adults. In contrast, esterase-5 appeared early in larval development, increasing with larval age to a peak shortly before pupation. Activity then declined during the pupal phase, but increased again during development of the pharate adult, and following eclosion, remained relatively stable in the imago. The tissue-specific expression of these two proteins also differed dramatically. The esterase-4 protein was expressed at high levels in the carcass, whereas esterase-5 was found in haemolymph and fat body.

3. DEVELOPMENTAL STUDIES OF THE ß-ESTERASES OF *D. BUZZATII*

At the time these data were reported, similar studies were being done in another member of the *mulleri* subgroup, *D. buzzatii* (East, 1982). Detailed analysis of the pattern of expression of the ß-esterases of *D. buzzatii* yielded results similar to those reported by Zouros *et al.* (1982) for *D. mojavensis*. Figure 1 summarises the ontogenic expression of

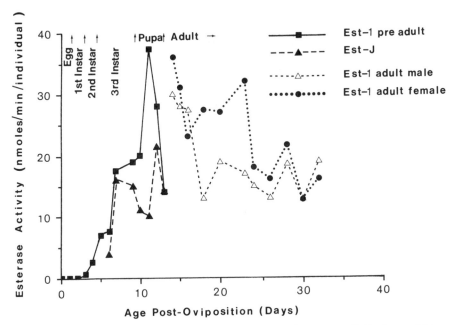

Figure 1. Ontogenic expression of the ß-esterases of *D. buzzatii*.

the two ß-esterases of *D. buzzatii*. Esterase-1 is homologous
to esterase-5 of *D. mojavensis* and esterase-J is homologous to
esterase-4.   The esterase-1 enzyme is present in all developmental stages, first appearing during embryogenesis, rising to
a peak in third larval instar and then declining during pupal
development. As in *D. mojavensis*, activity of esterase-1 in
the adult is relatively constant. Esterase-J activity appears
abruptly about half-way through the third larval instar, rises
rapidly to a peak just prior to pupation, then declines during
pupal development and is not detectable in the adult insect.

The pattern of tissue-specific expression of the enzymes
in *D. buzzatii* also matches that seen for the ß-esterases of
D. mojavensis, confirming the homology of the proteins in
these two species.   The expression of the enzymes is summarised in Table I.   Some effort was expended in trying to
determine more precisely the cellular distribution of
esterase-J in the carcass.   Histochemical staining of tissue
whole-mounts indicated the presence of non-specific esterases
in every cell type examined.   The application of a variety of
inhibitors failed to differentiate further classes of esterase
in these preparations, so attempts were made to identify
esterase isozymes by tissue dissection and electrophoresis on
high resolution cellulose acetate strips.   Esterase-J was not
detected in dissected muscle preparations, but was present in
"stripped" carcass comprising cuticle, cuticular hypodermis
and histoblasts.   It could not be detected in imaginal discs
and its presence in oenocytes could not be unambiguously
determined, although these cells stained strongly positive for
non-specific esterases in wholemount preparations and cryo-
sectioned material.

Table I
Tissue distribution of the ß-esterases of *D. buzzatii*
detected by electrophoresis and histochemical staining

	Esterase activity	
Tissue	Est-1	Est-J
Whole larva	+++	+++
Oenocyte	-	+/-
Stripped carcass (epidermis and histoblasts)	-	++
Haemolymph	+++	-
Imaginal discs	-	-
Brain	-	-
Brain with imaginal discs and ring gland	+/-	+/-
Fat body	+/-	-
Gut	-	-

+++ strong activity,   ++ moderately strong activity,
+/- weak activity,   - no detectable activity.

In summary, from biochemical and genetic data, there appears to be a tandemly duplicated ß-esterase locus which is widespread in members of the *mulleri* subgroup of the *repleta* species group. Histochemical data do not support the existence of this duplication in the related *hydei* subgroup (Zouros *et al.*, 1982) but immunological tests suggest that proteins homologous to the duplicated esterases are present in *D. eohydei* (Pen *et al.*, 1984). Consequently, the phylogenetic distribution of this duplication remains unclear. However, in those species where the duplication is detected, the two genes show extremely divergent patterns of developmental regulation. These esterases therefore provide an excellent opportunity to study the molecular events which have occurred as the duplicated loci have evolved novel patterns of temporal and spatial expression. In an attempt to exploit this experimental system, we have undertaken the molecular cloning of this duplicated esterase locus.

4. MOLECULAR CLONING OF THE DUPLICATION

The gene encoding the most abundant ß-esterase of *D. melanogaster* has been recently isolated and the sequences of a cDNA (Oakeshott *et al.*, 1987) and genomic clones (Collet *et al.*, 1989) are available. When the deduced amino acid sequence of the *D. melanogaster Est-6* cDNA was compared with the amino terminal peptide sequence data of Pen *et al.* (1986b) for the duplicated esterases of *D. mojavensis*, a remarkable degree of sequence homology was observed. The esterase-6 sequence is identical to one or other of the two *D. mojavensis* proteins at 27 of 36 residues. DNA sequence analysis of genomic clones from *D. melanogaster* (Collet *et al.*, 1989) revealed the existence of a previously unsuspected duplication of the ß-esterase locus in this species, and this duplication is found in other members of the *melanogaster* species subgroup (Oakeshott, pers. comm.).

The discovery of two esterase loci in *D. melanogaster*, coupled with the high level of sequence similarity between esterase-6 and the two *D. mojavensis* esterases suggested that the duplication in these divergent branches of the *Drosophila* phylogeny was homologous (Collet *et al.*, 1989).

Preliminary studies using the esterase-6 cDNA to probe Southern blots of *D. buzzatii* genomic DNA under conditions of reduced stringency of hybridisation (30% formamide, 6X SSC at 37°C) and washing (2X SSC at 50°C) revealed discrete bands of hybridisation over a low level of background. On this basis the *D. melanogaster* clone was used as an hybridisation probe to screen a *D. buzzatii* genomic library. Two independent recombinant bacteriophages were recovered and plaque purified from this screen. One of these, designated λDbE1, was selected for further analysis. A restriction map of the genomic insert in this phage is given in Figure 2.

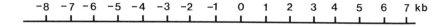

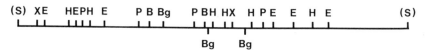

Figure 2. Restriction map of genomic clone λDbE1. Restriction enzyme abbreviations are as follows; B, *Bam* HI, Bg, *Bgl* II, E, *Eco* RI, H, *Hind* III, P, *Pst* I, S, *Sal* I, X, *Xho* I. (The *Sal* I sites shown in brackets are derived from the arms of the EMBL3 vector).

5. TRANSCRIPTIONAL ORGANISATION

Preliminary mapping of the transcriptional organisation of this genomic region was undertaken using two probes. One of these, the full length cDNA from *D. melanogaster*, was used to determine the regions with homology to esterase coding domains. The second probe was a 0.9 kb genomic fragment from the *Est-6* homologue of *D. simulans*, which comprised approximately 300 bp of coding sequence from the 5' end of the *Est-6* gene and a further 600 bp of non-transcribed flanking sequence. It was hoped that the use of this probe would enable the direction of transcription to be inferred from the pattern of hybridisation. The results of these hybridisation experiments are summarised in Figure 3.

Two broad domains showing homology to *Est-6* cDNA were identified. One of these, corresponding to a 5.2 kb *Hind* III fragment, provided a very strong hybridisation signal. Further analysis of this fragment with the *D. simulans* 5' specific probe indicated two separate regions of hybridisation, suggesting the existence of two coding domains within this fragment. This 5.2 kb *Hind* III fragment was digested into smaller fragments using a variety of restriction endonucleases and these were subcloned into the plasmids pTZ18U and pTZ19U. Single stranded DNA for sequence analysis was prepared from these plasmid subclones by infection with the helper phage M13K07.

In addition to the strong hybridisation signal observed with the *Est-6* cDNA to the 5.2 kb *Hind* III fragment, this probe also identified restriction fragments originating from the left-hand end of the genomic clone, in a region from -5.8 kb and extending to the end of the clone. Within this region two fragments were detected which hybridised to the *D. simulans* 5' specific probe (Fig. 3). This result raises the possibility that there may be other esterase genes in this region of the *D. buzzatii* genome. The observation that hybridisation of the *Est-6* cDNA to DNA from this region gave a weaker hybridisation signal than did fragments originating from the 5.2 kb *Hind* III fragment suggests that the DNA sequence homology in this region is somewhat less. It is possible that these are pseudogenes or remnants of a partial duplication which did not give rise to a complete gene. Alternatively,

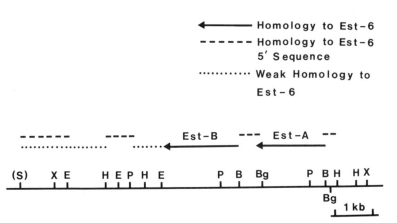

Figure 3. Restriction map of a segment of *D. buzzatii* genomic clone λDbE1 showing sites of hybridisation of the *D. melanogaster Est-6* cDNA and *D. simulans* 5'-specific clones. (Restriction enzymes are abbreviated as shown in Fig. 2).

there may be one or more active esterase genes whose products have not yet been identified. In *D. melanogaster* only two regions of esterase homology were found in a length of 20 kb and both of these appear to be transcribed (Collet *et al.*, 1989). However, in *D. pseudoobscura*, another member of the Sophophora subgenus, three regions of esterase homology have been identified (Richmond *et al.*, Chapter 15, this volume). Further studies, especially at the DNA sequence level will be required to clarify the situation in *D. buzzatii*.

6. DNA SEQUENCE ANALYSES

Preliminary DNA sequence data are presented below (Figs 4 and 5) for part of the region showing homology to the *D. melanogaster Est-6* cDNA probe. Initial studies have concentrated on a genomic region extending from the *Xho* I site at position 0 on the restriction map (Fig. 2), leftwards to the *Eco* RI site at position -5.3 kb. Approximately 4.9 kb of sequence has been obtained in two continuous stretches of 2035 bp (from the *Xho* I site at position 0, extending leftwards) and a further block of 2838 bp downstream, terminating at the *Eco* RI site (position -5.3 kb, Fig. 2). These two sequences are separated by a gap of approximately 500 - 600 bp, for which no sequence is currently available. Although these sequence data are preliminary and the majority require confirmation by sequencing in the opposite orientation, a number of useful observations can be made from the results obtained to date.

The preliminary transcriptional mapping with the two *Est-6* probes indicated that these two sequenced regions might harbour a duplicated esterase locus. Dot matrix homology comparisons were made between these two sequences, and extensive homology was found in one domain of each sequence. Both

```
CTCGAGCATATTTCTTAAACCATGCTTGCTTAGCTTTTGCCTACACTGCGGCATATACTTAATGCGAACTAATTCATTTATATACACACAGTCCATC  99

AGCTGGCATGTGTTGCCCTGCCCTGTCCCTTTACTCTTGCTCTATCATCATCATAATCTGCTTCATCGTTGATTAGAACGTTGATAATGGCATCGCTTA  198

TTTGTCTGGGACTTGGTCATTGACTGCTCCCCAGTTGTTGATGATACTAGTTGGGACCCGAGCCCCGGAGCATTCTATGGGATTAATGAAGCTTTTATC  297

AGTGGCTGCGATCGTTCGCTGAGCTGCTGATGATTAAGCAGCGAAGCCTGAGCTTATTGTATCCCTTAGGAATATACAAAAATATATTGTATCGCCATC  396

CATCACTTCGTCCAACTTCGCCATCTCGAAAGCCAGTGAAATCTGGAATATATACTGCACTACACTAAACTGCACACAATATATAGCGTGTTTCTACTT  495

CATATATCCGAGTGAAAAACCTCTCAGGTGTCATGAGTAATTGACTAATTTGAAACGAAATTAAACCAAATTTCAATTTTTGGTTTGGTTTCCAAGTGC  594

TAGATTGTTTCGAAACTAAACTCTATTAAGGCAACAAAAGCCCTCTGCCCCTCACGTGTCAAAGTCTTAAAGTTATTGTGCAATTTCGAAGAAGATTCA  693

GCTGAAACCAATTAGATTATCTCATGCGCGCGAATAAATTAAAGCTTGAAATGGCCATAAAAGGCAATGATTCTCTATGGGCGAAACAGTCGGAGTTGG  792
                                             ‾‾‾‾‾‾‾‾‾‾              ‾‾‾‾‾‾

AACTCCGTCTGATACATCCAGTTTACTATGAATACTCTAATTTTGGCAGTCGTGTTGGCTTTAGCCTCCACGGCAGCTGCCTTATACGATCCGCTACTC  891
                      MetAsnThrLeuIleLeuAlaValValLeuAlaAlaLeuAlaSerThrAlaAlaAlaLeuTyrAspProLeuLeu

GTCGAACTGGCCAATGGAGATCTACGTGGACGCGATAAAGGATTTTACTACAGCTTCGAGTCGATACCTTATGCACTGCCGCCCATCaATGACTTGCGC  990
ValGluLeuAlaAsnGlyAspLeuArgGlyArgAspLysGlyPheTyrTyrSerPheGluSerIleProTyrAlaLeuProProIleAsnAspLeuArg

CTGGAGGATCCAGTCCCATATACGGAAAGATGGAGTAACACCTTTGATGCCACCAAGCCGCCAACTGAGTGTCTACAATGGAGCCAAACCATTGCGCAG  1089
LeuGluAspProValProTyrThrGluArgTrpSerAsnThrPheAspAlaThrLysProProThrGluCysLeuGlnTrpSerGlnThrIleAlaGln

CCAAACAAGTTGACTGGCAGCGAAGACTGTCTGACCGTCAGTGTCTACAAGCCCAAGAATCACAGCCGCATCGAATTTCCGGTTGTTGCAAATATTTTT  1188
ProAsnLysLeuThrGlySerGluAspCysLeuThrValSerValTyrLysProLysAsnHisSerArgIleGluPheProValValAlaAsnIlePhe

GGCGGACGGTGGACCTTCGGCAGTCCGCTTGATGATGGAGTTGAGCACTTTATGTATAGAGGCAATGTCATAGTGGTGAAGATCAACTATAGAGTGGGT  1287
GlyGlyArgTrpThrPheGlySerProLeuAspAspGlyValGluHisPheMetTyrArgGlyAsnValIleValValLysIleAsnTyrArgValGly

CCACTTGGATTTTTAAGCACCGGTGACAAGGAACTTCCTGGAAACTATGGTCTCAAGGATCAGCGTGTCGCAATCCAATGGATCAAGCAGAATATAGAT  1386
ProLeuGlyPheLeuSerThrGlyAspLysGluLeuProGlyAsnTyrGlyLeuLysAspGlnArgValAlaIleGlnTrpIleLysGlnAsnIleAsp

CGATTTGGTGGAGATCCAGAGAACATAATTCTTCTTGGATTCGGTACAGGCGGCTCATCCGTCCACCTGCAGCTGATGCACAAGGATATGGAGAAGGTA  1485
ArgPheGlyGlyAspProGluAsnIleIleLeuLeuGlyPheGlyThrGlyGlySerSerValHisLeuGlnLeuMetHisLysAspMetGluLysVal

GTTAAGGGTGGCATCTCAATTAGTGGAACAGCCACATCGCCCTTTGCAGTTCAGTCCAGCGGACAAGAGGTGGCATTTCGATACGCTAGAATATTGGGA  1584
ValLysGlyGlyIleSerIleSerGlyThrAlaThrSerProPheAlaValGlnSerSerGlyGlnGluValAlaPheArgTyrAlaArgIleLeuGly

TGTGACAGTCGCAAAGCATCGACTGAACTGAAGGAATGCCTGAAGAAAATAGCAGCTGACGTTTTCGTTAGTGCCTTGAAACATCTTCAAGTCTTTGAC  1683
CysAspSerArgLysAlaSerThrGluLeuLysGluCysLeuLysLysIleAlaAlaAspValPheValSerAlaLeuLysHisLeuGlnValPheAsp

TATGTACTTTTTGGCGGCACTTTAGTCCAGTCATCGAGTCCTGATTCGGCTAAACCATTTCTAACCGAATTTCCTATGGCTAGCATCAGGAGTGCAAAA  1782
TyrValLeuPheGlyGlyThrLeuValGlnSerSerSerProAspSerAlaLysProPheLeuThrGluPheProMetAlaSerIleArgSerAlaLys

TCTGCCCAAGTACCTTGGTTGGCCAGCTATACCACGGAGAATGGCATATATAACGCCGCTCTATTGCTGAAAAGGCACTCCAATGGGAAAGCGAAAATT  1881
SerAlaGlnValProTrpLeuAlaSerTyrThrThrGluAsnGlyIleTyrAsnAlaAlaLeuLeuLeuLysArgHisSerAsnGlyLysAlaLysIle

GAACAGCTCAAATCTCGATGGAATGAGCTAGCCCATACTTCTTTTCATATCCTTACTTATGGGAAACGCTCTGAGCGGGATAGCCATTCCCGGGAAACTT  1980
GluGlnLeuLysSerArgTrpAsnGluLeuAlaHisThrSerPheHisIleLeuThrTyrGlyLysArgSerGluArgAspSerHisSerArgLysLeu

AAGCAGCAATACCTCGGCTATAGGAACTTCAGCGTGGAAAACTACTTCGATGTAC  2035
LysGlnGlnTyrLeuGlyTyrArgAsnPheSerValGluAsnTyrPheAspVal
```

Figure 4. Nucleotide sequence obtained from the upstream (*Est-A*) gene, including the 5' untranslated flanking region, and the amino acid sequence of the extended open reading frame identified in this region. Putative TATA and mRNA cap site homologies are underlined.

```
CCTTAATCTCACATTAATCATACTAGAAAACTGACTAGTTTATTTTATTTATCGCCATGAAAATTTTCAATATATTTATAAAGAACTGAAATAACAAAA  99

TCAAACACTTCATTTTTATTGTTATTTTTCTAAACAAAGCAAAATAAAGTCAAGGAAAAACCGAACGAATAACTGACTTGAGAATAACTTAAACGAAAT  198

AAATATATTACATTAAAAATAAGTTTGCAGTTAATGATCATACATATTTGTAGACTTGTAGTTGAGCGAAGGGTTGAACAGAGGTTTGGATTGATTAGC  297

TTGGTATTAAGTTCTTATTAATATACAAGTTTAGATCTGCTTACGCAAAAATATTTTAAACAAGTTTAAGCCATAAATAAACAGGCGTGCTTTACCAAA  396

ACTTGTAATAACTCTTCTCTGCACTCTTTAGAATAGGTTTTGCTTATAAATAAATTCATAAGCTAGTTATACGATTTTATAGTACATAAAGTCACAACT  495

TTACAAATGGCGATTAGCTCTTGAAAATCGCAACCGCTAATAAGTCAGCTCAGTTGCTAGCCCTTTGACCGCAAAGAGACAGTGGAGCAAGTGTGTCAA  594

ACTCATTTATATATACATAAATATAAATCGCATTTTTACTTCCATAGAGAAAGCCATCGGAACGAATCCGGTTTCACTTTATCTTCTATGTGCGAATAA  693

ATAGGCAACGATTGCCTTTGTTTAAGACAGTTTGACTTGGAGCGCATCATGAACATACTCGACGCGACACTCATTTTGTTCTCAGCATGTGCAGCATCT  792
                                                 MetAsnIleLeuAspAlaThrLeuIleLeuPheSerAlaCysAlaAlaSer

GCCTGGAACGATCCGCTACTCGTCGAGCTGGCCAATGGAGAGCTGCGTGGTCGCGACAATGGCATTTACTATAGCTACGAGTCGATACCTTATGCACAG  891
AlaTrpAsnAspProLeuLeuValGluLeuAlaAsnGlyGluLeuArgGlyArgAspAsnGlyIleTyrTyrSerTyrGluSerIleProTyrAlaGln

CCGCCCATCAATGAGTTGCGTCTGGAGGATCCAGTTTCATTTACTGGAAGGTGGGAGCATACCTTTGATGCCACCACACCGGCAACCGAATGCCTGCAA  990
ProProIleAsnGluLeuArgLeuGluAspProValSerPheThrGlyArgTrpGluHisThrPheAspAlaThrThrProAlaThrGluCysLeuGln

TTGAGTCAGTTCATCCAACAGCCTGACAAATTAACGGGCAGCGAGGACTGCCTGACCGTCAGCATTTTCAGGCCTAAAAATGCTACGCGAGAGAGCTTC  1089
LeuSerGlnPheIleGlnGlnProAspLysLeuThrGlySerGluAspCysLeuThrValSerIlePheArgProLysAsnAlaThrArgGluSerPhe

CCGGTCGTGGCAAGCATTTTCGGTGGAGCATGGAACTTCGGCAGTTCGCTTGATGATGCAGCTGACAACTTCATGGACAGCGGCAATGTAATAGTTGTG  1188
ProValValAlaSerIlePheGlyGlyAlaTrpAsnPheGlySerSerLeuAspAspAlaAlaAspAsnPheMetAspSerGlyAsnValIleValVal

AAGATCAACTACAGAGTGGGACCGCTTGGCTTTTTAAGCACCGGCGACAATGTACTTCCTGGAAACTATGGTCTCAAGGATCAGCGTCTCGCAATTCAA  1287
LysIleAsnTyrArgValGlyProLeuGlyPheLeuSerThrGlyAspAsnValLeuProGlyAsnTyrGlyLeuLysAspGlnArgLeuAlaIleGln

TGGATCAAGCAGAATATTGCTCGTTTCGGTGGAGATCCAGAGAACATAATTCTTCTTGGTTTCGGTGCAGGCGGCTCTGCCGTCCACCTGCAGCTGATG  1386
TrpIleLysGlnAsnIleAlaArgPheGlyGlyAspProGluAsnIleIleLeuLeuGlyPheGlyAlaGlyGlySerAlaValHisLeuGlnLeuMet

CACAAGGATATGGAGAAGATGGTCAAGGGTGGCATCTCAATTAGTGGAACAGCCTTCTCCCCCTTTGCTTTGCAAGGAAATGGACGCGAAGACGCTTTC  1485
HisLysAspMetGluLysMetValLysGlyGlyIleSerIleSerGlyThrAlaPheSerProPheAlaLeuGlnGlyAsnGlyArgGluAspAlaPhe

CGATTTGCCAAAATGTTGGGTTGCAACCGAATGAAGTCTTCATCTGAACTGAAAGACTGCTTCAAAAAGACAATGGCCAACCAACTCGTCAGAGCCGTT  1584
ArgPheAlaLysMetLeuGlyCysAsnArgMetLysSerSerSerGluLeuLysAspCysPheLysLysThrMetAlaAsnGlnLeuValArgAlaVal

AAGAAATTCCAAGTCTTTGAAAATATACCATCTGCTGTTTTTCGACCAGTTGTGGAGTCCCCTGATTCGGAAAAACCATTCCTCACCGAGCTTCCCCTC  1683
LysLysPheGlnValPheGluAsnIleProSerAlaValPheArgProValValGluSerProAspSerGluLysProPheLeuThrGluLeuProLeu

GAAACTATAAAAAGTGGAAGATATGCCCAAGTACCTTGGTTGGCCAGCTACACCACGCAGAATGGCATCTATAACACGGCAATCTTGCTACAGAAGGAT  1782
GluThrIleLysSerGlyArgTyrAlaGlnValProTrpLeuAlaSerTyrThrThrGlnAsnGlyIleTyrAsnThrAlaIleLeuLeuGlnLysAsp

GCGAATGGAAAAGAAATAATTTCACAGCTAAACTCGCGCTGGAATGAGCTAGCTCCGTATTTTTTATATTATCCAAAGGATATGAATAAATTTGAGAAA  1881
AlaAsnGlyLysGluIleIleSerGlnLeuAsnSerArgTrpAsnGluLeuAlaProTyrPheLeuTyrTyrProLysAspMetAsnLysPheGluLys

AATTTTCATTCTCAGAAACTAAAGCTGCAATATCTAGGATTTAGGAACTTCAGCGTCGAAAACTATTTCGATGTACAGCGCATGCTTACCAATGAACTG  1980
AsnPheHisSerGlnLysLeuLysLeuGlnTyrLeuGlyPheArgAsnPheSerValGluAsnTyrPheAspValGlnArgMetLeuThrAsnGluLeu

TTTAAGAATGGCGTTGAGATGGCCATAGATGCGCAACGAAAGCATGGAAGCAGTCCCGTTTATGCTTATGTCTACGACAATCCCGCTGATAAGTCAATC  2079
PheLysAsnGlyValGluMetAlaIleAspAlaGlnArgLysHisGlySerSerProValTyrAlaTyrValTyrAspAsnProAlaAspLysSerIle

GGACAGTTGCTTTCGAGACGAAAGGACCTGTTCTTGGGTAAAGTATTTCGTTAAGGCTGCCAAGCAGCTGCTTCAACTGATAATTAATATTCAGCTATT  2178
GlyGlnLeuLeuSerArgArgLysAspLeuPheLeu

CACAGGCACGGGCTTGGGTGACGATTTCTATTTATTGATGAACAATCCTCTGCGTGGACCACTGCGACCTGACGAGCAACTGGTTTCTTGGAAGCTGGT  2277
    GlyThrGlyLeuGlyAspAspPheTyrLeuLeuMetAsnAsnProLeuArgGlyProLeuArgProAspGluGlnLeuValSerTrpLysLeuVal

CAAGATGGTGGAGCAATTCGCCCAAAGCAGAAACTTGGTATATGATGACTGCGTATTCCCAGACAACCGGGGCCAAGAGGGCAGTTCCAGCTGGTGGCC  2376
LysMetValGluGlnPheAlaGlnSerArgAsnLeuValTyrAspAspCysValPheProAspAsnArgGlyGlnGluGlySerSerSerTrpTrpPro

ATCAAACGTACTCATTGCGAACTATTGGAGACAAGACACTTGCCGGAAACCGTGTCTGACGATTTGCCCCCATCTCCAGTAGTAATATACTAGTCCAAT  2475
SerAsnValLeuIleAlaAsnTyrTrpArgGlnAspThrCysArgLysProCysLeuThrIleCysProHisLeuGln

GTTTATCAACCTATCCATACATCAATCCATCACAATGAGTCTATGTAGGGACTTTTGGTAAATAAATCATTGCATTATGAGAAAAACTCCAAACAAAAC  2574

ATTTTTCTTTTGTATTTCGATATTTATGTAACCTAATAATAGAAAAGTATATATATAAGCAGTCTGATGGGTTTCTAAGTCCAAATTCCAATTGTGATA  2673

TCTCAACAACAAACCAAAACCTTTTAAGTACTTAATACGAACCAACTATCAGTAAGTTAAGGCAAAAAAAAAAAGAAAGAAAAAGTATTTAAATGATTG  2772

ATTCAAACTCATGAAGAGCTTGCCATTCATACACTCTTTTTACTCTATACTCTGTCACAGAATTC  2837
```

Figure 5. Nucleotide sequence from the downstream (*Est-B*) gene plus 5' and 3' untranslated flanking sequences. Amino acid sequence is given for the presumptive exon regions. Putative TATA and mRNA cap site homologies are underlined; also underlined are the consensus splice site sequences of the putative intron.

sequences were then examined for the presence of extended open
reading frames (ORF) and one long ORF was identified in each,
commencing in the regions of nucleotide sequence homology
detected by the dot matrix analysis. These extended ORFs were
translated and the protein sequences derived from them are
indicated in Figures 4 and 5. Available sequence from *D. buzzatii* permits the comparison of approximately 410 amino acids
extending from the amino terminal end of each protein. The
first 16 amino acids of both proteins are highly hydrophobic
and by comparison with the *D. melanogaster* esterases and
acetylcholinesterase (Hall and Spierer, 1986) are likely to
constitute a signal peptide which is presumably removed by
proteolytic cleavage to generate the mature esterase protein.
Excluding the presumptive signal peptides, the remaining amino
acids are 69% identical, and if conservative substitutions are
taken into account the degree of homology between these duplicated proteins is much higher. The existence of these long
ORFs, coupled with the high degree of amino acid homology is
stongly suggestive that we are indeed analysing the ß-esterase
duplication of *D. buzzatii* and that the products encoded by
these ORFs are active esterase proteins. A comparison of the
N-terminal peptide sequence of esterase-4 and esterase-5 of *D.
mojavensis* (Pen *et al.*, 1986b) with the comparable regions of
the two *D. buzzatii* proteins is given in Figure 6. These comparisons indicate very high similarity between species, and if
conservative substitutions are considered, the proteins have
74 % homology (average of the four possible pairwise comparisons).

In the absence of other experimental data it is not possible to relate the products of the two *D. buzzatii* genes to
known esterase proteins in this species. For this reason they
are referred to as *Est-A* and *Est-B* in the following discussion. Although we have not yet successfully sequenced all of
the upstream gene (*Est-A*), the sequence currently available
indicates that the two coding regions are separated by at
least 740 bp of DNA. This is substantially more than the 197
bp which separates the *Est-6* and *Est-P* genes of *D.
melanogaster*. Whether there has been an expansion of the
intergenic region in the *repleta* species or a contraction in
the *melanogaster* subgroup cannot be determined from these
results.

For the downstream gene (*Est-B*), more extensive data are
available. Further analysis of this sequence revealed a presumptive intron commencing at codon 459. This putative intron
is 67 bp in length and occurs at the same position in the
sequence as the introns of the *Est-6* and *Est-P* genes of *D.
melanogaster*. Although the identification of this stretch of
nucleotides as an intron is strictly hypothetical, being based
on analogy with the *D. melanogaster* genes, the putative donor
and acceptor splice sites in the *D. buzzatii* sequence are
flanked by short stretches of sequence highly homologous to
the consensus U1 snRNP splice signal (Lewin, 1987) 5'-
AG|GTAAGT.....6(Py)NCAG|N-3' as indicated in Figure 5.

If the sequences of the two esterase proteins of *D. buzzatii* are aligned with those of *D. melanogaster*, six spatially

				(ALA)	PRO	LEU	LEU	VAL	GLU	*D. mj.* Est-4
(SER)	ALA	ALA	ALA	ASP	PRO	LEU	ILE	VAL	GLU	*D. mj.* Est-5
ALA	ALA	LEU	TYR	ASP	PRO	LEU	LEU	VAL	GLU	*D. b.* Est-A
LEU	PRO	GLY	THR	SER	PRO	LEU	LEU	VAL	GLU	*D. b.* Est-B
LEU	PRO	ASN	GLY	LYS	LEU	ARG	GLY	ARG	ASP	*D. mj.* Est-4
LEU	PRO	ASN	GLY	LYS	VAL	(ARG)	GLY	(ARG)	ASP	*D. mj.* Est-5
LEU	ALA	ASN	GLY	ASP	LEU	ARG	GLY	ARG	ASP	*D. b.* Est-A
LEU	ALA	ASN	GLY	GLU	LEU	ARG	GLY	ARG	ASP	*D. b.* Est-B
ASN	(GLU)	GLY	TYR	TYR	(GLU)	ALA	GLU	(LEU)	ILE	*D. mj.* Est-4
ASN	GLU	GLY	TYR	TYR	(GLU)	ALA	GLU	(GLY)	ILE	*D. mj.* Est-5
LYS	GLY	PHE	TYR	TYR	SER	PHE	GLU	SER	ILE	*D. b.* Est-A
ASN	GLY	ILE	TYR	TYR	SER	TYR	GLU	SER	ILE	*D. b.* Est-B
PRO	(LYS)	ALA	(ASP)	PRO	PRO	VAL	GLY	(ASP)	LEU	*D. mj.* Est-4
PRO	(ARG)	ALA	GLU	PRO	(PRO)	VAL	GLY			*D. mj.* Est-5
PRO	TYR	ALA	LEU	PRO	PRO	ILE	MET	THR	CYS	*D. b.* Est-A
PRO	TYR	ALA	GLN	PRO	PRO	ILE	ASN	GLU	LEU	*D. b.* Est-B

Figure 6. Comparison of amino terminal peptide sequences of the Esterase-4 and Esterase-5 proteins of *D. mojavensis* (*D. mj.*) and the equivalent sequences deduced for the two proteins from *D. buzzatii* (*D. b.*).

conserved cysteine residues are observed in the esterase-B gene product and four of these are conserved in that region of the esterase-A gene for which sequence is available (Figure 7). These cysteine residues have been implicated in formation of the tertiary structure of esterase proteins (MacPhee-Quigley *et al.*, 1986).

Taken together, these separate observations; (a) a duplicated locus with homology to *Est-6* of *D. melanogaster*, (b) extended ORFs encoding polypeptides of high homology to each other and to the comparable amino terminal peptide sequence of the duplicated esterase-4 and esterase-5 proteins of *D. mojavensis*, (c) the presence of a putative intron in one of the *D. buzzatii* genes in the same position as the introns of the *D. melanogaster* genes and (d) the precise spatial conservation of functionally important cysteine residues, strongly suggest that we are analysing the duplicated esterase locus of *D. buzzatii* which encodes the esterase-1 and esterase-J proteins.

Clearly the DNA sequence must be completed and verified before we can proceed. Provided we can demonstrate, by the use of suitable hybridisation probes, that these genomic regions are transcribed into RNA, it seems likely that our initial objective, the cloning of the ß-esterase duplication from *D. buzzatii* has been achieved.

7. SEQUENCE COMPARISONS

Comparison of the nucleotide sequences between the duplicated loci in *D. buzzatii* indicated substantial homology
between the two genes in the coding region and essentially no
evidence of any appreciable homology outside this domain.
This is in marked contrast to the alcohol dehydrogenase duplication found in the *mulleri* subgroup species, where a 400 bp
region of 5' untranslated flanking sequence was found to be
75% similar between the duplicated loci, including some identical or near identical blocks of sequence (Atkinson *et al.*,
1988). The lack of sequence similarity in the 5' flanking
regions of the esterase genes when compared with the *Adh* genes
may simply reflect the relative age of the duplications. The
Adh duplication is restricted to members of the *mulleri*
species subgroup and perhaps some other closely related
species (Batterham *et al.*, 1984), whereas the data presented

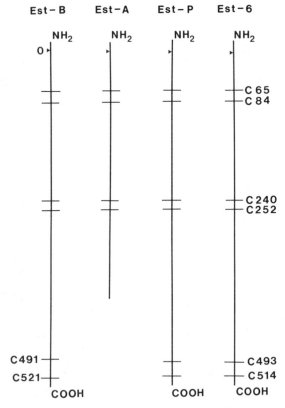

Figure 7. Spatial conservation of cysteine residues in four *Drosophila*
esterases. The sequences are aligned from the presumptive amino termini of
the mature, secreted form of the proteins (o).

here suggest that the esterase duplication may be at least as old as the genus *Drosophila*.

Computer assisted searches for short, highly conserved, domains in the non-coding segments of the *D. buzzatii* sequence identified potential TATA boxes and appropriately positioned sequences with good homology to insect mRNA cap-site consensus sequences of the type identified by Hultmark *et al.* (1986). These putative regulatory sequences are indicated in Figures 4 and 5.   In addition, several other short, highly conserved sequence motifs were identified in a region extending several hundred base pairs 5' to each of the coding domains (data not shown).   These sequences are not at all conserved in their spatial relationships to one another or to the putative mRNA transcription start sites, and therefore probably are not rem-nants of the original duplication event.   They may be *cis*-reg-ulatory elements, but further speculation is not warranted in the absence of functional tests of their significance.

Within the putative coding domains there is extensive nucleotide sequence homology, and as noted earlier, this is reflected in a high degree of sequence similarity at the amino acid level.   The homology between the two *D. buzzatii* sequences and the amino terminal peptide sequence of the duplicated ß-esterases of *D. mojavensis* was discussed previ-ously.   A dot matrix homology comparison of the duplicated esterase products of *D. melanogaster* and *D. buzzatii* is pre-sented in Figure 8.   This analysis was done to detect short regions of high similarity (a window of 20 amino acids with a match criterion of at least 16 residues).   If less stringent homology criteria are employed it is possible to demonstrate that all four proteins share homology over their entire sequence length (data not shown).   However, if the search is restricted to domains of very high homology (Fig. 8), then for the four interspecific comparisons, homology is confined to a short region centred around residue 180.   Figure 9 shows the alignment of the first 250 amino acids of the four proteins. Inspection of these sequences reveals extensive homology throughout the total length, but the highly conserved domain (identified in Fig. 8) is of particular interest.   This region of the protein has been implicated in the catalytic mechanism of esterases (Myers *et al.*, 1988).   Included in this sequence is an absolutely conserved dodecapeptide (see Fig. 9) contain-ing an aspartate residue thought to participate in the charge relay system of proton transfer which occurs during ester hydrolysis (Sikarov *et al.*, 1987).

The octapeptide which normally contains the active centre serine residue involved in proton transfer also is included in this highly conserved region.   The two sequences from *D. buz-zatii* do not agree at all well with the consensus sequence derived for this region of esterase proteins (Gly-Glu-Ser-Ala-Gly-Gly-Ala-Ser), and neither do they contain the active centre serine, although they are reasonably similar to each other.   We do not currently have an adequate explanation for this result.   It is possibly a sequencing error, and until this region of the sequence can be independently verified,

this must be considered the most plausible explanation for the
lack of homology to the conserved octapeptide.

Although the data presented here are only preliminary, one
further observation of interest may be made from the diagrams
in Figure 8.   The four orthologous comparisons revealed that
the only highly conserved domain was restricted to the region
discussed above.   However, the two paralagous comparisons also
illustrated in Figure 8 suggest that genes within a species
are more similar to each other than presumptive orthologous
homologues are between species. This appears to be the case in
both *D. melanogaster* and *D. buzzatii*.   This result is suggestive of concerted evolution of the duplicated genes within
each species.   There is a growing body of data which suggests
that concerted evolution may be relatively common in eukaryote
gene families (Payant *et al.*, 1988) and the studies currently
being undertaken on duplicated esterase loci should contribute
further to this topic.

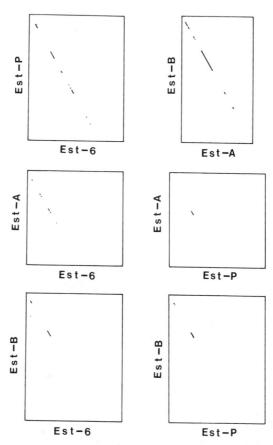

Figure 8. Dot matrix comparison of the amino acid sequences of the duplicated esterases of *D. melanogaster* and *D. buzzatii* using a search window of
20 and a match criterion of 16 or better.

8. SUMMARY

Drosophila buzzatii possesses two ß-esterase isozymes which differ dramatically in their patterns of temporal and spatial expression. Evidence from a variety of studies of other members of the *mulleri* subgroup indicates that these enzymes are the products of a duplicated locus. Recent molecular studies of the *Est-6* gene of *D. melanogaster* have demonstrated that it also is one product of a previously unidentified duplication. Results presented in this paper suggest that the event leading to the duplication of these esterase genes may have preceded the divergence of the Sophophora and Drosophila sub-genera.

The patterns of developmental expression of the duplicated genes in *D. buzzatii* differ considerably from that of *Est-6* of *D. melanogaster*. Molecular characterisation of the genes from

```
MNYVGLGLIIVLSCLWLGSNASDTDDPLLVQLPQGKLRGRDNGSYYSYES    (50)
 MSIFKRLLCLTLLWIAALESEA DPLIVEITNGKIRGKDNGLYYSYES
  MNTLILAVVLALASTAAALYDPLLVELANGDLRGRDKGFYYSFES
  MNILDATLILFSACAASAWNDPLLVELANGELRGRDNGIYYSYES


IPYAEPPTGDLRFEAPEPYKQKWSDIFDATKTPVACLQWDQFTPGANKLV   (100)
IPYAEHPTGALRFEAPQPYSHHWTDVFNATQSPVECMQWNQFINENNKLM
IPYALPPINDLRLEDPVPYTERWSNTFDATKPPTECLQWSQTIAQPNKLT
IPYAQPPINELRLEDPVSFTGRWEHTFDATTPATECLQLSQFIQQPDKLT


GEEDCLTVSVYKPKNSKRNSFPVVAHIHGGAFMFGAAWQNGHENVMREGK   (150)
GDEDCLTVSIYKPKKPNRSSFPVVVLLHGGAFMFGSGSIYGHDSIMREGT
GSEDCLTVSVYKPKNHSRIEFPVVANIFGGRWTFGSPLDDGVEHFMYRGN
GSEDCLTVSIFRPKNATRESFPVVASIFGGAWNFGSSLDDAADNFMDSGN


FILVKISYRLGPLGFVSTGDRDLPGNYGLKDQRLALQWIKQNIASFGGEP   (200)
LLVVKISYRLGPLGFASTGDRHLPGNYGLKDQRLALQWIKKNIAHFGGMP
VIVVKINYRVGPLGFLSTGDKELPGNYGLKDQRVAIQWIKQNIDRFGGDP
VIVVKINYRVGPLGFLSTGDNVLPGNYGLKDQRLAIQWIKQNIARFGGDP
                     ‾‾‾‾‾‾‾‾‾

QNVLLVGHSAGGASVHLQMLREDFGQLARAAFSFSGNALDPWVIQKGARG   (250)
DNIVLIGHSAGGASAHLQLLHEDFKHLAKGAISVSGNALDPWVIQQGGRR
ENIILLGFGTGGSSVHLQLMHKDMEKVVKGGISISGTATSPFAVQSSGQE
ENIILLGFGAGGSAVHLQLMHKDMEKMVKGGISISGTAFSPFALQGNGRE
      ‾‾‾‾‾‾‾‾
```

Figure 9. Alignment of protein sequences of the duplicated esterases of *D. melanogaster* and *D. buzzatii*. Sequences are aligned in the order; *D. melanogaster* Est-6, Est-P, *D. buzzatii* Est-A, Est-B (top to bottom).

these species provides an opportunity to study the evolution
of both enzyme structure and function and developmental regulation.   Based on the homology between the *D. melanogaster*
proteins and those of another member of the *mulleri* subgroup,
D. mojavensis, an *Est-6* cDNA clone was used to screen a *D.
buzzatii* genomic DNA library.   Two recombinant bacteriophages
were recovered from this screen and one of these,   DbE1 has
been analysed in further detail.   Homology to the *D.
melanogaster* sequence was identified in restriction fragments
spanning 8 kb of this cloned genomic fragment and a 5.2 kb
*Hin*d III fragment was identified which apparently harboured
two esterase genes.   This fragment was subcloned and partial
DNA sequence data have been obtained.

Analysis of the sequences obtained for approximately 4.9 kb
of genomic DNA revealed the existence of two long open reading
frames with extensive homology to each other.   The protein
sequences derived from these ORFs were highly homologous to
the esterases of *D. melanogaster* and to the amino terminal
peptide sequences of the duplicated ß-esterases of *D. mojavensis*.

Cysteine residues believed to be essential for the formation of the active tertiary structure of esterases were found
to be conserved in their spatial distribution in the *D. buzzatii* sequences.   In one of the *D. buzzatii* genes for which complete sequence was obtained, a putative intron was identified
in the same location as the intron of the *D. melanogaster*
esterase genes.

It seems most likely that we have isolated a duplicated
esterase locus homologous to the ß-esterase duplication of *D.
melanogaster*.   However, the existence of further esterase
coding sequences downstream from this duplication, and in
close proximity to it, is inferred from the hybridisation
experiments using the *D. melanogaster Est-6* cDNA probe.   We
are currently attempting to extend the cloned genomic DNA to
encompass these sequences.   Finally, paralagous and orthologous comparisons of the esterase coding regions of *D.
melanogaster* and *D. buzzatii* suggest that genes within a
species are more similar to each other than are presumably
orthologous homologues between species.   If this result is
confirmed by further sequence analysis then it suggests that
these duplications are being subjected to some form of concerted evolution.

ACKNOWLEDGEMENTS. We are greatly indebted to Dr. John Oakeshott
and Ms Jill Karotam for their gifts of the *D. melanogaster
Est-6* cDNA and the *D. simulans* 5' genomic clones respectively.
Our thanks also to John Oakeshott, Rollin Richmond and members
of their laboratories for their interest in this work, for
many stimulating discussions and for access to unpublished
data.   This work was supported by grants under the University
of New England - CSIRO Collaborative Research Scheme to Dr.
P.D. East and Dr. J.G. Oakeshott and in part by Grant No.
A18830670 from the ARC to PDE and Prof. J.S.F. Barker.

References

Atkinson, P. W., Mills, L. E., Starmer, W. T., and Sullivan, D. T., 1988, Structure and evolution of the *Adh* genes of *Drosophila mojavensis*, *Genetics* **120**:713-723.

Baker, W. K., 1980, Evolution of the alpha-esterase duplication within the *montana* subphylad of the *virilis* species group of *Drosophila*, *Genetics* **94**:733-748.

Baker, W. K., and Kaeding, E. A., 1981, Linkage disequilibrium at the alpha-esterase loci in a population of *Drosophila montana* from Utah, *Am. Nat.* **117**:804-809.

Batterham, P., Chambers, G. K., Starmer, W. T., and Sullivan, D. T., 1984, Origin and expression of an alcohol dehydrogenase gene duplication in the genus *Drosophila*, *Evolution* **38**:644-657.

Collet, C., Nielsen, K. M., Russell, R. J., Karl, M., Oakeshott, J. G., and Richmond, R. C., 1989, Molecular analysis of duplicated esterase genes in *Drosophila melanogaster*, *Manuscript submitted*.

Devonshire, A., 1977, The properties of a carboxylesterase from the peach potato aphid, *Myzus persicae* (Sulz), and its role in conferring insecticide resistance, *Biochem. J.* **167**:675-683.

East, P. D., 1982, Non-specific esterases of *Drosophila buzzatii*, in: *Ecological Genetics and Evolution: The Cactus-Yeast-Drosophila Model System* (J.S.F. Barker, and W.T. Starmer, eds.), Academic Press Australia, Sydney. pp. 323-338

Hall, L. M. C., and Spierer, P., 1986, The *Ace* locus of *Drosophila melanogaster*: structural gene for acetylcholinesterase with an unusual 5' leader, *EMBO J.* **5**:2940-2954.

Hedrich, H. J., and von Deimling, O., 1987, Re-evaluation of LG V of the rat and assignment of 12 carboxylesterases to gene clusters, *J. Hered.* **78**:92-96.

Hultmark, D., Klemenz, R., and Gehring, W. J., 1986, Translational and transcriptional control elements in the untranslated leader of the heat-shock gene hsp22, *Cell* **44**:429-438.

Lewin, B., 1987, *Genes III*, John Wiley and Sons, New York.

MacPhee-Quigley, K. T., Vedvick, T., Taylor, P., and Taylor, S. S., 1986, Profile of the disulphide bonds in acetylcholinesterase, *J. biol. Chem.* **260**:13565-13570.

Mouches, C., Pasteur, N., Berge, J. B., Hyrien, O., Raymond, M., de Saint Vincent, B. R., de Silvestri, M., and Georghiou, G. P., 1986, Amplification of an esterase gene is responsible for insecticide resistance in a California *Culex* mosquito, *Science* **233**:778-780.

Myers, M., Richmond, R. C., and Oakeshott, J. G., 1988, On the origins of esterases, *Mol. Biol. Evol.* **5**:113-119.

Oakeshott, J. G., Collet, C., Phillis, R. W., Nielsen, K. M., Russell, R. J., Chambers, G. K., Ross, V., and Richmond, R. C., 1987, Molecular cloning and characterization of esterase-6, a serine hydrolase of *Drosophila*, *Proc. natn. Acad. Sci. USA* **84**:3359-3363.

Payant, V., Abukashawa, S., Sasseville, M., Benkel, B. F., Hickey, D. A., and David, J., 1988, Evolutionary conservation of the chromosomal configuration and regulation of amylase genes among eight species of the *Drosophila melanogaster* species subgroup, *Mol. Biol. Evol.* **5**:560-567.

Pen, J., Rongen, H. A. H., and Beintema, J. J., 1984, Purification and properties of Esterase-4 from *Drosophila mojavensis*, *Biochim. biophys. Acta* **789**:203-209.

Pen, J., Schipper, A., Rongen, H. A. H., and Beintema, J. J., 1986a, Differences in specificity and catalytic efficiency between allozymes of Esterase-4 from *Drosophila mojavensis*, *Mol. Biol. Evol.* **3**:366-373.

Pen, J., van Beeumen, J., and Beintema, J. J., 1986b, Structural comparison of two esterases from *Drosophila mojavensis* isolated by immunoaffinity chromatography, *Biochem. J.* **278**:691-699.

Peters, J., 1982, Nonspecific esterases of *Mus musculus*, *Biochem. Genet.* **20**:585-606.

Sikarov, J-L., Krejci, E., and Massoulie, J., 1987, cDNA sequences of *Torpedo marmorata* acetylcholinesterase: primary structure of the presursor of a catalytic subunit; existence of multiple 5'-untranslated regions. *EMBO J.* **6**:1865-1873.

Zouros, E., 1973, Genic differentiation associated with the early stages of speciation in the *mulleri* subgroup of *Drosophila*, *Evolution* **27**:601.

Zouros, E., and van Delden, W., 1982, Substrate-preference polymorphism at an esterase locus of *Drosophila mojavensis*, *Genetics* **100**:307-314.

Zouros, E., van Delden, W., Odense, R., and van Dijk, H., 1982, An esterase duplication in *Drosophila*: Differences in expression of duplicate loci within and among related species, *Biochem. Genet.* **20**:929-942.

Zouros, E., and Krimbas, C. B., 1970, A case of duplication of an esterase locus in the olive fruit fly *Dacus oleae*, *Isozyme Bull.* **3**:44.

CHAPTER 21

The Evolution of Adh *Expression in the* Repleta *Group of* Drosophila

DAVID T. SULLIVAN, PETER W. ATKINSON,
CYNTHIA A. BAYER, and MARILYN A. MENOTTI-RAYMOND

1. INTRODUCTION

The mechanisms by which gene expression is controlled in time, in a cell type specific manner and in response to physiologically active agents is currently under intense study using a variety of strategies. We have utilized an approach to this problem which offers several attractive features. This involves the identification of specific variation in a pattern of gene expression within a group of related species followed by identification of the DNA sequence elements which operate to control expression of the gene using functional tests. Subsequently, a comparison of the nucleotide sequence of these regions from each member of the species set should lead to hypotheses about specific nucleotides having a role in the control of at least one aspect of the gene's pattern of expression. This hypothesis depends on the assumptions that sequence elements which are demonstrably conserved in different species are similar due to selection for a function, and that sequence differences may be regarded as candidates for the basis of the expression differences. Each of these can be tested by appropriate *in vitro* mutagenesis studies.

A by-product of this approach is that the role of changes in regulatory information with respect to evolution may be assessed. In addition the type of nucleotide sequence alterations which are the basis of these genetic changes may become apparent.

2. *ADH* GENE DUPLICATION

Several years ago it was noted that species of the mulleri and hydei subgroups of the repleta group contain a duplication of the gene which encodes alcohol dehydrogenase (Oakeshott *et al.*, 1982, Batterham *et al.*, 1982). The strategy used in

DAVID T. SULLIVAN, PETER W. ATKINSON, CYNTHIA A. BAYER, and MARILYN A. MENOTTI-RAYMOND * Department of Biology, Syracuse University, Syracuse, NY 13210, USA.

Ecological and Evolutionary Genetics of Drosophila, Edited by
J.S.F. Barker *et al.*, Plenum Press, New York, 1990

these studies to identify species whose genome contained the *Adh* duplication was to compare the ADH isozyme patterns of extracts of larvae and adults. The general result is that larvae are found to contain multiple bands of ADH activity while adults have a single ADH. This immediately implied that the two *Adh* genes are not coordinately controlled and suggests that this system would be an attractive one in which to investigate the evolution of genetic information involved in the control of gene expression. Thus, this gene-enzyme system offered the possibility of identifying specific nucleotide sequence elements involved with specific aspects of the stage or tissue specificity of gene expression and elucidation of the significant genetic changes which occurred during the evolution of the gene duplication.

Success in attaining these goals depended on several experimental results. First, it was essential to verify that the interpretation of the ADH gel patterns was correct. This was accomplished both biochemically and genetically. Batterham *et al.* (1983a), purified three ADH proteins from extracts of *D. mojavensis* larvae and demonstrated kinetic differences between them consistent with the three forms being the homodimeric products of two loci and an interlocus heterodimer. The variation in electrophoretic mobility between the forms was noted both in crude extracts and in purified enzyme preparations. Mills *et al.* (1986) using electrophoretic variant alleles of both *Adh* genes demonstrated that the functional genes are closely linked.

It was also important to verify that this *Adh* duplication is of relatively recent evolutionary origin so that nucleotide sequence analyses would offer a way to uncover the events leading to specific expression patterns. This was accomplished when Batterham *et al.* (1984) analyzed a large number of species and demonstrated that the duplication is present only in the mulleri and hydei subgroups. Furthermore it appears that most if not all species of these subgroups contain duplicate *Adh* genes. Exceptions may represent secondary events, i.e., loss of a gene or possibly extensions of existing phylogenetic interpretations beyond the point of high confidence. We have extended the observations of Batterham *et al.* (1984) to resolve some of the uncertainties previously reported. A summary of currently available information on the distribution of the *Adh* duplication within the repleta group is presented in Figure 1. In addition, we have indicated the criteria used to decide whether a given species contains the *Adh* duplication. These range from a comparison of larval-adult isozyme patterns to the complete nucleotide sequence of the region containing *Adh* loci.

Initially, we had expected that stage specific isozyme comparisons might misclassify a species because two isozymes might have the same net charge. This could result in larvae appearing to have a single gene while in reality there would be two. Surprisingly, this seems not to have been a common problem. There is only one species, *D.stalkeri,* that appears to have a single gene using isozyme comparison but has been found to have multiple genes at the DNA level. We have ana-

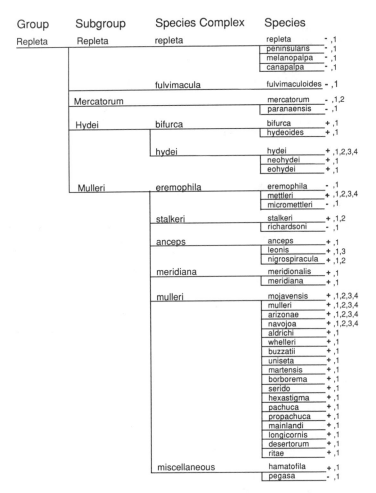

Figure 1. Distribution of the *Adh* duplication within the repleta group of *Drosophila*, + duplication present, - no evidence for *Adh* duplication. Criteria for conclusion: 1, larval - adult isozyme pattern comparison. 2, restriction map based on Southern blots of genomic DNA. 3, restriction map of cloned DNA. 4, nucleotide sequence determination of *Adh* genomic region.

lyzed flies from several populations of *D.stalkeri* and never found multiple ADH forms in larvae. However, Southern blots have clearly indicated that the *D. stalkeri* genome contains the *Adh* duplication (Sullivan and Weaver, unpublished). In addition, larvae and adults of all species of the eremophila species complex have a single ADH isozyme. However, the nucleotide sequence of the *Adh* region of *D. mettleri* reveals that this species does have the *Adh* duplication but one of these genes may be a pseudogene. Presumably, other species of the eremophila species complex will also be found to have

duplicate *Adh* genes. We have defined the genes and their corresponding isozymes based on their developmental pattern of
expression. ADH-1 is found in early larvae and disappears
during pupation or soon after emergence. ADH-2 may or may not
be present in early larvae but is found in late third instar.
This isozyme is the principal isozyme found in adults. Another
surprising result of the isozyme analysis is the observation
that in the set of over 30 species which we have examined,
ADH-2 is always a more basic protein than ADH-1. The reason
for this remains obscure but is likely to be related to some
aspect of protein function or turnover.

3. MOLECULAR CHARACTERIZATION OF *ADH* LOCI IN THE REPLETA GROUP

The initial molecular analysis of an *Adh* locus from a
species which has the *Adh* duplication was conducted by Fischer
and Maniatis (1985), who sequenced and analyzed the transcription of the *Adh* region of *D. mulleri*. These studies revealed
that the region contained three *Adh*-like genes. There is a
pseudogene, *Adh*-2 and *Adh*-1 arranged in that order within
approximately 9 kb. Subsequently, the nucleotide sequences of
the *Adh* regions of *D. mojavensis* (Atkinson *et al.*, 1988) and
D. hydei (Menotti-Raymond and Sullivan, in preparation) have
been determined. The organization of *Adh* region of these
species is generally similar to that of *D. mulleri*.

Expression of the *Adh* genes of *D. mojavensis* was found to
be similar to that of *D. mulleri*. *Adh*-1 is expressed in first
and second instar larvae exclusively. During third instar
Adh-1 continues to be expressed and expression of *Adh*-2
begins. *Adh*-2 expression occurs in the same cells as *Adh*-1,
with the exception of the gut which only has *Adh*-1 expression.
Adults of species with the *Adh* duplication only express *Adh*-2
with a specific exception of *D. mojavensis* and *D. arizonae*
which express high levels of *Adh*-1 in nurse cells of ovaries
(Batterham *et al.*, 1983b). This developmental pattern is
strikingly reminiscent of an aspect of control which occurs in
species of *Drosophila* with only one *Adh* gene. It has been
demonstrated that the *Adh* gene of *D. melanogaster* has two promoter regions (Benyajati *et al.*, 1983). One of these, located
close to the coding region and referred to as the proximal
promoter, is utilized in larval cells. The second, located
about 800 nucleotides upstream and referred to as the distal
promoter, is utilized in late third instar fat bodies and all
adult tissues which contain ADH (Savakis *et al.*, 1986). The
structure of the *Adh* locus as typified by *D. melanogaster*
appears to represent the general structure of the *Adh* gene in
the genus *Drosophila*. Of note is a strict correlation between
cell types which express *Adh*-1 in species having two functional *Adh* genes and proximal promoter usage in cells of
species having a single *Adh* gene. Similarly, there is a cell
type correlation between usage of the distal promoter and *Adh*-
2 expression. Consequently, during the evolution of the
repleta group a series of steps leading from one *Adh* gene to
the locus structure depicted in Figure 2 must have occurred.

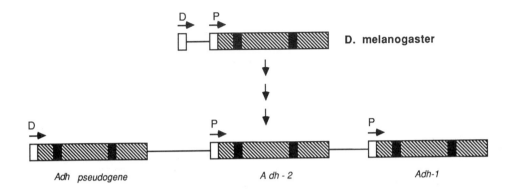

D. mojavensis

Figure 2. Comparison of the *Adh* locus of a species with one gene as typified by *D. melanogaster* with a species having the *Adh* duplication as typified by *D. mojavensis*. Promoter designation: D, distal promoter, P, proximal promoter. The designation of the *Adh*-2 promoter as P is based on sequence homologies.

4. EVOLUTION OF THE *Adh* LOCUS

Atkinson *et al.* (1988) compared the extent of nucleotide substitutions within the *Adh* genes of *D. mojavensis* and between *D. mojavensis* and *D. mulleri*. This comparison led to the proposal of a model which predicted a series of steps during the evolution of the present locus structure. The first event, a duplication, led to a gene with only the proximal promoter and coding region, and resulted in this gene positioned 3' to the original gene. The second event was a duplication of the 3' gene and subsequently the inactivation of the ancestor of the original gene to yield a pseudogene. Using this model and observations on the *Adh* expression patterns in species of the mulleri and hydei subgroups, we have focused on three specific regulatory differences. First, we have investigated the molecular basis of the differences in the expression of *Adh*-1 in the ovary of *D. mojavensis* and *D. arizonae* as compared to *D. mulleri* and *D. navojoa*. Second, we are studying the evolution of control of *Adh*-2 expression. This problem arises because the model of locus evolution of Atkinson *et al.* (1988) and other sequence comparisons reveal that the upstream region of *Adh*-2 is homologous to a proximal promoter, yet the gene is expressed with a developmental program analogous to a distal promoter. Third, and possibly related to the second problem, we are attempting to account for the observation that two expression patterns of *Adh*-2 have evolved in species having the duplication.

5. EXPRESSION OF *Adh* GENES

D. mojavensis and *D. arizonae* have high *Adh*-1 expression in the nurse cell cytoplasm of ovarian tissue while their close relatives *D. mulleri* and *D. navojoa* have no ADH-1 in the

ovary.   Species hybrids between *D. mojavensis* and *D. mulleri*
have ADH-1 of *D. mojavensis* in the ovaries while the *Adh-2*
genes of both species are expressed in adult fat body and
other ADH containing tissues (Mills *et al.*, 1986).   This
strongly suggests that ovarian *Adh-1* expression is under control of *cis*-acting elements.   In order to identify which
sequence elements are responsible for ovary expression, a
series of experiments using P-element mediated transformation
of the *Adh-1* gene of *D. mojavensis* into *D. melanogaster* hosts
was conducted.   The set of constructs in the vectors pUChsneo
(Steller and Pirrotta, 1985) and pDM24 (Mismer and Rubin,
1987) used for these analyses are diagrammed in Figure 3.   The
largest construct contains 13.5 kb of the *D. mojavensis Adh*
region.   *Adh-1* tissue expression in *D. melanogaster Adh* null
mutant larvae transformed with this *D. mojavensis* fragment is
essentially identical to that observed in larvae of *D.
mojavensis*.   This same result is obtained in transformants
whose *Adh-1* upstream regions are truncated to position -1519,
-469, -252 or -62, respectively (Fig. 3).   However in adult
ovaries, *Adh-1* expression does not occur in flies transformed
with *Adh-1* having a deletion to position -62 but does occur
when flies are transformed with the full length constructs or
the other deletion constructs.   Therefore it appears that
there is a sequence element(s) located between -259 and -62
which is responsible for the control of ovary expression.
Comparison of the nucleotide sequences of *D. mojavensis*, *D.
mulleri* and *D. navojoa* within this region reveals only one
region in which *D. mojavensis* differs from both of the other
species.   This region is between -65 and -75 and is underlined
in Figure 4.   This region is a prime candidate for a control
region involved in ovary expression.

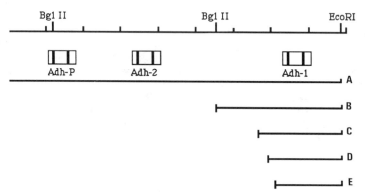

Figure 3.   Diagrammatic representation of the *D. mojavensis Adh* locus and
the deletion constructs used in P element mediated transformation using ADH
null mutant strains of *D. melanogaster* as hosts.   Constructs are: A, 13.5
Kb *Eco* R1 fragment of a clone of *D. mojavensis* DNA.   B, A clone which
contains 1519 nucleotides upstream of the transcription start site. C,
includes 469 nucleotides upstream of the transcription start site. D,
includes 252 nucleotides upstream of the transcription start site. E,
includes 62 nucleotides upstream of the transcription start site.

```
MOJ1  CGAAGACAACATAAAACAAACAAAATAGTCATCCTGAGTCGAAGTAAAACATATGGAATACGTGAAA
MUL1  CGAAGACAAAGTAAAACGAACAAA TAGTCATCCTGAGTCGAAGTAAAACATATGGAATACGAGAAA
NAV1  CGAGCGAAAAGTAAAACGAACAAA TAGTCATCCTGAGTCGAAGTAAAACATATGGAATACGTGGAA


MOJ1  CCGGCATGAGTTGCGTAGAATTTCGGAAAAAAACCAAACAGAAACGAATCAAAACAAAACGCCGAAA
MUL1  CCGGCATGAGTTGCGTAGAATTT GG                        AAAACAAAACACCGAA
NAV1  CCGGCATGAGTTGCGTAGAATTTCGGAAAAAAACCAAACAGAAGCGAATCAAAACAAAACGCCGAAA


MOJ1  ATTCTACTCGAAATGGCGGCAAAGTGGACGTCGGCGCCGACTGCG--GCCTTCGTTATTGATAAGCC
MUL1  ATTCTACTCGAACTGGCGGCAAAGTCGACGTCGACGCCGACCGCGCGGCCAGTGGTATTGATAAGAC
NAV1  ATTCTACTCGAACTGGCGGCAAAGTGGACGTCGACGCCGACTGCGCGGCCTGTGATGTTGATAAGCC
```

Figure 4. Alignment of the nucleotide sequence upstream of the *Adh*-1 genes of *D. mojavensis* (-252 to -55) (MOJI); *D. mulleri* (-226-55) (MULI); and *D. navojoa* (-257 to -58) (NAVI). Contained within the box is a region of dissimilarity which may be related to ovary expression.

The second problem related to the evolution of *Adh* expression is the developmental specificity of *Adh*-2 expression in *D. mojavensis*. The region of about 350 nucleotides upstream of the *Adh*-2 coding region has a high degree of sequence similarity in interspecific comparisons, implying sequence conservation due to function. In addition, it is clear that this 5' upstream region of *Adh*-2 is homologous to a proximal-like promoter (Atkinson *et al.*, 1988). Proximal promoters generally direct *Adh* expression in early larval tissues. However, *Adh*-2 is expressed in late third instar and adults, i.e., as if the gene had a distal promoter. In other words, the upstream region of *Adh*-2 is homologous to a proximal promoter but appears analogous to a distal promoter.

P-element mediated transformation has also been used to localize the essential control regions for *Adh*-2. A series of plasmid constructs, diagrammed in Figure 5, were injected into *D. melanogaster*, transformed lines obtained and the resultant adults assayed for *Adh*-2 expression by gel electrophoresis. Construct B, which has 549 nucleotides upstream of the transcription start site and contains all of the region of high interspecific sequence conservation, is expressed at barely detectable levels. A second construct (C) extends upstream of *Adh*-2 for 2000 nucleotides to a *Bgl* II site located in intron 1 of the pseudogene. Similar to transformants containing construct B, these transformants express *Adh*-2 at extremely low levels. In contrast, construct A which includes sequences upstream of the pseudogene results in transformants in which *Adh*-2 is expressed at high levels. The most active of these transformants have ADH activities approaching normal *D. mojavensis* levels. Of particular interest is construct D in which the *Bgl* II fragment containing the *Adh*-2 gene and 2000 nucleotides upstream is inverted with respect to the sequences upstream of the pseudogene. Transformants containing this construct express *Adh*-2 at levels which are indistinguishable from those with construct A. These data strongly suggest that high expression levels of *Adh*-2 are brought about by an element upstream of the pseudogene which has the properties of an

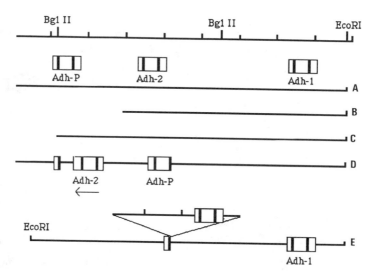

Figure 5. Diagrammatic representation of the *D. mojavensis* *Adh-2* locus and constructs used to assess *Adh-2* expression using P element mediated transformation into ADH null mutants of *D. melanogaster*. A, a 13.5Kb *Eco* R1 fragment of a clone of *D. mojavensis* DNA. B, A clone which contains 549 nucleotides upstream of the transcription start site. C, a clone which contains 2000 nucleotides upstream of the transcription start site. D, a clone which is comparable to A except the 4.4 Kb *Bgl* II fragment is inverted. E, a clone comparable to A with the 4.4 Kb *Bgl* II fragment deleted.

enhancer. A suggestion as to the possible nature of this enhancer can be obtained by inspecting the sequence upstream of the pseudogene of *D. mojavensis*. A comparison of this region and the distal promoter region of *D. melanogaster* is shown in Figure 6. There is a high degree of sequence similarity over several stretches. We suggest that these regions which are in the control region of the remnants of the distal promoter, upstream of the pseudogene are able to extend developmental specificity to the *Adh-2* gene further downstream. This would impart a stage specific expression pattern to *Adh-2* appropriate for a distal promoter.

We have attempted to determine whether this enhancer element could change the developmental regulation of *Adh-1*. Construct E (Fig. 5) has the *Bgl* II fragment containing *Adh-2* removed, thereby bringing the enhancer closer to *Adh-1* and eliminating the intervening gene. Transformant adult males were examined for *Adh-1* expression and were found to contain

```
MEL-D:-545 TCAAT T    GCGACCGCGAGCAACAACAGATCTTTCTACACTTCTCCTTGCTAT GC TTGACATTCAC AA GGT  -478.....//...-31 TATTTAAGGAGCTGCG AAG
HYD-P:-438 TCAATTT    GGGACCACAACAAA   CAAGATTCTTGACTATGTC             TATTGCTTGACATTCACTGAA GGT  -373.....//..-154.TATTTAAAGCTCTGCCCAAA
MUL-P:-520 TCAATTTGCTGCGACCACAACAAAATAACAAAGTTCTTGACTACGTC             GATTGCTTGACATTCAGTGAAAGT  -447.....//..-184.TATTTAAGGCGCTGCC AAG
MOJ-P:-490 TCAATTT    GCGACCACAACAAAATAACAAAGTTCTTGACTACGTC             GATTGCTTGACATTCAGTGAA GGT  -420.....//..-178.TATTTAAGC GCTGCCCGAG

CONSENSUS: TCAATTT...GCGACCACAACAAAACAACAA..TTCTTGACTACGTC...TATTGCTTGACATTCAGTGAA.GGT........//......TATTTAAGGCGCTGCC AAG
                     G    C T C T   AC  TTTC   G                                                    ACAT   G G A
```

Figure 6. Sequence in the areas which contain regions of high sequence similarly from the distal promoter regions of the *D. melanogaster Adh* locus (MEL-D) with the regions upstream of the pseudogenes of *D. hydei* (HYD-P), *D. mulleri* (MUL-P) and *D mojavensis* (MOJ-P). The numbers refer to the number of nucleotides upstream from known or putative start points of transcription.

ADH-1. Normally *Adh*-1 is not expressed in adult males and the
atypical expression might be due to action of the enhancer on
Adh-1. However, the interpretation of this result is complicated. As controls we examined adult males of the lines
obtained by transformation with the constructs shown in Figure
3. Surprisingly, *Adh*-1 expression is found in adult males in
all of the *Adh*-1 transformed lines but not in the transformants having both the *Adh*-2 and *Adh*-1 genes. These results
add an additional level of complexity and suggest the presence
of an *Adh*-1 negative control element upstream of -1500 of the
Adh-1 transcription start site. Conceivably, this negative
element could be the *Adh*-2 gene, i.e., transcription of a gene
upstream of *Adh*-1 may inhibit *Adh*-1 transcription. Clearly,
additional experiments are needed in order to understand the
details of the control of *Adh* expression.

We have also undertaken another approach to the evolution
of *Adh*-2 expression. We have studied the timing of *Adh*-2
expression in a number of species of the hydei and mulleri
subgroups (Menotti-Raymond and Sullivan, in preparation). Two
distinct developmental patterns are found. One is that typi-

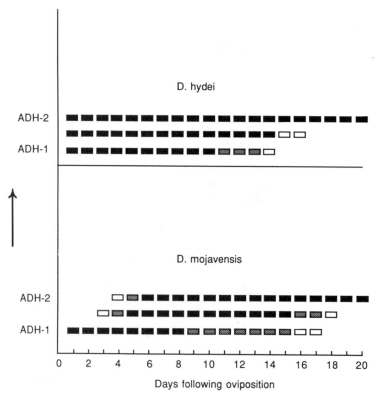

Figure 7. Diagrammatic comparison of the ADH isozyme profile found during
the development of *D. hydei* and *D. mojavensis*. Puparium formation occurs
on day 7 and eclosion occurs on days 12 to 13. Filled, hatched and open
boxes represent the relative intensity of the ADH activity, filled
representing the highest intensity. The band of ADH with electrophoretic
mobility between ADH-2 and ADH-1 is the heterodimer.

fied by *D. mojavensis*, in which *Adh-2* expression begins in late third instar and continues through adult stages. This pattern is found in all species of the mulleri species complex and also *D. hamatofila*. The second pattern is typified *D. hydei* and is also found in species of the meridiana, anceps and possibly stalkeri species complexes. In species with this pattern, *Adh-2* is first expressed early, either during late embryogenesis or first instar larva, and continues through larval, pupal and adult stages. The two *Adh-2* expression patterns are diagrammed in Figure 7.

The *D. hydei* expression pattern is what would be expected if the *Adh* locus consisted of two genes, one with both a proximal and distal promoter, i.e., like *D. melanogaster Adh* and a second gene with only a proximal promoter. This is a likely intermediate step in the evolution of the *D. mojavensis* locus (Atkinson *et al.*, 1988). Surprisingly, the nucleotide sequence reveals that the structure of the *D. hydei* locus is quite similar to that of *D. mojavensis* with a pseudogene, *Adh-2* and *Adh-1* arranged in that order (Menotti-Raymond and Sullivan, in preparation). Therefore the difference in the developmental expressions of *Adh-2* between *D. hydei* and *D. mojavensis* is not due to a substantial structural difference between the *Adh* regions of the two species. Rather this difference is likely to be due to subtle changes in nucleotide sequence. Given the existing phylogeny of the repleta group (Wasserman, 1982), we feel it is likely that the *D. mojavensis* pattern of expression is the derived pattern. If so, then a reasonable hypothesis as to what has happened during evolution would be that a specific aspect of *Adh-2* expression has been lost, i.e., the ability to be expressed in the early larval instars. If this hypothesis is correct, then it is likely that the molecular basis for this expression change can be uncovered through structural comparisons of the *Adh* regions coupled with P-element transformation.

6. SUMMARY

We have studied the evolution and expression pattern of the *Adh* genes of species of the mulleri and hydei subgroups. Species from these subgroups contain duplicate *Adh* genes which are the result of events occurring early in their lineage. We have used an evolutionary approach to identify sequence elements involved with specific aspects of the control of gene expression. Our strategy has been to identify candidate elements by species sequence comparison and then test the role of these sequences by P element mediated transformation into ADH deficient *D. melanogaster* hosts. In many species of the mulleri and hydei subgroups, the *Adh* locus contains three genes arranged in order of a pseudogene, *Adh-2* and *Adh-1*. *Adh-1* expression is usually confined to first, second and early third instar larvae. However, in two species *D. mojavensis* and *D. arizonae Adh-1* expression also occurs in nurse cells of ovaries. Deletion analysis of the *D. mojavensis* gene has localized the sequences responsible for this to a region upstream of -62 and downstream of -257. There have been several interesting events in the evolution of *Adh-2* expression.

Its promoter region is homologous to a proximal or larval promoter yet the pattern of Adh-2 expression in *D. mojavensis* and
D. mulleri is analogous to that of a distal or adult promoter.
The developmental specificity for Adh-2 expression is controlled by sequence elements upstream of the pseudogene which
probably once controlled that gene when it was able to encode
a protein.

ACKNOWLEDGEMENTS.   Research conducted in our laboratory and
reported here was supported by USPHS grant GM-31857.

References

Atkinson, P. W., Mills, L. E., Starmer W. T., and Sullivan, D. T., 1988,
    Structure and evolution of the Adh genes of *Drosophila mojavensis*,
    Genetics **120**:713-723.
Batterham, P., Starmer, W. T., and Sullivan, D. T., 1982, Biochemical
    genetics of the alcohol longevity response of *Drosophila mojavensis*,
    in: *Ecological Genetics and Evolution: The Cactus-yeast-Drosophila
    Model System* (J. S. F. Barker, and W. T. Starmer, eds), Academic Press
    Australia, Sydney, pp. 307-321.
Batterham, P., Chambers, G. K., Starmer, W. T., and Sullivan, D. T., 1984,
    Origin and expression of an alcohol dehydrogenase gene duplication in
    the genus *Drosophila*, *Evolution* **38**:644-657.
Batterham, P., Gritz, E., Starmer, W. T., and Sullivan, D. T., 1983a, Bio-
    chemical characterization of the products of the Adh loci of *Drosophila
    mojavensis*, *Biochem. Genet.* **21**:871-883.
Batterham, P., Lovett, J., Starmer, W. T., and Sullivan D., 1983b, Differ-
    ential regulation of duplicate alcohol dehydrogenase genes in *D.
    mojavensis*, *Devl Biol.* **96**:346-354.
Benyajati, C., Spoerel, H., Haymerle, H., and Ashburner, M., 1983, The mes-
    senger RNA for Adh in *D. melanogaster* differs in its 5' end in differ-
    ent developmental stages, *Cell* **33**:125-133.
Fischer, J., and Maniatis, T., 1985, Structure and transcription of the
    Drosophila mulleri alcohol dehydrogenase genes, *Nucl. Acids Res.*
    13:6899-6917.
Mills, L. E., Batterham, P., Alegre, J., Starmer, W. T., and Sullivan, D.
    T., 1986, Molecular genetic characterization of a locus that contains
    duplicate Adh genes in *Drosophila mojavensis* and related species,
    Genetics **112**:295-310.
Mismer, D., and Rubin, G. M., 1987, Analysis of the promoter of the ninaE
    opsin gene in *Drosophila melanogaster*, *Genetics* **116**:565-578.
Oakeshott, J. G., Chambers, G. K., East, P. D., Gibson, J. B., and Barker,
    J.S.F., 1982, Evidence for a genetic duplication involving alcohol
    dehydrogenase genes in *Drosophila buzzatii* and related species, *Aust.
    J. biol. Sci.* **35**:73-84.
Savakis, C., Ashburner, M., and Willis, J. H., 1986, The expression of the
    gene coding for alcohol dehydrogenase during the development of
    Drosophila melanogaster, *Devl Biol.* **114**:194-207.
Steller, M., and Pirrotta, V., 1985, A transposable P vector that confers
    selectable G418 resistance to *Drosophila* larva, *EMBO J.* **4**:107-171.
Wasserman, M.,1982, Evolution of the repleta group, in: *Genetics and Biol-
    ogy of Drosophila*, Vol. 3B (M. Ashburner, H. L. Carson, and J. N.
    Thompson, eds), Academic Press, New York, pp. 61-139.

CHAPTER 22

Evolutionary Relationships of Enzymes in the Genus Drosophila

GLEN E. COLLIER, ROSS J. MACINTYRE, AND DAVID FEATHERSTON

1. INTRODUCTION

While basking in the dawn of the age of DNA sequence data, this is an opportune time for us, as *Drosophila* evolutionary biologists, to take both a retrospective and prospective look at what the now traditional analyses of protein evolution have told us, and still can tell us, about the history of our favorite genus. The genus *Drosophila* is an extraordinarily rich testing ground for the evolutionary potential of molecular data. It is old enough so the evolutionary rates of change of conserved molecules can be measured, yet many speciation events are comparatively recent. Hence, the more ephemeral events involved in genome reorganization can be traced as well. This chapter will summarize and expand on the the conclusions of an earlier effort (MacIntyre and Collier, 1986), and, present new data on the evolutionary rates of change of several enzymes. In addition, we will examine in more detail the remarkable correspondence between allozyme and chromosome based phylogenies of several species groups.

2. EVOLUTION OF SPECIFIC ENZYMES

A recent compilation of the evolutionary rates of change of several *Drosophila* proteins is presented in Table I. Most of this information has come from immunological studies, primarily involving the technique of microcomplement fixation. The rates are presented in terms of unit evolutionary periods (UEP), or the time in millions of years for two homologous proteins to diverge in amino acid sequence by one percent. In calculating UEP's, the number of amino acid substitutions are first determined directly (from aligned sequences) or indirectly (relying on the immunological distances between proteins differing by a known number of amino acid substi-

_____

GLEN E. COLLIER * Department of Biological Sciences, Illinois State University, Normal, IL 61761, USA; ROSS J. MACINTYRE, AND DAVID FEATHERSTON * Section of Genetics and Development, Cornell University, Ithaca, NY 14853, USA.

Ecological and Evolutionary Genetics of Drosophila, Edited by
J.S.F. Barker *et al.*, Plenum Press, New York, 1990

Table I

Rate of evolutionary change (UEP) and average heterozygosities in several
Drosophila proteins

Enzyme	UEP	H[a]	1/UEP	H/(1 - H)
ACPH	3[b]	0.259	0.333	0.350
ADH	3.5[c]	0.102	0.286	0.114
LSP	4[d]	0.123[j]	0.250	0.140
SOD	11[e]	0.018	0.091	0.018
6PGDH	25[f]	0.045	0.040	0.047
GPDH	36[g],31[h]	0.028	0.028	0.029
ARGK	59[i]	0.003[i]	0.017	0.003

[a]Except for LSP and ARGK, values were taken from the surveys reviewed by
MacIntyre and Collier, 1986.
[b]MacIntyre *et al.*, 1978.
[c]Calculated from data in Fisher and Maniatis, 1985.
[d]Beverley and Wilson, 1984.
[e]Ayala, 1986.
[f]Reinbold and Collier, 1990.
[g]Collier and MacIntyre, 1977.
[h]Von Kalm *et al.*, 1989; Arai *et al.*, 1988.
[i]Collier, 1990.
[j]Band *et al.*, 1984; Singh and Coulthart, 1982.

Table II

Immunological distance values (± s.e.) for *Drosophila* arginine kinase

Antigen source	Antiserum mel.	vir.	imm.	Subgenus	Radiation
melanogaster	----	6±0.1	12±2.0	*Sophophora*	
virilis	7±0.4	----	7±0.1	*Drosophila*	*virilis-repleta*
immigrans	13±2.0	5±0.1	----	*Drosophila*	*immigrans-Hirtodrosophila*
yakuba	0±0.6	10±0.2	16±0.5	*Sophophora*	
persimilis	9±1.0	20±1.0	15±0.6	"	
pseudoobscura	10±0.7	21±0.5	18±2.0	"	
paulistorum	8±0.3	14±0.8	20±0.7	"	
willistoni	12±0.8	17±0.1	14±0.3	"	
emarginata	10±0.2	15±0.8	14±3.0	"	
robusta	12±0.1	0±0.3	4±0.2	*Drosophila*	*virilis-repleta*
paramelanica	15±0.1	1±2.0	12±2.0	"	
mercatorum	9±1.0	14±0.8	11±0.5	"	
arizonensis	12±0.6	10±0.3	23±2.0	"	
funebris	8±0.1	2±0.3	13±0.8	"	
nasuta	10±0.5	12±1.0	6±0.9	"	*immigrans-Hirtodrosophila*
Zaprionus vittiger	7±0.6	4±0.2	4±0.2	----	"
cardini	12±0.8	4±1.0	2±1.0	*Drosophila*	"
pallidipennis	12±3.0	0±0.9	7±2.0	"	"
pinicola	8±0.7	4±0.3	4±0.3	"	"
duncani	17±0.2	17±2.0	14±1.0	*Hirtodrosophila*	"

tutions). These numbers must then be corrected for parallel substitutions and reversions at each site. This correction generally uses a Poisson based estimator (e.g., see Dickerson, 1971), and its effect is logarithmically proportional to the observed number of differences between any two sequences. Finally the UEP is determined by dividing the corrected percent amino acid difference into the time since the two species donating the sequences which were compared most likely diverged from a common ancestor. Approximate rate constancy can be assessed if at least several proteins from different phylads are examined, via the so called relative rate test.

2.1. Arginine Kinase

Since so many of the UEP's reported in Table I were derived from immunological studies, we thought it would be useful, using arginine kinase as a paradigm, to review briefly the methodology, and show how the results are converted to percent amino acid sequence differentials (for a fuller description, see Champion *et al.*, 1974). Purified proteins from several species are used as antigens and the antisera against each protein are pooled from several rabbits. Thus, most epitopes are detected by these pooled polyclonal antisera. Immunological distances, which measure the concentration difference between homologous and heterologous antisera needed to produce equivalent reactions, should be reciprocal if the data are to be useful in evolutionary comparisons. Table II shows the very good reciprocity when antisera against three different *Drosophila* species' arginine kinases are tested. The percent standard deviation in these tests is 8.2. The remaining immunological distances in Table II represent so called "single distance" measurements, which, when averaged provide the data shown in Table III. In general, the distances are small, but agree with the basic evolutionary history of the three radiations, i.e., *Sophophora, virilis - repleta* and *immigrans - Hirtodrosophila*. They also indicate, by the relative rate tests, that arginine kinase is evolving at roughly the same rate in the three radiations.

How are these immunological distances converted to estimates of amino acid sequence differences? In this regard, allozymes have proven to be extremely useful. If the antisera

Table III
Average immunological distance of arginine kinase between major groups within the genus *Drosophila*

| Group | Antiserum | | |
	mel.	vir.	imm.
Subgenus *Sophophora*	7	15	16
Subgenus *Drosophila*			
virilis-repleta radiation	11	4	12
immigrans-Hirtodrosophila radiation	11	7	5

Table IV
Immunological distance differences for arginine kinase allozymes

Antigens compared	Antiserum		
	mel.	*vir.*	*imm.*
ARK[A] *vs* ARK[B] from D. melanogaster	2.2±0.06[a]	3.3±0.50	4.2±0.75

[a]Average (± s.e.) of four determinations.

react differentially with allozymic variants of the antigen, then the contribution of single amino acid differences to the total immunological distances can be measured. For example, all three antisera against arginine kinase can distinguish between the very rare and the common electrophoretic variants from *D. melanogaster* (see Table IV). Similarly, with regard to αglycerophosphate dehydrogenase (αGPDH), eight allozymic pairs were distinguished by the antisera against the enzyme from *D. melanogaster* with a mean ID difference of five units (and a range of 4-9). That substitutions are additive in their effect on immunological distance can be seen in Figure 1, taken from Collier and MacIntyre (1977). Benjamin *et al.* (1984) also document the additive effect of amino acid differences on immunological distances with a variety of antigens and taxa. Very recently, we were able to compare the number of amino acid substitutions directly determined from the sequences of αGPDH from *D. melanogaster* (Von Kalm *et al.*, 1989) and *D. virilis* (Arai *et al.*, 1988) with those estimated by the ID measurements. The seven amino acid substitutions predict a UEP of 31 million years, quite comparable to the 36 million years which was estimated from the immunological data (see Table I).

Given the estimate of a single amino acid substitution's effect on the immunological distances between arginine kinases (Table IV), and the generally small distances measured between the enzyme from distantly related *Drosophila* species (Table II), it is clear that arginine kinase is very slowly evolv-

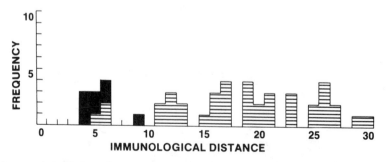

Figure 1. Distributions of immunological distances between *Drosophila* αglycerophosphate dehydrogenases as estimated with the antisera against the enzyme from *Drosophila melanogaster* (from Collier and MacIntyre, 1977).

ing.   Its average UEP is 59 million years; the calculations leading to that estimate are shown in Table V. Note that we are using the divergence times of 46 million years for the separation of the *virilis-repleta* and *immigrans-Hirtodrosophila* phylads and 62 million years for the *Sophophora-Drosophila* separation.   Beverley and Wilson (1984) estimated these dates after using several Dipteran outgroups while generating an evolutionary rate of change for a larval hemolymph protein.   Interestingly these dates are also supported by Spicer's recent analysis (1988) of *Drosophila* proteins as revealed by two dimensional gel electrophoresis and by his reevaluation of the data on αGPDH from Collier and MacIntyre (1977).   Even if the more recent dates of 36 and 52 million years are used for the two basic separations during the evolution of the genus, arginine kinase is still one of the most conserved macromolecules examined to date by *Drosophila* molecular evolutionists (see Table I).

2.2. 6-Phospho-Gluconate Dehydrogenase

The more rapidly evolving enzyme, 6 phospho-gluconate dehydrogenase (6PGDH), has recently been studied in the *Drosophila virilis* species group (Reinbold and Collier, 1990).   With five antisera, immunological distances between all 12 *virilis* group species and four outgroup Drosophilids were measured.

Table V
Calculation of Unit Evolutionary Period (UEP) for *Drosophila* arginine kinase

Branchpoint[a]	Time since common ancestor[b]	Average ID[c]	Amino acid substitutions		UEP
			Approx. No.[d]	No./100 residues[e]	
$D_V - D_I$	46 my	12	2.9	0.76	61 my
$D_I - D_V$	46 my	7	2.1	0.55	84 my
$S - D_I$	62 my	16	3.8	1.00	62 my
$S - D_V$	62 my	15	4.5	1.20	52 my
$D_I - S$	62 my	11	5.0	1.32	47 my
$D_V - S$	62 my	11	5.0	1.32	47 my
				Average UEP =	59 my

[a]D_V refers to members of the *virilis-repleta* radiation of the subgenus *Drosophila*.
D_I refers to members of the *immigrans-Hirtodrosophila* radiation of the subgenus *Drosophila*.
S refers to members of the subgenus *Sophophora*.   The group represented by species used as antigen sources is indicated first followed by the group represented by the antisera.
[b]Divergence times taken from Beverley and Wilson (1984).
[c]From Table II.
[d]Estimated by dividing the average ID by the immunological distance difference in Table VI.
[e]Assumes the polypeptide contains 380 residues.

With the Fitch-Margoliash tree building algorithm, the ten reciprocal measurements produced a dendrogram of the five species donating antigens entirely congruent with Lynn Throckmorton's phylogeny (Throckmorton, 1982) which is based primarily on cytogenetic data. When the additional species are added to the tree by the method of Beverley and Wilson (1982), the dendrogram is found to differ slightly from Throckmorton's tree of the 12 species. Both dendrograms are shown in Figure 2. When the tree based on the distances between the *virilis* group species and the four outgroup species is rooted, three major differences become apparent; (1) *D. kanekoi* clusters in the *virilis* phylad, (2) *D. a. americana* is joined to *D. virilis*, not to *D. a texana* and (3) *D. novamexicana* joins more closely to *D. virilis* than to *D. lummei*. Nevertheless, the coefficient of distortion between the two dendrograms is 0.44 and the probability that their similarity is due to chance is 0.003 (see below for a fuller description of how the two topologies are compared).

It is interesting that the average ID's between the outgroup species to members of the two virilis group phylads, i.e., the *montana* and *virilis* phylads, were not different. These data are summarized in Table VI. Thus, by these relative rate tests 6PGDH is evolving at the same apparent rate in both phylads, even though the species in the *montana* phylad have diverged ecologically, i.e., to cooler climates and new food sources and are more polymorphic with regard to allozymes and inversions. The *montana* phylad is also more speciose, with stronger reproductive isolation between the constituent species.

Finally, the data in Table VII indicate a UEP for 6PGDH of 25 million years, based on the immunological distance difference between two *Pgd* allozymes in *D. melanogaster*. If one immunological distance unit in 6PGDH accumulates every 2.6 million years (Table VII), and since the average distance between the species of the different phylads is approximately 6.0, then the *virilis* and *montana* phylads diverged about 15 million years ago, or in the middle of the Miocene, as suggested by Throckmorton (1982).

Since allozyme surveys have been conducted within several *Drosophila* species for the proteins listed in Table I, estimates of average heterozygosity for each are available in the literature. These are shown in the column labeled *H* in Table I. Skibinski and Ward (1981, 1982) suggest that the rate of evolutionary change in a protein will be related to its level of intrapopulation heterozygosity by the formula.

$$D = \frac{Ht}{2\ Ne\ (1-H)}$$

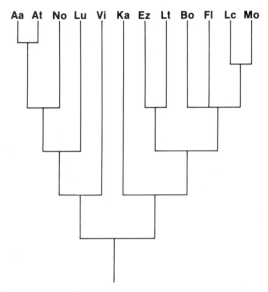

Figure 2a. Phylogeny of the *Drosophila virilis* species group redrawn from Throckmorton (1982). Abbreviations: Aa=*Drosophila americana americana*, At=*D. americana texana*, No=*D. novamexicana*, Lu=*D. lummei*, Vi=*D. virilis*, Ka=*D. kanekoi*, Ez=*D. ezoana*, Lt=*D. littoralis*, Bo=*D. borealis*, Fl=*D. flavomontana*, Lc=*D. lacicola*, and Mo=*D. montana*.

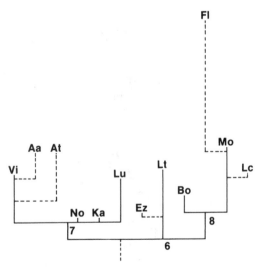

Figure 2b. Fitch-Margoliash tree of the species in Figure 2a as determined by the immunological distances measured with antisera against 6 phosphogluconate dehydrogenase. Seven legs (dashed lines) attached on the basis of best fit. Negative legs are indicated by a single vertical dash. Otherwise vertical distances are drawn to scale.

Table VI
Immunological distance from four outgroup species to members of the *virilis*
and *montana* phylads as measured with antisera against 6-phosphogluconate
dehydrogenase

Outgroup species	Average ID (± se)[a] from species in the *virilis* phylad	Average ID (± se) from species in the *montana* phylad
D. paramelanica	16 ± 8	18 ± 10
D. funebris	21 ± 13	12 ± 9
D. immigrans	24 ± 5	27 ± 9
D. melanogaster	25 ± 7	33 ± 5

[a]n = 3.

where D is genetic distance, H is average heterozygosity, Ne
is the effective population size and t is time. The UEP
should thus be related to H as follows:

$$1/UEP \ \alpha \ D/t = \frac{1}{2 \ Ne} \ \frac{H}{1-H} \ \text{ or } 1/UEP \ \alpha \ \frac{H}{1-H}$$

The regression on the data in Table I yields a coefficient
of determination of 0.84 with a slope of 0.92. These data
strongly support the hypothesis of Skibinski and Ward that, in
general, highly polymorphic protein systems will evolve more
rapidly.

3. THE EVOLUTION OF *DROSOPHILA* SPECIES GROUPS BASED ON PHYLOGENIES CONSTRUCTED WITH GENETIC DISTANCES

Since the pioneering work of Hubby and Throckmorton (1965),
many papers have appeared which survey and analyze allozymes
from groups of related species (reviewed in Avise and Aquadro,
1982; Powell, 1975; MacIntyre and Collier, 1986). With the

Table VII
Calculation of unit evolutionary period (UEP) for 6-phosphogluconate
dehydrogenase from *Drosophila*

Branchpoint	Divergence		Amino acid subs.		
	Time	ID	No.[a]	No./100[b]	UEP
virilis-immigrans	46 my	24.0	9.6	1.96	23.5
virilis-melanogaster	62 my	27.6	11.0	2.25	27.5
				Average	25.5

[a]Based on a difference of 2.5 ID units between *Pgd*[3] versus *Pgd*[8]
from *D. melanogaster*.
[b]Based on a polypeptide of 491 residues calculated by dividing the average
subunit molecular weight of 54,000 (Williamson *et al.*, 1980) by the
average molecular weight of an amino acid of 110.

assumption that enzymes catalyzing the same reaction in different species are homologous and that electrophoretic mobility is a reasonable indication of structural identity, the data from such interspecific surveys can be used to estimate the "genetic distance" between the species studied (Nei, 1971, 1972). These distance values can be analyzed by any of several algorithms to construct dendrograms; diagrams which may reflect the evolutionary relationships of these species. Such use of distance estimates in general, and immunological and genetic distance estimates in particular, have been severely criticized in recent years (Farris, 1981, 1985; Swofford, 1981; Throckmorton, 1978). These criticisms, as well as alternative analyses of electrophoretic data, are reviewed by Buth (1984) and Berlocher (1984). As a result, the reliability of genetic distance based trees has been called into question. Despite cogent replies to a number of these criticisms (Felsenstein, 1984, 1985; Nei *et al.*, 1983), the question remains, "How good are phylogenetic inferences derived from electrophoretic comparison of proteins?"

The genus *Drosophila* offers several distinct advantages for addressing this question. First, a relatively large number of species groups within the genus have been surveyed by these methods. Moreover, additional data are available from other sources for **independently** inferring phylogenetic relationships for many of these same groups. Primary among these sources are cytological data, for which salivary gland chromosome analyses are invaluable. Other data are available on comparative morphology, behavior, interspecific hybridization tests, and zoogeography. Thus, an obvious strategy for addressing this question is to choose species groups of *Drosophila* for which both kinds of data are available and to compare the phylogenetic topologies derived from each data type. We here report such comparisons of phylogenetic trees derived from 22 allozyme surveys of nine species groups of the genus *Drosophila*. This analysis updates and expands the data presented earlier (MacIntyre and Collier, 1986).

3.1. Sources of Data

Table VIII lists the interspecific allozyme surveys of *Drosophila* species for which other data are available for phylogenetic inference. In some cases, e.g., the *obscura* group species, data from more than one allozyme study were used to construct phylogenetic trees. Included in this Table are the number of loci scored and the method used to calculate genetic distance. The nature of the independent data sets (cytological, behavioral, etc.) and the references from which they were drawn are also listed.

The reference trees for the *bipectinata* subgroup (Bock, 1971), *melanogaster* subgroup (Lemeunier and Ashburner, 1976), *mesophragmatica* group (Brncic *et al.*, 1971), *mulleri* subgroup (Wasserman, 1982), *virilis* group (Throckmorton, 1982) and *planitibia* group (Carson *et al.*, 1970) are based primarily upon analyses of salivary gland polytene chromosome inversion polymorphisms. These analyses consider inversions as derived

Table VIII
Allozyme surveys of *Drosophila* species groups

		Allozyme surveys			Taxonomic data			Data for reference tree	
No.	Reference	No. of Loci	Metric	No. of OTU's	Subgenus	Species group	Subgroup	References	Data type
1.	Yang *et al.*, 1972	23	S	6	Sophophora	*melanogaster*	*bipectinata*	Bock, 1971	Sexual isolation tests; chromosome homologies in interspecific hybrids.
2.	Triantaphyllidis *et al.*, 1978	6	% shared electromorphs	6	"	"	*montium*	None	
3.	Eisses *et al.*, 1979	18	I	6	"	"	*melanogaster*	Lemeunier and Ashburner, 1976	Cytology; hybrid sterility tests
4.	Gonzales *et al.*, 1982	55	I	3	"	"	"	"	"
5.	Lakovaara *et al.*, 1972	27	% shared electromorphs	11	"	*obscura*	*obscura*	Buzzati-Traverso and Scossiroli, 1955; Dobzhansky, 1970[d]	Morphology; cytology
6.	Lakovaara *et al.*, 1976	21	"	21	"	"	*obscura* and *affinis*	Buzzati-Traverso and Scossiroli, 1955; Dobzhansky, 1970; Miller and Sanger 1968; Krimbas and Loukas 1984	Morphology; cytology hybridization tests
7.	Prakash, 1977	43	I	3	"	"	*obscura*	Buzzati-Traverso and Scossiroli, 1955; Dobzhansky and Powell, 1975	Morphology; cytology
8.	Marinkovic *et al.*, 1978	6	I	5	Sophophora	*obscura*	*obscura*	Buzzati-Traverso and Scossiroli, 1955; Dobzhansky and Powell, 1975	Morphology; cytology
9.	Gonzalez *et al.*, 1983	72	I	2	"	"	"	"	"
10.	Cabrera *et al.*, 1983	68	I	5	"	"	"	"	"
11.	Ayala *et al.*, 1974	36	I	7	"	*willistoni*	"	Spassky *et al.*, 1971	Morphology; cytology; behavior; ecology, zoogeography
12.	Nair *et al.*, 1971	24	S	6	Drosophila	*mesophragmatica*	"	Brncic *et al.*, 1971	Cytology
13.	Zouros, 1973	17	I	4	"	*repleta*	*mulleri*	Wasserman, 1982	Cytology
14.	Richardson and Smouse, 1976	7	i,i[b]	11	Drosophila	*repleta*	*mulleri*	Wasserman, 1982	Cytology
15.	Throckmorton (pers. comm.)	9	I	11	"	*virilis*	*virilis* and *montana*	Throckmorton, 1982	Cytology
16.	Johnson *et al.*, 1975	10	S	16	"	picture-winged	*planitibia*	Carson *et al.*, 1970	Cytology; sexual isolation tests
17.	Ayala, 1975	31	I	8	"	picture-winged; white tipped; *crassifemur*	"	Ayala, 1975	Cytology; morphology
18.	Carson *et al.*, 1975	14	I	2	Drosophila	picture-winged	*planitibia*	Carson *et al*, 1970	Cytology, sexual isolation tests
19.	Sene and Carson, 1977	25	I	2	"	picture-winged	"	Carson *et al*, 1970	Cytology, sexual isolation tests
20.	Craddock and Johnson, 1979	12	S	4	"	"	"	Craddock and Johnson, 1979	Cytology
21.	Onishi *et al.*, 1983	-[c]	I	18	Sophophora and Drosophila	*melanogaster virilis*	several	Lemeunier and Ashburner, 1976; Lee, 1971; Throckmorton,1982	Cytology; Sexual isolation tests; zoogeography
22.	Spicer, 1988	-[d]	I	8	"	several	several	"	"

[a] For groups or subgroups surveyed repeatedly, references for the reference tree are given only for the first study listed.
[b] A unique distance estimate based upon average electrophoretic mobility of each gene product.
[c] A combination of 27 allozymes analyzed by conventional electrophoretic techniques and approximately 100 abundant proteins detected by two dimensional gel electrophoresis.
[d] 135 protein "spots" were scored in this study. The UPGMA tree (Fig. 5) was used for our comparison.

character states and are therefore cladistic in nature. The resulting patterns of relationships are unrooted networks. Additional data from sexual isolation tests, hybrid sterility tests, observations of interspecific chromosomal asynapsis, metaphase karyotypes, and zoogeography were used to root these networks. The reference tree for the *willistoni* group is based upon a composite of information on the morphology, hybrid sterility tests, cytology, and zoogeography (Spassky *et al*., 1971). The reference tree for the *obscura* group is based upon UPGMA analysis of the morphological differences tabulated by Buzzati-Traverso and Scossiroli (1955). These relationships are purely phenetic and as such are the least reliable reference tree of the set. The relationships of three sets of triads of sibling species within this group were established by analysis of polytene chromosome inversion polymorphisms (Dobzhansky, 1970; Miller and Sanger, 1968; Krimbas and Loukas, 1984). The relationships of species groups within the genus have been assessed by a cladistic analysis of an extensive set of morphological characters and zoogeography (Throckmorton, 1968, 1975).

3.2. Genetic Distance Estimates

The studies in Table VIII differ in the method used to estimate genetic distance. Either "percent shared electromorphs", Rogers' S (Rogers, 1972), or Nei's I (Nei, 1972) have been used, except for Richardson and Smouse (1976). These authors define an "average mobility" for each gene product surveyed. This value is a weighted average determined by the contribution of each electromorph and its frequency in the population. Fortunately, the three commonly used genetic distance estimates are highly correlated, indeed they are virtually interchangeable (Ayala *et al*., 1974; Ayala, 1975; Nevo, 1974; see MacIntyre and Collier, 1986 for fuller discussion of this point). Thus, for constructing trees we use "D", following Ayala (1975), which is either -lnI or -lnS or -ln% shared electromorphs.

3.3. Construction of Phylogenetic Trees

Most of the analyses of allozyme-based genetic distance estimates (Table VIII) have used the unweighted pair group method (UPGMA) of Sneath and Sokal (1973). Despite the vigorous debate of the relative merits of alternative algorithms for calculating trees (Farris, 1981; Prager and Wilson, 1978; Swofford, 1981; Nei *et al*., 1983), the UPGMA method appears adequate for our purpose of comparing data types, although it necessarily assumes a constant rate of change of the genetic distance metric.

3.4. Comparison of Phylogenetic Trees

For the purpose of this analysis we assume that the trees based upon information other than electrophoretic data are correct. This is necessary as well as conservative. It is necessary because we wish to assess the suitability of the genetic distances for phylogenetic reconstructions. It is conservative because instances in which the reference trees are in error and the genetic distance based trees are more correct will only increase the number of poor fits between trees based upon the different kinds of data.

We have used two complementary methods to compare dendrograms constructed from matrices of genetic distances with those constructed from matrices of alternative types of data. The first follows, with some modifications, Camin and Sokal (1965) and Farris (1973). A "coefficient of distortion" is calculated which estimates the number of steps, on a per cluster basis, to convert one dendrogram to the other. A convenient property of this method is that it is easy to identify which part(s) of a genetic distance dendrogram contributes to a large value for the coefficient of distortion. In other words, it is easy to identify which OTUs are clustered "improperly". The data sets used to generate the discordant trees, for these OTUs in particular, can then be scrutinized.

Remembering that the reference tree to which others are compared may not represent the "true" phylogeny, it would be helpful to pool reasonably congruent data sets to give more accurate estimates. An unfortunate aspect of the coefficient of distortion metric to date is that it offers no way to determine when data sets, and the trees they generate, are "reasonably congruent"; i.e., its statistical properties are unknown. In order to remedy this problem, we have empirically determined the statistical distribution of randomly generated coefficients of distortion for trees with four, five and seven OTUs, and will soon publish distributions for trees with six and eight through 11 OTUs.

To develop these statistical distributions, consider trees with five OTUs. There are 105 different ways to relate five OTUs in bifurcating trees; 60 permutations of topology a[1] (the "ladder"), 30 of topology b[2] and 15 of topology c[3]. There are 10,920 (= 105 x 104) pairs of trees to compare and thus generate coefficients of distortion; a complete statistical distribution would involve tabulating them all and assigning them to percentile groups. We chose to estimate the statistical distribution with 100 pairs of trees. Twenty trees were generated with a random number table such that the sequence of OTUs (12345, 41352, etc.) for a tree was random and the tree's form (a, b, or c) was both random and proportionate. Thus, Table IX, a 10 x 10 matrix of coefficient of distortion values, has 11 trees of form a (≈ 20 x 60/105), six of form b,

---

[1] Topology a

[2] Topology b

[3] Topology c

and three of form **c**. These coefficient of distortion values range from 0.056 to 1.25, with a mean of 0.738; their distribution by percentile is shown in Figure 3. The important value for present purposes is that under which only 5% of the randomly generated coefficient of distortion values fall. In a comparison of two trees with five OTUs, a coefficient of distortion value ≤ 0.167 will occur with only 5% probability by chance.

For seven OTU trees, with 108,045,630 (= 10,395 x 10,394) possible pairs and 11 forms (**a** through **k**), 289 pairs were compared in a 17 x 17 matrix of randomly generated trees. The coefficient of distortion values range from 0.07 to 1.767, with a mean of 1.434. Figure 4 shows these distributed by percentile, with the value at 5% being 0.533. Note that this value is appreciably higher than the one for five OTUs. We anticipate the 5% values will continue to increase as statistical distributions for trees with more OTUs are produced.

Consider the 100 coefficient of distortion values in Table IX, generated for the case of five OTUs. There are six different "classes" of comparisons, based on the topologies in each pair, and these classes occur at different frequencies: **a** x **a** - 30 times, **a** x **b** - 33, **a** x **c** - 17, **b** x **b** - 9, **b** x **c** - 9, and **c** x **c** - 2. The classes involving topology **a** are the most frequent simply because this topology occurs most often. How-

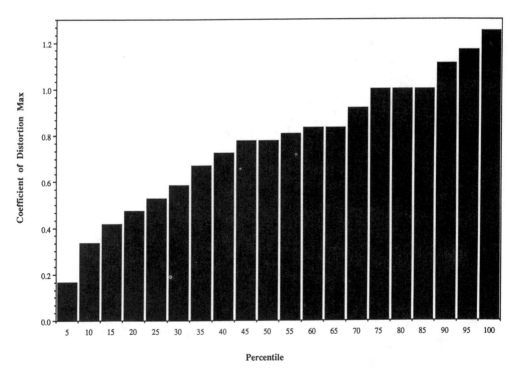

Figure 3. Distribution by percentile of randomly generated coefficients of distortion for five OTU's.

Table IX

100 Coefficient of Distortion (C.D.) values from comparisons using 20 randomly generated "phylogenies" with five OTUs. Mean Coefficient of Distortion values are shown for each row and column, as well as an overall mean

OTU order Topology	23541 a	15432 a	25431 c	15234 c	34152 a	21345 a	34152 b	42531 b	53412 a	43125 b	Mean C.D. for rows
15324, b	0.806	0.222	1.110	0.611	0.806	0.806	1.250	0.333	0.472	1.250	0.767±0.363
45132, a	0.944	0.167	1.000	0.889	0.500	1.000	1.000	1.000	0.417	1.000	0.792±0.310
43125, c	0.778	0.944	0.667	1.000	0.194	0.139	0.583	1.000	0.778	0.583	0.667±0.305
21345, b	0.472	0.722	1.110	0.778	0.806	0.056	1.250	0.833	0.722	1.250	0.800±0.362
41253, b	0.722	0.667	1.110	1.110	0.667	0.472	1.170	0.917	0.167	1.170	0.817±0.377
31524, a	0.667	0.472	1.060	0.556	0.722	0.528	1.000	0.833	0.472	1.000	0.731±0.229
34125, a	0.778	0.806	0.667	1.060	0.560	0.500	0.500	0.833	0.722	0.500	0.642±0.272
25314, a	0.222	0.833	0.611	0.556	1.000	0.667	0.833	1.000	0.583	0.833	0.714±0.236
45312, a	0.806	0.417	0.778	1.060	0.333	1.000	0.833	1.000	0.167	0.833	0.722±0.308
13542, a	0.917	0.333	1.060	0.778	0.417	0.778	0.917	0.833	0.333	0.917	0.728±0.267
Mean C.D. for columns	0.711 ±0.217	0.558 ±0.273	0.917 ±0.210	0.839 ±0.215	0.555 ±0.304	0.595 ±0.323	0.933 ±0.257	0.858 ±0.201	0.483 ±0.221	0.933 ±0.257	0.738±0.293

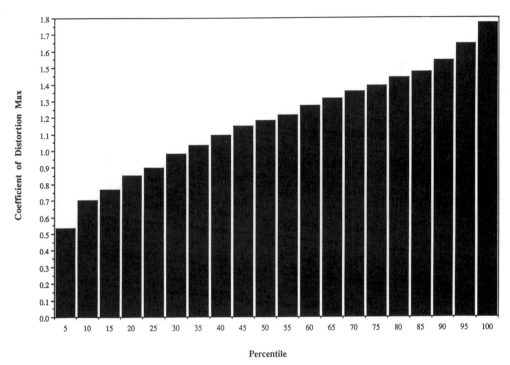

Figure 4. Distribution by percentile of randomly generated coefficients of distortion for seven OTU's.

ever, while the mean coefficient of distortion value for five OTU trees is estimated as 0.738, the mean values from each of the classes are not equivalent. Figure 5 shows the frequency distribution of coefficient of distortion values for each of the classes and the mean value by class. The smallest mean values occur in classes with topology **a**, and these classes are also the most abundant. As trees with more OTUs are analyzed, however, the relative proportion of topology **a** decreases. With six OTUs, topology **a** is one of six possible topologies, and occurs with 38% frequency, with seven OTUs, it is one of 11 and 24%, and with eight OTUs, it is one of 23 and 15%. Thus, as higher order trees are analyzed, the relative proportion of "non-bushy" topologies like **a** decreases more slowly, and the distribution of coefficient of distortion values shifts to greater values more slowly. It is likely that the coefficient of distortion value at the 5% level will effectively reach a plateau at 11 OTUs.

The second method (Penny *et al.*, 1982) converts two trees to be compared to unrooted networks. The number of internal links in common between the two networks, i.e., those which divide the OTUs into identical subsets, is used to calculate a "difference" (d) between the two networks. The advantage of this approach is that the exact probabilities of randomly choosing two networks with a given value of d for four through 11 OTUs have been calculated (Table 1 in Penny *et al.*, 1982).

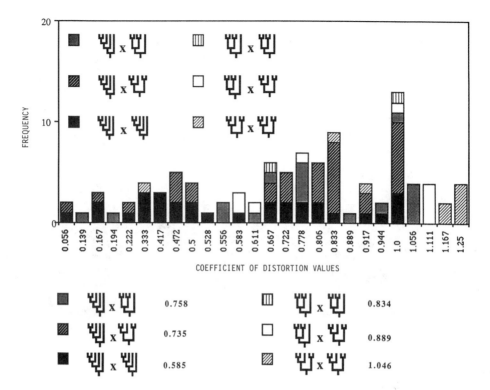

Figure 5. Frequencies of randomly generated coefficients of distortion for five OTU's, arranged by comparisons of the three basic topologies.

We have estimated the probabilities for larger numbers of OTU's by extrapolation (MacIntyre and Collier, 1986).

The two methods complement one another in that the first allows one to identify which portion of a tree is responsible for a large coefficient of distortion while the second provides the probability of obtaining a particular d value by chance. These methods are fundamentally different in the way in which they compare trees. There is no obvious, simple relationship between coefficient of distortion and d for the same set of trees. Thus, we report the results of both methods of analyses in Table X.

3.5. Analysis of the Data

A more complete presentation of the data along with the dendrograms is found in MacIntyre and Collier (1986). Three subsequent additions have been made to these data. First, a cytological study of the salivary gland polytene chromosomes of *D. gaunche*, *D. madeirensis*, and *D. subobscura* (Krimbas and Loukas, 1984) has allowed the placement of the former two species on the *subobscura* branch of the reference tree of this subgroup. Previously (MacIntyre and Collier, 1986) we took

EVOLUTION OF ENZYMES IN *DROSOPHILA*   435

Table X

Coefficients of distortion and d values (Penny *et al.*, 1982) for comparison
of dendrograms constructed from genetic distances and dendrograms
constructed from other types of data

Study[a]	No. of OTU's	Coefficient of Distortion	d	Probability of d, given n OTU's
1	6	0.29	2	5.7×10^{-2}
3,4	6	0	0	9.5×10^{-3}
5[b]	8	0.07	3.3	4.7×10^{-3} (UPGM)
5[c]		0.20	3.3	4.7×10^{-3} (Farris)
6[d]	8	0.11	1.3	6.7×10^{-4}
6[e]	7	0.55	5.3	1.6×10^{-1}
8	5	0.17	0	6.7×10^{-2}
10	5	0.67	2	2.7×10^{-2}
11	7	0.78	5.3	1.6×10^{-1}
12	6	0	0	9.5×10^{-3}
14	11	0.67	10.6	8.4×10^{-3}
15	12	0.24	6	4.5×10^{-6} [f]
16	16	1.90	25.5	6.5×10^{-1} [f]
17	8	0.07	1.3	6.7×10^{-4}
20	6,4	0.31	3.3	1.8×10^{-1}
21	17	0.09	6	1.1×10^{-10} [f]
22	8	0.09	4	6.5×10^{-3}

[a]See Table VIII for references.
[b]For the genetic distance based tree in Lakovaara *et al.* (1972).
[c]For the genetic distance based tree derived by Farris (1973).
[d]For the genetic distance based tree of *obscura* subgroup species.
[e]For the genetic distance based tree of *affinis* subgroup species.
[f]Estimated. Table 1 in Penny *et al.* (1982) gives probabilities of obtaining d values for topologies up to 11 OTUs. For each of these d values we plotted probability versus number of OTUs. These fell on straight lines. Each of these were extended to estimate the probability of a given d value for these larger numbers of OTUs.

the conservative approach of attaching these two species to the base of the tree derived from the morphological data (Buzzati-Traverso and Scossiroli, 1955). Second, Ohnishi *et al.* (1983) examined the variation in approximately 100 abundant proteins by two dimensional gel electrophoresis for species in subgroups of the *melanogaster* species group of the subgenus *Sophophora* and within the *virilis* species group of the subgenus *Drosophila*. The genetic distance estimates based on these proteins were used to construct a composite dendrogram of the species from the several subgroups and groups. Thirdly, Spicer (1988) also examined and analyzed two dimensional gel patterns of 135 protein "spots" in seven species from the subgenus *Drosophila*, and the sophophoran species, *D. melanogaster*. We compared the UPGMA tree he constructed to Throckmorton's phylogeny of the eight species.

This empirical analysis (Table X) of the phylogenetic value of genetic distance based trees has revealed instances in which the topological similarity of "molecular" dendrograms and traditional phylogenies is quite good and in at least two

cases, is perfect.   There are other cases where the correspondence is not particularly impressive, but in general, there
are many more good genetic distance based trees than bad ones.
In fact, there are only four cases out of 17 in which the
coefficient of distortion is greater than 0.4 and p(d) is
greater than 0.057.

For two of these four cases, the failure of only two of the
species to form primary clusters in the genetic distance based
dendrogram is responsible for virtually all of the coefficient
of distortion (MacIntyre and Collier, 1986).   Further, one of
these two cases involves species of the *obscura* group.   For
these it must be remembered that the soundness of the reference dendrogram which is based only upon morphological comparisons (Buzzati-Traverso and Scossiroli, 1955) is problematical.

The remaining two cases (Richardson and Smouse, 1976; Johnson *et al.*, 1975) involve greater discrepancies between the
genetic distance based dendrogram and the respective reference
dendrogram.   In both cases rather few loci were scored, *viz.*,
seven and 10 respectively.   Even so, in the study of the *mulleri* subgroup the coefficient of distortion would have been
near 0.3 if the genetic distance estimate between *D. longicornis* and *D. propachuca* had been as small as that between *D.
tira* and *D. ritae* or that between *D. aldrichi* and *D. wheeleri*.
Further, it is possible that these allozyme data would have
yielded a better dendrogram had genetic distance been calculated in one of the more traditional ways rather than by the
unique method formulated by these authors.   Thus, the major
discrepancy between a genetic distance based dendrogram and a
more traditional dendrogram is for the sixteen species of the
planitibia subgroup surveyed by Johnson *et al.* (1975).   Unlike
the other large coefficients of distortion discussed above,
this one cannot be explained by the failure of a small number
of OTU's to form a primary cluster.   In this case, over 90%
of the coefficient of distortion is due to discordances in four
clusters which involve seven of the sixteen species (MacIntyre
and Collier, 1986).   In each of these cases, species which are
included in primary or secondary clusters of the reference
tree, cluster only in the ultimate or penultimate cluster of
the genetic distance based tree.

Upon close inspection, however, the survey of Johnson *et
al.* (1975) differs from the others in Table VIII in another
important respect.   Allozymic variation for 11 of the 16
species was based upon population sample sizes of eight to
1290 (mean=208).   For the remaining five species very few
individuals were examined.   The electromorphs represented by
these individuals were assumed to be the most common electromorphs for each of these species.   Genetic distances between
these five and the remaining 11 species were calculated as per
cent shared electromorphs.   For the few loci scored, failure
to correctly identify the most common electromorph could have
disastrous consequences for the topology of a genetic distance
based tree.   More importantly, for closely related species
having the same common allozymes, the identity and frequency
of the rare allozymes are the determinants of tree topology.

This important information is lost when only a few individuals are surveyed. Indeed, all five of these species were among the seven that contributed to the large coefficient of distortion discussed above.

Accordingly, these five species were omitted from the genetic distance matrix, and the topology was recalculated for the remaining 11 species. This reduced the coefficient of distortion to 0.46 and p(d) to 9.5×10^{-3}. If the DNA hybridization data of Hunt *et al.* (1981) are used to resolve one of the trifurcations in the reference tree, the coefficient of distortion is further reduced to 0.32 and p(d) to 9×10^{-4}. Most of the remaining coefficient of distortion is due to another, unresolved trifurcation in the reference tree.

The data in Tables VIII and X allow us to ask more than just whether there is good correspondence between genetic distance based trees and the respective reference trees. Several properties of these surveys can be examined to determine what factors may affect this correspondence.

Three primary factors, number of OTU's studied, number of loci scored, and the heterozygosities of the loci scored, may contribute to a fit between a genetic distance based dendrogram and a reference dendrogram. With regard to the number of OTUs, as they increase the number of possible trees increases dramatically (Felsenstein, 1978). Therefore, the probability of errors in estimating the true tree similarly increases (e.g., Tateno *et al.*, 1982). Secondly, the accuracy of the genetic distance estimate can be a primary determinant of correct tree topology. The variance around an estimate of D is more sensitive to the number of loci examined than to the number of individuals surveyed from a species (Nei, 1975, 1978; Mueller and Ayala, 1982). Therefore, Nei (1978) and Nei *et al.* (1983) recommend that 60 or more loci be scored to minimize the variance of D.

The possible importance of a third determinant of genetic distance based topologies, the average heterozygosities of the loci scored, is not as immediately obvious, and therefore, deserves fuller explanation. The mutation rate of the loci surveyed has a striking effect upon the probability of error in estimating phylogenetic trees as shown in the simulation studies of Nei *et al.* (1983). It is certain that mutation rate (and therefore the rate of change) varies over loci; thus if these rates could be estimated, the sample of loci surveyed could be adjusted to increase the accuracy of trees constructed from allozyme data. Skibinski and Ward (1981, 1982) as well as ourselves (see above) have demonstrated a clear relationship between per locus rates of divergence and per locus average heterozygosity for proteins. Thus, as a first approximation, this measure of extant variation may be used to classify loci as to their relative rates of change.

Table XI reports correlation coefficients between these three attributes of the allozyme studies analyzed and the measures of topological similarity between the genetic distance based dendrograms and the reference dendrograms. Further, the correlation coefficients between the different measures of topological similarity are reported. In both of these cases (coefficient of distortion *vs* d and coefficient of distortion *vs* p(d)), r was near one indicating that these methods measure similar degrees of distortion between the topologies compared. Of the remaining comparisons, only the correlation coefficients between number of OTU's and coefficient of distortion and between number of OTU's and d were significantly different from zero. The average heterozygosity over all loci in any given study (whether based upon the average heterozygosities of each locus or upon the heterozygosity estimated by the individual studies) was negatively correlated with both coefficient of distortion and d, but not significantly. The number of loci surveyed had only a very weak correlation with d and the coefficient of distortion. This is surprising given the importance of the accuracy of the genetic distance estimate and the theoretical role that increasing the number of loci has in reducing the variance of this estimate.

Given the range of heterozygosities seen for the *Drosophila* proteins surveyed (and the inferred range of rates of change for these proteins) and the effect rates of change can have on estimating phylogenies in simulation studies (Nei *et al.*, 1983), the effect of the number of loci surveyed may be compromised by the different rates of change among these loci. This possibility was assessed by analyzing two allozyme surveys (Cabrera *et al.*, 1983; Ayala, 1975) and asking if one or more subset of loci defined by average heterozygosity could more accurately define the topology determined by the total set of loci than other subsets. These analyses are summarized in Figures 6 and 7 in which the coefficient of distortion between the topology of trees generated by sets of five loci are compared to the tree generated by the total set of loci. In both cases, a subset of loci of relatively low heterozygosity more accurately predicted the topology generated by the total set of loci. Although there were few loci in the highest heterozygosity class in both surveys, this class was also reasonably accurate in predicting this topology. Subsets of loci in other heterozygosity classes produced trees which were

Table XI
Correlation coefficients (r) between properties of genetic distance data sets and coefficients of distortion, d values, or probability of d

N	Comparisons			r	
16	Coefficient of distortion	vs.	d	0.92	**
16	Coefficient of distortion	vs.	probability of d	0.92	**
16	No. of OTUs	vs.	coefficient of distortion	0.66	**
16	No. of OTUs	vs.	d	0.89	**
16	No. of "loci" scored	vs.	coefficient of distortion	-0.07	
16	No. of 'loci' scored	vs.	d	-0.39	
16	H^a	vs.	coefficient of distortion	-0.27	
16	H^a	vs.	d	-0.26	

** Significantly different from zero at 0.01 level.
[a]Average heterozygosity over all loci in a given study.

more discordant. This suggests that the effect of the number of loci scored can be compromised by the different rates of change of the individual loci. Unfortunately, there are too few data sets available in which sufficient loci are surveyed to determine the generality of this observation.

In summary, we take issue with those who claim that genetic distance based trees are not or cannot be "good" trees (Throckmorton, 1978; Farris, 1981). We have analyzed an extensive set of allozyme surveys for species groups of *Drosophila* for which independent data are available for estimating phylogenetic relationships. We have quantified the differences between dendrograms based upon genetic distance and dendrograms based upon other types of data and have assessed the likelihood that these similarities are due to chance. In general, we find the differences to be small or, if relatively large, to be due to only one or a few discordant genetic distance estimates.

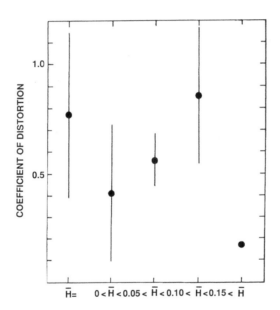

Figure 6. Coefficients of distortion for trees generated by subsets of five loci from the survey of Cabrera *et al.* (1983). Loci were assigned to one of five classes based upon average heterozygosity (H). The loci included in each class and the number (N) of sets of five loci sampled for each class are as follows; H=0 - CAT, DIA-1, EST-1, EST-4, FUM, HK-5, ME, Pt-1, Pt-4 (N=38); 0<H<0.05 - AK-3, ALD, DIA-2, EST-11, EST-13, GOT-1, GPDH, HBDH, HK-3, MDH, ODH, PHI, PYR, TO, XDH, Pt-2, Pt-7, Pt-8, Pt-10, Pt-12, Pt-5 (N=24); 0.05<h<0.10 - AR, G6PDH, IDH, PEP-2, PEP-3, 6PGDH, PGM-1 (N=12); 0.10<H<-0.15 - ADH, AMY, DIA-4, COT-2, HK-1, LAP-4, PEP-1 (N=12); 0.15<H - ACPH, EST-7, EST-8, APH-3. The tree generated by all loci in the survey was used as the reference tree. ● - Average of the coefficients of distortion for trees generated by sets of five loci within a given heterozygosity class. S.D. is indicated by the vertical line through each point.

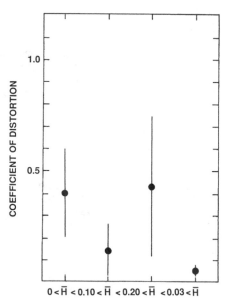

Figure 7. Coefficients of distortion for trees generated by subsets of five loci from the survey of Ayala (1975). Loci were assigned to one of five classes based upon average heterozygosity (H). The loci included in each class and the number (N) of sets of five loci sampled for each class are as follows; 0<H<0.10 - ADH, FUM, GOT, GPHD, HBDH, HK-3, MDH, ODH, PGI, TO TPI-2 (N=17); 0.10<H<0.20 - ADK-2, AO-2, APH, EST-4, IDH, LAP-5, ME-1, 6PGDH (N=13); 0.20<H<0.30 - ACPH, ADK-1, G3PDH, HK-1, HK-2, PGM (N=6); 0.30<H - ALD-1, ALD-2, AO-1, EST-1, G6PDH, XDH (N=6). The tree generated by all loci in the survey was used as the reference tree. ● - Average of the coefficients of distortion for trees generated by sets of five loci within a given heterozygosity class. S.D. is indicated by the vertical line through each point.

4. PROSPECTUS

There is no doubt that molecular data will assume an ever more prominent role in helping us understand the history of the genus *Drosophila*. These data will include the results from single copy sequence hybridizations, although the resolution of this technique limits the studies to species which diverged from a common ancestor between 1 and 20 million years ago. Genetic distances from allozyme surveys are also useful to define the histories of the members of single species groups. On the other hand, slowly evolving enzymes like αGPDH or 6PGDH, and the proteins detected by two dimensional gel electrophoresis can be used to assess the relationships between species groups and the major radiations within the genus. Very slowly evolving proteins like αGPDH and arginine kinase will help to place the genus *Drosophila* in a dendrogram of Dipteran families and perhaps to assess the evolutionary relationships of the insect orders themselves. For example, data from Collier and MacIntyre (1977) and Von Kalm *et al.* (1989) demonstrate that the structure of αGPDH is conserved

within the order Diptera, and the more slowly evolving arginine kinase can be detected immunologically in other major
groups in the insecta (see Table XII).

Given the demonstrated promise of these more traditional
ways of measuring interspecific variation in genes and their
products, what does the future, i.e., the expected flood of
DNA sequence information, hold in store?   Clearly, there is
reason to be optimistic.   Not only are DNA sequences relatively easy to obtain and are three times richer than amino
acid sequences in terms of information content, the evolution
of the structure, regulation and organization of individual
genes can now be precisely traced.   In this regard, Saitou and
Nei (1986) have dramatically shown how powerful DNA sequence
information can be with regard to resolving trifurcations in
phylogenetic reconstructions.   For example, even when the two
divergences are within 10% of each other in the time period,
the "correct" branching pattern can be discerned (at the 95%
confidence level) with only 4 Kb of sequence.   This depends
upon the rate of nucleotide substitution, and to some extent
upon the tree building algorithm used in the construction of
dendrogram.

There has been one recurring objection to DNA sequence
data, *viz.*, that the studies generally do not include enough
"alleles" from a single species.   Hence, a difference between
two species could have existed as a polymorphism in the common
ancestor, and thus would affect in an unknown way estimates of
the rate of change of the gene and/or its product.   Now, with
the advent of polymerase chain reaction technology, this
objection may quickly be obviated.   We surely stand on the
threshold of a golden age in evolutionary biology.   Yet we
must be mindful that the beauty and the power of evolutionary
studies lie in the comparative approach.   No single technique
can ever tell us the whole story or allow us to unambiguously
reconstruct a phylogenetic history.   Every method, every data
set has its limitations, and not to consider all the information when trying to trace the past of the genus *Drosophila*
would be not only foolish but could even lead to erroneous
reconstructions.   A combination of data from cytogenetic, morphological, biogeographical, and molecular studies (e.g., see
Lachaise *et al.*, 1988) can produce internally consistent or

Table XII

Immunological distance between *Drosophila* arginine kinase and other
arthropod arginine kinases

Heterologous antigen source		Immunological distance (mean ± s.d., n=3)
Hymenoptera	(*Specius speciosa*)	66 ± 1
Coleoptera	(*Pelidnota* sp.)	71 ± 5
Orthoptera	(*Acheta domesticus*)	83 ± 6
Odonta	(*Calopteryx maculata*)	89 ± 5
Thysanura	(*Lepisma saccharina*)	99 ± 7
Arachnida	(*Achaeranea tepidariorum*)	110 ± 4

concensus phylogenies, which are not only testable but ones in which we, as evolutionary biologists can have quiet confidence.

5. SUMMARY

This chapter provides an analysis of evolution in the genus *Drosophila*, as deduced from 1) immunological distance data for specific enzymes, and 2) genetic distance data for allozymes. Immunological distance data shows that while different enzymes evolve at different rates, the rate of change for any given enzyme is relatively constant. This rate constancy predicates the general rule that more rapidly evolving enzymes are useful for resolving the divergence times of younger phylads. In addition, as in vertebrates, the more rapidly evolving enzymes are those with higher levels of intrapopulation heterozygosity.

Given the controversy regarding the validity of phylogenies constructed from genetic distance values, we compared these *Drosophila* species group phylogenies with those based on morphological, cytological, hybridization and/or zoogeographical data. The quantitation of differences between the two kinds of phylogenetic trees involved the use of two methods: the "Coefficient of Distortion" and Penny's "Difference" metrics. In 17 comparisons, only four show statistically significant differences between the genetic distance dendrograms and traditional phylogenies, and these four are readily explained by small sample sizes of both enzymes and individuals. Genetic distance-based phylogenies are most accurate when 1) the number of taxa involved is kept under fifteen, 2) the number of enzymes scored is maximized, and 3) the enzymes used to compute the genetic distance are from the moderately and rapidly evolving sub-sets (as estimated from the average heterozygosities of the enzymes). When these three features are optimized, phylogenies from genetic distance estimates closely match those from other, more traditionally accepted, datasets.

References

Arai, K., Tominaga, H., Yokote, Y., and Narise, S., 1988, The complete amino acid sequence of cytoplasmic glycero-3-phosphate dehydrogenase from *Drosophila virilis*, *Biochem. biophys. Acta* 953:6-13.

Avise, J. C., and Aquadro, C. F., 1982, A comparative summary of genetic distances in the vertebrates: Patterns and correlations, *Evol. Biol.* 15:151-186.

Ayala, F., Tracey, M., Hedgecock, D., and Richmond, R. C., 1974, Genetic differentiation during the speciation process in *Drosophila*, *Evolution* 28:576-592.

Ayala, F., 1975, Genetic differentiation during the speciation process, *Evol. Biol.* 8:1-75.

Ayala, F. J., 1986, On the virtues and pitfalls of the molecular evolutionary clock, *J. Hered.* 77:226-235.

Band, H. T., Band, R. N., and Ives, P. T., 1984, The existence of LSP-1 Beta[S] in *Drosophila melanogaster* natural populations in two northern states, *Biochem. Genet.* 22:551-566.

Benjamin, D. C., Berzofsky, J. A., East, I. J., Gurd, F. R. N., Hannum, C., Leach, S. J., Margoliash, E., Michael, J. G., Miller, A., Prager, E. M., Reichlin, M., Sercrz, S., Smith-Gail, J., Todd, P. E., and Wilson, A. C., 1984, The antigenic structure of proteins: a reappraisal, *Ann. Rev. Immun.* **2**:67-101.

Berlocher, S. H., 1984, Insect molecular systematics, *A. Rev. Ent.* **298**:403-433.

Beverley, S. M., and Wilson, A. C., 1982, Molecular evolution in *Drosophila* and higher Dipteran. I. Microcomplement fixation studies of a larval hemolymph protein, *J. Mol. Evol.* **18**:251-264.

Beverley, S. M., and Wilson, A. C., 1984, Molecular evolution in *Drosophila* and higher Dipteran. II. A time scale for fly evolution, *J. Mol. Evol.* **21**:1-13.

Bock, I. R., 1971, Intra and interspecific chromosomal inversions in the *Drosophila bipectinata* species complex, *Chromosoma* **34**:206-229.

Brncic, D., Nair, P. S., and Wheeler, M. R., 1971, Cytotaxonomic relationships within the *mesophragmatica* species group of *Drosophila*, *Univ. Tex. Publs* **7103**:1-16.

Buth, D. G., 1984, Application of electrophoretic data in systematic studies, *Annu. Rev. Ecol. & Syst.* **15**:501-522.

Buzzati-Traverso, A., and Scossiroli, R., 1955, The "obscura" group of the genus *Drosophila*, *Adv. Genet.* **7**:47-92.

Cabrera, V., Gonzalez, A. M., Larruga, J. M., and Gullon, A., 1983, Genetic distance and evolutionary relationships in the *Drosophila obscura* group, *Evolution* **37**:675-689.

Camin, J., and Sokal, R., 1965, A method for deducing branching sequences in phylogeny, *Evolution* **19**:311-326.

Carson, H. L., Hardy, D. E., Spieth, H. T., and Stone, W. S., 1970, The evolutionary biology of the Hawaiian Drosophilidae, in: *Essays in Evolution and Genetics in Honor of Theodosius Dobzhansky* (M. K. Hecht, and W. S. Steere, eds), Appleton Century Crofts, New York, pp. 437-543.

Carson, H., Johnson, W., Nair, P., and Sene, F., 1975, Allozymic and chromosomal similarity in two *Drosophila* species, *Proc. natn. Acad. Sci. USA* **72**:4521-4525.

Champion, A. B., Prager, E. M., Watcher, D., and Wilson, A. C., 1974, Microcomplement fixation, in: *Biochemical and Immunological Taxonomy of Animals* (C. A. Wright, ed.), Academic Press, New York, pp. 397-416.

Collier, G., 1990, Evolution of arginine kinase within the genus *Drosophila*, *J. Hered.* **81**(3) (In press).

Collier, G. E., and MacIntyre, R. J., 1977, Microcomplement fixation studies on the evolution of α-glycerophosphate dehydrogenase within the genus *Drosophila*, *Proc. natn. Acad. Sci. USA* **74**:684-688.

Craddock, E., and Johnson, W., 1979, Genetic variation in Hawaiian *Drosophila*. V. Allozymic diversity in *Drosophila sylvestris* and its homosequential species, *Evolution* **33**:137-155.

Dickerson, R., 1971, The structure of cytochrome-c and the rates of molecular evolution, *J. Mol. Evol.* **1**:26-45.

Dobzhansky, Th., 1970, *Genetics of the Evolutionary Process*, Columbia Univ. Press, New York.

Dobzhansky, Th., and Powell, J., 1975, *Drosophila pseudoobscura* and its American relatives, *D. persimilis* and *D. miranda*, in: *Handbook of Genetics*, Vol. 3 (R. C. King, ed.), Plenum, New York, pp. 357-387.

Eisses, K., Van Dijk, H., and Van Delden, W., 1979, Genetic differentiation within the melanogaster species group of the genus *Drosophila* (Sophophora), *Evolution* **33**:1063-1068.

Farris, J. S., 1973, On comparing the shapes of taxonomic trees, *Syst. Zool.* **22**:50-54.

Farris, J. S., 1981, Distance data in phylogenetic analysis, *in: Advances in Cladistics* (V. A. Funk, and D. R. Brooks, eds.), The New York Botanical Garden, Bronx, New York, pp. 3-23.

Farris, J. S., 1985, Distance data revisited, *Cladistics* 1:67-85.

Felsenstein, J., 1978, The number of evolutionary trees, *Syst. Zool.* 27:27-33.

Felsenstein, J., 1984, Distance methods for inferring phylogenies: A justification, *Evolution* 38:16-24.

Felsenstein, J., 1985, Phylogenies from gene frequencies: A statistical problem, *Syst. Zool.* 43:300-311.

Fisher, J. A., and Maniatis, T., 1985, Structure and transcription of the *Drosophila mulleri* ADH gene, *Nucl. Acids Res.* 13:6899-6917.

Gonzalez, A., Cabrera, V., Larruga, J., and Gullon, A., 1982, Genetic distance in the sibling species *Drosophila melanogaster*, *D. simulans* and *D. mauritiana*, *Evolution* 36:517-522.

Gonzalez, A. M., Cabrera, V., Larruga, J., and Gullon, A., 1983, Molecular variation in insular endemic *Drosophila* species of the macronesian archipelago, *Evolution* 37:1128-1140.

Hubby, J., and Throckmorton, L., 1965, Protein differences in *Drosophila*. II. Comparative species, genetics and evolutionary problems, *Genetics* 52:203-215.

Hunt, J. A., Hall, T. J., and Britten, R. J., 1981, Evolutionary distances in Hawaiian *Drosophila* measured by DNA reassociation, *J. Mol. Evol.* 17:361-367.

Johnson, W., Carson, H., Kaneshiro, K., Steiner, W., and Cooper, M., 1975, Genetic variation in Hawaiian *Drosophila*. II. Allozymic differentiation in the *D. planitibia* subgroup, in: *Isozymes*, Vol. IV (Clement Markert, ed.), Academic Press, New York, pp. 563-584.

Krimbas, C. B., and Loukas, M., 1984, Evolution of the obscura group *Drosophila* species. I. Salivary chromosomes and quantitative character in *D. subobscura* and two closely related species, *Heredity* 53:469-482.

Lachaise, D., Cariou, M. L., David, J., Lemunier, F., Tsacas, L., and Ashburner, M., 1988, Historical biogeography of the *Drosophila melanogaster* species group, *Evol. Biol.* 22:159-225.

Lakovaara, S., Saura, A., and Falk, C., 1972, Genetic distance and evolutionary relationships in the *Drosophila obscura* group, *Evolution* 26:177-184.

Lakovaara, S., Saura, A., Lankinen, P., Pohjola, L., and Lokki, J., 1976, The use of isoenzymes in tracking evolution and classifying Drosophilidae, *Zool. Scr.* 5:173-179.

Lee, T. J., 1971, Frequency of races in males of *D. auraria* in natural populations, *Drosoph. Inf. Serv.* 47:125-126.

Lemeunier, F., and Ashburner, M. J., 1976, Relationships within the *melanogaster* species subgroup of the genus *Drosophila* (*Sophophora*). II. Phylogenetic relationships between six species based upon polytene chromosome banding sequences, *Proc. R. Soc. B.* 193:275-294.

MacIntyre, R. J., Dean, M. R., and Batt, G., 1978, Evolution of acid phosphatase-1 in the genus *Drosophila*, Immunological studies, *J. Mol. Evol.* 12:121-142.

MacIntyre, R. J., and Collier, G. E., 1986, Evolution of *Drosophila* proteins, in: *The Genetics and Biology of Drosophila*, Vol. 3E (M. Ashburner, H. L. Carson, and J. N. Thompson, Jr., eds.), Academic Press, New York, pp. 39-146.

Marinokovic, D., Ayala, F., and Andjelkovic, M., 1978, Genetic polymorphism and phylogeny of *D. subobscura*, *Evolution* 32:164-173.

Miller, D. D., and Sanger, W., 1968, Salivary gland chromosome variation in the *D. affinis* subgroup. II. Comparison of C-chromosome patterns in *D. athabasca* and five related species, *J. Hered.* 59:322-327.

Mueller, L. D., and Ayala, F. J., 1982, Estimation and interpretation of genetic distance in empirical studies, *Genet. Res.* **40**:127-137.

Nair, P., Brncic, D., and Kojima, K., 1971, Isozyme variations and evolutionary relationships in the *mesophragmatica* species group of *Drosophila*, *Univ. Tex. Publs* **7103**:17-28.

Nei, M., 1971, Interspecific gene differences and evolutionary time estimated from electrophoretic data on protein identity, *Am. Nat.* **105**:385-398.

Nei, M., 1972, Genetic distance between populations, *Am. Nat.* **106**:283-292.

Nei, M., 1975, *Molecular Population Genetics and Evolution*, American Elsevier Publishing Co., Inc., New York.

Nei, M., 1978, Estimation of average heterozygosity and genetic distance from a small number of individuals, *Genetics* **89**:583-590.

Nei, M., Tajima, F., and Tateno, Y., 1983, Accuracy of estimated phylogenetic trees from molecular data, *J. Mol. Evol.* **19**:153-170.

Nevo, E., Kim, Y. J., Shaw, C. R., and Thaeler, Jr., C. S., 1974, Genetic variation, selection and speciation in *Thomomys talpoides* pocket gophers, *Evolution* **28**:1-23.

Ohnishi, S., Kawanishi, M., and Watanabe, T., 1983, Biochemical phylogenies of *Drosophila*: protein differences detected by two-dimensional electrophoresis, *Genetica* **61**:55-63.

Penny, D., Foulds, L. R., and Hendy, M. D., 1982, Testing the theory of evolution by comparing phylogenetic trees constructed from five different protein sequences, *Nature, Lond.* **297**:187-201.

Powell, J., 1975, Protein variation in natural populations of animals, *Evol. Biol.* **8**:79-113.

Prager, E. M., and Wilson, A. C., 1978, Construction of phylogenetic trees for proteins and nucleic acids: Empirical evaluation of alternative matrix methods, *J. Mol. Evol.* **11**:129-142.

Prakash, S., 1977, Genetic divergence in closely related sibling species *Drosophila pseudoobscura*, *D. persimilis* and *D. miranda*, *Evolution* **31**:14-23.

Reinbold, S., and Collier, G., 1990, Molecular systematics of the *Drosophila virilis* species group (*Diptera, Drosophilidae*), *Ann. ent. Soc. Am.* **83**(3) (In press).

Richardson, R., and Smouse, P., 1976, Patterns of molecular variation. I. Interspecific comparisons of electromorphs in the *Drosophila mulleri* complex, *Biochem. Genet.* **14**:447-466.

Rogers, J. S., 1972, Measures of genetic similarity and genetic distance, *Univ. Tex. Publs* **7213**:145-153.

Saitou, N., and Nei, M., 1986, The number of nucleotides required to determine the branching order of three species with special reference to the Human Chimpanzee Gorilla divergence, *J. Mol. Evol.* **24**:189-204.

Sene, F., and Carson, H., 1977, Genetic variation in Hawaiian *Drosophila*. IV allozymic similarity between *D. silvestris* and *D. heteroneura* from the island of Hawaii, *Genetics* **86**:187-198.

Singh, R. S., and Coulthart, M. B., 1982, Genic variation in abundant soluble proteins of *Drosophila melanogaster* and *Drosophila pseudoobscura*, *Genetics* **102**:437-453.

Skibinski, D. O. F., and Ward, R. D., 1981, Relationship between allozyme heterozygosity and rates of divergence, *Genet. Res.* **38**:71-92.

Skibinski, D. O. F., and Ward, R. D., 1982, Correlations between heterozygosity and evolutionary rates of protein, *Nature, Lond.* **298**:490-492.

Sneath, P. R., and Sokal, R., 1973, *Numerical Taxonomy: The Principles and Practice of Numerical Classification*, W.H. Freeman, San Francisco.

Spassky, B., Richmond, R. C., Perez-Salas, S., Pavlovsky, O., Mourao, C. A., Hunter, A. S., Hoenigsberg, H., Dobzhansky Th., and Ayala, F. J., 1971, Geography of the sibling species related to *Drosophila willistoni*, and of the semispecies of the *Drosophila paulistorum* complex,

Evolution **25**:129-143.

Spicer, G., 1988, Molecular evolution among some *Drosophila* species groups as indicated by two-dimensional electrophoresis, *J. Mol. Evol.* **27**:250-260.

Swofford, D. L., 1981, On the utility of the distance Wagner procedure, in: *Advances in Cladistics* (V. A. Funk, and D. R. Brooks, eds), N.Y. Botanical Garden Press, pp. 25-44.

Tateno, Y., Nei, M., and Tajima, F., 1982, Accuracy of estimated phylogenetic trees from molecular data, *J. Mol. Evol.* **18**:387-404.

Throckmorton, L., 1968, Concordance and discordance of taxonomic characters in *Drosophila* classification, *Syst. Zool.* **17**:355-387.

Throckmorton, L., 1975, The phylogeny, ecology and geography of *Drosophila*, in: *Handbook of Genetics*, Vol. 3 (R.C. King, ed.), Plenum, New York, pp. 421-469.

Throckmorton, L. H., 1978, Molecular phylogenetics, in: *Beltsville Symposia in Agricultural Research*, Vol. 2 (J. A. Romberger, R. H. Foote, L. Knutson, and P. D. Lintz, eds), Allenhend, Osmond & Co., Montclair, New York; John Wiley and Sons, New York, pp. 221-239.

Throckmorton, L., 1982, The *virilis* species group, in: *The Genetics and Biology of Drosophila*, Vol 3B (M. Ashburner, H. L. Carson, and J. N. Thompson, Jr., eds), Academic Press, New York, pp. 227-296.

Triantaphyllidis, C., Panourgias, J., and Scouras, Z., 1978, Isozyme variation and phylogenetic relationships among six species of the *montium* subgroup of the *Drosophila melanogaster* species group, *Genetica* **48**:223-227.

Von Kalm, L., Weaver, J., DeMarco, J., MacIntyre, R., and Sullivan, D., 1989, Multiple isoforms of *Drosophila* glycero-3-phosphate dehydrogenase are generated by differential splicing of 3' exons, *Proc. natn. Acad. Sci. USA* (submitted).

Wasserman, M., 1982, Evolution of the *repleta* group, in: *The Genetics and Biology of Drosophila* Vol. 3B (M. Ashburner, H. L. Carson, and J. N. Thompson, Jr., eds), Academic Press, New York, pp. 61-140.

Williamson, J. H., Krochko, D., and Geer, B. W., 1980, 6-Phosphogluconate dehydrogenase from *Drosophila melanogaster*. I. Purification and properties of the A isozyme, *Biochem. Genet.* **18**:87-101.

Yang, S., Wheeler, L., and Bock, I., 1972, Isozyme variation and phylogenetic relationships in the *Drosophila bipectinata* species complex, *Univ. Tex. Publs* **7213**:213-227.

Zouros, E., 1973, Genetic differentiation associated with early stages of speciation in the *mulleri* subgroup of *Drosophila*, *Evolution* **27**:601-621.

CHAPTER 23

Genetic Differentiation Among Island Populations and Species of Cactophilic Drosophila in the West Indies

W. B. HEED, A. SÁNCHEZ, R. ARMENGOL, AND A. FONTDEVILA

1. INTRODUCTION

In 1954, Marvin Wasserman, in his initial cytological analysis of the repleta species group of *Drosophila*, discovered there were six "mulleri-like" species united and set apart from the others by one fixed inversion common to all six forms and five other inversions shared among the species in various combinations, even though in the homozygous condition. Also of special interest, which was not considered an issue at the time, was the fact that three of the species, when hybridized, showed no inversions in the salivary chromosomes; *D. mulleri*, *D. aldrichi* and *D. wheeleri* were homosequential in banding pattern (Wasserman, 1954).

By 1962 six additional "mulleri-like" species were identified and the term "mulleri complex" was coined (Wasserman, 1962). By 1975 the term "phyletic cluster" was introduced by Richardson *et al.* (1975) and Richardson and Smouse (1976) to denote four sets of sibling species from North America in the mulleri complex for an ANOVA classification on patterns of electrophoretic mobility. Wasserman and Koepfer (1977, 1979) subsequently incorporated the cluster terminology into their cytological studies. By 1982 the martensis and buzzatii clusters had been clearly identified on cytological grounds and thought to be an extension of the mulleri complex into South America (Fontdevila, 1982; Wasserman, 1982a, 1982b).

Recent concentrated collecting in the dry areas of the West Indies and the Andean chains in northern South America has increased the number of homosequential species in the mulleri cluster to six (Wasserman, 1990). The additional species are: *D. mayaguana* from the West Indies and two undescribed forms, "from Venezuela" and "from Peru." With the addition of two more undescribed species from the West Indies, closely related to but not homosequential with *D. mayaguana*, and known tem-

---

W.B. HEED * Department of Ecology and Evolutionary Biology, University of Arizona, Tucson, AZ 85721, USA; *A. SÁNCHEZ, R. ARMENGOL, AND A. FONTDEVILA* * Departament di Genética i de Microbiologia, Universitat Antónoma de Barcelona, Bellaterra-Barcelona, Spain.

porarily as *D. SB+* and *D. SB-5*, there now exists a mulleri cluster consisting of eight species (Wasserman, 1990). This cluster is here and of necessity divided into three subclusters (Table I) based on the present report and the report of Armengol (1986).

The present report provides details on electrophoretic studies on the three species in the mayaguana subcluster from the West Indies. It was made possible by two major expeditions to those islands in 1983 and 1984 by a number of investigators interested in the Cactus-Yeast-*Drosophila* model system (Barker and Starmer, 1982) and by subsequent starch gel electrophoresis conducted on the laboratory strains while one of us (W.B.H.) was a guest at the Universitat Autònoma de Barcelona in 1987-88. The study is an extension of the large electrophoretic survey by Armand Sánchez (1986) of the eight South American species in the martensis and buzzatii clusters as well as the two species in the stalkeri complex from the West Indies (Table I). It is also an extension of the elec-

Table I
The 18 species of the repleta group analyzed electrophoretically at the Universitat Autónoma de Barcelona, Unitat de Genética

I. Mulleri cluster (Tropical America)

    A. Aldrichi subcluster (Tropical America)
       1. *D. aldrichi* (U.S., Mexico, C. America, S. America)
       2. *D. wheeleri* (U.S., Mexico)

    B. Mulleri subcluster (Tropical America)
       3. *D. mulleri* (U.S., Mexico, W. Indies)
       4. *D. "from Peru"* (Peru)
       5. *D. "from Venezuela"* (Venezuela)

    C. Mayaguana subcluster (West Indies)
       6. *D. mayaguana* (Bahamas, G. Antilles, G. Cayman, Tortola)
       7. *D. SB+* (Cuba, Hispañola, Navassa)
       8. *D. SB-5* (Cuba, Hispañola, Jamaica)

II. Martensis cluster (South America)
       9. *D. martensis* (Colombia, Venezuela)
      10. *D. starmeri* (Colombia, Venezuela)
      11. *D. uniseta* (Colombia, Venezuela)
      12. *D. venezolana* (Venezuela)

III. Buzzatii cluster (South America)
      13. *D. buzzatii* (Brazil, Argentina, Bolivia)
      14. *D. serido* (Brazil, Argentina)
      15. *D. koepferae* (Bolivia, Argentina)
      16. *D. borborema* (Brazil)

IV. Stalkeri complex (U.S., West Indies)
      17. *D. stalkeri* (Florida, Bahamas, Caymans, G. Antilles,
              except Puerto Rico)
      18. *D. richardsoni* (Lesser Antilles, Puerto Rico)

trophoretic analysis by Rosa Armengol (1986) of the other five species in the mulleri cluster from tropical America (Table I). Armengol's study is of particular value since it must be realized that inversions in the larval salivary chromosomes become powerless as phylogenetic determinants in the presence of a group of species which are homosequential.

Parenthetically, the two South American clusters have recently been determined to be cytologically distinct from the clusters whose members are found chiefly in North America (Alfredo Ruiz, pers. comm; Wasserman, 1990). Therefore Wasserman (1990) has placed them in a new complex, the martensis complex, while the mulleri cluster thankfully remains in the mulleri complex.

The two undescribed forms from the West Indies, D. SB+ and D. SB-5, are of extreme interest in part because they carry a number of inversions, in contrast to the other six species in the mulleri cluster (Wasserman, 1990) and because they occur sympatrically but utilize different host plants (cacti) in several of the islands. Furthermore, the two species are very closely related. David Grimaldi and the senior author are collaborating on the descriptions of D. SB+ and D. SB-5.

2. MATERIALS AND METHODS

2.1. Collections

Listed below are the species used in the analysis along with the locality (numbered as in the tables and figures), date of collection, means of capture and number recorded (many or all of which were used to establish the cultures of the specimens analyzed electrophoretically).

D. mayaguana - (1) Fond Parisien, Haiti, May 6-7, 1982, banana bait, 32, (2) Hatillo, Dominican Republic, May 11-12, 1982, banana bait, 58, (3) Beef Island, Tortolas (B.W.I.), May 17-18, 1982, Ex: Agave leaf, 10, (4) Conception Island, Bahamas, Nov. 18, 1983, banana bait, 5, Ex: Opuntia stricta, 4, (5) Grand Cayman Island (U.K.), Nov. 27-28, 1983, banana bait, 139, (6) Great Inagua Island, Bahamas, Nov. 19-20, 1983, banana bait, 980, Ex: Cephalocereus sp., 88, (7) Great Inagua Island, Bahamas, Nov. 19-20, 1983, banana bait [see (6)], Ex: Opuntia stricta, 48.

D. SB+ - (10) Fond Parisien, Haiti, May 6-7, 1982, banana bait, many, Ex: Opuntia moniliformis, 1,542, (13) Navassa Island (U.S.) Nov. 21, 1983, banana bait, 213, Ex: Opuntia stricta, 195.

D. SB-5 - (8) Fond Parisian, Haiti, May 6-7, 1982, banana bait, many, Ex: Stenocereus hystrix, 720, (9) Port Henderson, Jamaica, Nov. 22-23, 1983, banana bait, 781, (11) Fond Parisien, Haiti, May 6-7, 1982, same as above, (12) Pont Beudet, Haiti, May 9, 1982, fruit and cactus bait, 11, Ex: Stenocereus hystrix, 80, (14) Palisades airport, Jamaica, Nov. 22-23, 1983, banana bait, 225.

Figure 1 identifies the location of the collecting localities listed above. Other localities in Jamaica and Hispañola (not illustrated) have been successfully sampled for the species in the mayaguana subcluster by banana baiting on three occasions by Marvin Wasserman (1985, 1986 and 1987). The Cuban Guantanamo Bay United States Naval Station is identified in Figure 1 since the three species in the mayaguana subcluster were collected there and host plant records obtained by William T. Starmer and the senior author November 15-18, 1988. However none of the specimens and their descendents obtained after the first two collecting trips in 1983 and 1984 have been subjected to electrophoretic analysis.

2.2. Starch Gel Electrophoresis

The basic procedure for making the gel is based on Smithies (1955). It is started with 22.5 g hydrolyzed starch (Sigma, S-4501) heated to almost boiling in 250 ml of the correct buffer in which the air is evacuated to eliminate bubbles and instantly poured into a plexiglass mold of 430 x 170 x 30 mm. The mold has serpentine refrigeration that circulates water at 4°C and the gel is run on the desk top. Specimens are ground individually in 7 mm holes in a Teflon block in 30 ml distilled water and kept on ice while they are slotted in a row near the cathode base of the gel with 6 x 6 mm Whatmann (3 MM) filter papers. The terminal filters are marked with bromophenol in order to visualize the progress of migration. The electrodes are placed in a pan of 500 x 80 x 80 mm with 750 ml of the proper buffer and are connected to a Heath power supply. The duration of the run is from 4-6 hours, depending on the system analyzed.

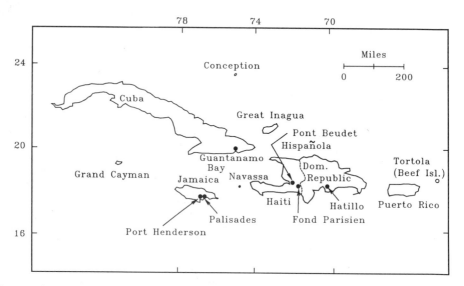

Figure 1. Islands and localities where collections were made for the mayaguana subcluster. *D. mayaguana* has been collected on all of the islands illustrated except Navassa and Puerto Rico. Mayaguana Island in the Bahamas (not shown) is the type locality for this species (Vilela, 1983). *D. SB+* has been collected in Cuba, Hispañola and Navassa. *D. SB-5* has been collected in Cuba, Hispañola and Jamaica.

With two filter papers per specimen it is possible to analyze four enzyme systems by duplicating the 22-25 slots that
make up the first one half of the gel. After the run the gel
is cut in half and sliced, providing four surfaces with iden-
ical sequences of specimens. If a similar unit is run simultaneously, it is possible to analyze 44-50 specimens in one
day. However, since every fifth slot is occupied by a control, a total of 32-40 unknowns can be analyzed. The final
size of the gel that is to be stained is 215 x 100 x 2.5 mm
since the upper and lower edges are cut and discarded. The
gels are incubated in their respective stains at 37°C until
the bands become distinct. The staining is terminated by the
addition of 100 ml of a mixture of methanol: water: acetic
acid of 5:5:1 for eight hours. The significant gels were
washed, wrapped and stored at 4°C. Table II lists the 15
enzymes studied.

2.3. Staining Methods

Adenylate kinase-2: 180 mg glucose, 40 mg $MgCl_2$, 40 mg ADP,
15 mg NADP, 50 mg MTT, 80 units G6PDH, 320 units hexokinase,
100 ml tris-HCl 0.05M pH 7.1. Two mg PMS added in absence of
light. Adult females utilized. Three bands sometimes appear.
The middle band was measured since it was always present, even
in pupae. All specimens monomorphic. Hexokinase bands mimic
Adk-2 bands in all specimens. This data was deleted. *Alcohol
dehydrogenase-1* and *2*: 2 ml isopropanol, 20 mg NBT, 15 mg
NAD, 100 ml tris-HCl pH 8.6. One mg PMS added in absence of

Table II

Enzymes analyzed in the three species of the mayaguana subcluster in the
West Indies. Buffer systems for the gel and tray are referenced for each
enzyme. Locations of the 15 loci by chromosome are listed

Enzyme	Locus	Buffer[a]	Chromosome
Adenylate kinase-2	Adk-2	II	X
Alcohol dehydrogenase-1	Adh-1	I	3
Alcohol dehydrogenase-2	Adh-2	I	3
Aldehyde oxidase-1	Ao-1	III	2
Esterase-1	Est-1	III	2
Fumarase-1	Fum-1	III	X
Glucose-6-phosphate dehydrogenase	G-6-pdh	III	X
α-Glycerophosphate dehydrogenase	α-Gpdh	III	3
Isocitrate dehydrogenase	Idh-1	II	4
Malate dehydrogenase	Mdh-1	II	3
Octanol dehydrogenase-1	Odh-1	I	2
Peptidase-1	Pep-1	IV	?
Peptidase-2	Pep-2	IV	2
Phosphoglucomutase-1	Pgm-1	I	4
Xanthine dehydrogenase-1	Xdh-1	III	2

[a]Buffer I: Discontinuous tris-citrate (Poulik, 1957). Buffer II: Continuous
tris-citrate pH 7.1 (Ayala *et al.*, 1972). Buffer III: Continuous tris-
borate-EDTA, pH 9.0 (Shaw and Prasad, 1970). Buffer IV: Continuous tris-
borate-EDTA (Loukas *et al.*, 1979).

light. The gel was saturated with NAD by adding 100 mg NAD to 10 ml of gel buffer before evacuation. Larvae and pupae were utilized. Heterozygotes detectable. However, when both loci are homozygous an intermediate band is present which is the heterodimer. *Aldehyde oxidase-1*: 2 ml benzaldehyde, 15 mg NAD, 25 mg MTT, 100 ml tris-HCl pH 8.6. Two mg PMS added in absence of light. Vapors kept separate from other gels during incubation. Adults utilized. Highly variable. Seven classes of "heterozygotes" scored. *Esterase-1*: 60 mg Fast Garnet-GBC, 1.5 ml of ANA (200 mg ß-naphylacetate in 10 ml acetone agitated in 10 ml water), 100 ml phosphate buffer (0.1M) pH 8.0. After incubation gel submerged in 0.5M boric acid for 15 min. Young adults stain best. Five classes of heterozygotes scored. *Fumarase-1*: 50 mg fumaric acid, 20 mg NBT, 25 mg NAD, 100 units of malate dehydrogenase, 100 ml tris-HCl pH 8.6. 1 mg PMS added in absence of light. Adult females scored. Heterozygotes show a second band, outside bands not evident. *Glucose-6-phosphate dehydrogenase*: 100 mg glucose-6-phosphate, 15 mg NADP, 25 mg MTT, 100 ml tris-HCl pH 8.6. Two mg PMS added in absence of light. Thirty mg of NADP was dissolved in 10 ml of gel buffer and added to the gel before evacuation. Young adult females utilized. Usually only single bands appeared. *α-Glycerophosphate dehydrogenase*: 20 mg DL-α-glycerophosphate, 15 mg NAD, 25 mg MTT, 100 ml tris-HCl pH 8.6. Two mg PMS added in absence of light. Adults utilized. All specimens monomorphic. *Isocitrate dehydrogenase-1*: 20 mg MTT, 20 mg NADP, 20 mg MgCl$_2$, 150 mg sodium isocitrate, 100 ml tris-HCl pH 8.6. Two mg PMS added in absence of light. Pupae and adults utilized. All specimens monomorphic. When tris-citrate is the buffer, malic enzyme mimics IDH bands exactly in all specimens. When Poulik is the buffer, malic enzyme bands are faster but inconsistent and so ME data was deleted. *Malate dehydrogenase-1*: 50 mg L-malic acid, 20 mg NAD, 20 mg MTT, 100 ml tris-HCl pH 8.6. Two mg PMS added in absence of light. Pupae utilized. All specimens monomorphic. *Octanol dehydrogenase-1*: 2 ml octanol-ethanol (20%) added to 100 ml tris-HCl pH 8.6 and agitated for two hours. Adding several drops of TWEEN-80 reduces agitation time. 20 mg NBT, 25 mg NAD and 3 mg PMS added in absence of light. Adults utilized. Other systems appear below this system. All specimens monomorphic. *Peptidase-1 and 2*: The stain was prepared in an agar solution. 120 mg agar in 15 ml buffer (1 g NaHPO$_4$ 12 H$_2$O in 100 ml water, pH 7.6). The mixture was agitated over heat until the boiling point. Two mixtures were made that were added to the agar solution when it was cooled to 45°C:(1) 12 mg L-leucyl-L-tyrosine, 6 mg snake venom and 12 mg peroxidase in 15 ml buffer, (2) 15 mg 3-amino-N-ethyl-carbazole in 1.7 ml n-n-dimethyl-formanide. Gel and electrode buffer was 0.18M Tris: 0.1M boric acid: 0.004M EDTA. The agar solution was poured onto the bottom slice of the gel within a 1 cm wide frame made from the upper slice of the gel and allowed to solidify before incubating for 30-60 min. The reading must be made before the bands become diffuse. Adults were utilized. Bands very readable. *Phosphoglucomutase-1*: 10 mg NADP, 200 mg MgCl$_2$, 600 mg glucose-1-phosphate, 50 mg EDTA, 20 mg MTT, 80 units G6PDH, 100 ml tris-HCl 0.1M pH 7.1. Two mg PMS added in absence of light. Pupae utilized. Heterozygotes show 2 bands. *Xanthine dehydrogenase-1*: 100 ml hypoxanthine in 100 ml Tris-HCl 0.05 M pH 7.5 heated to boiling and cooled. 15 mg KCl, 25 mg NAD, 20 mg MTT added. Two mg PMS added in absence of light. Pupae utilized. Highly variable. Heterozygotes very difficult to read because of absence of outside bands.

2.4. Other Enzymes Not Scored

Both ME and HK mimicked other systems as explained above. APH and ACPH bands would not migrate. When left longer in the stain the bands became diffuse. Both 6-GPDH and EST-2 did not stain on a consistent basis although different buffers were attempted. The LAP system produced too many bands and so interpretation was not possible.

2.5. Allozyme and Electromorph Homology

Homology was determined when specimens of a control species with a previously determined allozyme were matched with specimens of one or more strains in the mayaguana subcluster. The two, or more, bands must migrate precisely the same distance from the origin in order to be considered homologous. In the usual case a control species was placed in every fifth slot of the gel. More controls were added when necessary. Homology or nonhomology was scored on the majority of the 55 allozymes and electromorphs from the 15 loci analyzed in the mayaguana subcluster by comparison with one or more of the 10 species in the martensis and buzzatii clusters and the stalkeri complex (Table I) as analyzed by Sánchez (1986).

However, comparisons with the other five members of the mulleri cluster, as listed in Table I and analyzed by Armengol (1986), proved to be more difficult. As a consequence only 12 loci in the mayaguana subcluster had allozymes and electromorphs that could be homologized with the species in the remainder of the mulleri cluster. The three enzyme systems that could not be matched, chiefly because of the necessity for different buffers, were EST-1, XDH-1 and G6PDH. The latter enzyme had bands that migrated much slower than those in the mayaguana subcluster and so homology was not important in this case. However, EST-1 and XDH-1 allozymes showed much overlap with those from D. mayaguana, D. SB+ and D. SB-5 and so could not be deciphered.

Richardson and Smouse (1976) have correctly outlined the problem of homology in regards to electrophorectic surveys and the authors are fully aware of the ambiguities that arise by simply matching bands. Even so, and especially at the current level of analysis, the bands can only be matched.

2.6. Analyses of Data

The frequencies of the allozymes and electromorphs for each enzyme were entered into BIOSYS-1, a FORTRAN IV computer program (Swofford and Selander, 1981, University of Illinois).

Absolute frequencies, as calculated from the gels, were used to determine Nei's (1972) genetic distance and identity coefficients. Equal frequencies were employed to determine Rogers' (1972) genetic distance coefficients. Wright's F-statistics (Wright, 1965, 1978; Nei, 1977) based on absolute allozyme frequencies were also generated by the program and are incorporated. A single classification ANOVA program was kindly prepared by Mauro Santos from Nei's genetic identity matrix. Distance Wagner trees (Farris, 1972; Swofford, 1981) were computed on the basis of Rogers' genetic distance coefficients. The Goodness of Fit statistics, which measure the degree of deviation from the genetic distance matrix with the projection of the OTU's in the final tree, are based on Farris (1972), Prager and Wilson (1976) and Fitch and Margoliash (1967). The cophenetic correlation coefficient is a patristic distance measurement. In addition the data were kindly subjected to a Wagner parsimony analysis by Mike Sanderson and Mark Porter (University of Arizona) using PAUP version 2.4 (Swofford, 1985). All autapomorphies and constant characters were eliminated while inversion data, kindly supplied by Marvin Wasserman, were included but coded as reversible and unweighted characters. In order to compare the most parsimonious solutions from PAUP with the topologies from the distance Wagner trees, the MacClade program was used (version 1.0; copyright 1986, W. Maddison).

3. RESULTS

3.1. Identification of Electromorphs and Allozymes

The *Adk-2* locus is fixed for electromorph 99 in the 14 cultures of the mayaguana subcluster. The middle band is scored when there are three bands. The slow band is scored when only two bands appear. The slow band in the hexokinase system gives the same readings as the ADK-2 enzyme. The controls were *D. venezolana* for allozyme 99, with which the cultures were homologized, *D. buzzatii* for 100 and *D. martensis* for 101. Allozyme 99 is also found in *D. starmeri*, *D. uniseta* and *D. stalkeri*.

The *Adh-1* locus produced six different electromorphs (presumptive allozymes, see below). Table III shows that *D. mayaguana* differs from *D. SB+* and *D. SB-5* in all cultures except the one from Conception Island which shares allozyme 102 and 100 with the *D. SB-5* strain from Pont Beudet, Haiti. These two strains are also similar by the fact that no heterozygotes could be detected in either strain. The other strains of *D. mayaguana* are very different from the one from Conception Island. While the culture from Grand Cayman Island is fixed for allozyme 85, all other strains are homozygous for 93. An expected 1:2:1 ratio was realized when flies from Grand Cayman were crossed in reciprocal fashion to those from

Table III

Allozyme and electromorph frequencies for the 10 variable loci in laboratory cultures of the mayaguana subcluster

(N = number of individuals analyzed)

Locus	Culture Number[a]													
	1	2	3	4	5	6	7	8	9	10	11	12	13	14
ADH-1														
102	---	---	---	0.13	---	---	---	---	---	---	---	0.08	---	---
100	---	---	---	0.87	---	---	---	1.00	---	0.55	1.00	0.92	0.13	---
96	---	---	---	---	---	---	---	---	---	---	---	---	0.87	---
95	---	---	---	---	---	---	---	---	---	---	---	---	---	1.00
93	1.00	1.00	1.00	---	---	1.00	1.00	---	1.00	0.45	---	---	---	---
85	---	---	---	---	1.00	---	---	---	---	---	---	---	---	---
N	11	14	8	23	19	12	10	7	7	10	9	20	20	15
ADH-2														
99	1.00	1.00	1.00	1.00	---	1.00	1.00	1.00	1.00	1.00	1.00	1.00	1.00	1.00
98	---	---	---	---	1.00	---	---	---	---	---	---	---	---	---
N	12	7	10	21	19	8	10	7	7	10	9	23	20	4
AO-1														
103	---	---	---	---	---	---	---	---	---	0.08	---	---	---	0.09
102	---	0.29	---	0.03	---	0.01	---	0.05	0.61	0.69	0.32	0.52	0.52	---
101.5	---	---	---	---	---	---	---	---	---	---	0.08	0.09	0.15	0.09
101	0.93	0.71	0.50	0.34	0.30	0.07	---	---	0.39	0.19	0.49	0.35	0.33	0.82
100.5	---	---	---	---	---	0.15	---	---	---	---	0.05	---	---	---
100	0.07	---	0.34	---	0.67	0.49	0.23	0.05	---	0.04	0.06	0.04	---	---
99.5	---	---	0.09	0.62	0.03	0.14	0.07	0.06	---	---	---	---	---	---
99	---	---	0.07	---	---	0.14	0.70	0.74	---	---	---	---	---	---
98	---	---	---	---	---	---	---	0.10	---	---	---	---	---	---
N	15	7	20	28	39	73	30	42	18	26	37	23	42	11

Culture Number[a]

Locus	1	2	3	4	5	6	7	8	9	10	11	12	13	14
EST-1														
104	---	0.07	---	---	---	---	---	0.06	---	---	---	---	---	---
103	0.75	0.60	0.38	---	0.89	0.05	---	0.61	0.92	---	---	0.59	---	---
102	0.25	0.33	0.62	0.14	0.11	0.26	0.79	0.31	0.08	0.19	---	---	0.60	0.35
101	---	---	---	0.86	---	0.46	0.21	0.02	---	0.81	1.00	0.41	0.34	0.65
100	---	---	---	---	---	0.23	---	---	---	---	---	---	---	---
"NULL"	---	---	---	---	---	---	---	---	---	---	---	---	0.06	---
N	32	15	37	42	19	28	19	54	20	29	10	11	54	17
FUM-1														
103	---	---	---	---	---	---	---	0.11	---	---	---	---	---	---
102	---	---	---	---	---	---	---	0.27	---	---	---	---	---	---
100	---	---	---	---	---	---	---	0.06	---	---	---	---	---	---
98	1.00	1.00	1.00	1.00	1.00	1.00	1.00	0.56	1.00	1.00	1.00	1.00	1.00	1.00
N	15	6	17	32	11	13	8	27	15	17	13	7	28	5
G6PDH														
104	---	0.08	---	0.09	0.91	---	---	---	---	---	---	---	---	---
103	0.77	0.54	0.94	0.23	0.03	---	0.74	---	0.15	0.35	0.21	0.15	0.68	0.25
102.5	0.16	0.38	---	0.66	0.03	---	0.26	0.50	0.15	0.65	0.76	0.15	0.29	0.75
102	0.07	---	0.06	0.03	0.02	1.00	---	0.50	0.70	---	0.02	0.70	0.03	---
101	---	---	---	---	---	---	---	---	---	---	---	---	---	---
N	43	13	16	35	29	15	23	32	40	26	42	40	31	12
PEP-1														
102	1.00	1.00	1.00	1.00	0.25	1.00	1.00	1.00	1.00	1.00	1.00	1.00	1.00	1.00
101	---	---	---	---	0.28	---	---	---	---	---	---	---	---	---
100	---	---	---	---	0.43	---	---	---	---	---	---	---	---	---
99	---	---	---	---	0.04	---	---	---	---	---	---	---	---	---
N	5	6	7	9	46	4	6	18	17	7	12	4	18	5

Culture Number[a]

Locus	1	2	3	4	5	6	7	8	9	10	11	12	13	14
PEP-2														
104	---	---	---	---	---	---	---	1.00	1.00	1.00	1.00	1.00	0.97	1.00
101	1.00	1.00	1.00	1.00	1.00	1.00	1.00	---	---	---	---	---	0.03	---
N	5	6	7	5	38	9	10	18	17	7	12	4	18	5
PGM-1														
105	1.00	1.00	1.00	---	1.00	1.00	0.62	---	---	---	---	---	---	---
103	---	---	---	1.00	---	---	0.38	---	---	---	---	---	---	---
100	---	---	---	---	---	---	---	1.00	1.00	0.96	1.00	1.00	0.92	1.00
97	---	---	---	---	---	---	---	---	---	0.04	---	---	0.08	---
N	8	17	13	14	15	26	25	10	5	12	8	8	12	6
XDH-1														
105	---	---	---	---	---	---	---	---	---	0.07	---	---	---	---
104	---	---	---	---	---	---	---	---	---	0.79	0.29	0.60	0.13	---
103	---	---	---	0.07	---	---	---	0.14	0.50	0.14	0.29	0.13	0.65	0.14
102.5	---	---	---	---	---	---	---	---	---	---	---	---	0.09	---
102	---	0.17	---	0.40	0.08	0.04	---	0.66	0.50	---	0.43	0.27	0.13	0.86
101.5	---	---	---	---	---	---	---	0.18	---	---	---	---	---	---
101	1.00	0.83	1.00	0.53	0.92	0.58	1.00	---	---	---	---	---	---	---
100	---	---	---	---	---	0.38	---	0.02	---	---	---	---	---	---
N	10	6	5	30	26	12	5	23	6	14	7	15	23	14

[a]Culture Number: (1) *D. mayaguana*, Fond Parisien, Haiti, (2) *D. mayaguana*, Hatillo, Dominican Republic, (3) *D. mayaguana*, Beef Island, Tortolas, (4) *D. mayaguana* Conception Island, Bahamas, (5) *D. mayaguana*, Grand Cayman Island, (6) *D. mayaguana*, Great Inagua Island, Bahamas, (7) *D. mayaguana*, Great Inagua Island, Bahamas, (8) *D. SB-5*, Fond Parisien, Haiti, (9) *D. SB-5* Port Henderson, Jamaica, (10) *D. SB+*, Fond Parisien, Haiti, (11) *D. SB-5*, Fond Parisien, Haiti, (12) *D. SB-5*, Pont Beudet, Haiti, (13) *D. SB+*, Navassa Island, (14) *D. SB-5*, Palisades Airport, Jamaica

Great Inagua (culture No. 6).   The results were 10 93/93: 14
93/85: 5 85/85.   Chi square is 1.75.   Allozymes 102 and 85 are
new to our study.   Allozyme 100 is the common one in *D. buzza-*
tii and it has been homologized with this species. Allozyme 93
was homologized with 93 in "from Peru."

The two remaining strains that are polymorphic at the *Adh-1*
locus belong to *D. SB+* and in both cases heterozygotes were
easily determined.   Culture 10 from Fond Parisien, Haiti, is
polymorphic for allozymes 100 and 95.   The gels read 14 96/96:
5 96/100: 0 100/100.   This is in accordance with the Hardy-
Weinberg equilibrium if allozyme 96 were 0.87 and 100 were
0.13 in frequency.   Allozyme 95 has been homologized with 95
in *D. uniseta*.   All the species in the martensis cluster have
95 as does *D. richardsoni*.   Culture 13 from Navassa Island is
polymorphic for allozymes 100 and 96.   The gels read 2 95/95:
5 95/100: 3 100/100.   If the allozyme frequencies were 0.45
for 95 and 0.55 for 100, one would expect this distribution of
genotypes at equilibrium with 10 individuals.   The Navassa
Island culture probably shares 96 with *D. martensis* and *D.*
starmeri but it could not be homologized because of the low
frequency of 96 in these species.   The strains of *D. SB-5* have
allozymes 102, 100 and 95 mostly in the fixed condition possibly because of the low number of specimens analyzed.

The *Adh-2* locus shows the two allozymes 99 and 98 (Table
III). The Grand Cayman culture of *D. mayaguana* is fixed for 98
which is unique to our study.   Allozyme 99, fixed in the 13
other cultures, has also been found in a very low frequency
only in *D. buzzatii* from South America.   In effect, then,
allozyme 99 is probably also unique to our study in the sense
that it could be a separate mutation.

The *Ao-1* locus produced nine electromorphs (Table III).
However, only 46 double bands were identified from a total of
413 bands scored probably because of failure of clear separation making heterozygotes difficult to detect.   One strain
(No. 4) met the Hardy-Weinberg test.   If the 102 electromorph
is ignored, the banding read 13 99/99, 10 99/101 and 5
101/101, which is nonsignificantly different from expectation
for a locus with two alleles,.   However, strain No. 5 showed a
reading of 21 100/100, 10 100/101 and 7 101/101.   This is significantly different from expectation and the heterozygote is
less than expected.   The band at 99 was not included.   Later
analysis from three separate cultures were not significantly
different from expectation based on the Hardy-Weinberg test.
A pair mating of *D. mayaguana* from Grand Cayman Island was
scored as 12 100/100: 9 100/101: 1 101/101, on the basis of
two alleles.   Another pair mating of *D. mayaguana* from Great
Inagua Island (culture No. 6) was scored as 2 99/99: 5 99/100:
10 100/100: 1 100/101: 0 101/101: 0 99/101, on the basis of
three alleles.   A third pair mating representing *D. SB+* from
Fond Parisien was scored as 0 101/101: 1 101/102: 6 102/102: 3
102/103: 1 103/103: 0 101/103, on the basis of three alleles.
These data are not included in Table III but are presented
here to illustrate that a deficiency of heterozygotes does not
exist under the conditions of mated pairs.   The zero classes
make it difficult for precise statistical analyses.

In any event, there is a general separation of *D. mayaguana* from *D. SB+* and *D. SB-5* but very little separation of the latter two species (Table III). The only exception is culture No. 8 of *D. SB-5* from Fond Parisien, Haiti, which is similar to several strains of *D. mayaguana* by having a high frequency of electromorph 99 and lower frequencies of 98 and 99.5. All other cultures of *D. SB+* and *D. SB-5* do not have these three bands. The majority of the allozymes are shared with *D. stalkeri* and *D. richardsoni*, the buzzatii and martensis clusters and the mulleri subcluster in various arrangements except 99.5, 100.5 and 101.5 which were measured only in the mayaguana subcluster. Allozyme 103, however, is clearly unique to this subcluster. The controls used for the Ao-1 locus polymorphism were extensive. *D. serido* (from Brazil) and "from Venezuela" represented allozyme 98. *D. richardsoni* represented allozyme 99. *D. buzzatii*, *D. starmeri*, *D. mulleri* and "from Peru" represented allozyme 100. *D. wheeleri* and *D. martensis* represented allozyme 101. Allozyme 102 is a rare electromorph in *D. buzzatii* and *D. venezolana* and so no confirmation of homology was possible with the 102 in the mayaguana subcluster.

The *Est-1* locus produced five different electromorphs plus a probably null mutant discovered in *D. SB+* from Navassa Island (Table III). In general there is a large amount of overlap between all three species in the sharing of bands. In *D. mayaguana*, the strains from Conception Island and Great Inagua Island exhibit mostly bands 101 and 102 which is very similar to the two strains of *D. SB+* and the *D. SB-5* strain from Palisades, Jamaica. In contrast, the two cultures of *D. SB-5* from Fond Parisien, Haiti and Port Henderson, Jamaica, have a greater proportion of bands 102 and 103, which is more similar to the remaining cultures of *D. mayaguana*.

Hardy-Weinberg estimates were made on all strains and a strong deficiency of heterozygotes was present in all strains that were polymorphic. The heterozygotes were scored by the occurrence of light staining wide bands. Three examples are given here with the expectations in parenthesis: Culture No. 1, 7 (2.0) 102/102, 2 (12.0) 102/103, 23 (18.0) 103/103; culture No. 4, 35 (30.85) 101/101, 2(10.29) 101/102, 5(0.86) 102/102; culture No. 10, 22(19.033) 101/101, 3(8.93) 101/102, 4(1.04) 102/102. In the case of culture No. 13 with the null mutant the interpretation read as: 16 (5.84) 101/101, 5 (20.79) 101/102, 30 (18.52) 102/102, 6 (0.69) null/null, 0 (4.01) null/101, 0 (7.15) null/102. The null was recognized as the complete absence of a band for the ß-esterase stain but the presence of the slower migrating α-esterase band in each case.

The *EST-1* bands of the species in the mayaguana subcluster could not be homologized with the other species in the mulleri subcluster since different buffer conditions were necessary for mulleri species. The only consistent controls for *Est-1* were *D. buzzatii* for allozyme 100 and *D. uniseta* for 102. However, allozyme 100 is present in *D. stalkeri* and all species in the martensis and buzzatii clusters except *D. serido* from Brazil. Allozyme 102 is present in the four species of the martensis cluster and also in *D. buzzatii* and *D. serido* from Brazil. Allozyme 101 is present only in *D. venezolana* but this was not homologized with the mayaguana subcluster species. The remainder of the electromorphs in the mayaguana subcluster (103, 104, null) have not turned up in the eight species from South America nor in *D. stalkeri* or *D. richardsoni*.

The *Fum-1* locus is fixed at electromorph 98 in 13 cultures of the mayaguana subcluster (Table III). This has been homologized with *D. starmeri*, *D. venezolana* and *D. uniseta* where the allozyme is also fixed, or nearly so, in the three species. Allozyme 98 is also present in *D. buzzatii* but is very rare. Culture No. 8 of *D. SB-5* has three additional and faster migrating bands. This strain is not in Hardy-Weinberg equilibrium for the four electromorphs. A total of 22 homozygotes were scored but 11.03 were expected and 5 heterozygotes were scored but 15.97 were expected. *Fum-1* 100 is present in all species of the mulleri and aldrichi subclusters and the buzzatii cluster and *D. martensis*, *D. stalkeri* and *D. richardsoni*. *Fum-1* 102 is new to our study while 103 is present in *D. buzzatii* but in a very low frequency and therefore it could not be homologized.

A total of five electromorphs appeared in the *G6PDH* system. These were all single bands except for one individual in the Grand Cayman strain of *D. mayaguana* where a double band of electromorphs 101 and 103 was detected. In Table III *G6pdh* 102 and/or 102.5 are the common morphs in the strains from Conception and Great Inagua Islands of *D. mayaguana* and also in both cultures of *D. SB+* and three cultures of *D. SB-5*. The remainder of the *D. mayaguana* strains have a high frequency of *G6pdh* 103 which is present moderately in two strains of *D. SB-5*. Electromorphs 103 and 104 are new to our study. Morph 102 is fixed or almost fixed in all four species of the martensis cluster and also in *D. stalkeri* and *D. richardsoni*. G6PDH bands in species of the mulleri and aldrichi subclusters and the buzzatii cluster migrate more slowly although electromorph 101 is present in *D. koepferae* from Argentina and Bolivia.

The *α-Gpdh* locus is monomorphic for electromorph 100 as are practically all other species in the repleta group. The controls were *D. buzzatii*, *D. richardsoni*, *D. serido* and *D. aldrichi*.

The *Idh-1* locus is monomorphic for electromorph 99. The only other species that have this band are *D. starmeri* and *D. venezolana* with which the cultures of the mayaguana subcluster have been homologized.

The *Mdh-1* locus is monomorphic for electromorph 104. This band matches the band in *D. uniseta* (#3). *D. mulleri* and "from Venezuela" also have 104.

The *Odh-1* locus is fixed for electromorph 104 in the 14 cultures of the mayaguana subcluster. This electromorph was never homologized with any species. However, the band was consistently measured as 4 mm faster than *D. buzzatii* 100.

Other controls were *D. mulleri* which read 102 and *D. richardsoni* and *D. stalkeri* which read 103 each. *D. starmeri* and *D. serido* are reported to have electromorph 104 in high frequencies but unfortunately they were never used as controls.

The *Pep-1* locus was monomorphic at electromorph 102 for 13 of the 14 strains analyzed. The Grand Cayman culture of *D. mayaguana* had three additional morphs (Table III). Only four heterozygotes were scored in this polymorphism and so there is no Hardy-Weinberg confirmation in this case. A total of 42 "homozygotes" were observed where 14.92 were expected and 4 "heterozygotes" were observed where 31.09 were expected. Electromorph 102 has been homologized with the *D. mulleri* 102. Electromorph 101 has been homologized with 101 of *D. aldrichi* and *D. venezolana* (from Los Frijoles Island) and morph 100 with *D. buzzatii*. Morph 102 is present in all other species of the mulleri subcluster except *D. aldrichi* as well as *D. venezolana* from Los Frijoles. Morph 101 is present in all species of the martensis cluster as well as *D. aldrichi, D. wheeleri, D. stalkeri* and *D. richardsoni*. Morph 100 is in high frequency in all the species of the buzzatii cluster except *D. borborema*. It is also present in low frequencies in all species of the martensis cluster except *D. venezolana*. Electromorph 99 is fixed in *D. borborema* but the band in *D. mayaguana* was not matched with it.

The *Pep-2* locus shows a clear separation between the *D. mayaguana* cultures with electromorph 101 and those of *D. SB+* and *D. SB-5* with electromorph 104 (Table III). The Navassa Island culture of *D. SB+* had a single individual heterozygous for both morphs. Electromorph 101 was homologized with 101 of *D. richardsoni* and 104 was homologized with 104 of *D. aldrichi*. Morph 101 is present in *D. venezolana, D. stalkeri* and "from Venezuela" as well as *D. richardsoni*. Morph 104 is present in all species of the mulleri and aldrichi subclusters except *D. mulleri*. It is also present in *D. martensis* and *D. venezolana*.

The *Pgm-1* locus shows a clear separation between the *D. mayaguana* cultures and all of the remaining cultures with the faster electromorphs 105 and 103 (Table III). *D. SB-5* cultures are fixed for electromorph 100 while *D. SB+* cultures are polymorphic (100/97). The Conception Island strain differs from all other strains of *D. mayaguana* by being fixed for the 103 morph. One of the Great Inagua cultures (No. 7) is polymorphic (105/103). This strain is in Hardy-Weinberg equilibrium, 8 (105/105): 15 (105/103): 2 (103/103). The two *D. SB+* strains are also in Hardy-Weinberg equilibrium. The culture from Fond Parisien reads as 11 (100/100): 1 (100/97). The culture from Navassa Island, with additional data added after the analysis presented in Table III, reads as 13 (100/100): 6 (100/97): 1 (97/97). All of these tests are very close to expectation so there is little doubt that the four electromorphs are correctly identified as allozymes.

Pgm-1 105 is new to our study. *Pgm-1* 103 has been reported only in *D. buzzatii* in a low frequency. Since the two 103's were not run together, they may or may not be the same allozyme. The control for allozymes 105 and 103 was *D. vene-zolana* from La Blanquilla Island which was scored as *Pgm-1* 104. Its band migrated between the above two bands representing *Pgm-1* 105 and 103. *D. buzzatii* and *D. mulleri* served as controls for *Pgm-1* 100 and 97 respectively. *Pgm-1* 100 is also present in the members of the martensis cluster (except *D. uniseta*), *D. serido*, *D. wheeleri*, *D. stalkeri* and *D. richard-soni*. *Pgm-1* 97 is also present in "from Peru" and "from Venezuela." In summary, the 105 allozyme, and possibly the 103 allozyme, are diagnostic for *D. mayaguana*. *D. SB-5* and *D. SB+* share allozyme 100 with eight other species while *D. SB+* shares 97 only with *D. mulleri* and its closest relatives, "from Peru" and "from Venezuela."

The *Xdh-1* locus clearly separates out the *D. mayaguana* cultures since they have electromorph 101 which is absent in *D. SB+* and *D. SB-5* (Table III). The latter two species have electromorphs 102, 103 and 104 in various combinations and frequencies. Bands 101.5 and 102.5 were classified by measurements only. Very few heterozygotes were detected because of lack of outside bands. The controls were *D. buzzatii* for 100 (homologized), *D. borborema* and *D. uniseta* for 102 (homologized) and *D. serido* from Brazil for 103 (homologized). The mulleri and aldrichi subclusters could not be homologized with our flies. Electromorphs 101 and 105 are new to the martensis and buzzatii clusters and to *D. stalkeri* and *D. richardsoni*. The martensis cluster has allozymes 102 and 103 while the buzzatii cluster has these and also allozyme 100.

3.2. Apomorphic Electromorphs and Allozymes

When the mayaguana subcluster is considered as a unit there exist a varying number of apomorphic loci for it depending on the outgroup employed. Table IV, which is based on the data from Armengol (1986) and Sánchez (1986), shows the South American clusters and the stalkeri complex are "less isolated" from *D. mayaguana*, *D. SB+* and *D. SB-5* than are the five other species in the mulleri cluster. This anomaly cannot easily be explained, in light of the chromosome data, and the problem will be addressed in the Discussion. *D. buzzatii* is included in Table IV since this species is one of the outgroups in the Wagner trees.

In regard to the distinctness of *D. mayaguana*, the combination of *Pgm-1* 103 and 105 assure one apomorphic system while *Xdh-1* 101 has a probability of 0.74 of being diagnostic for any individual in the *D. mayaguana* strains (Ayala and Powell, 1972). These bands have not been detected in other species of the mulleri cluster nor the martensis and buzzatii clusters. They are not present in *D. stalkeri* nor *D. richardsoni*. By contrast *D. SB+* and *D. SB-5* share two allozyme systems which distinguish them from *D. mayaguana* but not from other species. *Pgm-1* 100 is found in eight other species and *Pgm-1* 97 in three other species, while *Pep-2* 104 is present in six other species. This latter morph has a probability of 0.97 of being diagnostic for any individual in the *D. SB+* strains. If the

Table IV

Apomorphic loci and their electromorphs for the three species in the mayaguana subcluster compared with other groups and *D. buzzatii*. Number of species in parenthesis. Based on Armengol (1986) and Sánchez (1986)

Outgroup	No. Loci	No. Electromorphs
Martensis cluster (4)[a]	1	2
Stalkeri complex (2)[b]	3	4
Buzzatii cluster (4)[c]	4	5
Mulleri subcluster (3)[d]	5	10
Aldrichi subcluster (2)[e]	6	16
D.buzzatii (1)[f]	6	11

[a]*Adh-2* 98, 99 (from 15 loci).

[b]*Adh-2* 98, 99; *Idh-1* 99; *Mdh-1* 104 (from 15 loci).

[c]*Adk-2* 99; *Idh-1* 99; *Mdh-1* 104; *Pep-2* 101, 104 (from 15 loci).

[d]*Adk-2* 99; *Adh-2* 98, 99; *G6pdh* 101, 102, 102.5 103, 104; *Idh-1* 99; *Odh-1* 104 (from 12 loci).

[e]Same as mulleri subcluster plus *Adh-1* 85, 93, 95, 96, 100, 102 (from 12 loci).

[f]Same as buzzatii cluster plus *G6pdh* 101, 102, 102.5, 103, 104; *Odh-1* 104 (from 15 loci).

Mexican and South American species are ancestral, then *D. mayaguana* is the more derived of the three species in the West Indies.

A similar pattern of differentiation is encountered within the islands of the West Indies. There exist two unique bands within *D. mayaguana* that are not present in any other species. *Adh-1* 85 and *Adh-2* 98 are monomorphic and unique to the culture from Grand Cayman Island. There are no similarly unique bands in *D. SB+* and *D. SB-5*. However, *Pgm-1* 97 is the only allozyme representative of both strains of *D. SB+* in the mayaguana subcluster even though it is also found in the fixed condition in *D. mulleri*, "from Peru" and "from Venezuela". Thus *D. mayaguana* is genetically more isolated between islands than are the other two species.

If the culture from Grand Cayman Island is the most isolated in *D. mayaguana*, then the culture from Conception Island must certainly be the least isolated and the one most closely allied with *D. SB+* and *D. SB-5* since it contains four loci that have similar electromorphs and allozymes and in similar frequencies to one or both of the two other species. They are *Adh-1*, *Est-1*, *G6pdh* and *Xdh-1* (in part) (Table III).

3.3. Genetic Distance Coefficients

Table V presents the matrix of genetic identity and genetic distance (Nei, 1972) between the 14 strains of the mayaguana subcluster. Table VI summarizes Nei's genetic distance coefficients between the three species in the mayaguana subcluster. It shows as expected, that *D. mayaguana* is more isolated from *D. SB+* and *D. SB-5* than the latter two species are from

Table V

Matrix of genetic identity of Nei (1972) above diagonal and genetic distance of Nei (1972) below diagonal for 15 loci from *D. mayaguana* (Nos 1 - 7) and the two undescribed species *D. SB+* (Nos 10, 13) and *D. SB-5* (Nos 8, 9, 11, 12, 14)

Population	1	2	3	4	5	6	7	8	9	10	11	12	13	14
D. mayaguana														
1. F. Parisien, Haiti	****	0.988	0.979	0.741	0.825	0.877	0.872	0.620	0.673	0.619	0.623	0.668	0.650	0.645
2. Hatillo, D. Republic	0.012	****	0.972	0.759	0.807	0.914	0.902	0.651	0.695	0.655	0.647	0.688	0.693	0.666
3. Beef Isl., Tortola (A.[a])	0.022	0.029	****	0.745	0.828	0.890	0.904	0.619	0.645	0.621	0.613	0.648	0.655	0.633
4. Conception Isl. (O.s.)	0.299	0.276	0.294	****	0.633	0.774	0.834	0.761	0.675	0.760	0.809	0.771	0.715	0.737
5. G. Cayman Isl.	0.192	0.215	0.188	0.457	****	0.726	0.707	0.527	0.575	0.507	0.506	0.565	0.523	0.511
6. G. Inagua Isl. (C.)	0.131	0.090	0.117	0.257	0.320	****	0.941	0.654	0.623	0.679	0.662	0.651	0.708	0.653
7. G. Inagua Isl. (O.s.)	0.137	0.103	0.101	0.182	0.347	0.061	****	0.686	0.624	0.669	0.643	0.643	0.709	0.652
D. SB+*/D. SB-5														
8. F. Parisien, Haiti (S.)	0.478	0.429	0.480	0.273	0.641	0.425	0.377	****	0.852	0.847	0.886	0.913	0.841	0.828
9. P. Henderson, Jamaica	0.396	0.364	0.438	0.393	0.554	0.473	0.471	0.160	****	0.876	0.846	0.897	0.857	0.918
*10. F. Parisien, Haiti (O.m.)	0.480	0.423	0.476	0.275	0.679	0.387	0.402	0.166	0.132	****	0.954	0.955	0.898	0.890
11. F. Parisien, Haiti (S.)	0.473	0.435	0.490	0.211	0.680	0.412	0.442	0.121	0.168	0.047	****	0.964	0.880	0.897
12. P. Beudet, Haiti (S.)	0.404	0.374	0.434	0.260	0.571	0.429	0.442	0.091	0.109	0.046	0.037	****	0.877	0.863
*13. Navassa Isl. (O.s.)	0.431	0.367	0.424	0.335	0.649	0.345	0.343	0.174	0.154	0.107	0.128	0.132	****	0.859
14. Palisades, Jamaica	0.439	0.406	0.458	0.306	0.671	0.426	0.428	0.189	0.085	0.117	0.108	0.147	0.151	****

[a]C. = bred from *Cephalocereus* sp.; O.s. = bred from *Opuntia stricta*; S. = bred from *Stenocereus hystrix*; O.m. = bred from *Opuntia moniliformis*; A. = bred from *Agave*.

each other.   In fact the genetic distance between the two
species of 0.125 is no greater than the mean distance between
the five strains of D. SB-5 (0.122).

Tables VII and VIII illustrate the differentiation between
islands in the mayaguana subcluster on the basis of Nei's
coefficients. Since there is little allozyme and electromorph
differentiation between D. SB+ and D. SB-5, the data are
pooled in Table VII.   In this table, within island D's are
slightly lower than those between islands which are similar to
one another and of very modest proportions.

In Table VIII, however, one can see an increasing dissimilarity from the Bahamas southward through Hispañola (and Tor-
tola) and then westward to Grand Cayman, making a semicircular
pattern around Cuba (Fig. 1).   The genetic distance between
the ends of the semi-circle, Conception and Great Cayman
Islands, is as great as that between D. mayaguana as a whole
and the other two species (Table VI).   The reality of this
pattern is reinforced when three geographic subdivisions of D.
mayaguana are compared individually to D. SB+ and D. SB-5
(Table IX).   This illustrates once again that D. mayaguana is
more highly differentiated.   It is also now quite evident that
the culture from Conception Island of D. mayaguana has the
lowest genetic distance with the other two species, as was
concluded above on incomplete qualitative data.

Table X presents the matrix of Rogers' genetic distance
coefficients (1972).   These data were used primarily to construct Wagner trees (see below).   However, other useful information may be extracted from the table in the form of genetic
similarities (1 - genetic distance) for comparison with
outside work and also there emerges a slightly refined
geographic pattern to the one previously discussed in that the
three strains from the Bahama Islands (Conception and Great
Inagua) can be more readily grouped together than with Nei's
genetic distance coefficients (Table XI).   For example, when
the strains of D. mayaguana are individually compared and
ranked by mean genetic similarity to pooled D. SB+ and D. SB-
5, the first three strains to emerge are from the Bahamas.
When the opposite comparison is made, the first two strains in
the ranking belong to D. SB+.

3.4. Wright's F-Statistics

Wright's hierarchical F-statistics for the mayaguana subcluster are presented in Table XII.   The data on D. SB+ and D.
SB-5 are pooled.   A total of eight of the 10 variable loci are
variable for D. mayaguana.   Fum-1 and Pep-2 are monomorphic
and were exempt from the analysis.   Similarly, Adh-2 and Pep-1
have no variation in D. SB+ and D. SB-5 and were not included.
Demes translates to cultures and regions to islands in Table
XII.   Two divisions were set up in each case.   For D.
mayaguana the first division includes Conception and Great
Inagua Islands.   For the other two species, Jamaica resides in
the first division.   The F values are arranged facing each

Table VI
Mean (and ranges) of Nei's genetic distance coefficients in the mayaguana
subcluster of the mulleri cluster. Number of cultures indicated

	1	2	3
1. *D. mayaguana* (7)	.182 (.012-.457)		
2. *D. SB+* (2)	.430 (.275-.679)	.107 (.107-.107)	
3. *D. SB-5* (5)	.440 (.211-.680)	.125 (.046-174)	.122 (.037-.189)

Table VII
Mean (and ranges) of Nei's genetic distance coefficients among the
geographic subdivisions of the pooled data of *D. SB+* and *D. SB-5*. Number of
cultures indicated

	1	2	3
1. Jamaica (2)	.085 (.085-.085)		
2. Hispañola (4)	.141 (.108-.189)	.085 (.037-.166)	
3. Navassa (1)	.153 (.151-.154)	.135 (.107-.174)	---

Table VIII
Mean (and ranges) of Nei's genetic distance coefficients among the
geographic subdivisions of *D. mayaguana*. Number of cultures indicated

	1	2	3	4
1. Conception (1)	---			
2. Great Inagua (2)	.220 (.182-.257)	.061 (.061-.061)		
3. Hispañola- Tortola (3)	.290 (.276-.299)	.113 (.090-.137)	.021 (.012-.029)	
4. Grand Cayman (1)	.457	.334 (.320-.347)	.198 (.188-.215)	---

Table IX
Mean (and ranges) of Nei's genetic distance coefficients for three
subdivisions of *D. mayaguana* with *D. SB+* and *D. SB-5*. Number of cultures
indicated

D. mayaguana	*D. SB+*	*D. SB-5*
Conception (1)	.305	.289
	(.275-335)	(.211-.393)
Great Inagua (5)		
Hispañola & Tortola	.408	.433
	(.343-.480)	(.364-.490)
Grand Cayman (1)	.664	.623
	(.649-.679)	(.554-.680)

other for direct comparison with the differences listed for
each level between them. *D. mayaguana* has higher values on all
levels except for "cultures in islands." This exception is no
doubt caused by the variation in the four cultures from Hispañola in *D. SB+* and *D. SB-5*. The two comparisons of interest
are the culture effect (cultures in total) and the island
effect (islands in total). There is a substantial amount of
between culture variation in both groups of flies, even though
D. mayaguana is higher, but there is very little between
island variation in *D. SB+* and *D. SB-5*. *D. mayaguana* has a
strong island effect as expected.

3.5. Genetic Identity Coefficients

Another method to demonstrate that the strains of *D. SB+*
and *D. SB-5* are a tightly knit group compared to those of *D.
mayaguana* is by single classification ANOVA tests on the
genetic identity coefficients of Nei (1972) (Table V). The
mean genetic identity of the pooled seven strains of *D. SB+*
and *D. SB-5* is 0.886 and the F_s = 1.9252 which is nonsignificant. Of the total variance, 86.6% lies within strains. For
the seven *D. mayaguana* strains, the mean similarity is 0.839
while the F_s = 3.2197 which is significant at the 5% level. In
this case 73.0% of the variance lies within strains.

In the interstrain comparisons of the *D. SB+*/*D. SB-5* v. *D.
mayaguana*, F_s = 0.1333 which is nonsignificant, and 100% of the
variance component lies within strains. However, when the *D.
mayaguana* strains are compared to *D. SB+*/*D. SB-5* strains, F_s =
37.2518 which is significant at the 0.1% level, and 83.8% of
the variance component lies between strains. Clearly the one
set of strains is more heterogeneous than the other. The
strains responsible for the heterogeneity in *D. mayaguana* are
of course the ones from Conception and Grand Cayman Islands.
The mean genetic identity between *D. SB+*/*D. SB-5* and *D.
mayaguana* is 0.649. As expected the isolated Grand Cayman
strain is less than this (0.531) while the least isolated

Table X

Matrix of genetic distance coefficients of Rogers (1972) for 15 loci from *D. mayaguana* (Nos 1 - 7), *D. SB-5* (Nos 8, 9, 11, 12, 14), *D. SB+* (Nos 10, 13), and *D. buzzatii* (No. 15). Allozymes and electromorphs given equal frequency when locus is polymorphic

Population	1	2	3	4	5	6	7	8	9	10	11	12	13	14	15
D. mayaguana															
1. F. Parisien, Haiti	****														
2. Hatillo, D. Republic	0.108	****													
3. Beef Island, Tortola	0.043	0.119	****												
4. Conception Island	0.243	0.220	0.243	****											
5. G. Cayman Island	0.240	0.250	0.245	0.306	****										
6. G. Inagua Island	0.128	0.108	0.139	0.230	0.277	****									
7. G. Inagua Island	0.119	0.176	0.123	0.184	0.320	0.148	****								
D. SB+*/D. SB-5															
8. F. Parisien, Haiti	0.370	0.354	0.369	0.302	0.453	0.329	0.325	****							
9. P. Henderson, Jamaica	0.291	0.275	0.310	0.276	0.388	0.323	0.344	0.207	****						
*10. F. Parisien, Haiti	0.313	0.315	0.327	0.236	0.413	0.296	0.256	0.185	0.176	****					
11. F. Parisien, Haiti	0.360	0.344	0.369	0.260	0.436	0.325	0.328	0.147	0.191	0.160	****				
12. P. Beudet, Haiti	0.302	0.303	0.322	0.226	0.402	0.303	0.319	0.164	0.134	0.158	0.099	****			
*13. Navassa Island	0.290	0.264	0.298	0.203	0.367	0.265	0.268	0.237	0.222	0.152	0.169	0.182	****		
14. Palisades, Jamaica	0.344	0.333	0.355	0.265	0.429	0.319	0.287	0.189	0.086	0.123	0.163	0.153	0.208	****	
D. buzzatii															
15. (P.I.)	0.823	0.812	0.816	0.766	0.771	0.789	0.805	0.718	0.796	0.766	0.779	0.770	0.752	0.795	****

Table XI
Mean (and ranges) of Rogers' genetic similarity coefficients for three
subdivisions of *D. mayaguana* with *D. SB+* and *D. SB-5*. Number of cultures
indicated

D. mayaguana	SB+	SB-5
Bahamas (3)	.746	.697
	(.704-.797)	(.656-.774)
Hispañola- Tortola (3)	.690	.667
	(.635-.736)	(.630-.725)
G. Cayman (1)	.611	.578
	(.587-.633)	(.547-.612)

strain (Conception Island) is greater (0.747). In fact, and
as discovered using genetic distances, there is no difference
between the mean I for Conception Island versus either *D.
SB+/D. SB-5* (0.747) or the other six strains of *D. mayaguana*
(0.748).

3.6. Distance Wagner Trees

Finally it is appropriate to attempt to place the 14 cultures in the mayaguana subcluster in proper phylogenetic relationship among themselves and with outside groups. Remembering
that cytological information is available on all 14 strains,
and also in regard to almost any conceivable outside group,
the task is to identify the trees that are congruent with
inversion data. Briefly put, this has been possible to accom-
plish within the mayaguana subcluster but only when *D. buzzatii* and *D. aldrichi* are the outgroups. However, *D. buzzatii*
is not directly related on the inversion data to the mayaguana
cluster while *D. aldrichi* is homosequential with *D. mayaguana*.

Figures 2 and 3 represent distance Wagner trees, with *D.
buzzatii* as an outgroup, based on 15 loci using Rogers'
genetic distance coefficients (1972) as presented in Table X.
The coefficients are based on equal frequencies of the
allozymes when they occur in polymorphic loci. This procedure
hopefully evens out many differences between strains that may
not be too meaningful. Figure 4 is a Wagner tree based on 12
loci and *D. aldrichi* is the outgroup. The conditions otherwise are the same as above. The chief difference between Figures 2 and 3 is that *D. SB+* is paraphyletic to *D. SB-5* in
Figure 2, with the strain from Haiti being closer to *D. SB-5*,
while it is monophyletic in Figure 3. The choice here is for
the paraphyletic situation since the Goodness of Fit statistics are slightly better and the tree is slightly shorter.

Cytologically there is a slight distinction between the two
trees also in favor of the paraphyletic interpretation. The
following data were kindly supplied by Marvin Wasserman (Table

Table XII

F-statistics for the mayaguana sub-cluster (15 loci)

		D. mayaguana 7 Demes, 5 Islands (Regions), 2 Divisions (Bahamas _vs_ Others)			_D. SB+/D. SB-5_ 7 Demes, 3 Islands (Regions), 2 Divisions (Jamaica _vs_ Others)			
X	Y	Variance Component	F_{xy}	Difference	Variance Component	F_{xy}	X	Y
Deme	Island	0.44003	0.214	.109	0.99099	0.323	Deme	Island
Deme	Division	1.93720	0.546	.148	1.37418	0.398	Deme	Division
Deme	Total	1.90558	0.542	.172	1.22069	0.370	Deme	Total
Island	Division	1.49717	0.422	.311	0.33318	0.111	Island	Division
Island	Total	1.46556	0.417	.347	0.22970	0.070	Island	Total
Division	Total	-0.03161	-0.009	.038	-0.15349	-0.047	Division	Total

Table XIII

Chromosome constitution (in genomes) of *D. SB+* and *D. SB-5*. Standard sequence (ST) includes the following fixed inversions: Xabc, 2abcfg, 3abc, 4, 5. The 5th chromosome is fixed for inversion r (5r) in *D. SB-5*. (from Wasserman, 1990)

Species	Island	X_{ST}	X_z	2_{ST}	$2n^8$	$2r^8$	$2n^8r^8$	$2n^8l^8$	$2m^8*$	20^8**	$2s^8***$	3_{ST}	$3n^2$
D. SB+	Navassa	12	0	0	12	0	0	0	0	0	0	12	0
	Hispañola	58	0	3	8	3	P	39	0	0	0	58	0
D. SB-5	Hispañola	278	0	0	0	0	0	0	257	12	9	3	275
	Jamaica	0	54	0	0	0	0	0	61	1	0	61	1

P = Present, $2m^8* = 2n^8l^8m^8$, $20^8** = 2n^8l^8m^8o^8$, $2s^8*** = 2n^8l^8m^8s^8$

XIII).   The two inversions of interest are $2n^8$, a single inversion on the second chromosome, and $2n^81^8$, the single inversion plus an adjacent inversion so that $2n^81^8$ is inherited as a unit.   The single inversion, $2n^8$, is fixed in all strains from Navassa Island of *D. SB+* and heterozygous with $2n^81^8$ in the strains from the Cul de Sac region of Hispañola still representing *D. SB+*.   The strain from Fond Parisien is located in this region.   All *D. SB-5* strains, whether from Hispañola or Jamaica, are fixed for $2n^81^8$ ($2n^8$ is not present).   The tree in Figure 2 permits the inversions to arise in sequence ($2n^8$ -> $2n^81^8$) rather than simultaneously, as would be a necessary interpretation in Figure 3.   Furthermore, *D. SB+* is cytologically more ancestral than *D. SB-5* also by having the standard (ST) chromosome within its repertoire of second chromosome homologues in Hispañola.   The ST chromosome is the one that is homosequential in the six other species of the mulleri cluster.

D. SB-5 is clearly monophyletic in Figures 2 and 3 as are the two strains from Jamaica within the species.   Inversion 5r is fixed on the fifth chromosome in all the strains of *D. SB-5*, while inversion Xz, on the X chromosome, is fixed in all strains from Jamaica (Table XIII).   Thus any Wagner tree that did not show these monophylies would not be accepted on cytological grounds.   For instance, when either *D. martensis, D. starmeri* or *D. venezolana* is used as an outgroup, either one or the other monophylies is disrupted.   None of these three South American species cytologically fits with the mayaguana subcluster but many allozymes and electromorphs are shared between the strains from Venezuela and those from the West Indies as was discussed previously and illustrated in Table IV.

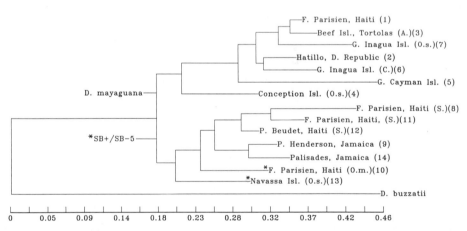

Figure 2. Distance Wagner Tree of 14 cultures based on 15 loci in the mayaguana subcluster. The cultures of *D. SB+* are marked with an asterisk. *D. buzzatii* is the outgroup.   Letters and numbers following the strains same as in Table V. Allozymes and electromorphs of polymorphic loci given equal frequencies. Tree length = 1.961. Goodness of fit statistics : Farris "F" = 3.602, Prager and Wilson "F" = 10.539, percent standard deviation of Fitch and Margoliash = 19.879 and the cophenetic correlation = 0.970.

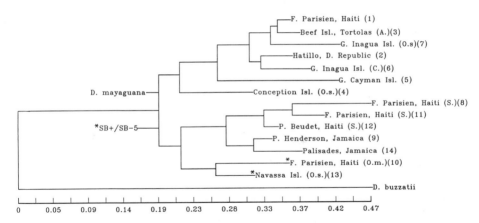

Figure 3. Distance Wagner Tree of 14 cultures based on 15 loci in the mayaguana subcluster. The cultures of *D. SB+* are marked with an asterisk. *D. buzzatii* is the outgroup. Letters and numbers following the strains same as in Table V. Allozymes and electromorphs of polymorphic loci given equal frequencies. Tree length = 1.990. Goodness of fit statistics : Farris "F" = 5.373, Prager and Wilson "F" = 15.720, percent standard deviation of Fitch and Margoliash = 32.910 and the cophenetic correlation = 0.966.

The only other species that was examined for an outgroup was *D. aldrichi* (Figure 4). Once again *D. SB+* is paraphyletic to *D. SB-5* but there is a trichotomy here involving the two Jamaican strains. However, the chromosome data permits the placement of the Jamaican strains with the other *D. SB-5* strains and this is also illustrated in the trees in which *D. buzzatii* is the outgroup. Figure 4 has an added attraction, however, by showing an early divergence in *D. mayaguana* of the three Bahama strains from the islands of Conception and Great Inagua. This is of course geographically satisfying and is similar to the data presented in Table XI. Reference to the three figures shows that the tree with *D. aldrichi* as an outgroup is the shortest and it also has the highest cophenetic correlation. Comparisons must be made cautiously, however, since the trees in Figure 2 and 3 are based on 15 loci while the one in Figure 4 is based on 12, as discussed previously.

3.7. Phylogenetic Analysis using Parsimony

The results of the parsimony analysis (PAUP) in which the allozymes and inversions are either present or absent showed similarities to parts of all three Wagner trees (Figures 2, 3 and 4). The outgroup was *D. aldrichi* and *D. wheeleri* (allozyme data from Armengol, 1986). Eleven trees were equally parsimonious and the tree length was 114 steps. The consistency index was 0.49. A total of 19 taxa with 56 characters were employed. Six of these characters were the inversions $2n^8$, $2n^81^8$, $2n^81^8m^8$, $3n^2$, 5r and Xz. The concensus tree indicated that all three species are monophyletic as in Figure 3. It takes one extra step to make *D. SB+* paraphyletic as in Figures 2 and 4. The MacClade program demonstrated, however,

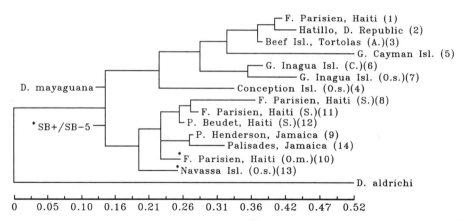

Figure 4. Distance Wagner Tree of 14 cultures based on 12 loci in the mayaguana subcluster. The cultures of *D. SB+* are marked with an asterisk. *D. aldrichi* is the outgroup. Letters and numbers following the strains same as in Table V. Allozymes and electromorphs of polymorphic loci given equal frequencies. Tree length = 1.875. Goodness of fit statistics: Farris "F" = 5.188, Prager and Wilson "F" = 14.607, percent standard deviation of Fitch and Margoliash = 29.471 and the cophenetic correlation = 0.976.

that the paraphyly was more parsimonious with inversions $2n^8$ and $2n^81^8$ but that monophyly was more parsimonious with allozymes *Pgm* 97 and *Xdh* 104. The latter allozyme is also shared with two strains of *D. SB-5*. In *D. mayaguana*, the Bahama Island strains are closely related and can be in the same positions as in Figure 4 without additional steps.

The ingroup topology did not change by substituting the mulleri subcluster for the aldrichi subcluster (allozyme data from Armengol, 1986). However, *D. buzzatti* as an outgroup, produced a consistently different topology that was not congruent with the others. Thus PAUP is more sensitive than the Wagner trees in this respect. When PAUP was run without the inversion data and with the mulleri and aldrichi subclusters as outgroups, *D. mayaguana* remained monophyletic but *D. SB+* and *D. SB-5* became unresolved in the strict concensus tree. There were 114 equally parsiminious trees in this case. This demonstrates the necessity for the inversion data in order to resolve the relationships within *D. SB+* and *D. SB-5* using PAUP.

3.8. Genetic Distance Dendrograms

The relations of eight species in the mulleri cluster are shown in a dendrogram in Figure 5 based on Nei's genetic distances from the absolute allozyme and electromorph frequencies of 12 loci. The data are an amalgamation of Armengol's data (1986) and that of the present report. The species are probably in correct relative relation to one another. The surprising developments here are the large genetic distances between subclusters and the small genetic distances within

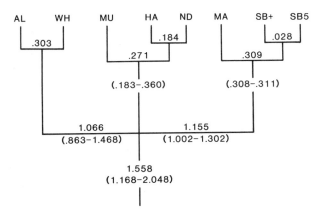

Figure 5. Dendrogram of the eight species in the mulleri cluster based on mean values of 12 loci using Nei's (1972) genetic distance coefficients. Ranges of mean values in parentheses. Coefficients based on actual frequencies of allozymes and electromorphs . AL = *D. aldrichi*, WH = *D. wheeleri*, MU = *D. mulleri*, HA = "from Peru", ND = "from Venezuela", MA = *D. mayaguana*. The genetic distance of 1.558 separates the aldrichi subcluster (left) from the mayaguana subcluster (right). Data from Armengol (1986) and the present study.

them.   Figure 6 shows two partial dendrograms of the same species each with the maximum number of loci analyzed.   As can be seen the distances within clusters have increased to more realistic statistics.   Even so there is a hierarchical pattern of genetic distances between twin species (0.125 – 0.321), within subclusters (0.431-0.449) and between the three subclusters (>1.0).

4. DISCUSSION

4.1. Previous Studies

Considering the large number of species in the repleta species group, electrophoretic studies aimed at deciphering species relationships have been limited until very recently. Zouros (1973) worked with four species using 16 loci while

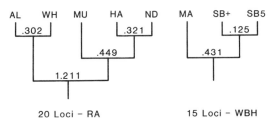

Figure 6. Two partial dendrograms of the eight species in the mulleri cluster based on mean values of 20 loci (left), from Armengol (1986), and 15 loci (right), from the present study, using Nei's (1972) genetic distance coefficients. Otherwise as in Figure 5.

Richardson *et al.* (1975, 1977) and Richardson and Smouse (1976) analyzed 12 species with seven enzyme systems. These studies were generally in agreement with Wasserman's inversion phylogeny of the mulleri complex, as recently demonstrated by McIntyre and Collier (1986), and both sets of studies emphasized the role of host plant specificity for creating differences in allozyme patterns. This point will be discussed below. In addition, Zouros demonstrated a nested hierarchy of D values among populations, subraces, races, sibling species and nonsiblings in *D. mojavensis*, *D. arizonensis*, *D. mulleri* and *D. aldrichi*, but the values were very low for each category. This point will also be discussed below.

In 1976, Barker and his group initiated allozyme studies with *D. buzzatii* in Australia (Barker and Mulley, 1976). This study was climaxed by a very detailed analysis of temporal and microgeographic variation in the species (Barker *et al.*, 1986). Evidence for selection and inbreeding was found and the importance of individual *Opuntia* pad rots was emphasized, as was the probability that habitat selection was occurring. But of primary interest for the present report was the analysis of allozymes of *D. buzzatii* in Brazil and Argentina (Barker *et al.*, 1985). This publication, which is the result of collections (1977-1979) by Sene and his group in São Paulo, represents the first available data on allozyme polymorphism in South American populations. Of the 31 loci assayed, 14 were variable in at least one locality. Subsequent work in South America on natural populations of *D. buzzatii* and the remaining species in the buzzatii cluster, as well as the martensis cluster, was conducted by Sànchez and Fontdevila in Barcelona (Sànchez, 1986; Fontdevila *et al.*, 1988) The remainder of this account is outlined and brought up to data in the Introduction.

4.2. Outgroup Determination

The history of the mulleri cluster is also reviewed in the Introduction. This cluster presents a most interesting picture of genetic similarities and differences among the eight species in relation to their geographic distributions (Table I, Figures 5 and 6). Six of the eight species have restricted distributions either in North or South America while the other two species, *D. aldrichi* and *D. mulleri*, are more widespread. A very intriguing geographic pattern exists among the three species in the mulleri subcluster. The presently known distribution of *D. mulleri* extends from Nuevo Leon and Tamaulipas, Mexico, northward and eastward through Texas and then sporadically to Florida. The species is also present in many islands in the West Indies. Sturtevant (1921) reported *D. mulleri* from Tegucigalpa, Honduras. The two closely related species of *D. mulleri*, however, are found in the Andean mountain chains of Peru and Venezuela, respectively. It is most likely that this large disjunction between *D. mulleri*, "from Peru" and "from Venezuela" represents one of the most extreme degrees of isolation of closely related species in the repleta species group and it is most significant since it encompasses both the North American and South American continents.

For this reason, and until more information is available, it is reasonable to consider the mulleri subcluster to be ancestral to both the aldrichi and mayaguana subclusters. As described previously, the mulleri and aldrichi subclusters are interchangeable as outgroups to the West Indian species in the PAUP analysis. However, reference to Table IV shows how tenuous this relationship might be (at least in a phenetic sense) in the absence of the cytological information, for it indicates a greater sharing of West Indian alleles with the South American clusters than with either the mulleri or aldrichi subcluster.

The probable reason behind this similarity, aside from the possibility of convergence, resides in the selection of the 15 loci that could be utilized in the mayaguana subcluster. Since a total of 22 enzyme systems were employed in the South American forms by Sánchez (1986) and since the other seven systems were attempted for the West Indian flies but could not be analyzed for various reasons as explained in the Materials and Methods, the 15 loci become a biased subsample of the original 22 loci. In addition it must be recalled that none of three species in the martensis cluster, when used as an outgroup to the West Indian flies, produced a Wagner tree congruent with the inversion sequences. The same may be said for *D. buzzatii* when it was employed as an outgroup in the PAUP analysis. Further consideration of the evolutionary relationships among the complexes, clusters, and subclusters in Table I is not practical at this time.

However, it is interesting to note that the outgroup for the mulleri cluster, in its entirety, should be the mojavensis cluster (Richardson *et al.*, 1975) located chiefly in western Mexico and southwestern U.S. and recently reviewed by Ruiz *et al.* (1990). Thus the relation between the clusters of cactophilic *Drosophila* in North America and in South America are not immediately evident. Even so, it is noteworthy that, with more recent mammalian fossil data, several different time levels during the Tertiary and Quaternary can be elected in which to traverse the two continents and in conditions probably suitable for the spread of cacti (Marshall, 1988; Webb, 1978). The last savannah corridor between the Americas was 12,000 to 10,000 years ago (Marshall, 1988).

4.3. Sources of Error

Errors in the present analysis may have arisen from several sources. The first and most critical source of error, aside from technical problems of staining and separation of bands in the gel, resides in the low number of loci examined. Comparisons among outgroups in the absence of outside information could give rise to erroneous relationships due a sampling bias as discussed above. However, comparisons of genetic differentiation within and between the species in the mayaguana subcluster should be on firmer ground. But even in this case, a larger number of loci with which to make comparisons would have been preferable. The highly variable enzymes may be grouped according to those that are strongly diagnostic (fixed differences) for strains but not especially species (ADH-1), and those that have low diagnostic powers (AO-1, EST-1, G-6-PDH, XDH-1). Even so, these five enzymes are responsible for larger values of genetic distance calculated from low genetic identities than the mean values across all enzymes simply because in many of the paired comparisons for each enzyme, an electromorph is present in one species but absent in the other. Of the remaining 10 enzymes with low or no variation between strains, PEP-2 and PGM-1 are the only ones strongly diagnostic for species. The remaining eight enzymes average out to high genetic identities between species, thus producing lower genetic distances than the mean values across all enzymes. The extent to which the genetic distances within and between species would be modified by an increase in the number of enzymes analyzed remains to be determined. It is significant that four of the five highly variable enzymes are known also as variable substrate enzymes (Kojima *et al.*, 1970). G-6-PDH is a glucose metabolizing enzyme.

Another source of error is inherent in a possible culture effect. Comparisons have been made between laboratory strains of several years duration and they were initiated by a varying number of founders. Rare alleles may have been lost or even fixed. Thus differentiation between cultures could have increased. Table XII illustrates considerable differentiation among cultures in *D. mayaguana*. The weighted average F_{DT} among eight variable loci equals 0.542. The non-hierarchical F-statistics (not included in Table XII) shows a range for this fixation index of 0.223 to 1.0 among the loci. The average variance for the index is 0.238. An interesting comparison which matches these statistics with other *Drosophila* data was generated by Wright (1978) who compared the allelic frequencies at 10 of the most strongly divergent loci in *D. willistoni* and *D. equinoxialis*. The fixation index was 0.533 (0.258 - 0.927) with a variance of 0.294. Wright noted that in all 10 pairs of loci the predominant allele was different in the two species. Therefore it is possible that there may be an artificially high deme and island effect in *D. mayaguana* (Table XII) but to what degree remains an open question. The same consideration applies to *D. SB+* and *D. SB-5* but in this case there is very little differentiation between islands (Table XII) showing that the cultures from these islands did not significantly drift apart. The weighted average F_{DT} for eight loci for the two species combined equals 0.370 (0-0.784) and the average variance equals 0.152. This index is considerably lower than the index for *D. mayaguana*. The fixation indexes for islands in divisions and islands in total were 0.111 and 0.070, respectively (Table XII). These indexes are four to six times lower than the ones for *D. mayaguana*. The possibility that common selective laboratory regimes accounted for the presence of similar alleles in similar frequencies does not explain the large differences discovered between several of the *D. mayaguana* cultures. Finally the issue of possible mixing of cultures is ruled out since phenotypic differences, as discussed below, identify each culture according to species, while inversion differences even identify each culture as to their island of origin (for *D. SB+* and *D. SB-5*).

Possible errors in species identification are discussed below.

4.4. Species Identification

From the electrophoretic data in the present report, one may rightly question the accuracy of the species identification since *D. mayaguana* has accumulated rather large genetic distances between certain cultures, while *D. SB+* and *D. SB-5* could be considered one species based solely on the small genetic distance between their cultures. However, there is little doubt that these three species are valid biological species for the following reasons. Even though *D. mayaguana* has formed two body color races, no other significant morphological differences have been identified. The Bahamas flies are lighter than the flies from the more southern islands. Furthermore, all crosses between island cultures were fertile (Heed and Armengol, unpublished). It appears as though electrophoretic differentiation between islands has preceded speciation events yet to occur in *D. mayaguana*. Nei's genetic identity ranges as low as 0.633 for intraspecific variation in the species (Table V). The breeding sites for *D. mayaguana* are known from Conception Island, where a few individuals have been reared from *Opuntia stricta*, and Great Inagua Island, where a moderate number of specimens have been reared from *O. stricta* and also the columnar cactus, *Cephalocereus* sp. (prob. *C. millspaughii*).

In contrast to the *D. mayaguana* cultures, *D. SB+* and *D. SB-5* differ in sexual characteristics including the shape of the genital claspers and the shape of the toe of the genital arch. They also differ in characters that have ecological significance such as the length of the pupal horns and shape of the larval mouthhooks. In addition the shape of the egg guide (ovipositor) differs between the two species. This difference could have ecological as well as sexual significance. Therefore it is possible, in theory, to distinguish *D. SB+* and *D. SB-5* on the phenotypes of the adults, larvae and/or pupae. Fertility between the species has been measured by Wasserman (1990). Male sterility has been detected in one cross, but not in the reciprocal test, while in another cross, in which a third culture replaced one of the original cultures, the hybrids of both sexes were fertile in reciprocal tests. The fertility data serves to underscore the close genetic similarity between the species. However there are species specific inversion differences as seen in Table XIII. It appears, that in this case, a speciation event has preceded electrophoretic differentiation since Nei's genetic distance between *D. SB+* and *D. SB-5* is only 0.125 (Table VI).

4.5. Significance of Host Plant Selection

It is most significant that *D. SB+* and *D. SB-5* inhabit different host cacti, especially where the two species are broadly sympatric. In Cuba, *D. SB+* has been reared from the columnar cactus, *Cephalocereus sp.* (prob. *C. brooksianus*), and from the tree opuntia, *O. moniliformis*, while *D. SB-5* has been reared from the columnar cactus, *Stenocereus hystrix*. In Hispañola, *D. SB+* has been reared abundantly from *O. moniliformis* while many *D. SB-5* specimens emerged from *S. hystrix*. These differences in host plant association may account for the differences in the ecological characters as well as the characters associated with sexual differences between the two species, if the different cactus rots act as mating platforms for each species, as they probably do. However, and at least for the loci analyzed in the present report, the host plant differences do not account for greater electrophoretic differentiation as proposed for several other species in the mulleri complex by Zouros (1973) and Richardson *et al.* (1977) and summarized and emphasized by MacIntyre and Collier (1986). In fact Armengol (1986) discovered that the genetic distance between *D. mulleri* and *D. aldrichi* is much greater than originally claimed in 1973 by Zouros (1.05 and 0.124, respectively). The discrepancy is explained by the different enzymes analyzed in each study. Zouros investigated 16 enzymes and proteins while Armengol examined 20 enzymes. The two studies overlapped by a total of only eight systems, which, parenthetically, showed very similar D values (Armengol, 1986). In retrospect, it was Richardson et al. (1975) who originally demonstrated the distinctiveness of *D. mulleri* in relation to *D. aldrichi* (and *D. wheeleri*) in their analysis of the mulleri complex. This information should lower the "coefficient of distortion" in the mulleri complex as analyzed by MacIntyre and Collier (1986) and Collier et al. (Chapter 22, this volume). Thus the fact that *D. mulleri* and *D. aldrichi* breed together in several species of *Opuntia* cacti in Texas, and that they are homosequential in gene order of the polytene chromosomes, has little effect on allozyme differentiation.

With the West Indian flies, it is more probable that *D. SB+* and *D. SB-5* are either more recently evolved than *D. mayaguana* (but see below) and/or that there has been hybridization between the two forms in the past. The inversion data do not indicate hybrids are frequent in present time. These conclusions are very similar to the ones presented by Sene and Carson (1977) in their description of the genetic similarities between *D. silvestris* and *D. heteroneura* in Hawaii. Nei's genetic identity coefficient, as recorded by Sene and Carson, was 0.961 within *D. silvestris,* 0.949 within *D. heteroneura,* and 0.939 between the two species based on 25 loci. Comparative figures for *D. SB+* and *D. SB-5* may be obtained from Table III that show I = 0.898 within *D. SB+*, I = 0.886 within D. SB-5, and I = 0.883 between the two species based on 15 loci. Aside from the fact that the West Indian species are not as closely related as the Hawaiians and that they do not breed in the same host plant, other details are strikingly similar. There are no electrophoretically fixed differences either within or between the species and the genetic variation between and within localities and species is very similar in the West Indian material as in the Hawaiian material. Since both sets of species are very closely related it is probably the recency of common ancestry (but see below) that accounts for these small electrophoretic differences (rather than the presence of common selective factors acting on *D. silvestris* and *D. heteroneura* which occupy at times the same host plant). Separate host plants in the West Indies probably serve the same function as separate lek sites in Hawaii in so far as assisting the maintenance of behavioral differences between truly closely related species that have the potential to hybridize. This is the significance of strict host plant selection for neospecies. The suggestion, however, remains to be verified by direct observation for the Caribbean material. The conclusion concerning the absence of a noticeable host plant effect on allozyme differentiation is similar to the conclusion described by Fontdevila *et al.* (1988) in which they report, from as yet unpublished data, the lack of correspondence between ecological differentiation and protein evolution in an "exhaustive study" by Moritz Benado and his group for the species in the martensis cluster in Colombia and Venezuela.

4.6. Geographic Genetics

The three species in this study are subdivided into 14 strains (cultures) which yield 91 paired comparisons in Table V for each of Nei's genetic identity and genetic distance coefficients. Measurement of mean genetic identity within the three species combined is 0.855 (0.633-0.988) from 32 pairwise comparisons. According to Thorpe (1982), a value of I=0.85, or above, is typical of within species values. Obviously the mayaguana complex is a borderline case. The species that makes it so is *D. mayaguana* which has a mean genetic identity of 0.839 (0.633-0.988) and it falls below Thorpe's criterion. The mean genetic identity between the three species, as a group, is 0.689 (0.506-0.955) from 59 pairwise comparisons. This value is typical of the difference between species found by Thorpe ($\bar{I}$ = 0.540 ± 0.168 S.D.). However the mean identity between *D. SB+* and *D. SB-5* with 10 paired comparisons is 0.884 (0.841-0.955). This value is clearly at the very high end of the scale.

Inquiry into the effect of islands on genetic distance yields interesting results but in the last analysis, this measurement no doubt also depends upon the history of the group (Nei, 1987). Listed below are values of Nei's mean genetic distance for all strains and all species arranged hierarchically. The numbers in parenthesis are ranges while the numbers in square brackets represent the number of paired comparisons.

Within species/within islands: 0.068 (0.012-0.121) [6]
Within species/between islands: 0.183 (0.022-0.457) [26]
Between species/within islands: 0.341 (0.046-0.480) [11]
Between species/between islands: 0.394 (0.117-0.680) [48]

An island effect on D is very noticeable within species but it is relatively weak between species. One reason for this difference resides in the different histories of the strains from the various islands. The most interesting culture in this respect is the one derived from Conception Island since it has strong affinities not only for other strains of *D. mayaguana* but also for the species pair of *D. SB+/D. SB-5*, as discussed previously and shown in Table IX.   Inspection of Table III shows the Conception Island strain shares various allozymes and electromorphs with other *D. mayaguana* cultures (i.e. *Pep-2* 101) and with *D. SB+/D. SB-5* cultures (i.e. *Adh-1* 102, 100; *Xdh-1* 103).   The culture from Conception Island also has had a semi-independent evolution since it is fixed for *Pgm-1* 103, which is present in only one other strain, also from the Bahamas (No. 7).   Thus the Conception Island culture, among the 14 analyzed in the present report, comes closest to the ancestral type for the mayaguana subcluster and it may be considered at present to be a segregant from that ancestor. Also, as seen in Figure 4, the Bahamas strains of *D. mayaguana*, in general, are most closely related to the ones of the *D. SB+/D. SB-5* species pair.   Therefore, that the early evolution of this subcluster occurred in the Bahama Islands must be considered as a realistic hypothesis.   For a discussion of present-day vertebrate relictual distribution patterns resulting from changing climatic and eustatic sea level events in the Bahama Archipelago during the Pleistocene, see Pregill and Olson (1981) and Olson and Pregill (1982).   For a discussion of relictual distribution patterns in the family Drosophilidae, among the islands in the West Indies compared to other areas, see Grimaldi (1988).

However, in regard to the present report, the Bahama Islands are only part of this history since neither *D. SB+* nor *D. SB-5* have been collected there.   Furthermore, there is no host plant in the Bahamas for *D. SB-5*, since no *Stenocereus hystrix*, nor a *Stenocereus* of any species, has ever been reported from these islands (Britton and Millspaugh, 1962). The significant evolutionary events for *D. SB+* and *D. SB-5* probably occurred in Hispañola where the full range of ancestral inversion types exist (Table XIII) and where *S. hystrix* is plentiful.   The inversion sequences are derived in Jamaica for *D. SB-5* and in Cuba for both species, from present information, in comparison to the Hispañola strains.   The culture from Navassa Island of *D. SB+* probably represents an early migrant of this species from Hispañola.

The possibility that all three species in the mayaguana subcluster originated in Hispañola may also be considered a possibility.   The large genetic distance between the cultures from Grand Cayman and Conception Islands (D=0.457) would be subdivided if different dispersal events were initiated from an original *D. mayaguana* population in Hispañola.   The mean genetic distance between the Hispañola cultures and the one from Conception Island is 0.288.   The mean genetic distance between the Hispañola cultures and the one from Grand Cayman is 0.204.   However, since the Conception Island culture converges on the other two species in genetic identity while the Grand Cayman culture diverges from them, the interpretation of exactly how this might have happened is not easily forthcoming.   In any case, under this model the Conception Island flies would be considered a relictual marginal population that derived its identity from an early and now extinct *D. mayaguana* population in Hispañola.

In summary, *D. mayaguana* has diverged considerably as a species in comparison to most other organisms while the D. *SB+*/D. *SB-5* species pair shows very little divergence. One explanation for this phenomenon may reside in the large difference in island size inhabited by each one. Except for Navassa Island, *D. SB+* and *D. SB-5* inhabit only the largest islands in the Caribbean area, while *D. mayaguana* inhabits both the largest and among the smallest islands in the area. Furthermore Conception and Grand Cayman Islands are surrounded by deep sea which means they did not merge with other islands during times of decreased eustatic sea levels during the Pleistocene. Whether or not these islands, and other nearby islands, were submerged during times of glacial retreats in the Pleistocene remains an open question at the moment. According to Pregill and Olson (1981), as recorded from the literature on the Bahamas area, 65,000 years ago the sea level was evidently about 29 to 32 feet above the present level while 17,000 years ago it was depressed by about 384 feet. It is interesting to note that Conception Island peaks out at 60 feet above sea level as does Grand Cayman. Moreover Mayaguana Island has areas above sea level of 80 to 100 feet, while Cat Island, 60 miles north of Conception, has several areas over 100 feet with the highest point being 160 feet above sea level. Furthermore, the central limestone plateau of Cayman Brac, 90 miles east of Grand Cayman, reaches up to 140 feet on the eastern end of the island and it is densely vegetated. Thus not only did island hopping occur more often in *D. mayaguana* but it occurred generally among much smaller islands than in *D. SB+* and *D. SB-5*. These conditions should promote a significant degree of genetic divergence.

Another explanation for the divergent genetic distances between the three species in the present study resides in the possibility that reinforcement for prezygotic isolation has increased the rate of speciation in *D. SB+* and *D. SB-5*, which are broadly sympatric in Hispañola. Coyne and Orr (1989) have postulated this effect in a review of the literature by plotting prezygotic isolation indexes against genetic distance with allopatric sets of species compared to sympatric sets. By using the mean and variance of the isolation index for sympatric forms as an indicator of species status also for allopatric forms, they determined the mean genetic distances under sympatry to be 0.31 and under allopatry to be 0.66.

Measurements of the degree of sexual isolation between allopatric and sympatric forms in the mayaguana subcluster remain to be carried out on a large scale. Preliminary tests by Wasserman (1990) were in the direction expected under female choice conditions. The degree of postzygotic isolation also remains to be tested on a large scale. In any event, the sympatric-allopatric model of speciation rates must be considered as a viable hypothesis to explain the results obtained in the mayaguana subcluster.

The sudden origin of inversions in *D. SB+* from a line of descent devoid of chromosomal rearrangements is extraordinary. The genome of the two species may have become unstable early in their history possibly by a series of hybridization events

and subsequent introgression as described by Naveira and Fontdevila (1985) for *D. buzzatii* and *D. serido*. The reviews by
McDonald (1983) and Fontdevila (1988) are especially pertinent
here. However, in the present analysis, little hybrid sterility has evolved. Thus in addition to being generally
restricted to the large islands, and possibly having undergone
a quickened rate of speciation, these twin species could have
hybridized in the past, which also, of course, would aid in
explaining their very similar genetic identity.

5. SUMMARY

     The mulleri and martensis complexes of cactophilic
species in the repleta species group of *Drosophila* continues
to be one of the more valuable groups of flies for evolutionary studies. In recent years, the islands in the Caribbean
area have become better known for these "mulleri flies," and
their ecology, due to the efforts of a number of investigators
(Heed, 1989; Starmer *et al.*, 1987; Starmer *et al.*, 1990;
Wasserman, 1990). The present work on the genetic relationships among three species in the mayaguana subcluster in the
West Indies was initiated to take advantage of the large electrophoretic survey recently completed on the martensis and
buzzatii clusters in South America (Sánchez, 1986) and the
mulleri and aldrichi subclusters in North and South America
(Armengol, 1986). Even though *D. buzzatii* was the standard
species for comparison in all of these studies, a large number
of species of known allozyme patterns were also utilized for
comparative material in the present study.

     The investigation is particularly interesting not only
because all of the 14 strains originated from islands but
because the islands differ greatly in size and complexity.
Furthermore the host cacti have been identified from a number
of islands. The study may be divided, first, into the ascertainment of the phylogenetic relationships of the three
species and, second, into the analysis of the contrasting
genetic divergences discovered within and among the species.
The phylogenetic relationships were determined on the basis of
equal frequencies of 55 electromorphs among 15 enzyme loci
based on Rogers' genetic distance coefficients and from this
the distance Wagner trees were computed. In addition a
cladistic PAUP analysis based on 56 characters of electromorphs and inversions was performed. The topologies of the
two trees are very similar when either the aldrichi or mulleri
subclusters are employed as the outgroup. The major split is
between *D. mayaguana* and the two undescribed neospecies, *D.
SB+* and *D. SB-5*. The latter two species express very little
differentiation. The preferred Wagner trees indicate *D. SB+*
to be paraphyletic to *D. SB-5* while the preferred clade shows
each species as monophyletic. Close genetic similarity
between the mayaguana subcluster and the martensis cluster
(especially *D. venezolana* and *D. starmeri*), even though a phenetic measurement, remains the only outstanding contradiction.
It may be the result of a sampling bias or the singular possibility of convergence. If it is the result of close phyloge-

netic relatedness, then the inversion differences between the two groups must be re-examined.

The genetic divergences and similarities can be summarized by Nei's genetic identity coefficents and Wright's fixation indexes. The strains of *D. mayaguana* analyzed in the present study originated from the Bahamas (3), the Virgin Islands (1), Grand Cayman Island (1) and Hispañola (2). Nei's mean identity for all strains is 0.839 (0.633-0.988). This illustrates considerable differentiation as does the fixation index (F_{DT}) of 0.542 (0.223- 1.0). *D. SB+* strains originated from Hispañola (1) and Navassa Island (1). *D. SB-5* strains originated from Hispañola (3) and Jamaica (2). Nei's mean identity for all strains of both species is 0.886 (0.828-0.964) while the fixation index is 0.370 (0-0.784). The species are pooled in this case since they are so similar. Nei's I between species is 0.884 (0.841-0.955).

Thus *D. mayaguana*, as a species, has been modified on the enzyme level to a greater value than the two neospecies. This may be a matter of age, with *D. mayaguana* somewhat older than the other species, or it may be an allopatric island effect since five of the seven strains originate from islands very much smaller than any of the Greater Antilles. The fixation index for islands (regions) in the total array (F_{RT}) is 0.417 for *D. mayaguana* while it is only 0.070 for *D. SB+* and *D. SB-5* combined. The explanation may be that both phenomena have occurred, since the Conception Island strain of *D. mayaguana* shows a strong genetic identity with the two neospecies, illustrating an almost intermediate phylogenetic position for this strain. Furthermore the cytological characteristics of *D. mayaguana* lie with the outgroup species in the aldrichi and mulleri subclusters by having no heterozygous inversions and by being homosequential with them.

One remaining question concerns the very close genetic similarity between two species which differ, however slightly, in a number of morphological characters, are polymorphic for a number of inversions (see below), exist on separate host plants, and are completely sympatric on the two largest islands in the Caribbean, Cuba and Hispañola. Obviously speciation has preceded enzyme differentiation in this case and it appears that the sympatric condition may be the underlying ingredient which instigated the event. Reference is made to the "Coyne-Orr effect" in which it was implicated that differentiating sympatic populations arrive at species status about twice as fast as populations in allopatry, as measured by Nei's genetic distance coefficient. Of course the assumption here is that allozyme differentiation runs in clock-like fashion, which has yet to be demonstrated in *Drosophila*. Even so, reinforcement of sexual isolation in areas of sympatry, however achieved (i.e. by natural and/or sexual selection), once again becomes a matter of considerable interest. A detailed study of the pre- and postzygotic isolation between *D. SB+* and *D. SB-5* is in order. In any event, the coincidence of morphological species differences with host plant differences between *D. SB+* and D. SB-5 leads to the presumption that the two phenomena are related, as detailed by Ruiz and Heed (1988) for the mojavensis cluster in western Mexico.

The final point to be made concerns the inversion discovered by Wasserman in *D. SB+* and *D. SB-5*. The former species has four gene sequences, i.e. three heterozygous inversions, while the latter species has a total of eight gene sequences, two of which were fixed from *D. SB+*. It is believed the genomes of these species must have been disrupted by an extraordinary event, possibly by hybridization, at a critical stage in the speciation process. Thus there remain several explanations for the close electrophoretic similarity between the two species. Reinforcement may increase the rate of the speciation process while hybridization would decrease the genetic distance between them. The antithetical nature of these two mechanisms, however, demands that they almost certainly would have to operate independently of each other.

ACKNOWLEDGMENTS. This work was accomplished while W. B. H. spent a sabbatical leave at the Universitat Autónoma de Barcelona, Bellaterra (Barcelona), Spain. He remains greatly indebted to Armand Sànchez and Rosa Armengol for their expertise in the laboratory and the computer room and for the use of their data from the Ph.D. and Tesis de Licenciature, respectively; to Antonio Fontdevila for hosting the project and his aid on a number of problems; to Alfredo Ruiz for his unfailing support and friendship and for cytological confirmation of the cultures, as well as pointing out the morphological differences of the immature stages of the two neospecies; and to Mauro Santos for statistical analysis and numerous insightful discussions.

Marvin Wasserman generously donated his as yet unpublished data on the inversions and hybridization in the mayaguana subcluster. Tom Starmer was instrumental in initiating the two Caribbean trips on the ORV Cape Florida. Lt. Bob Hines and D.P.C. Michael Potts hosted Starmer and Heed at the U.S. Naval Station, Guantanamo Bay, Cuba. Their support is gratefully acknowledged. Mike Sanderson and Mark Porter (University of Arizona) very kindly analyzed the data from a cladistic viewpoint.

The project was supported by the U.S.-Spain Joint Committee for Scientific and Technical Cooperation of the National Science Foundation, Division of International Programs (No. IBP-8509031).

References

Armengol, R. M., 1986, *Relaciones filogéneticas en el cluster* mulleri *de* Drosophila, Tesis de Licenciatura, Universidad Autónoma de Barcelona, (Bellatera), Barcelona.

Ayala, F. J., and Powell, J. R., 1972, Allozymes as diagnostic characters of sibling species of *Drosophila*, *Proc. natn. Acad. Sci.* USA 69:1094-1096.

Ayala, F. J., Powell, J. R., Tracey, M. L., Mauro, C. A., and Pérez-Salas,

S., 1972, Enzyme variability in the *Drosophila willistoni* group. IV. Genic variation in natural populations of *Drosophila willistoni*, *Genetics* **70**:113-139.

Barker, J. S. F., and Mulley, J. C., 1976, Isozyme variation in natural populations of *Drosophila buzzatii*, *Evolution* **30**:213-233.

Barker, J. S. F., and Starmer, W. T., 1982, *Ecological Genetics and Evolution: The Cactus-Yeast-Drosophila Model System*, Academic Press Australia, Sydney.

Barker, J. S. F., Sene, F. M., East, P. D., and Pereira, M. A. Q. R., 1985, Allozyme and chromosomal polymorphism of *Drosophila buzzatii* in Brazil and Argentina, *Genetica* **67**:161-170.

Barker, J. S. F., East, P. D., and Weir, B. S., 1986, Temporal and microgeographic variation in allozyme frequencies in a natural population of *Drosophila buzzatii*, *Genetics* **112**:577-611.

Britton, N. L., and Millspaugh, C. F., 1962, *The Bahama Flora*, Hafner Publ. Co., New York.

Coyne, J. A., and Orr, H. A., 1989, Patterns of speciation in *Drosophila*, *Evolution* **43**:362-381.

Farris, J. S., 1972, Estimating phylogenetic trees from distance matrices, *Am. Nat.* **106**:645-668.

Fitch, W. M., and Margoliash, E., 1967, Construction of phylogenetic trees, *Science* **155**:279-284.

Fontdevila, A., 1982, Recent developments on the evolutionary history of the *Drosophila mulleri* complex in South America, in: *Ecological Genetics and Evolution: The Cactus-Yeast-Drosophila Model System* (J. S. F. Barker, and W. T. Starmer, eds), Academic Press Australia, Sydney, pp. 81-95.

Fontdevila, A., 1988, The evolutionary potential of the unstable genome, in: *Population Genetics and Evolution* (G. de Jong, ed.), Springer-Verlag, Berlin, pp. 251-263.

Fontdevila, A., Pla, C., Hasson, E., Wasserman, M., Sánchez, A., Naveira, H., and Ruiz, A., 1988, *Drosophila koepferae*: a new member of the *Drosophila serido* (Diptera: Drosophilidae) superspecies taxon, *Ann. ent. Soc. Am.* **81**:380-385.

Grimaldi, D. A., 1988, Relicts in the Drosophilidae (Diptera), in: *Zoogeography of Caribbean Insects* (J. K. Liebherr, ed.), Cornell Univ. Press, Ithaca, pp. 183-213.

Heed, W. B., 1989, Origin of *Drosophila* of the Sonoran Desert II. In search for a founder event, in: *Genetics, Speciation, and the Founder Principle* (L. V. Giddings, K. Y. Kaneshiro, and W. W. Anderson, eds), Oxford Univ. Press, New York, pp. 253-278.

Kojima, K., Gillespie, J., and Tobari, Y. N., 1970, A profile of *Drosophila* species' enzymes assayed by electrophoresis. I. Number of alleles, heterozygosities, and linkage disequilibrium in glucose-metabolizing systems and some other enzymes, *Biochem. Genet.* **4**:627-637.

Loukas, M., Krimbas, C. B., Mavragani-Tsipidou, P., and Kastristsis, C. D., 1979, The genetics of allozyme loci in *Drosophila subobscura* and its photographic chromosome maps, *J. Hered.* **70**:17-26.

Marshall, L. G., 1988, Land mammals and the great American interchange, *Amer. Sci.* **76**:380-388.

McDonald, J. F., 1983, The molecular basis of adaptation: a critical review of revelant ideas and observations, *Annu. Rev. Ecol. & Syst.* **14**:77-102.

MacIntyre, R. J., and Collier, G. E., 1986, Protein evolution in the genus *Drosophila*, in: *The Genetics and Biology of Drosophila,* vol. 3e (M. Ashburner, H. L. Carson, and J. N. Thompson, eds), Academic Press, London, pp. 39-146.

Naveira, H., and Fontdevila, A., 1985, The evolutionary history of *Drosophila buzzatii* IX. High frequencies of new chromosome rearrangements induced by introgressive hybridization, *Chromosoma* **91**:87–94.

Nei, M., 1972, Genetic distance between populations, *Am. Nat.* **106**:283–292.

Nei, M., 1977, F-statistics and analysis of gene diversity in subdivided populations, *Ann. Hum. Genet.* **41**:225–233.

Nei, M., 1987, *Molecular Evolutionary Genetics*, Columbia Univ. Press, New York.

Olson, S. L., and Pregill, G. K., 1982, Introduction to the paleontology of Bahaman vertebrates, in: *Fossil Vertebrates from the Bahamas* (S. L. Olson, ed.), Smithsonian Contribution to Paleontology No. 48, Smithsonian Inst. Press, Washington, D.C., pp. 1–7.

Poulik, M. D., 1957, Starch gel electrophoresis in a discontinuous system of buffers, *Nature* **180**:1477–1479.

Pregill, G. K., and Olson, S. L., 1981, Zoogeography of West Indian vertebrates in relation to Pleistocene climatic cycles, *Annu. Rev. Ecol. & Syst.* **12**:75–98.

Prager, E. M., and Wilson, A. C., 1976, Congruency of phylogenies derived from different proteins. A molecular analysis of the phylogenetic position of cracid birds, *J. Mol. Evol.* **9**:45–57.

Richardson, R. H., Richardson, M. E., and Smouse, P. E., 1975, Evolution of electrophoretic mobility in the *Drosophila mulleri* complex, in: *Isozymes, IV: Genetics and Evolution* (C. L. Markert, ed.), Academic Press, New York, pp. 533–545.

Richardson, R. H., and Smouse, P. E., 1976, Patterns of molecular variation. I. Interspecific comparisons of electromorphs in the *Drosophila mulleri* complex, *Biochem. Genet.* **14**:447–466.

Richardson, R. H., Smouse, P. E., and Richardson, M. E., 1977, Patterns of molecular variation. II. Associations of electrophoretic mobility and larval substrate within species of the *Drosophila mulleri* complex, *Genetics* **85**:141–154.

Rogers, J. S., 1972, Measures of genetic similarity and genetic distance, *Univ. Tex. Publs* **7213**:145–153.

Ruiz, A., and Heed, W. B., 1988, Host-plant specificity in the cactophilic *Drosophila mulleri* species complex, *J. Anim. Ecol.* **57**:237–249.

Ruiz, A., Heed, W. B., and Wasserman, M., 1990, The evolution of the *mojavensis* cluster of cactophilic *Drosophila* with descriptions of two new species, *J. Hered.* **81**:30–42.

Sánchez, A., 1986, *Relaciones filogenéticas en los clusters* buzzatii y martensis *(grupo* repleta*) de* Drosophila, Tesis Doctoral, Universidad Autónoma de Barcelona (Bellaterra), Barcelona.

Sene, F. M., and Carson, H. L., 1977, Genetic variation in Hawaiian *Drosophila*. IV. Allozymic similarity between *D. silvestris* and *D. heteroneura* from the island of Hawaii, *Genetics* **86**:187–198.

Shaw, C. R., and Prasad, R., 1970, Starch gel electrophoresis of enzymes. A compilation of recipes. *Biochem. Genet.* **4**:297–318.

Smithies, O., 1955, Zone electrophoresis in starch gels: group variations in the serum proteins of normal human adults, *Biochem. J.* **61**:629–641.

Starmer, W. T., Lachance, M., and Phaff, H. J., 1987, A comparison of yeast communities found in necrotic tissue of cladodes and fruits of *Opuntia stricta* on islands in the Caribbean Sea and where introduced into Australia, *Microb. Ecol.* **14**:179–192.

Starmer, W. T., Lachance, M., Phaff, H. J., and Heed, W. B., 1990, The biogeography of yeasts associated with decaying cactus tissue in North America, the Caribbean, and Northern Venezuela, *Evol. Biol.* **24**:253–296.

Sturtevant, A. H., 1921, *North American species of* Drosophila, Carnegie Inst. of Wash., Publ. 301, Washington, D.C., pp. 1–150.

Swofford, D. L., 1981, On the utility of the distance Wagner procedure, in: *Advances in Cladistics* (V. A. Funk, and D. R. Brooks, eds), N.Y. Botanical Garden Press, New York, pp. 25–44.

Swofford, D. L., 1985, *PAUP: Phylogenetic analysis using parsimony. User's manual*, Illinois Natural History Survey, Champaign, Illinois.

Swofford, D. L., and Selander, R. B., 1981, BIOSYS-1: A FORTRAN program for the comprehensive analysis of electrophoretic data in population genetics and systematics, *J. Hered.* **72**:281-283.

Thorpe, J. P., 1982, The molecular clock hypothesis: biochemical evolution, genetic differentiation and systematics, *Annu. Rev. Ecol. & Syst.* **13**:139-168.

Vilela, C. R., 1983, A revision of the *Drosophila repleta* species group (Diptera, Drosophilidae), *Rev. Bras. Ent.* **27**:1-114.

Wasserman, M., 1954, Cytological studies on the *repleta* group. *Univ. Tex. Publs* **5422**:130-152.

Wasserman, M., 1962, Cytological studies of the *repleta* group of the genus *Drosophila*, V. The *mulleri* subgroup, *Univ. Tex. Publs* **6205**:85-118.

Wasserman, M., 1982a, Evolution in the *repleta* group, in: *The Genetics and Biology of* Drosophila, vol. 3b (M. Ashburner, H. L. Carson, J. N. Thompson, eds), Academic Press, London, pp. 61-140.

Wasserman, M., 1982b, Cytological evolution in the *Drosophila repleta* species group, in: *Ecological Genetics and Evolution: The Cactus-Yeast*-Drosophila *Model System* (J. S. F. Barker, and W. T. Starmer, eds), Academic Press Australia, Sydney, pp. 49-64.

Wasserman, M., 1990, Cytological evolution of the *Drosophila* repleta species group, in: *Inversion Polymorphism in* Drosophila (J. R. Powell, and C. B. Krimbas, eds), C.R.C. Press, Boca Raton, Fla. (in press).

Wasserman, M., and Koepfer, H. R., 1977, Phylogenetic relationships among *Drosophila longicornis, Drosophila propachuca* and *Drosophila pachuca*, a triad of sibling species, *Genetics* **87**:557-568.

Wasserman, M., and Koepfer, H. R., 1979, Cytogenetics of the South American *Drosophila mulleri* complex: the martensis cluster. More interspecific sharing of inversions, *Genetics* **93**:935-946.

Webb, S. D., 1978, A history of savanna vertebrates in the new world. Part II: South America and the great interchange, *Annu. Rev. Ecol. & Syst.* **9**:393-426.

Wright, S., 1965, The interpretation of population structure by F- statistics with special regard to systems of mating, *Evolution* **19**:395-420.

Wright, S., 1978, *Evolution and the Genetics of Populations, vol. 4: Variability within and among natural populations*, Univ. Chicago Press, Chicago.

Zouros, E., 1973, Genic differentiation associated with the early stages of speciation in the *mulleri* subgroup of *Drosophila*, *Evolution* **27**:601-621.

Contributors

Virginia Aberdeen Department of Biology , Syracuse University, Syracuse, New York 13210, USA

J. Ruben Abril Department of Biological Sciences, University of Denver, Denver, Colorado 80208, USA

R. Armengol Departament di Genética i de Microbiologia, Universitat Antónoma de Barcelona, Bellaterra-Barcelona, Spain

Peter W. Atkinson Department of Biology, Syracuse University, Syracuse, New York 13210, USA

Louise Baker Department of Genetics, University of Melbourne, Parkville, Victoria 3052, Australia

J.S.F. Barker Department of Animal Science, University of New England, Armidale, New South Wales 2351, Australia

Philip Batterham Department of Genetics, University of Melbourne, Parkville, Victoria 3052, Australia

Cynthia A. Bayer Department of Biology, Syracuse University, Syracuse, New York 13210, USA

James P. Brady Department of Biology and Institute for Molecular and Cellular Biology, Indiana University, Bloomington, Indiana 47405, USA

Glen E. Collier Department of Biological Sciences, Illinois State University, Normal, Illinois 61761, USA

Mira M. Dumancic CSIRO Division of Entomology, G.P.O. Box 1700, Canberra, A.C.T. 2601, Australia

Peter East Department of Animal Science, University of New England, Armidale, New South Wales 2351, Australia

Abigail Elizur Department of Biochemistry, University of Adelaide, South Australia 5000, Australia

William J. Etges Department of Ecology and Evolutionary Biology, University of Arizona, Tucson, Arizona 85721, USA. Present address: Department of Zoology, University of Arkansas, Fayetteville, Arkansas 72701, USA

David Featherston Section of Genetics and Development, Cornell University, Ithaca, New York 14853, USA

James C. Fogleman Department of Biological Sciences, University of Denver, Denver, Colorado 80208, USA

A. Fontdevila Departament di Genética i de Microbiologia, Universitat Antónoma de Barcelona, Bellaterra-Barcelona, Spain

Geoffrey G. Foster CSIRO Division of Entomology, G.P.O. Box 1700, Canberra, A.C.T. 2601, Australia

Anne Y. Game CSIRO Division of Entomology, GPO Box 1700, Canberra, A.C.T. 2601, Australia

Billy W. Geer Department of Biology, Knox College, Galesburg, Illinois 61401, USA

Anne Graham Department of Animal Science, University of New England, Armidale, New South Wales 2351, Australia

Ygal Haupt Walter and Eliza Hall Institute of Medical Research, PO Royal Melbourne Hospital, Victoria 3050, Australia

Marion J. Healy CSIRO Division of Entomology, G.P.O. Box 1700, Canberra, A.C.T. 2601, Australia

Philip W. Hedrick Department of Biology, Pennsylvania State University, University Park, Pennsylvania 16802, USA

W. B. Heed Department of Ecology and Evolutionary Biology, University of Arizona, Tucson, Arizona 85721, USA

Pieter W.H. Heinstra Department of Population and Evolutionary Biology, University of Utrecht, The Netherlands

Ary A. Hoffmann Department of Genetics and Human Variation, La Trobe University, Victoria 3083, Australia

Hope Hollocher Department of Biology, Washington University, St. Louis, Missouri 63130, USA

Antony J. Howells Department of Biochemistry, Faculty of Science, Australian National University, Canberra, A.C.T. 2601, Australia

John Jaenike Department of Biology, University of Rochester, Rochester, New York 14627, USA

J. Spencer Johnston Department of Entomology, Texas A & M University, College Station, Texas 77843, USA

Ann M. Kapoun Department of Biology, Knox College, Galesburg, Illinois 61401, USA

Susan Lawler Department of Biology, Washington University, St. Louis, Missouri 63130, USA

Ross J. MacIntyre Section of Genetics and Development, Cornell University, Ithaca, New York 14853, USA

Therese Ann Markow Department of Zoology, Arizona State University, Tempe, Arizona 85287, USA

Stephen W. McKechnie Department of Genetics and Developmental Biology, Monash University, Clayton, Victoria 3168, Australia

John A. McKenzie Department of Genetics, University of Melbourne, Parkville, Victoria 3052, Australia

Marilyn A. Menotti-Raymond Department of Biology, Syracuse University, Syracuse, New York 13210, USA

Chris Moran Department of Animal Husbandry, University of Sydney, New South Wales 2006, Australia

Michael Murphy School of Biological Sciences, University of Sydney, New South Wales 2006, Australia

Karen M. Nielsen Department of Biology and Institute for Molecular and Cellular Biology, Indiana University, Bloomington, Indiana 47405, USA

Sharon O'Donnell Department of Genetics and Human Variation, La Trobe University, Victoria 3083, Australia

John G. Oakeshott CSIRO Division of Entomology, GPO Box 1700, Canberra, A.C.T. 2601, Australia

P.A. Parsons Division of Science and Technology, Griffith University, Nathan, Brisbane. Present address: Department of Zoology, University of Adelaide, Adelaide, South Australia 5000, Australia

Rollin C. Richmond Department of Biology and Institute for Molecular and Cellular Biology, Indiana University, Bloomington, Indiana 47405, USA

Jennifer L. Ross Department of Genetics and Developmental Biology, Monash University, Clayton, Victoria 3168, Australia

Robyn J. Russell CSIRO Division of Entomology, G.P.O. Box 1700, Canberra, A.C.T. 2601, Australia

A. Sánchez Departament di Genética i de Microbiologia, Universitat Antónoma de Barcelona, Bellaterra-Barcelona, Spain

Elizabeth M. Snella Department of Biology and Institute for Molecular and Cellular Biology, Indiana University, Bloomington, Indiana 47405, USA

William T. Starmer Biology Department, Syracuse University, Syracuse, New York 13210, USA

David T. Sullivan Department of Biology, Syracuse University, Syracuse, New York 13210, USA

John A. Sved School of Biological Sciences, University of Sydney, New South Wales 2006, Australia

Richard G. Tearle Department of Biochemistry, University of Adelaide, South Australia 5000, Australia

Alan R. Templeton Department of Biology, Washington University, St. Louis, Missouri 63130, USA

Eric C. Toolson Department of Biology, University of New Mexico, Albuquerque, New Mexico 87131, USA

Adam Torkamanzehi Department of Animal Husbandry, University of Sydney, New South Wales 2006, Australia

Kerrin L. Turney Department of Genetics and Developmental Biology, Monash University, Clayton, Victoria 3168, Australia

Aleid Van der Zel Department of Biology, Knox College, Galesburg, Illinois 61401, USA

Gaye L. Weller CSIRO Division of Entomology, G.P.O. Box 1700, Canberra, A.C.T. 2601, Australia

Gillian Whitington Department of Animal Science, University of New England, Armidale, New South Wales 2351, Australia

Author Index

Organism Index

Subject Index

Abdominal bristle number, 104–107, 109, 112

Abnormal abdomen (*aa* syndrome), 18–34, 165

Absorption, insecticide: *see* insecticide, absorption

Acetaldehyde, 234–235, 237–239, 247

Acetaldehyde dehydrogenase (ALDH), 237

Acetic acid, 133–137, 139–140, 234, 237–238, 247

Acetoin, 133–140

Acetone, 133–140, 232

Acetylcholinesterase: *see* esterases, acetylcholinesterase

Activity
behavioral, 76, 80–81
variation, 359–360, 375–376, 378–379

Adaptation, 37, 287–288
by natural selection, 162
demographic, 31
to desiccation, 31
to young adult age structure, 31

Adaptive evolution, 101, 103, 114

Adaptive infiltration, 8

Additive genetic variance (V_A), 111–113

Adenylate energy charge (AEC), 79

ADH (alcohol dehydrogenase), 232–247, 333, 359–360, 407–410, 416, 420
ADH-1, 409–410, 416
ADH-2, 410, 416
ADH-F, 239–240
ADH-S, 239–240

duplication, 407–411

genes, 334
Adh, 238, 240–244, 247–248, 333, 359–360, 407–417
Adh-1, 410–414, 416–417, 451, 454–455, 463, 482
Adh-2, 410–414, 416–418, 451, 455, 458, 463, 465
Adh-pseudogene, 410–415, 417
Adh^F, 238, 240, 242, 244
Adh^{f6n}, 236, 242
Adh^{n2}, 232, 235, 243
Adh null, 412
Adh^S, 238, 240, 244

isozyme, 408–410, 416

locus,
evolution of, 411
pathway, 234–237, 240, 243, 245, 247

null, 232–235, 243, 247, 414

variants, 230

Adk-2 (adenylate kinase-2 gene), 451, 454, 463

Adult experience
early, 221

AED: *see* anterior ejaculatory duct

Age structure, 25–27, 29, 31
adult 33–34

Alcohol
dehydrogenase: *see* ADH
metabolism, 234
secondary, 238
short chain, 234
tolerance, 232, 247

Aldehyde
dehydrogenase, 237
gem-diol form, 238

Alditol acetates, 128